Physical Optics

Physical Optics

S. A. AKHMANOV
S. YU. NIKITIN
Department of Physics
M. V. Lomonosov Moscow State University

CLARENDON PRESS • OXFORD
1997

This book has been printed digitally in order to ensure its continuing availability

OXFORD
UNIVERSITY PRESS

Great Clarendon Street, Oxford OX2 6DP

Oxford University Press is a department of the University of Oxford.
It furthers the University's objective of excellence in research, scholarship,
and education by publishing worldwide in

Oxford New York

Auckland Bangkok Buenos Aires Cape Town Chennai
Dar es Salaam Delhi Hong Kong Istanbul Karachi Kolkata
Kuala Lumpur Madrid Melbourne Mexico City Mumbai Nairobi
São Paulo Shanghai Singapore Taipei Tokyo Toronto

with an associated company in Berlin

Oxford is a registered trade mark of Oxford University Press
in the UK and in certain other countries

Published in the United States
by Oxford University Press Inc., New York

First published 1997
Reprinted 1999, 2002

A catalogue record for this book is available from the British Library

Library of Congress Cataloging in Publication Data
Akhmanov, S. A. (Sergei Alexandrovich)
Physical optics / S. A. Akhmanov. S. Yu. Nikitin.
Includes bibliographical references and index.
1. Physical optics. I. Nikitin, S. Yu. III. Title.
QC395.2A33 1997 535'.2—dc20 96-48895 CIP

ISBN 0-19-851795-5

Preface

I write this preface with emotion since my teacher and co-author Sergei Akhmanov is no longer with us. This untimely death took place after unsuccessful surgery, and his death has been a major tragedy for the Moscow academic community and a personal loss to all of us, his colleagues and friends. Professor S. Akhmanov's scientific work had a profound impact on the development of physical optics. Perhaps even more profound was the impact of his outstanding personality on his colleagues and students.

This book is based on the lecture course that has been given at the Physics Department of M. V. Lomonosov Moscow State University during recent years. It is written primarily as a textbook for university students of physics. It is my hope that the book can also be useful to teachers and to professional scientists working in the area of optics research and its diverse applications.

The main purpose of the book has been to provide a comprehensive and up-to-date coverage of the fundamentals of physical optics as one of the central divisions of contemporary physics. The text draws extensively on the concepts of classical and quantum radiation theories, non-linear physics, statistics, and Fourier transformation forming the bedrock of the present-day physical science. At the same time a great deal of attention has been devoted to such important topics as lasers, non-linear optics, new methods of optical spectroscopy, as well as analysis, transformation, and synthesis of optical fields.

The text consists of 23 chapters. Within each chapter the theoretical material is supported by experimental and laboratory evidence, numerous illustrations and tables. Organized in this way, the volume can be used by students as a fundamental source of knowledge and by instructors and researchers as a reference guide.

I am pleased to thank my colleagues Prof. Nikolai Koroteev, Prof. Yuri Dyakov, Prof. Mikhail Vorontsov, and Prof. Victor Vysloukh, who have contributed much to the development of the Physical Optics course at the Physics Department of M. V. Lomonosov Moscow State University.

Special thanks to Prof. Nicolaas Bloembergen, Nobel Prize Laureate (Harvard University, USA), Prof. Herbert Walther (Max-Planck-Institut für Quantenoptik, Germany), Prof. Anatoliy Chirkin (Moscow State University), and Mrs. Maya Akhmanova for the idea to preserve Akhmanov's legacy to the world of physics also in the English language.

I am indebted to Dr. Yuri Atanov for his invaluable help with translating the text into English, to Dr. Victor Zadkov for typesetting and preparation of the camera-ready copy, and to Dr. Alexei Lukashev for preparing the illustrations.

I should like to thank the staff of Oxford University Press and Dr. Tatyana Sokolova, the OUP-Moscow Project manager, for their help and support in the production of this book.

Moscow, S.N.
September 1996

Contents

PART II LIGHT EMISSION AND OPTICAL OSCILLATORS

PART III INTERFERENCE, DIFFRACTION, AND COHERENCE

Part I

The electromagnetic theory of light

1

Optical waves in vacuum

The electromagnetic nature of light and Maxwell's theory. Hertz's experiments. Maxwell's equations. The wave equation. A plane wave. A spherical wave. Modulated waves and radiation of real sources of light. Spectral expansion of the optical field. The superposition principle.

1.1 The electromagnetic nature of light and Maxwell's theory

The question of the nature of light constitutes one of the central problems in physical optics. Attempts to solve this problem have always been in the mainstream of physical research. Many great discoveries in physics have been directly or indirectly associated with efforts to gain a more penetrating insight into the nature of light.

By the end of the 19th century physicists possessed a variety of facts testifying to the electromagnetic nature of light. Such facts were obtained experimentally by Faraday, who observed the effects of a magnetic field on the light propagation in substances, by Lebedev, who measured the pressure of light, and by Hertz, who was the first to demonstrate the existence of electromagnetic waves. Some other research on the interaction of light and matter could also be mentioned.

One of the first arguments in favour of the electromagnetic nature of light was the coincidence of the velocity of electromagnetic waves, calculated by Maxwell, with the velocity of light.

In 1849 Fizeau measured the velocity of light and obtained the value 315 000 $\mathrm{km\,s^{-1}}$. In 1857 Weber and Kohlrausch measured the electrodynamic constant c, equal to the ratio of the electromagnetic and electrostatic units of charge, and obtained the value $c = 310\,800\ \mathrm{km\,s^{-1}}$. The existence of electromagnetic waves followed from the set of equations that Maxwell proposed in 1861 to describe the electromagnetic field. The propagation velocity of the waves was determined from the value of the electrodynamic constant. Maxwell noticed that the value of c found by Weber and Kohlrausch was very close to the velocity of light measured by Fizeau. This observation made him suggest that light is indeed an *electromagnetic wave.*

1.1.1 The velocity of light

Measuring the velocity of light offered quite a challenge because of its extremely large value, equal to about 300 000 $\mathrm{km\,s^{-1}}$. The history of measurements of the velocity of light is fascinating (see, for example, Jaffe 1961). Table 1.1 summarizes

Table 1.1 The velocity of light in vacuum.

Year	Author	Method	Result ($km\,s^{-1}$)	Uncertainty
1676	Roemer	Jupiter satellites	214 000	
1726	Bradley	Aberration of stars	301 000	
1849	Fizeau	Gear	315 000	
1862	Foucault	Light beam deflection by rotating mirror	298 000	±500
1872	Cornu	Gear	298 500	±900
1874	Cornu	Light beam deflection	300 400	±800
1878	Michelson	The same	300 140	±700
1879	Michelson	The same	299 910	±50
1882	Michelson	The same	299 853	±60
1924	Michelson	Mirror prism	299 796	±30
1973	Evenson	Laser	299 792.4562	$\pm 10^{-3}$

the basic measurement techniques and the results obtained. The advent of the laser made it possible to perform very accurate measurements of the velocity of light. According to the latest data, the velocity of light in vacuum is equal to

$$c = 299\,792\,458 \pm 1.2\,\mathrm{m\,s^{-1}}. \tag{1.1}$$

1.1.2 The electrodynamic constant

The electrodynamic constant c has the sense of the ratio of the electromagnetic and electrostatic units of charge. The method of measurement of this quantity can be explained by the example of the Weber and Kohlrausch experiment.

One can find the charge of a condenser of known capacity in electrostatic units by measuring the potential difference across its electrodes with the help of an electrometer. On the other hand, the same charge can be measured in electromagnetic units. For this purpose one should discharge the condenser through the coil of a ballistic galvanometer and measure the angle of deviation of its mirror. The ratio of the values thus obtained yields the ratio of the units of charge.

The results of experimental measurements of the electrodynamic constant c (see Table 1.2) clearly show that it coincides with the velocity of light in vacuum.

Table 1.2 The ratio of the electromagnetic and electrostatic units of charge.

Year	Author	Value in vacuum ($km\,s^{-1}$)
1857	Weber and Kohlraush	310 800
1868	Maxwell	284 300
1880	Stoletov	298 000
1898	Fabry and Perot	299 870

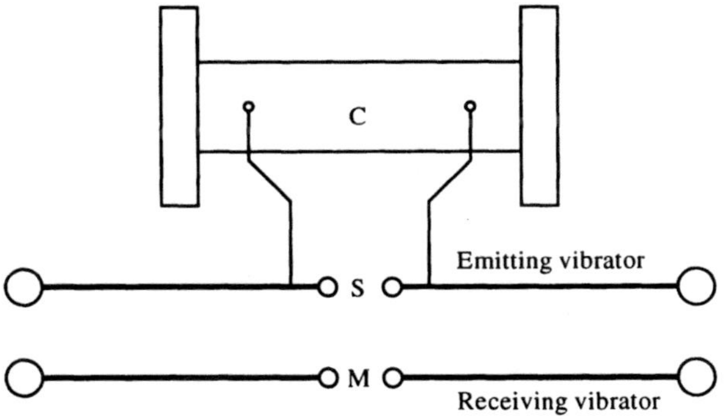

Fig. 1.1 A diagram of Hertz's apparatus. C, Induction coil; S, spark gap; M, micrometer.

1.2 Hertz's experiments

Maxwell's prediction of the existence of electromagnetic waves needed experimental confirmation. The necessary evidence was provided by Hertz, who succeeded in generating and receiving electromagnetic waves, as well as carrying out other studies of their properties.

In his experiments with the spark discharge of a powerful induction coil, Hertz observed ultra-fast oscillations of electric current (with a period of about 10^{-8} s). Later he found that a high-frequency alternating current flowing in a straight conducting wire is capable of inducing a similar current in another conductor located at some distance from the first one. That was how electromagnetic waves were discovered.

In one of the experiments Hertz made use of electric oscillations generated by a vibrator consisting of two identical metal rods separated by the spark gap (Fig. 1.1). Both rods of the vibrator were charged from the induction coil until the potential difference reached the breakdown value. The resulting spark short-circuited the gap and produced high-frequency attenuating electric oscillations in the vibrator. Simultaneously, it emitted an electromagnetic wave into surrounding space. The emergence of the wave was registered by the incidence of a spark in the micrometer of a receiving vibrator.

To prove the identical nature of optical and electromagnetic waves, Hertz ventured to demonstrate the reflection, refraction, and polarization of electromagnetic waves. Using parabolic mirrors made of zinc tin, he succeeded in forming a beam of electromagnetic waves, similar to the optical beam. With the help of a moulded-resin 1.2-ton prism, 1.5 m high, Hertz achieved a deflection of the obtained beam, which was the analogy of refraction of optical rays in a glass prism. Finally, he could also obtain convincing evidence of polarization of electromagnetic waves using a wire net. Summarizing his research performed in 1886-1889, Hertz wrote: "... it seems quite probable to me that the experiments described prove the identity of light, thermal rays and electrodynamic wave motion".

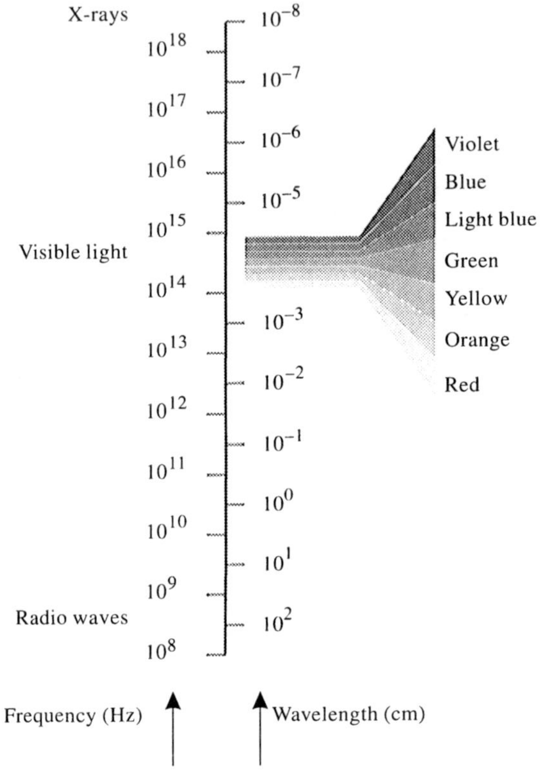

Fig. 1.2 The optical range on the scale of electromagnetic waves.

Thus, the identical nature of light and electricity was proved experimentally. From that time, optics could be included into electrodynamics, just as acoustics had been incorporated into mechanics.

Figure 1.2 shows the place of the optical range on the scale of electromagnetic waves. This range spans wavelengths from 0.4 μm to 0.7 μm (or frequencies from 400 to 750 THz).

1.3 Maxwell's equations

The set of equations for an electromagnetic field was proposed by Maxwell in the middle of the 19th century as a result of generalization of the data gained in experiments with electric charges, currents, and magnets. Further investigations showed that Maxwell's equations are much more profound and go far beyond the facts and notions on which they were based. The equations have proved to be relativistically invariant. They adequately describe high-frequency alternating electromagnetic fields, including optical waves. The theory of electromagnetic wave radiation by moving charges, as well as the theory of interaction of light and matter, can use Maxwell's equations as the starting point.

Maxwell's equations describing an electromagnetic field in vacuum can be written in the Gaussian system of units (which is convenient for optics) as

$$\mathrm{rot}\vec{E} = -\frac{1}{c}\frac{\partial \vec{H}}{\partial t}, \tag{1.2}$$

$$\mathrm{rot}\vec{H} = \frac{1}{c}\frac{\partial \vec{E}}{\partial t}, \tag{1.3}$$

$$\mathrm{div}\vec{E} = 0, \tag{1.4}$$

$$\mathrm{div}\vec{H} = 0. \tag{1.5}$$

Here $\vec{E}$ and $\vec{H}$ are the strengths of electric and magnetic fields and c is the electrodynamic constant, equal to the velocity of light in vacuum.

Equation (1.2) furnishes a mathematical formulation of the electromagnetic induction law. Equation (1.3) shows that an alternating electric field gives rise to the magnetic field. Equation (1.4) asserts the absence of a static electric field in vacuum. Equation (1.5) postulates the absence of magnetic charges.

1.4 The wave equation

Equations (1.2)–(1.5) enable a derivation of closed equations for the fields $\vec{E}$ and $\vec{H}$. Differentiating eqn (1.3) with respect to time and interchanging the succession of temporal and spatial derivatives, we obtain

$$\mathrm{rot}\left(\frac{\partial \vec{H}}{\partial t}\right) = \frac{1}{c}\frac{\partial^2 \vec{E}}{\partial t^2}.$$

Using eqn (1.2), we obtain

$$\mathrm{rot}\,\mathrm{rot}\vec{E} = -\frac{1}{c^2}\frac{\partial^2 \vec{E}}{\partial t^2}.$$

Making use of the differential operator relation known from vector analysis, we transform the left-hand side of the last equation into the form

$$\mathrm{rot}\,\mathrm{rot}\vec{E} = \mathrm{grad}\,\mathrm{div}\vec{E} - \Delta\vec{E}.$$

Here Δ is the Laplacian which, in Cartesian coordinates, takes the form

$$\Delta = \frac{\partial^2}{\partial x^2} + \frac{\partial^2}{\partial y^2} + \frac{\partial^2}{\partial z^2}. \tag{1.6}$$

Since free electric charges are absent in the case considered, i.e. $\mathrm{div}\vec{E} = 0$, we obtain the following equation for the vector of electric field strength:

$$\Delta\vec{E} - \frac{1}{c^2}\frac{\partial^2\vec{E}}{\partial t^2} = 0. \tag{1.7}$$

It is easy to make sure that the vector $\vec{H}$ satisfies the analogous equation

$$\Delta\vec{H} - \frac{1}{c^2}\frac{\partial^2\vec{H}}{\partial t^2} = 0. \tag{1.8}$$

Equations (1.7) and (1.8) are linear with respect to the field strength. Therefore they are equivalent to the set of scalar equations of the same form, containing only one component of electric (or magnetic) field strength. Indeed, let us write the vectors $\vec{E}$ and $\vec{H}$ using Cartesian components:

$$\vec{E} = \vec{x}_0 E_x + \vec{y}_0 E_y + \vec{z}_0 E_z, \quad \vec{H} = \vec{x}_0 H_x + \vec{y}_0 H_y + \vec{z}_0 H_z, \tag{1.9}$$

where $\vec{x}_0$, $\vec{y}_0$, and $\vec{z}_0$ are the unit vectors, directed along the x-, y- and z-axes of the Cartesian frame of reference. Then each of the components E_α or H_α ($\alpha = x, y, z$) satisfies the scalar equation

$$\Delta E_\alpha - \frac{1}{c^2}\frac{\partial^2 E_\alpha}{\partial t^2} = 0, \quad \Delta H_\alpha - \frac{1}{c^2}\frac{\partial^2 H_\alpha}{\partial t^2} = 0. \tag{1.10}$$

This is easy to prove by scalar multiplication of eqns (1.7) and (1.8) successively by $\vec{x}_0$, $\vec{y}_0$, and $\vec{z}_0$. Equations (1.7), (1.8), and (1.10) are called the wave equations. Their solutions yield the behavour characteristic of propagating waves.

1.5 A plane wave

Suppose that the arbitrary component of the field f (for example, E_α or H_α) depends only on one spatial coordinate and time, i.e. $f = f(z,t)$. Then eqn (1.10) becomes simplified to the form

$$\frac{\partial^2 f}{\partial z^2} - \frac{1}{c^2}\frac{\partial^2 f}{\partial t^2} = 0. \tag{1.11}$$

To find the solution of this equation, we shall write it as

$$\left(\frac{\partial}{\partial z} - \frac{1}{c}\frac{\partial}{\partial t}\right)\left(\frac{\partial}{\partial z} + \frac{1}{c}\frac{\partial}{\partial t}\right) f = 0$$

or

$$\frac{\partial^2 f}{\partial\xi\partial\eta} = 0, \tag{1.12}$$

where the variables $\xi = t - z/c$ and $\eta = t + z/c$ are introduced. Equation (1.12) is satisfied by the function

$$f(\xi,\eta) = f_1(\xi) + f_2(\eta)$$

or, in variables z and t, by the function

$$f(z,t) = f_1(t - z/c) + f_2(t + z/c), \tag{1.13}$$

where f_1 and f_2 are arbitrary (differentiable) functions of their arguments.

Equation (1.13) expresses the general solution of eqn (1.11). This solution describes the superposition of two waves. One of them propagates along the z-axis and the other in the opposite direction. The velocities of both waves are equal to c. Indeed, the perturbation f_1, located at point z_1 at moment t_1, arrives at moment t_2 at the point z_2 determined by the relation $t_1 - z_1/c = t_2 - z_2/c$. Then, if $t_2 > t_1$, we have $z_2 > z_1$, and the propagation velocity of the wave perturbation is $v = (z_2 - z_1)/(t_2 - t_1) = c$.

The functions $f_1 = f_1(z,t)$ and $f_2 = f_2(z,t)$ describe *plane waves*, because the wave perturbation has the same magnitude at all points of the infinite plane perpendicular to the propagation direction. The specific form of the functions f_1 and f_2 is governed by the boundary and initial conditions.

1.5.1 A plane harmonic wave

Now let us specify the spatial and temporal variation of the optical field. Consider the Cartesian component of the field $E_x(z,t)$. Let us assume that, for $z = 0$,

$$E_x(0,t) = A\cos\omega t, \tag{1.14}$$

i.e. the optical field strength obeys a harmonic law. Then, in accordance with eqn (1.13), a plane harmonic wave will propagate in the region $z \geq 0$:

$$E_x(z,t) = A\cos[\omega(t - z/c)] = A\cos(\omega t - kz) = A\cos[\omega t - \varphi(z)]. \tag{1.15}$$

In this expression A is the wave *amplitude*, ω is the *angular frequency* related to the *period* T and the *oscillation frequency* $\nu = 1/T$ as

$$\omega = 2\pi/T = 2\pi\nu, \tag{1.16}$$

$$k = \omega/c = 2\pi\nu/c = 2\pi/\lambda \tag{1.17}$$

is the *wave number*, and

$$\lambda = cT = c/\nu \tag{1.18}$$

is the *wavelength*. The quantity

$$\Phi = \omega t - kz \tag{1.19}$$

is referred to as the *total phase* of the wave and depends on t and z. The phase $\varphi(z) = kz$ varying with the distance z traversed by the wave is called the *phase shift*.

The locus of the equal phase values is referred to as the *wavefront*. In a plane harmonic wave it is a plane perpendicular to the direction of propagation.

1.6 A spherical wave

It is easy to see that eqn (1.10) is also satisfied by the waves

$$E_\alpha = E_\alpha(t, r), \quad H_\alpha = H_\alpha(t, r),$$

the field strengths of which depend only on one spatial variable; that is, the modulus of the radius vector

$$r = \sqrt{x^2 + y^2 + z^2}.$$

Such waves are called *spherical*. Just as in the case of the plane wave, the spherical wave is treated as a reference object, being of major importance in optics.

Let us find the solution of the scalar wave equation

$$\Delta f - \frac{1}{c^2}\frac{\partial^2 f}{\partial t^2} = 0 \tag{1.20}$$

in the form $f = f(t, r)$. For the spherically symmetric function f the Laplacian takes the form

$$\Delta f = \frac{1}{r}\frac{\partial^2}{\partial r^2}(rf). \tag{1.21}$$

Inserting eqn (1.21) into eqn (1.20), we obtain the equation

$$\frac{1}{r}\frac{\partial^2}{\partial r^2}(rf) = \frac{1}{c^2}\frac{\partial^2 f}{\partial t^2}. \tag{1.22}$$

Let us introduce an auxiliary function $F = rf$. Then eqn (1.22) changes to the form analogous to eqn (1.11),

$$\frac{\partial^2 F}{\partial r^2} = \frac{1}{c^2}\frac{\partial^2 F}{\partial t^2}, \tag{1.23}$$

and its general solution can therefore be represented by a superposition of two waves propagating in mutually opposite directions

$$F(r, t) = F_1(t - r/c) + F_2(t + r/c). \tag{1.24}$$

Coming back to the sought-for function f, we obtain

$$f(r, t) = \frac{F_1(t - r/c)}{r} + \frac{F_2(t + r/c)}{r}. \tag{1.25}$$

Equation (1.25) describes two spherical waves: one *diverging* from the origin of the coordinates and the other *converging* towards it (see Fig. 1.3).

1.6.1 A harmonic spherical wave

Suppose that an equiphase harmonic perturbation,

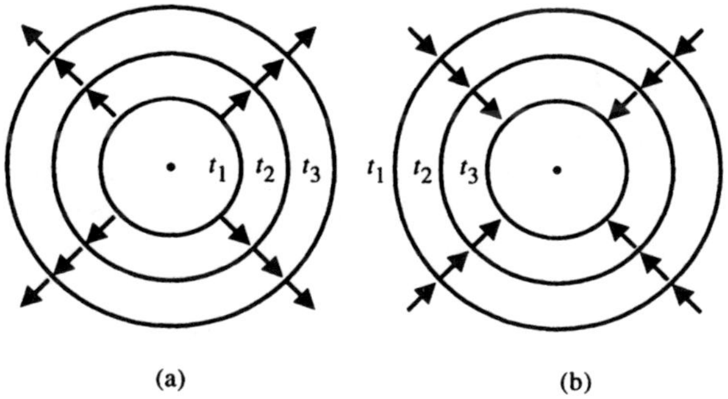

Fig. 1.3 Diverging (a) and converging (b) spherical waves. Wavefront positions are shown for sequential instants $t_1 < t_2 < t_3$.

$$f(r_0, t) = (A_0/r_0)\cos[\omega(t - r_0/c)], \tag{1.26}$$

is applied at all points on the sphere of radius r_0. Then the diverging wave excited by such a source can be written as

$$f(r, t) = (A_0/r)\cos[\omega(t - r/c)], \tag{1.27}$$

for $r > r_0$. In contrast to the plane wave, the amplitude A_0/r depends on the coordinate, with the phase and amplitude fronts representing spherical surfaces.

1.7 Modulated waves and radiation of real sources of light

Real sources of light emit radiation that at first glance has little in common with the reference waves considered earlier. The difference consists first of all in the character of the spatial variation of the wave perturbation. For example, the helium–neon laser beam constitutes neither a plane nor a spherical wave. For many laser, and especially non-laser, sources of light, the model of temporal harmonic variation of electric and magnetic fields ("a monochromatic wave"[1]) is also far from reality. The ruby laser radiation pattern is a vivid example of such a situation. Here we deal with a short light burst, or an optical pulse, which represents a non-harmonic wave process of finite duration.

How should one describe real optical beams and pulses? What are the laws governing their propagation in space? The detailed answer to these questions is given in terms of wave optics, to which Part III of this book is devoted. The properties of real wave processes can be found if one comprehends the laws of light emission as well as the physics of interference and diffraction. In this chapter, before undertaking such consistent examination, we shall discuss some simple models based on clear physical concepts. This will enable us to formulate problems arising when we describe optical beams and pulses. It turns out that

[1]Monochromatic means "single-colour". The origin of this term is associated with the fact that the human eye registers change in the visible radiation frequency as a change of colour.

such a description can be based on the notions of plane and spherical harmonic waves.

1.7.1 Quasi-plane waves

In practice it is impossible to devise the source of light generating the field described by eqn (1.14). The finite size, or as it is often called, the finite *aperture* of the source, leads to variation of the amplitude of optical vibrations in the plane perpendicular to the propagation direction; that is, to the emergence of a *spatially modulated* wave. In such a wave the values of the amplitude and phase depend on the coordinate, i.e. the field pattern drastically differs from that of a plane wave. However, there are cases in which the plane wave can be considered as more than a mathematical abstraction. Physical considerations indicate that if the aperture of a light source is equal to d, the wave process will have the properties of a plane wave not only for $d \to \infty$, but also for a finite but sufficiently wide aperture.

What does the term "wide aperture" mean in this case? Obviously, the value of d should be compared to a certain intrinsic spatial scale of the process. The natural scale for the optical wave is its wavelength λ. In the visible radiation range $\lambda \simeq 10^{-4}$ cm. The aperture of a typical laser beam is $d \approx 0.1$ cm. Consequently, $d/\lambda \approx 10^3 \gg 1$.

With the help of laser optical resonators as well as different optical collimators, one can produce wide-aperture optical beams the amplitude and phase of which vary slowly (on the scale of the wavelength) in the plane perpendicular to the direction of propagation. In such beams the wavefronts at large distances from the source do not differ much from the plane surface. Thus, we deal with *quasi-plane waves*. In this case the field at the exit aperture of the source can be written in the form

$$E_x(t, x, y, 0) = A(x, y)\cos \omega t. \tag{1.28}$$

Here the wavefront is plane, but the wave amplitude varies slowly in the x–y plane with the characteristic scale of the variation being equal to d.

One can expect that at $d \gg \lambda$ there should be a region, located at sufficiently long distance z ($z \gg \lambda$, $z > d$), in which the wave excited by the source (eqn (1.28)) remains quasi-plane:

$$E_x(t, x, y, z) = A(x, y)\cos(\omega t - kz). \tag{1.29}$$

The criterion of applicability of eqn (1.29) can be established only after solving eqn (1.10) with the boundary condition given by eqn (1.28). The further discussion of this problem will be resumed in Part III of this book. However, we shall give here a remarkable corollary describing the propagation of the so-called *Gaussian optical beam*.

Suppose that at $z = 0$ the phase wavefront is plane and the amplitude distribution is described by a Gaussian curve

$$E_x(t, x, y, 0) = A_0 \exp[-(x^2 + y^2)/d^2]\cos \omega t \tag{1.30}$$

or

$$E_x(t, r, 0) = A_0 \exp(-r^2/d^2) \cos \omega t, \tag{1.31}$$

where $r^2 = x^2 + y^2$ and d is the beam radius.

As shown in Part III, the approximate solution of the wave equation

$$\Delta E_x - \frac{1}{c^2}\frac{\partial^2 E_x}{\partial t^2} = 0 \tag{1.32}$$

with the boundary condition given by eqn (1.31) takes the form

$$E_x(t, r, z) = \tfrac{1}{2} A_c(r, z) \exp[i(\omega t - kz)] + \text{c. c.} \tag{1.33}$$

Here "c. c." means "complex-conjugate", and $A_c(r, z)$ is the complex wave amplitude, defined as

$$A_c(r, z) = \frac{A_0}{(1 - iz/z_d)} \exp\left[-\frac{r^2}{(1 - iz/z_d)d^2}\right], \tag{1.34}$$

where

$$z_d = kd^2/2. \tag{1.35}$$

Equation (1.34) can be rewritten as

$$A_c(r, z) = A \exp(-i\varphi), \tag{1.36}$$

where A and φ are real amplitude and phase of the wave:

$$A(r, z) = \frac{A_0}{\sqrt{1 + (z/z_d)^2}} \exp\left[-\frac{r^2/d^2}{(1 + z/z_d)^2}\right], \tag{1.37}$$

$$\varphi(r, z) = \frac{r^2}{d^2}\frac{z/z_d}{1 + (z/z_d)^2} - \arctan(z/z_d). \tag{1.38}$$

According to eqns (1.37) and (1.38) the distribution of the amplitude and the phase over the beam cross-section is governed by one parameter only, which is the ratio of the distance z traversed by the wave to the characteristic length z_d defined by eqn (1.35) and called the *diffraction length* of the optical beam. For $z \ll z_d$ the expression for the optical field simplifies and takes the form

$$E_x(t, r, z) = A_0 \exp(-r^2/d^2) \cos(\omega t - kz), \tag{1.39}$$

i.e. we have wave of the form given by eqn (1.29).

The higher the value of z_d, i.e. the higher the ratio d/λ, the longer is the distance z over which the phase wavefront remains plane. For example, the diffraction length $z_d = \pi d^2/\lambda - 5$ m for helium–neon laser radiation with a wavelength $\lambda = 0.63\ \mu$m and a beam radius $d = 0.1$ cm. By broadening the beam to $d = 1$ cm we obtain $z_d = 500$ m. Thus, the calculation results confirm the preliminary qualitative reasoning.

What is remarkable is that the Gaussian beam keeps its form when $z > z_d$. According to eqn (1.37), in this region only the width of the beam varies. The beam acquires, however, the properties of a spherical wave. It follows from eqns (1.37) and (1.38) that for $z \gg z_d$ the wave amplitude decreases proportionally with the distance traversed,

$$A(r, z) \simeq A_0 z_d / z \sim 1/z, \tag{1.40}$$

while the phase wavefront becomes spherical:

$$\varphi(r, z) \sim r^2/d^2. \tag{1.41}$$

The above-mentioned properties of the Gaussian beam are illustrated in Fig. 1.4, showing the evolution of the cross-section and the distortion of the wavefront of the beam as it propagates in free space.

1.7.2 Superposition of spherical waves and a randomly modulated wave

An observer located sufficiently far from a source of light perceives it as a luminous point. It is natural to expect that in this case the characteristics of the optical field at the observation point would be close to those of a spherical wave. The radiation of stars reaching the Earth is a typical example.

The results cited earlier help to trace the formation of a spherical wave for a specific radiation model; namely, for a Gaussian optical beam. By virtue of eqns (1.40) and (1.41) an observer located at a distance $z \gg z_d$ from the exit mirror of a helium–neon laser perceives the laser radiation as a spherical wave.

Individual atoms and molecules are elementary sources of optical radiation. Since the characteristic size of an atom is about 10^{-8} cm, the radiation of individual atoms and molecules can be treated as spherical waves even on microscopic scales.

Usually, we deal with the emission of a large number N of elementary oscillators. For instance, $N \simeq 10^{19}$ cm^{-3} in gas in normal conditions and $\simeq 10^{22}$ cm^{-3} in a condensed medium. That is why the radiation from conventional sources of light is a superposition of a large number of spherical waves:

$$E_x(t, x, y, z) = \sum_{i=1}^{N} (A_{0i}/r_i) \cos[\omega(t - r_i/c)], \tag{1.42}$$

where

$$r_i = [(x - x_i)^2 + (y - y_i)^2 + (z - z_i)^2]^{1/2}, \tag{1.43}$$

and x_i, y_i, and z_i are the coordinates of the ith oscillator (atom or molecule). A qualitative picture of the wave field of an ensemble of oscillators is given in

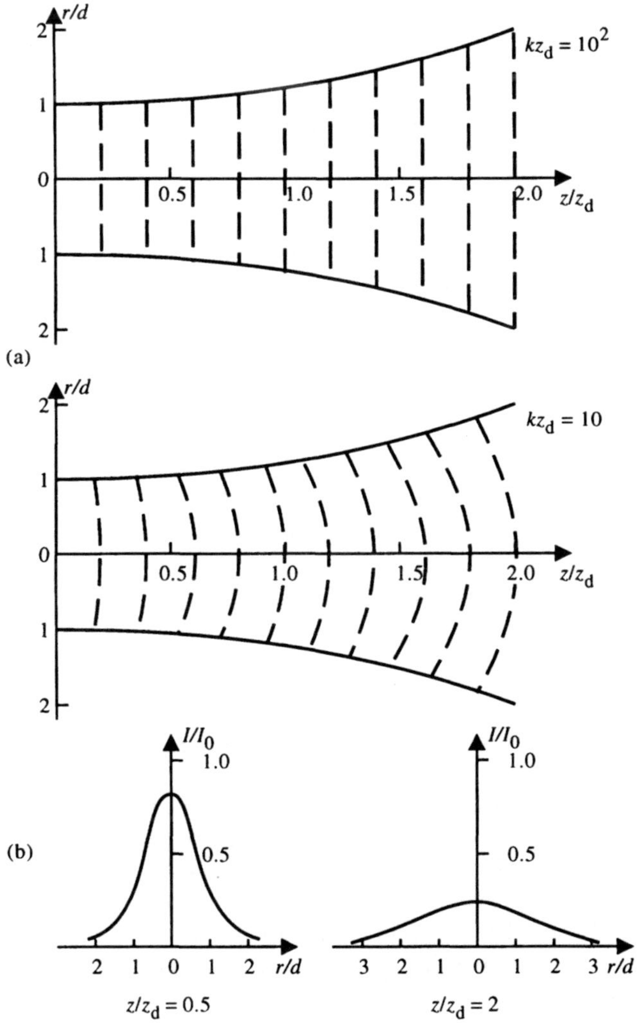

Fig. 1.4 The propagation of a Gaussian optical beam. (a) The beam contour in longitudinal section (solid curves) and wavefronts (dotted lines). (b) The radiation intensity profile in the beam cross-section.

Fig. 1.5, which shows the evolution of the spatial superposition of spherical waves. One can see that near the oscillators the field has the complicated structure of a spatially modulated wave. As the distance from the source of light grows, the spherical wavefronts of individual waves are seen to coincide over larger areas. Finally, at $z \to \infty$ the superposition of many spherical waves is well approximated by a single spherical wave.

How does one describe quantitatively the properties of a complicated wave

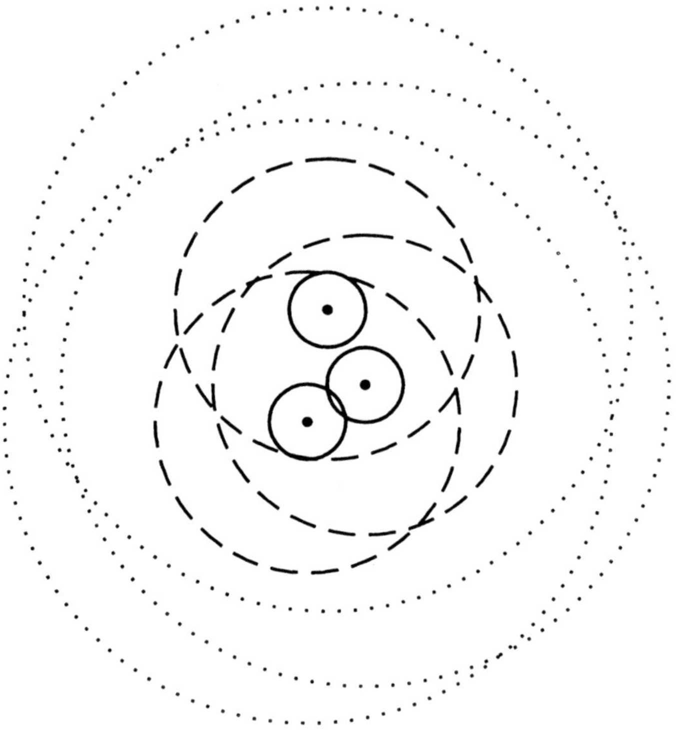

Fig. 1.5 The superposition of spherical waves.

field constituting a superposition of many spherical waves? This is one of the main problems of the physics of light emission, to which Part II of the book is devoted. Here we shall only remark that as soon as the behaviour of a large collection of atoms or molecules obeys *statistical* laws, the behaviour of the total optical field is also statistical. Hence, the notion of a *randomly modulated* wave appears in optics, i.e. a wave the amplitude and phase of which vary in a random, irregular way.

1.7.3 Quasi-harmonic waves

It has already been mentioned that in experiments we usually deal with optical pulses of finite duration, but not with an ideal harmonic wave. Real sources of light emit waves that are modulated not only in space, but also in time. Taking into account only the temporal modulation of the field and considering the distances $z \ll z_{\rm d}$ and the points located near the axis of the optical beam (x, $y \ll d$), we can replace eqn (1.28) by the expression

$$E_x(t, x, y, 0) = f(t) = A(t)\cos\omega t. \tag{1.44}$$

According to eqn (1.13) the source described by eqn (1.44) will excite a plane *amplitude-modulated* wave

$$E_x(t,z) = A(t - z/c)\cos[\omega(t - z/c)], \tag{1.45}$$

which is the solution of the unidimensional wave equation (1.11). In a more general case it is not only the amplitude, but also the phase of the vibrations of the source that vary in time:

$$E_x(t,0) = f(t) = A(t)\cos[\omega t - \varphi(t)]. \tag{1.46}$$

As a result, a wave having temporal modulation of both amplitude and phase is generated.

The variations of $A(t)$ and $\varphi(t)$ can be quite rapid. For instance, modern lasers are capable of generating optical pulses with a duration $\tau \sim 0.1$–1.0 ps. Nevertheless, in most cases the corresponding amplitude and phase variations remain slow on the scale of the period of optical vibrations, $T = 2\pi/\omega$. In the visible range $T \simeq 1$–10 fs, and even for picosecond pulses, the inequality $\tau \gg T$ holds. The properties of such slowly modulated waves resemble those of a harmonic wave. As a consequence, these waves are called *quasi-harmonic*.

By direct substitution one can verify that eqn (1.45) satisfies the wave equation (1.10). According to eqn (1.45) the form of temporal modulation of a plane wave remains as it propagates in a free space. Such behaviour of temporally modulated waves differs principally from that of spatially modulated waves. Nevertheless, as will be shown in Part IV of this book, the shape and duration of the optical pulse can vary substantially when it propagates in a material medium.

While on the subject of temporal modulation of optical waves, let us ponder over the following question: How close can one realize in practice an ideal monochromatic wave? On the face of it, it seems that we are dealing with a technical problem. Presumably, improvements in the stability of the parameters of a single-frequency continuous-wave laser can lead to the generation of optical vibrations that differ from harmonic waves by as little as one wishes. However, there is an ultimate limit to monochromaticity, defined by quantum fluctuations in emitting atoms and molecules. These fluctuations prove to be the reason for intrinsically unavoidable amplitude and phase modulations of the wave. It should be noted that radiation non-monochromaticity caused by quantum fluctuations is relatively small. The deviation of frequency from the average value does not exceed 100 Hz. Nevertheless, this deviation is essential in quantum standards of time and frequency and should be taken into consideration while performing precise optical experiments. In summary, we can say that the ideal monochromatic wave is just another abstraction. Like the ideal plane wave, it cannot be realized in practice.

1.8 Spectral expansion of the optical field

As explained in the previous section, in many cases the radiation properties of real sources of light are close to the properties of plane or spherical harmonic

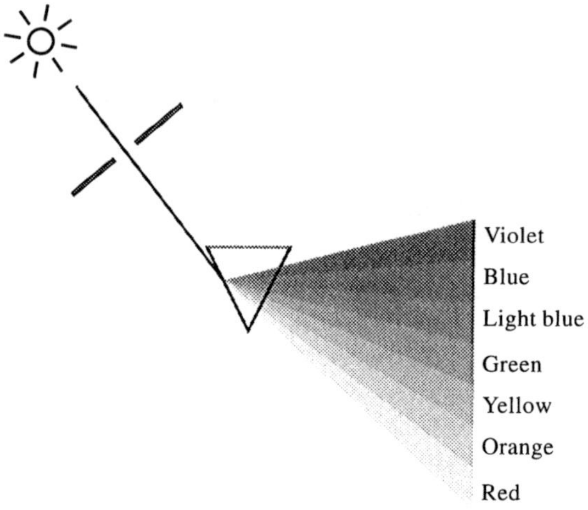

Fig. 1.6 A scheme for Newton's experiment.

waves. However, the standard waves are significant in other aspects. Any arbitrary optical field can be regarded as a superposition of standard waves. Expressed otherwise, it is possible to decompose an arbitrary optical field into a spectrum, i.e. to perform *spectral expansion.*

Expansion of optical beams and pulses into plane harmonic waves is of particular importance for optics. As a matter of fact, such an expansion not only proves to be a convenient mathematical operation, but is in fact *actually performed* in optical experiments. One of the classic experiments of this kind is the famous Newton's experiment in which sunlight was decomposed into a spectrum using a glass prism (see Fig. 1.6). Newton's conclusion that "...the sunlight consists of the rays of different refractability" can be easily translated into the mathematical language of spectral expansions. It means that the field of a plane non-monochromatic wave, $E_x(t, z)$, can be represented as a superposition of plane monochromatic waves.

Indeed, the arbitrary function $f(t)$ describing an optical perturbation at $z = 0$ can be written in the form of the Fourier integral

$$f(t) = \frac{1}{\pi}\int_0^\infty [a(\omega)\cos\omega t + b(\omega)\sin\omega t]d\omega = \frac{1}{\pi}\int_0^\infty f(\omega)\cos[\omega t - \varphi(\omega)]d\omega, \quad (1.47)$$

i.e. it can be decomposed into a set of harmonic vibrations or into the frequency spectrum

$$f_\omega = f(\omega)\cos[\omega t - \varphi(\omega)]. \quad (1.48)$$

The amplitudes of quadrature spectral components $a(\omega)$ and $b(\omega)$, or the spectral amplitude $f(\omega)$ and phase $\varphi(\omega)$ defining the frequency spectrum of the function $f(t)$, are calculated applying the inverse Fourier transformation

$$a(\omega) = \int_{-\infty}^{\infty} f(t)\cos\omega t\, dt, \quad b(\omega) = \int_{-\infty}^{\infty} f(t)\sin\omega t\, dt,$$
$$f(\omega) = \sqrt{a^2(\omega) + b^2(\omega)}, \qquad \tan\varphi(\omega) = b(\omega)/a(\omega). \tag{1.49}$$

Each harmonic component of the perturbation $f(t)$ specified at $z = 0$ excites a monochromatic optical wave

$$E_\omega(t, z) = f(\omega)\cos[\omega t - kz - \varphi(\omega)], \tag{1.50}$$

satisfying the wave equation. The total field, being the superposition of the waves given by eqn (1.50),

$$E_x(t, z) = \frac{1}{\pi}\int_0^{\infty} E_\omega\, d\omega. \tag{1.51}$$

also satisfies the wave equation. Equation (1.50) together with eqn (1.45) provides one more method of describing a non-monochromatic plane wave; namely, as a superposition of plane monochromatic waves. It is generally said that expression (1.45) describes a non-monochromatic wave in the *time domain*, while (1.51) describes that in the *spectral, or frequency, domain*.

Spectral expansions can naturally be generalized to wave beams, i.e. to spatially modulated waves. The wave beam can also be treated as a superposition of plane waves propagating in different directions. Different spectral components in such an expansion are characterized by the angles between the wave propagation direction and the coordinate axes. It is therefore conventional to define the *angular spectrum* of a spatially modulated wave. Remarkably, the expansion into the angular spectrum is actually performed in very simple experiments. For instance, the form of the angular spectrum of an optical beam is found from the illumination intensity distribution over a screen either removed sufficiently far from the source of light (at $z \gg z_\mathrm{d}$) or located in the focal plane of a lens. In such experiments, free space and a lens perform the same Fourier transformation with respect to the angular spectrum as does a glass prism with respect to the frequency spectrum. The detailed discussion of this issue will be presented in Part III of this book; its analysis will require more detailed views about the physics of propagation of spatially modulated waves.

1.9 The superposition principle

According to this principle, optical waves with different frequencies and propagation directions travel in vacuum independently of one another. There are simple experiments clearly illustrating the superposition principle. For instance, two observers may see different objects when looking through the same hole in the screen, and, generally speaking, the patterns that they see are by no means interrelated.

Mathematically, the superposition principle is a consequence of the linearity of the wave equation describing the propagation of optical waves in vacuum.

Indeed, if the fields E_1, E_2, ... are solutions of the wave equation, then the sum of the fields,

$$E = \sum_n E_n,$$

is its solution too. One can verify this by inserting in eqn (1.10) the expressions for plane waves of the form

$$E_{xn} = A_n \cos(\omega_n t - k_n z), \quad k_n = \omega_n / c.$$

In this case the wave equation disintegrates into a set of independent equations for individual waves.

Being trivial in electromagnetic theory, the superposition principle presented quite an obstacle to proponents of the corpuscular theory of light. It seemed that the corpuscles belonging to different optical beams should interact somehow or be scattered by one another. Even Huygens himself, the founder of wave theory, by no means regarded the superposition principle as self-evident. In his *Treatise on Light* Huygens wrote "...The most amazing property of light consists in that the rays coming from different, and even opposite, directions penetrate one through another without impeding their mutual actions". We shall see later that Huygens' doubts have serious grounds. Contemporary laser optics gives many examples of definite violations of the superposition principle for intense optical waves propagating in a medium. Quantum electrodynamics predicts violations of the superposition principle in vacuum (although for very high light intensities). For example, the scattering of light by light has to take place in superintensive optical fields in vacuum.[2] Therefore, the superposition principle in optics is by no means universal. We shall consider it to be valid for optical waves in vacuum. The limits of its applicability in material media are discussed in Part IV of this book.

Bibliography

Born, M. and Wolf, E. (1968). *Principles of optics.* Pergamon Press, Oxford.

Evenson, K. M., Wells, J. S., Petersen, F. R., Danielson, B. L., and Day, G. W. (1973). *Applied Physics Letters*, **22**, 192–195.

Hertz, H. R. (1887). *Annalen der Physik und Chemie*, **31**, 421–428.

Jaffe, B. (1961). *Michelson and the speed of light.* Heinemann, London.

Kittel, C., Knight, W. D., and Ruderman, M. A. (1965). *Mechanics.* McGraw-Hill, New York.

Letokhov, V. S. and Chebotaev, V. P. (1977). *Nonlinear laser spectroscopy.* Springer-Verlag, Berlin.

Maxwell, J. C. (1864). *Transactions of the Royal Society,* **CLV**.

Michelson, A. A. (1927). *Studies in optics.* University of Chicago Press, Chicago.

[2] To date, there are no experiments that confirm this prediction.

2

Polarization of light

The transverse nature of an optical wave. The polarization states of a plane harmonic wave—elliptical, circular, and linear polarizations. Non-monochromatic light—the natural polarization. Experimental methods of polarization measurement.

2.1 The transverse nature of an optical wave

The evolution of an arbitrary component of the vector $\vec{E}$ or $\vec{H}$ in a plane light wave is described by eqns (1.10) and (1.13). This is not, however, the only information that Maxwell's equations can provide. Using eqns (1.2)–(1.5), we can define the relationship between the magnitudes and directions of the vectors $\vec{E}$ and $\vec{H}$.

Consider a plane optical wave propagating along the z-axis. In such a wave, $\vec{E} = \vec{E}(z,t)$ and $\vec{H} = \vec{H}(z,t)$. We shall show that this wave is transverse, i.e. the field components of the wave are perpendicular to the propagation direction. Indeed, it follows from eqn (1.4) that

$$\mathrm{div}\vec{E} = \frac{\partial E_x}{\partial x} + \frac{\partial E_y}{\partial y} + \frac{\partial E_z}{\partial z} = 0.$$

Since in the considered wave $\partial E_x/\partial x = 0$ and $\partial E_y/\partial y = 0$, then $\partial E_z/\partial z = 0$, i.e. the component E_z does not vary in space. Writing the equation for the z component of the circulation of $\vec{H}$, we obtain

$$(\mathrm{rot}\vec{H})_z = \frac{\partial H_y}{\partial x} - \frac{\partial H_x}{\partial y} = \frac{1}{c}\frac{\partial E_z}{\partial t}.$$

Since $\partial H_y/\partial x = 0$ and $\partial H_x/\partial y = 0$, we have $\partial E_z/\partial t = 0$. Consequently, E_z is a constant independent of z and t. It should be put equal to zero because we consider only rapidly alternating fields. The similar conclusion can be drawn with respect to the longitudinal component H_z of the magnetic field. Thus, the components E_x, E_y and H_x, H_y of a plane electromagnetic wave propagating in free space along the z-axis differ from zero, whereas $E_z = H_z = 0$.

Maxwell's equations allow us to obtain the relationship between the magnitudes of the vectors $\vec{E}$ and $\vec{H}$ in an optical wave. From eqns (1.2) and (1.3), we obtain

$$\begin{aligned}(\mathrm{rot}\vec{E})_y &= \frac{\partial E_x}{\partial z} = -\frac{1}{c}\frac{\partial H_y}{\partial t}, \\ (\mathrm{rot}\vec{H})_x &= -\frac{\partial H_y}{\partial z} = \frac{1}{c}\frac{\partial E_x}{\partial t}.\end{aligned} \tag{2.1}$$

Similarly, we find the relationship between the components E_y and H_x:

$$\frac{\partial E_y}{\partial z} = \frac{1}{c}\frac{\partial H_x}{\partial t}, \quad \frac{\partial H_x}{\partial z} = \frac{1}{c}\frac{\partial E_y}{\partial t}. \tag{2.2}$$

Equations (2.1) and (2.2) describe two independent plane optical waves. Both waves propagate along the z-axis; one is characterized by the orthogonal field components E_x and H_y, and the other by E_y and H_x.

2.1.1 A plane harmonic wave

Let us consider a plane optical wave with the field configuration "E_x, H_y". Let

$$E_x = A\cos(\omega t - kz), \quad k = \omega/c. \tag{2.3}$$

Using eqns (2.1) and (2.3), we can easily show that

$$\frac{\partial H_y}{\partial t} = -c\frac{\partial E_x}{\partial z} = \frac{\partial E_x}{\partial t}, \quad \frac{\partial H_y}{\partial z} = -\frac{1}{c}\frac{\partial E_x}{\partial t} = \frac{\partial E_x}{\partial z},$$

whence it follows that the fields E_x and H_y can differ only by a constant. The optical field, however, does not contain any constant component and, consequently,

$$E_x = H_y = A\cos(\omega t - kz). \tag{2.4}$$

Thus, in a plane harmonic optical wave of "E_x, H_y" type both field components vary in phase according to a harmonic law (Fig. 2.1). Similarly, for the wave "E_y, H_x" we have

$$-E_y = H_x = A\cos(\omega t - kz). \tag{2.5}$$

2.2 The polarization states of a plane harmonic wave—elliptical, circular, and linear polarizations

The two waves E_x, H_y and E_y, H_x considered above differ one from another by the directions of the vectors $\vec{E}$ and $\vec{H}$, i.e. by the *polarization directions*. The waves described by eqns (2.1)–(2.5) are called *linearly polarized* because for a fixed value of z the tip of the vector $\vec{E}$ traces a straight line. The direction of the wave polarization vector is assumed to coincide with the direction of the $\vec{E}$ vector. Let us introduce the polarization plane by defining it as a plane containing the vector $\vec{E}$ and the unit vector $\vec{z}_0$ characterizing the wave propagation direction. Linear polarized waves are also referred to as *plane polarized waves*.

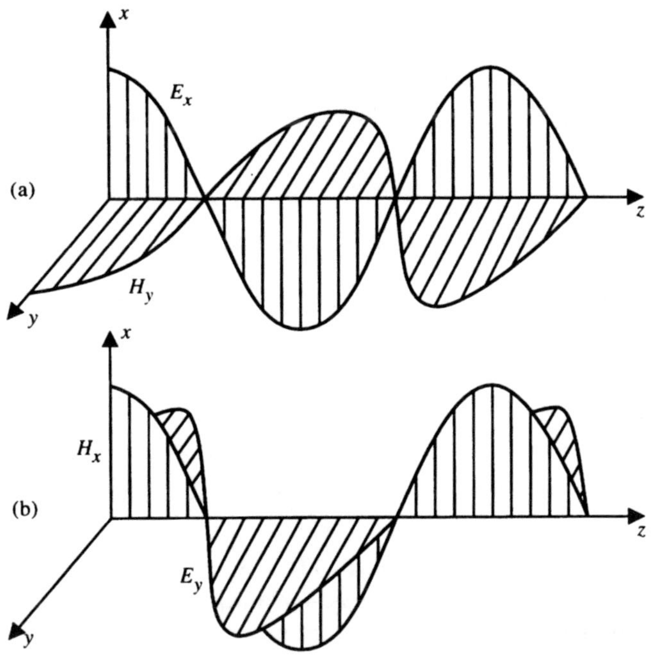

Fig. 2.1 Plane harmonic waves of type (a) E_x, H_y and (b) E_y, H_x.

The preference given in the formulation of this notion to the electric field vector is only a matter of definition. Both vectors $\vec{E}$ and $\vec{H}$ are "light vectors". However, when speaking about the interaction of light with matter, a certain precedence should be still given to the vector $\vec{E}$. In fact, the force exerted by an electromagnetic field on an electric charge is equal to $\vec{F} = e\vec{E} + (e/c)[\vec{v}, \vec{H}]$, and the action of the magnetic field at $v/c \ll 1$ is therefore much weaker than that of the electric field.

Either of the waves E_x, H_y or E_y, H_x satisfies the wave equation. Evidently, the sum of these waves satisfies it as well. This is a manifestation of the superposition principle for optical waves in vacuum. In the general case both of the components E_x and E_y of the harmonic wave are non-zero, and the electric field vector takes the form

$$\vec{E}(t, z) = \vec{x}_0 E_x(t, z) + \vec{y}_0 E_y(t, z). \tag{2.6}$$

We shall consider a plane wave the electric field components of which vary according to a harmonic law:

$$E_x(t, z) = A_1 \cos(\omega t - kz + \varphi_1), \quad E_y(t, z) = A_2 \cos(\omega t - kz + \varphi_2). \tag{2.7}$$

Let us write down the equation of the track traced out by the tip of the vector $\vec{E}$ in the plane $z =$ constant. To do this, we shall introduce an auxiliary variable

$\tau = \omega t - kz$ and rewrite eqn (2.7) as

$$E_x(t,z) = A_1(\cos\tau\cos\varphi_1 - \sin\tau\sin\varphi_1),$$
$$E_y(t,z) = A_2(\cos\tau\cos\varphi_2 - \sin\tau\sin\varphi_2);$$

whence

$$\frac{E_x}{A_1}\sin\varphi_2 - \frac{E_y}{A_2}\sin\varphi_1 = \cos\tau\sin(\varphi_2 - \varphi_1),$$
$$\frac{E_x}{A_1}\cos\varphi_2 - \frac{E_y}{A_2}\cos\varphi_1 = \sin\tau\sin(\varphi_2 - \varphi_1).$$

Taking the squares of the left-hand and right-hand sides of these equations and summing them, we find

$$\left(\frac{E_x}{A_1}\right)^2 + \left(\frac{E_y}{A_2}\right)^2 - 2\frac{E_xE_y}{A_1A_2}\cos(\varphi_2 - \varphi_1) = \sin^2(\varphi_2 - \varphi_1). \tag{2.8}$$

Equation (2.8) represents an ellipse. The ellipse is inscribed into a rectangle the sides of which are parallel to the x- and y-axes and equal to $2A_1$ and $2A_2$ (Fig. 2.2(a)). Hence, in the general case of propagation of a monochromatic optical wave the tip of its vector $\vec{E}$ traces out an ellipse in the plane z = constant. The vector of magnetic field strength behaves in a similar manner. Such a wave is called an *elliptically polarized wave.*

While tracing an elliptical curve, the tip of the vector $\vec{E}$ rotates either clockwise or anticlockwise within the plane z = constant. In optics these two cases are distinguished by defining *right-handed* polarization (when facing the source of the light—the vector $\vec{E}$ is observed to rotate clockwise) and *left-handed* polarization (when the vector rotates anticlockwise). We shall show that the rotation direction of the vector $\vec{E}$ is determined by the sign of the phase difference $\varphi = \varphi_2 - \varphi_1$. Let t_0 be the moment of time at which $\omega t_0 - kz + \varphi_1 = 0$. It follows from eqn (2.7) that at this moment

$$E_x(t_0, z) = A_1, \quad \dot{E}_y(t_0, z) = -\omega A_2\sin\varphi, \tag{2.9}$$

where a dot over a letter denotes time differentiation, i.e. $\dot{E}_y = dE_y/dt$. Equations (2.9) show that at the moment at which the tip of the vector $\vec{E}$ reaches the extreme right point of its trajectory (see Fig. 2.2(a)), $\dot{E}_y < 0$ for $0 < \varphi < \pi$ and $\dot{E}_y > 0$ for $-\pi < \varphi < 0$. Obviously, the former case corresponds to the right-handed polarized wave and the latter to the left-handed polarized wave. Now we shall consider some specific examples.

2.2.1 Linear polarization

If $\varphi = \varphi_2 - \varphi_1 = m\pi$, where $m = 0, \pm1, \pm2, \ldots$, the ellipse degenerates into a straight line described by the equation

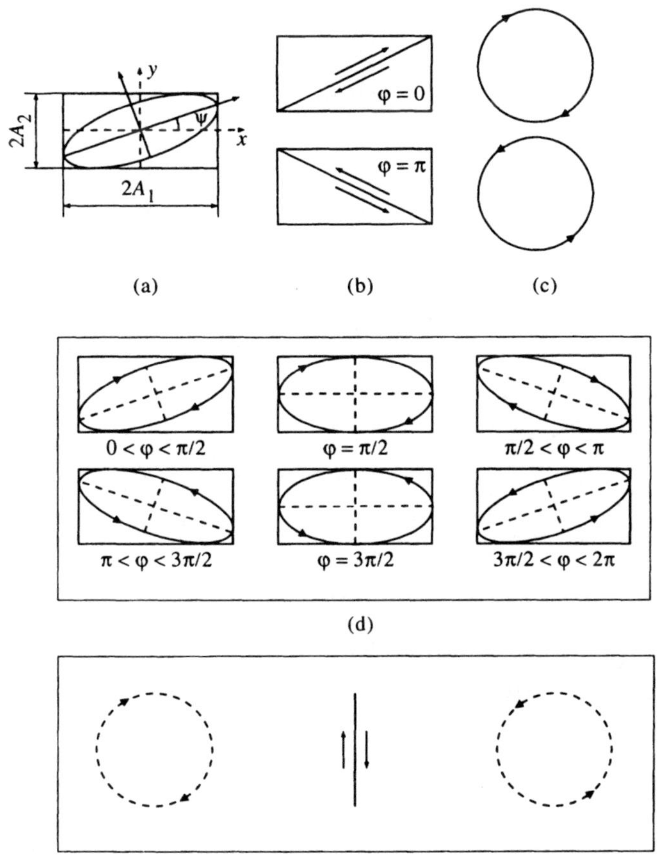

(e)

Fig. 2.2 The polarization states of a plane harmonic wave.

$$E_x/E_y = (-1)^m (A_1/A_2).$$

In this case the wave is linear polarized or plane polarized. Two possible polarization directions in a plane polarized wave, corresponding to $\varphi = 0$ and $\varphi = \pi$, are shown in Fig. 2.2(b).

2.2.2 Circular polarization

If $A_1 = A_2 = A$ and $\varphi = \varphi_2 - \varphi_1 = m\pi/2$, where $m = \pm 1, \pm 3, \pm 5, \ldots$, then the ellipse degenerates into a circle described by the equation

$$E_x^2 + E_y^2 = A^2.$$

Thus, the tip of the vector $\vec{E}$ (just as of the vector $\vec{H}$) traces out the circle clockwise or anticlockwise. This kind of polarization is referred to as *circular*.

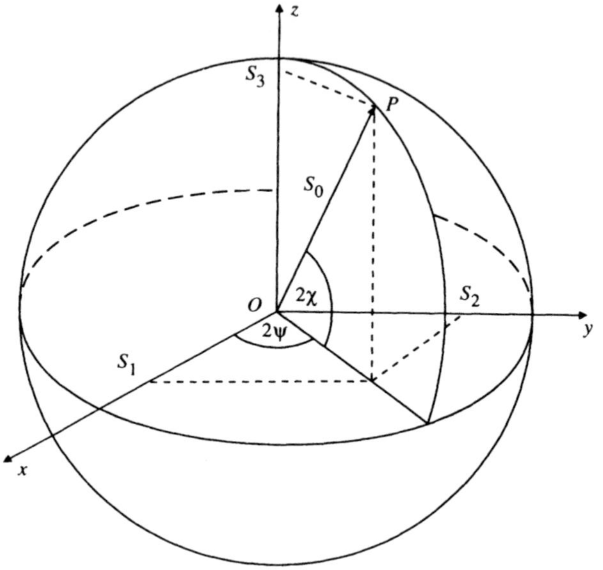

(f)

Right-handed and left-handed circular polarizations are distinguished. For right-handed polarization,

$$\varphi = \pi/2 + 2m\pi, \quad \vec{E}^{(+)} = \vec{x}_0 E_x^{(+)} + \vec{y}_0 E_y^{(+)}, \tag{2.10}$$

where

$$E_x^{(+)} = A\cos(\tau + \varphi_1), \quad E_y^{(+)} = -A\sin(\tau + \varphi_1). \tag{2.11}$$

For left-handed polarization,

$$\varphi = -\pi/2 + 2m\pi, \quad \vec{E}^{(-)} = \vec{x}_0 E_x^{(-)} + \vec{y}_0 E_y^{(-)}, \tag{2.12}$$

where

$$E_x^{(-)} = A\cos(\tau + \varphi_1), \quad E_y^{(-)} = A\sin(\tau + \varphi_1). \tag{2.13}$$

It follows from eqns (2.10)–(2.13) that

$$\vec{E} = \vec{E}^{(+)} + \vec{E}^{(-)} = \vec{x}_0 2A\cos(\tau + \varphi_1). \tag{2.14}$$

This means that the sum of right-handed and left-handed polarized waves yields a linearly polarized wave.

2.2.3 Stokes' parameters and Poincaré's sphere

In the general case, a plane monochromatic optical wave has right-handed or left-handed elliptical polarization. The polarization ellipse can be adequately characterized by three parameters, e.g. by the amplitudes A_1 and A_2 of orthogonal oscillations and by the phase difference φ, or by the ellipse's half-diameters a and b and the angle ψ between the x-axis and the major axis of the ellipse (see Fig. 2.2(a)). It is also convenient to describe an elliptically polarized wave by means of Stokes' parameters:

$$\begin{aligned} S_0 &= A_1^2 + A_2^2, \qquad S_1 = A_1^2 - A_2^2, \\ S_2 &= 2A_1A_2\cos\varphi, \quad S_3 = 2A_1A_2\sin\varphi. \end{aligned} \tag{2.15}$$

Only three of these parameters are independent, because the following identity is valid:

$$S_0^2 = S_1^2 + S_2^2 + S_3^2. \tag{2.16}$$

Introducing an auxiliary angle χ, defined by the equation

$$\tan\chi = \pm b/a, \tag{2.17}$$

where a and b are the half-diameters of the polarization ellipse, the "+" sign corresponds to the right-handed polarized wave, and the "−"sign to the left-handed polarized wave, we can derive the following relations for Stokes' parameters:

$$\begin{aligned} S_1 &= S_0\cos(2\chi)\cos(2\psi), \\ S_2 &= S_0\cos(2\chi)\sin(2\psi), \\ S_3 &= S_0\sin(2\chi). \end{aligned} \tag{2.18}$$

Equations (2.15)–(2.18) can be used as a basis for a visual geometric illustration of polarization. Stokes' parameters, S_1, S_2, and S_3, can be treated as the Cartesian coordinates of a point lying on a sphere of radius S_0. The angles 2χ and 2ψ are then the spherical angular coordinates of this point (Fig. 2.2(f)). The angle ψ characterizes the orientation of the polarization ellipse, while the angle χ denotes its ellipticity—that is, the ratio of diameters—and the direction of rotation. The sphere shown in Fig. 2.2(f) is called Poincaré's sphere because he was first to propose this geometric representation of polarization.

To summarize, we list the possible polarization states of a plane harmonic wave (illustrated in Fig. 2.2): (a) an elliptical polarized wave; (b) a linear polarized wave; (c) a circular polarized wave (right-handed and left-handed polarizations); (d) an elliptical polarized wave for various values of the phase difference φ of orthogonal field components; (e) a linear polarized wave as a combination of two circularly polarized waves with opposite directions of rotation; (f) Poincaré's sphere, representing the polarization of a plane harmonic wave.

2.2.4 Complex variable notation for an optical wave

Optical wave fields can be described in a concise and convenient way by using complex variable notation. Making use of Euler's formula,

$$\cos\alpha = \tfrac{1}{2}\exp(i\alpha) + \tfrac{1}{2}\exp(-i\alpha), \tag{2.19}$$

we can represent the electric field of a plane harmonic wave,

$$\vec{E} = \vec{x}_0 A_1 \cos(\omega t - kz + \varphi_1) + \vec{y}_0 A_2 \cos(\omega t - kz + \varphi_2), \tag{2.20}$$

in the form

$$\vec{E} = \tfrac{1}{2}(\vec{x}_0 \mathcal{E}_x + \vec{y}_0 \mathcal{E}_y)\exp[i(\omega t - kz)] + \text{c. c.} \tag{2.21}$$

Here the abbreviation "c. c." designates "complex conjugate"; the quantities $\mathcal{E}_x$ and $\mathcal{E}_y$ are called *complex amplitudes* and defined by the following formulae:

$$\mathcal{E}_x = A_1\exp(i\varphi_1), \quad \mathcal{E}_y = A_2\exp(i\varphi_2). \tag{2.22}$$

According to the definition, the modulus of the complex amplitude is equal to the real amplitude, and its argument to the real phase of the optical wave. For a linear polarized wave,

$$\vec{E} = \tfrac{1}{2}\vec{e}\mathcal{E}\exp[i(\omega t - kz)] + \text{c. c.}, \tag{2.23}$$

where $\mathcal{E}$ is the complex amplitude and $\vec{e}$ is the unit vector characterizing the polarization direction ("polarization vector"). If the wave is circularly polarized, then

$$\vec{e} = (\vec{x}_0 \pm i\vec{y}_0)/\sqrt{2}, \tag{2.24}$$

where the "+" and "−" signs correspond to the right-handed and left-handed rotations.

2.2.5 A wave vector

Let a plane harmonic optical wave propagate in an arbitrary direction set by the unit vector $\vec{n}$. Then constant-phase surfaces of the wave represent planes perpendicular to the vector $\vec{n}$ (Fig. 2.3). Let us introduce the wave vector

$$\vec{k} = \vec{n}\omega/c. \tag{2.25}$$

According to eqn (2.25), the vector $\vec{k}$ indicates the wave propagation direction, and its modulus is equal to the wave number $k = \omega/c$. Let ξ denote the distance traversed by the wave in the direction $\vec{n}$. By drawing a radius vector $\vec{r}$ from the origin of coordinates to an arbitrary point at the wavefront (see Fig. 2.3), we can write

$$\xi = \vec{n}\vec{r}. \tag{2.26}$$

Hereinafter the symbol $\vec{a}\vec{b}$ denotes the scalar product of vectors, i.e. $\vec{a}\vec{b} \equiv (\vec{a}, \vec{b})$. Using eqn (2.26), we obtain

$$k\xi = k\vec{n}\vec{r} = \vec{k}\vec{r}. \tag{2.27}$$

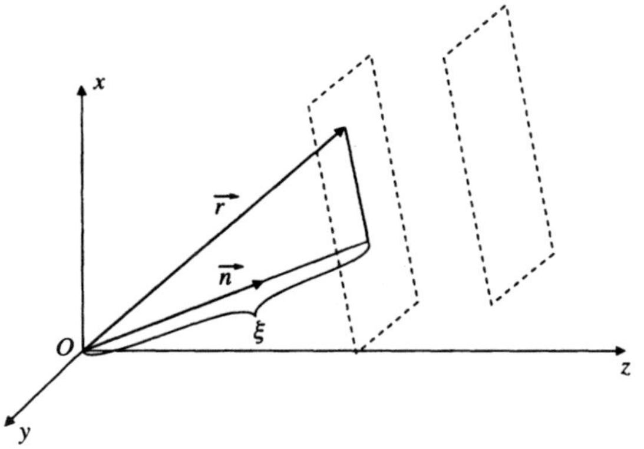

Fig. 2.3 A plane wave propagating in an arbitrary direction $\vec{n}$. The constant phase surfaces perpendicular to the vector $\vec{n}$ are shown.

Now the wave field can be presented in the form

$$\vec{E} = \tfrac{1}{2}\vec{\mathcal{E}} \exp[i(\omega t - \vec{k}\vec{r})] + \text{c. c.}, \tag{2.28}$$

where $\vec{\mathcal{E}} = \vec{e}\mathcal{E}$ and $\vec{e}$ is the polarization vector. The magnetic field can be written similarly:

$$\vec{H} = \tfrac{1}{2}\vec{\mathcal{H}} \exp[i(\omega t - \vec{k}\vec{r})] + \text{c. c.} \tag{2.29}$$

Now we can obtain some general relationships for a plane monochromatic optical wave. We shall use the nabla operator

$$\vec{\nabla} = \vec{x}_0 \frac{\partial}{\partial x} + \vec{y}_0 \frac{\partial}{\partial y} + \vec{z}_0 \frac{\partial}{\partial z}. \tag{2.30}$$

For an arbitrary vector field $\vec{a}(\vec{r})$,

$$\text{rot}\, \vec{a} = [\vec{\nabla}, \vec{a}], \quad \text{div}\, \vec{a} = (\vec{\nabla}, \vec{a}). \tag{2.31}$$

Using eqns (2.31), we can represent Maxwell's equations (1.2)–(1.5) in the form

$$\begin{aligned} [\vec{\nabla}, \vec{E}] &= -\frac{1}{c}\frac{\partial \vec{H}}{\partial t}, \quad (\vec{\nabla}, \vec{E}) = 0, \\ [\vec{\nabla}, \vec{H}] &= \frac{1}{c}\frac{\partial \vec{E}}{\partial t}, \quad (\vec{\nabla}, \vec{H}) = 0. \end{aligned} \tag{2.32}$$

It is easy to see that

$$\frac{\partial}{\partial \alpha} \exp[i(\omega t - \vec{k}\vec{r})] = -ik_\alpha \exp[i(\omega t - \vec{k}\vec{r})],$$

where $\alpha = x, y, z$. Making use of this identity together with eqn (2.28), we obtain

$$(\vec{\nabla}, \vec{E}) = \tfrac{1}{2}(-i\vec{k}, \vec{\mathcal{E}}) \exp[i(\omega t - \vec{k}\vec{r})] + \text{c. c.} \tag{2.33}$$

and

$$[\vec{\nabla}, \vec{E}] = \tfrac{1}{2}[-i\vec{k}, \vec{\mathcal{E}}] \exp[i(\omega t - \vec{k}\vec{r})] + \text{c. c.} \tag{2.34}$$

Finally,

$$\frac{\partial \vec{E}}{\partial t} = \tfrac{1}{2}(i\omega)\vec{\mathcal{E}} \exp[i(\omega t - \vec{k}\vec{r})] + \text{c. c.} \tag{2.35}$$

Similar formulae can also be written for the vector $\vec{H}$. Using them, we can represent eqns (2.32) in the form

$$\begin{aligned} [\vec{k}, \vec{\mathcal{E}}] &= \frac{\omega}{c}\vec{\mathcal{H}}, \quad (\vec{k}, \vec{\mathcal{E}}) = 0, \\ [\vec{k}, \vec{\mathcal{H}}] &= -\frac{\omega}{c}\vec{\mathcal{E}}, \quad (\vec{k}, \vec{\mathcal{H}}) = 0 \end{aligned} \tag{2.36}$$

or, taking into account eqn (2.25), as

$$\begin{aligned} [\vec{n}, \vec{\mathcal{E}}] &= \mathcal{H}, \quad (\vec{n}, \vec{\mathcal{E}}) = 0, \\ [\vec{n}, \vec{\mathcal{H}}] &= -\mathcal{E}, \quad (\vec{n}, \vec{\mathcal{H}}) = 0. \end{aligned} \tag{2.37}$$

Inserting eqns (2.37) for $\vec{\mathcal{E}}$ and $\vec{\mathcal{H}}$ into eqns (2.28) and (2.29), we obtain the following vector relations describing the structure of the field of a plane monochromatic optical wave:

$$\begin{aligned} \vec{E} &= -[\vec{n}, \vec{H}], \quad (\vec{n}, \vec{E}) = 0, \\ \vec{H} &= [\vec{n}, \vec{E}], \quad (\vec{n}, \vec{H}) = 0. \end{aligned} \tag{2.37a}$$

The relations $(\vec{n}, \vec{E}) = 0$ and $(\vec{n}, \vec{H}) = 0$ prove the transverse nature of the optical wave. According to eqns (2.37a), the right set of three orthogonal vectors $\vec{n}$, $\vec{E}$ and $\vec{H}$ characterize the plane harmonic optical wave propagating in an arbitrary direction $\vec{n}$ in vacuum (Fig. 2.4).

2.3 Non-monochromatic light—the natural polarization

In a plane monochromatic optical wave the electric field strength $\vec{E}$ is a regular function of coordinates and time. Such a wave is called a completely polarized or just *polarized* wave. Earlier we discussed in detail the polarization state of a plane harmonic wave. In the general case such a wave is elliptically polarized and the characteristics of the polarization ellipse are determined by the amplitudes and phases of the orthogonal components E_x and E_y of the optical field.

The finite aperture of real optical beams and the non-monochromaticity of light lead to distortions of this ideal picture. While the structure of laser light can

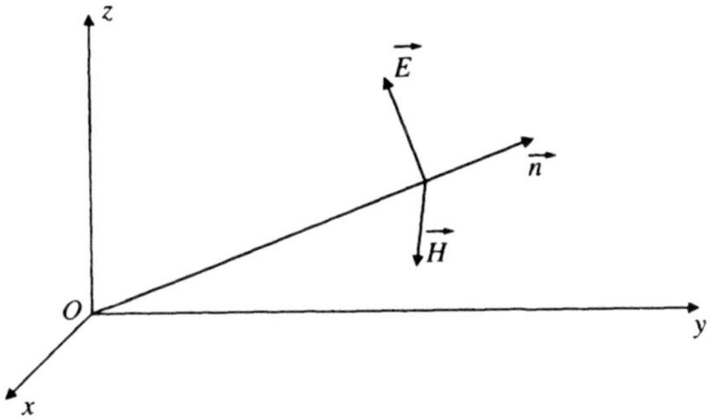

Fig. 2.4 The field structure of a plane optical wave.

be close to that of a polarized wave, the polarization of non-laser light exhibits rapid chaotic variations with time.

It is natural to treat the field variations of a non-monochromatic optical wave as a *random* process. For such a wave, the direction of the vector $\vec{E}$ in the wavefront plane varies randomly with time. If all directions of the vector $\vec{E}$ are equally probable, then the light is called non-polarized or *naturally polarized*. Sunlight or an incandescent electric lamp provide typical examples. On the other hand, when a predominant direction of the vector $\vec{E}$ exists, the light is called partially polarized.

The optical field of a plane non-monochromatic wave propagating along the z-axis takes the form

$$\vec{E} = \vec{x}_0 E_x + \vec{y}_0 E_y, \tag{2.38}$$

where

$$\begin{aligned} E_x &= \tfrac{1}{2}\mathcal{E}_x \exp[i(\omega t - kz)] + \text{c. c.}, \\ E_y &= \tfrac{1}{2}\mathcal{E}_y \exp[i(\omega t - kz)] + \text{c. c.} \end{aligned} \tag{2.39}$$

Treating the complex amplitudes $\mathcal{E}_x$ and $\mathcal{E}_y$ as random functions of time, we can introduce a *coherence matrix* of the optical wave:

$$\hat{\mathcal{J}} = \begin{pmatrix} \langle \mathcal{E}_x \mathcal{E}_x^* \rangle & \langle \mathcal{E}_x \mathcal{E}_y^* \rangle \\ \langle \mathcal{E}_y \mathcal{E}_x^* \rangle & \langle \mathcal{E}_y \mathcal{E}_y^* \rangle \end{pmatrix}, \tag{2.40}$$

where the angular brackets denote time averaging. The elements of this matrix can be measured experimentally (see below). The coherence matrix completely characterizes the polarization of a plane non-monochromatic optical wave.

2.4 Experimental methods of polarization measurement

The polarization of light is measured experimentally using anisotropic crystals. In this section we briefly touch upon the optics of anisotropic media. A detailed discussion of this topic is given in Part IV.

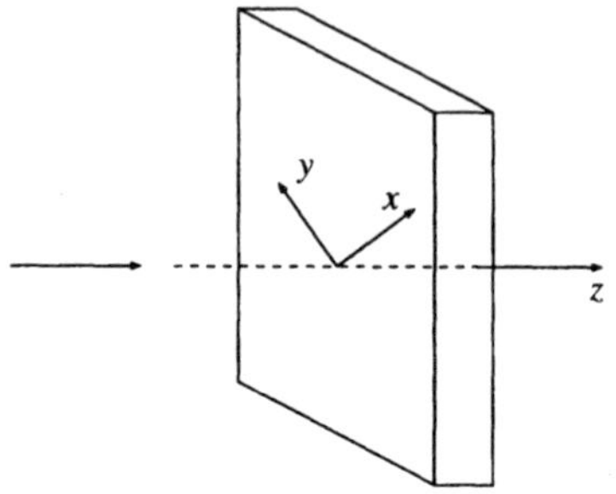

Fig. 2.5 Ordinary and extraordinary optical waves in an anisotropic crystal.

The anisotropy of a crystal lattice causes the propagation of a light wave in the crystal to depend on the polarization and propagation direction of that wave. Having entered the crystal, an optical wave with arbitrary polarization breaks down into two linear polarized waves with orthogonal polarization directions, the so-called "ordinary" and "extraordinary" waves. Generally, the propagation velocities of these waves are different. As the ordinary and extraordinary waves propagate, a phase shift $\Delta\varphi$ appears between them. This shift is proportional to the velocity difference and the distance covered.

2.4.1 Ordinary and extraordinary waves

There are two inherent polarization directions, x and y, for any direction z in the crystal (Fig. 2.5). Physically, both of these directions are distinguished because the optical waves linearly polarized along these directions keep their polarization states invariable as they propagate through the crystal. One of these waves is called *ordinary* and its velocity is independent of the direction of propagation. The other wave is referred to as *extraordinary* and its velocity is different for different propagation directions.

Unlike the ordinary or extraordinary waves, an arbitrarily polarized wave alters its polarization state as it travels in the crystal. Such a wave breaks down into ordinary and extraordinary waves travelling at different velocities.

In any crystal there is at least one direction for which the velocities of the ordinary and extraordinary waves are equal. Such a direction is called an optic axis of the crystal. Depending on the number of the axes, anisotropic crystals are subdivided into uniaxial and biaxial crystals. The uniaxial crystals, such as quartz and calcite, are usually applied in optical devices for polarization control. As follows from the definition, an arbitrarily polarized wave can propagate along the optic axis of the crystal, keeping its polarization constant. In other words, the properties of the crystal in the direction of its optic axis are identical to those of an isotropic medium. Conversely, anisotropy manifests itself most strongly when the wave propagates at right angles to the optic axis.

The phase shift arising between ordinary and extraordinary waves can be utilized for polarization control. For example, having passed through a crystal plate, a linear polarized light beam experiences a phase shift $\Delta\varphi = \pi/2$ between the

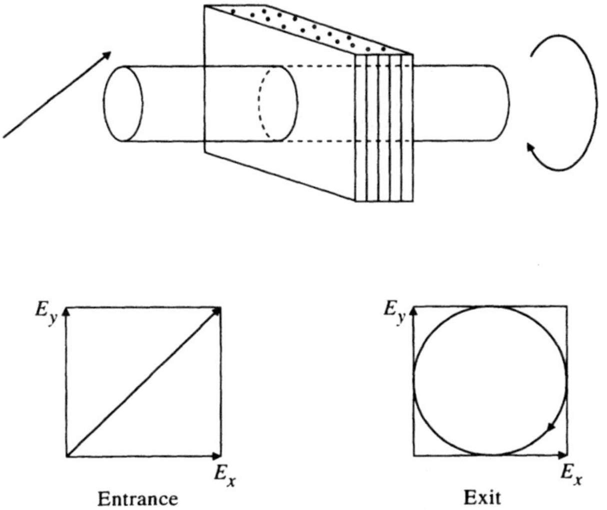

Fig. 2.6 The transformation of linear polarization of light to circular polarization.

field components E_x and E_y and turns into circularly polarized light. By placing another identical crystal plate in the path of this light, we will again obtain a linear polarized light, the polarization direction of which is perpendicular to the initial one. A proper choice of plate thickness can ensure the transformation of elliptically polarized light into light with linear or circular polarization, and vice versa.

2.4.2 Quarter-wave and half-wave plates

Let a linear polarized optical beam fall on a transparent anisotropic crystal so that the vector $\vec{E}$ is directed at an angle 45° to the x and y directions of the inherent polarizations of the wave in the crystal. Then the ordinary and extraordinary waves emerging at the entry have equal phases and amplitudes. The thickness of the crystal plate is chosen so that the phase difference between the ordinary and extraordinary waves becomes equal to $\pi/2$. The amplitudes of these waves thus remain equal as before and the light leaving the crystal is circularly polarized (Fig. 2.6).

The plate performing such a transformation is referred to as a *quarter-wave* plate because it introduces the phase difference $\Delta\varphi = \pi/2$ corresponding to a path difference equal to $\lambda/4$. Such plates are widely used in modern laser devices for the transformation of linear polarization of light to circular polarization and vice versa. *Half-wave* plates introducing a phase difference $\Delta\varphi = \pi$ are used for rotation of the polarization plane of a linear polarized optical wave through 90°.

2.4.3 Polarizers, analysers, and compensators

The absorption coefficients of some crystals, e.g. tourmaline, have widely differing values for ordinary and extraordinary waves. This causes a practically total absorption of one of the waves in a crystal plate only a few millimeters thick, which results in the remaining wave with linear polarization exiting the crystal. Accordingly, a tourmaline plate selects a linear polarized component from light with random polarization, i.e. the plate acts as a *polarizer* of light. Some materials having different absorption coefficients for light polarized in different directions (for example, polyvinylalcohol films impregnated by iodine) can be used to fabricate polaroid films. Such films are widely used in polaroids; that is, devices intended for selecting the linear polarized component with a definite polarization direction from light with random polarization.

The effect of total internal reflection can be also applied for the separation of ordinary and extraordinary waves, and consequently for the production of linear polarized light in anisotropic crystals possessing different refractive indices for ordinary and extraordinary waves. Various polarizing prisms based on this effect, i.e. the Glan prism, the Nicol prism, and others, are also used as polarizers.

Using a polarizer, we can determine the polarization direction of a linear polarized wave and make sure that the polarization is indeed linear. This is done by rotating the polarizer around the axis of the light beam and observing the intensity variations of the light exiting the crystal. If the output light intensity drops to zero for a certain polarizer orientation, then the initial beam is linearly polarized, and its polarization direction is orthogonal to the transmission direction, i.e. the "axis", of the polarizer. The polarizer performs as an *analyser* in such experiments. Figure 2.7 illustrates examples of some polarizers and analysers. These are (a) a tourmaline plate, (b) a polaroid, (c) a Glan prism, and (d) a Nicol prism. The letter "o" denotes an ordinary wave and letter "e" an extraordinary one. The arrow at the polaroid rim indicates the direction of the polaroid axis; that is, the polarization direction of the optical wave totally transmitted by the polaroid.

Optical devices (called *compensators*), which convert elliptical polarized light to linear polarized light, are used for the measurement of the polarization parameters of elliptically polarized light. The compensator is a plate made of two wedge-like anisotropic crystals, so that a displacement of one crystal relative to the other causes a change in the plate thickness. Such a device enables a continuous variation of the thickness of the anisotropic plate, which leads to a continuous variation of the phase difference $\Delta\varphi$ between the ordinary and extraordinary waves. Figure 2.8(a) shows Soleil's compensator. When used in the configuration shown in the figure, the compensator generates a phase shift $\Delta\varphi = (2\pi/\lambda)(n_e - n_o)(h_2 - h_1)$ between the ordinary and extraordinary waves. Gradual control of the phase shift is accomplished by displacing one compensator wedge relative to the other. Figure 2.8(b) illustrates a configuration of *Babinet's compensator* generating a phase shift $\Delta\varphi = (2\pi/\lambda)(n_e - n_o)(h_1 - h_2)$ between the ordinary and extraordinary waves. The phase shift is continuously monitored by displacing the optical beam, or the compensator, in a transverse direction. In the above expressions for the phase shift, the symbol n_o denotes the refraction index for the ordinary wave and n_e that for the extraordinary wave. The wedges used in Soleil's and Babinet's compensators are made of quartz, the refractive indices of

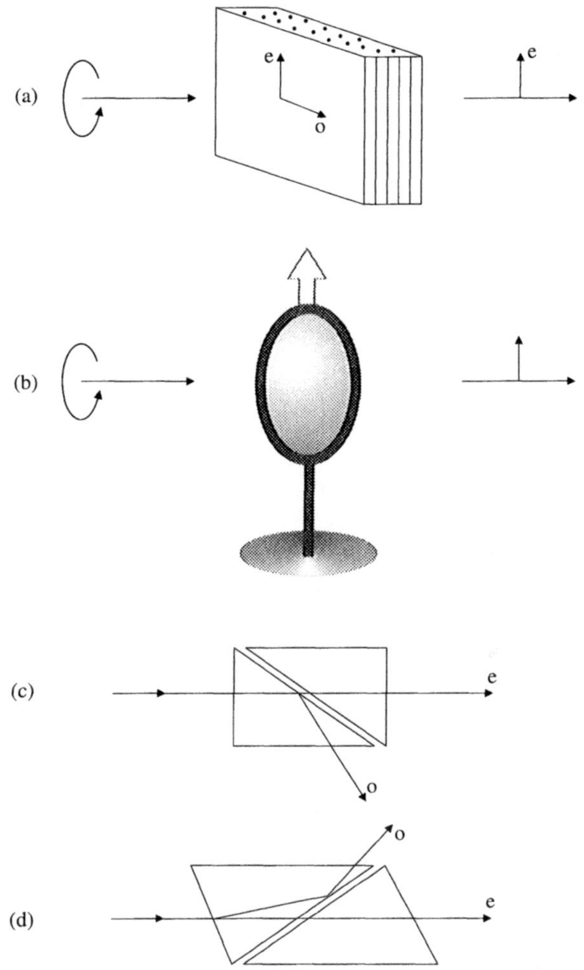

Fig. 2.7 Polarizers and analysers.

which are n_e=1.553 and n_o=1.544. The phase shifts generated by compensators are thus found by measuring the displacement of one compensator wedge relative to the other, or the displacement of the compensator as a whole with respect to the optical beam. Polarized light transmitted through a "compensator–analyser" system gives the characteristic pattern of alternating light and dark fringes in the beam cross-section, whereas the natural (non-polarized) light retains a homogeneous intensity distribution under the same conditions.

2.4.4 Polarization analysis of a plane monochromatic optical wave

A plane monochromatic optical wave is generally elliptically polarized. (Linear and circular polarizations can be regarded as partial cases of elliptical polar-

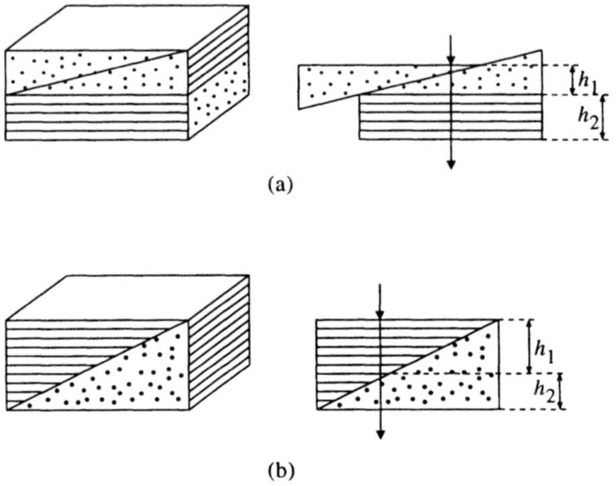

Fig. 2.8 Compensators.

ization.) Polarization analysis is thus reduced to experimental measurements of polarization ellipse parameters; that is, the spatial orientation of the ellipse and the ratios of its axes. These measurements can be carried out in the following way. While turning the analyser around the axis of the optical beam, we can observe variations in the intensity of the light transmitted by the analyser. Let us note the directions of the analyser axis corresponding to the maximum I_{max} and to the minimum I_{min} of intensity. Obviously, the former direction coincides with that of the ellipse's major axis and the latter with the direction of its minor axis. Since the light intensity is proportional to the square of the amplitude of the optical field (see Chapter 3), the ratio of the axes of the polarization ellipse can be evaluated as

$$a/b = \sqrt{I_{max}/I_{min}}.$$

2.4.5 Polarization analysis of non-monochromatic light

As we noted before, the field of a non-monochromatic optical wave is governed by random processes. The light is therefore characterized by mean values of various quantities: intensity, correlation function, etc. The averaging procedure can be carried out either statistically, i.e. by processing a great number of measurements, or temporally, i.e. by temporal integration. Note that the temporal integration invariably occurs in any real optical experiment due to the inertia of the measuring instrument.

Polarization parameters are measured with the aid of the analyser and the compensator. Suppose that an optical wave first passes through the compensator, which introduces the phase difference $\Delta\varphi = \varepsilon$ between the field components E_x and E_y (with the x- and y-axes being directed along the compensator edges) and then through the analyser, the transmission axis of which makes an angle θ with the x-axis (Fig. 2.9). The wave transmitted by the analyser is linearly polarized

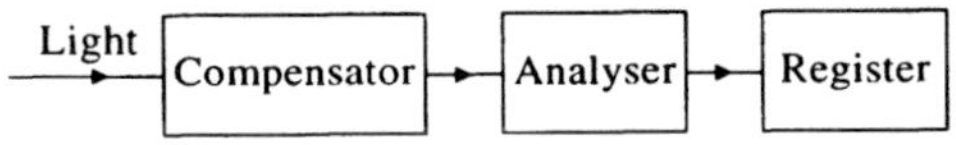

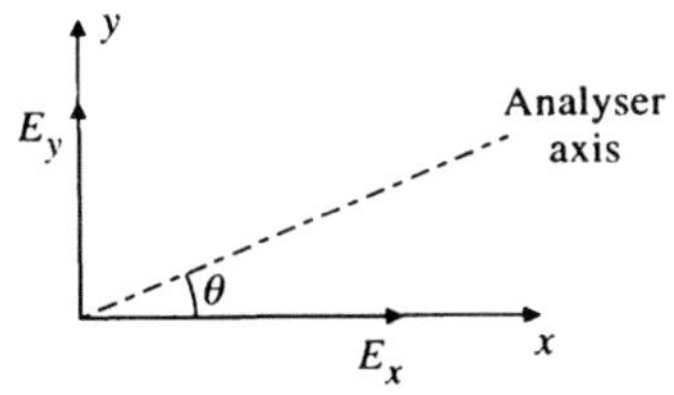

Fig. 2.9 Measurement of the polarization of non-monochromatic light.

and has an intensity

$$I = I(\theta, \varepsilon) = (c/8\pi)\langle|\mathcal{E}|^2\rangle, \tag{2.41}$$

where

$$\mathcal{E} = \mathcal{E}_x \cos\theta + \mathcal{E}_y e^{i\varepsilon} \sin\theta. \tag{2.42}$$

Here the angular brackets denote time averaging. The light intensity $I = I(\theta, \varepsilon)$ is measured by a measuring device (Fig. 2.9). Such measurements are repeated for different values of the parameters θ and ε. From the results of measurements one can derive quantitative conclusions about the polarization state of the optical wave and, specifically, calculate the elements of the coherence matrix.

Indeed, substituting eqn (2.42) into eqn (2.41), we obtain

$$I(\theta, \varepsilon) = (c/8\pi)(\mathcal{J}_{xx} \cos^2\theta + \mathcal{J}_{yy} \sin^2\theta + \mathcal{J}_{xy} e^{-i\varepsilon} \sin\theta\cos\theta + \mathcal{J}_{yx} e^{i\varepsilon} \sin\theta\cos\theta), \tag{2.43}$$

where $\mathcal{J}_{xx}$, $\mathcal{J}_{yy}$, $\mathcal{J}_{xy}$, and $\mathcal{J}_{yx}$ are the elements of the coherence matrix defined by eqn (2.40). Using eqn (2.43), it is easy to show that

$$\mathcal{J}_{xx} = \frac{8\pi}{c} I(0°, 0), \quad \mathcal{J}_{yy} = \frac{8\pi}{c} I(90°, 0),$$
$$\mathcal{J}_{xy} = \frac{8\pi}{c}\left\{\frac{1}{2}\left[I(45°, 0) - I(135°, 0)\right] + \frac{i}{2}\left[I(45°, \frac{\pi}{2}) - I(135°, \frac{\pi}{2})\right]\right\}, \tag{2.44}$$
$$\mathcal{J}_{yx} = \frac{8\pi}{c}\left\{\frac{1}{2}\left[I(45°, 0) - I(135°, 0)\right] - \frac{i}{2}\left[I(45°, \frac{\pi}{2}) - I(135°, \frac{\pi}{2})\right]\right\}.$$

Equations (2.44) express the coherence matrix in terms of experimentally measured quantities.

By varying the parameters θ and ε we can determine the value of

$$P = \frac{I(\theta,\varepsilon)_{\max} - I(\theta,\varepsilon)_{\min}}{I(\theta,\varepsilon)_{\max} + I(\theta,\varepsilon)_{\min}}, \tag{2.45}$$

which is called the *degree of polarization*. This quantity can be expressed via the elements of the coherence matrix. It follows from eqns (2.43) and (2.45) that

$$P = \sqrt{1 - \frac{4|\mathcal{J}|}{(\mathcal{J}_{xx} + \mathcal{J}_{yy})^2}}, \tag{2.46}$$

where

$$|\mathcal{J}| = \mathcal{J}_{xx}\mathcal{J}_{yy} - \mathcal{J}_{xy}\mathcal{J}_{yx}. \tag{2.47}$$

is the determinant of the coherence matrix.

Let us consider some partial cases.

Non-polarized (natural) light

This is the light for which

$$I(\theta,\varepsilon) = \text{constant} \tag{2.48}$$

for all values of θ and ε. It follows from eqn (2.48) that in this case

$$\mathcal{J}_{xx} = \mathcal{J}_{yy} = \tfrac{1}{2} I_0 \frac{8\pi}{c}, \quad \mathcal{J}_{xy} = \mathcal{J}_{yx} = 0, \tag{2.49}$$

where

$$I_0 = \frac{c}{8\pi}(\mathcal{J}_{xx} + \mathcal{J}_{yy})$$

is the intensity of the initial optical wave. Using eqns (2.46), (2.47), and (2.49), we find that $P = 0$. The same result follows from eqns (2.45) and (2.48).

Totally polarized light

Let

$$\mathcal{E}_x = A_1 \exp(i\varphi_1), \quad \mathcal{E}_y = A_2 \exp(i\varphi_2), \tag{2.50}$$

where A_1, A_2, φ_1, and φ_2 are constants. Substituting eqn (2.50) into eqn (2.40), we obtain

$$\hat{\mathcal{J}} = \begin{pmatrix} A_1^2 & A_1 A_2 e^{-i\varphi} \\ A_1 A_2 e^{i\varphi} & A_2^2 \end{pmatrix},$$

where $\varphi = \varphi_2 - \varphi_1$. In this case $|\mathcal{J}| = 0$ and $P = 1$.

Partially polarized light

The models of non-polarized light and of totally polarized light are idealizations. Real optical beams have a degree of polarization lying in the range $0 < P < 1$. Such light is called partially polarized. Note that partially polarized light can be treated as a superposition of non-polarized and totally polarized components.

Bibliography

Azzam, R. M. A. and Bashara, N. M. (1977). *Ellipsometry and polarized light.* North-Holland, Amsterdam.

Born, M. and Wolf, E. (1968). *Principles of optics.* Pergamon Press, Oxford.

Yariv, A. and Yeh, P. (1984). *Optical waves in crystals.* John Wiley, New York.

3

Light energy

The energy flux carried by an optical wave. Optical beams and pulses: energy, power, and intensity. The spatial and temporal compression of light energy. Basic concepts of photometry.

The capability to carry energy is one of the fundamental properties of light. The notion of electromagnetic energy is shown to emerge from Maxwell theory. The possibilities for spatial and temporal compression of optical energy are discussed and numerical estimates given.

3.1 The energy flux carried by an optical wave

The notion of energy of the optical field follows directly from Maxwell theory. Let us first consider a simple example of a plane optical wave and then generalize the result obtained to an arbitrary optical field.

Suppose that a plane light wave with the field components $E_x(z,t)$ and $H_y(z,t)$ propagates along the z-axis. Then, according to eqns (2.1), we can write

$$\frac{1}{c}\frac{\partial E_x}{\partial t} = -\frac{\partial H_y}{\partial z}, \quad \frac{1}{c}\frac{\partial H_y}{\partial t} = -\frac{\partial E_x}{\partial z}.$$

By multiplying the first of these equations by E_x and the second by H_y and summing up, we obtain the following relation:

$$\frac{1}{c}\left(E_x\frac{\partial E_x}{\partial t} + H_y\frac{\partial H_y}{\partial t}\right) = -\left(E_x\frac{\partial H_y}{\partial z} + H_y\frac{\partial E_x}{\partial z}\right). \tag{3.1}$$

After introduction of the field energy density (energy per unit volume),

$$w = \frac{1}{8\pi}(E_x^2 + H_y^2), \tag{3.2}$$

eqn (3.1) can be rewritten as

$$\frac{\partial w}{\partial t} = -\frac{\partial S}{\partial z}, \tag{3.3}$$

where

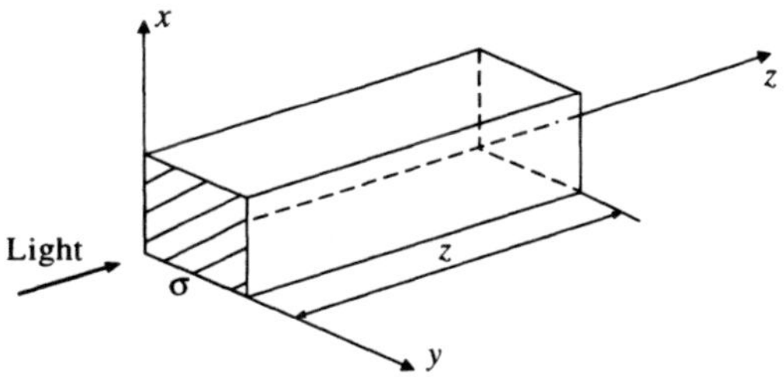

Fig. 3.1 The derivation of the energy conservation law for an optical wave.

$$S = \frac{c}{4\pi} E_x H_y. \tag{3.4}$$

Let us clarify the physical meaning of the parameter S. Consider the optical field within a finite region having the shape of a rectangular parallelepiped of height z and base area σ (see Fig. 3.1). Integrating eqn (3.3) over the parallelepiped volume, we obtain

$$\frac{\partial W}{\partial t} = -\sigma[S(t,z) - S(t,0)], \tag{3.5}$$

where

$$W = \int_V w \, dV \tag{3.6}$$

is the optical field energy inside the parallelepiped with volume V. Since we consider the optical field in vacuum, field energy variations within the volume can be caused only by the energy flow across its boundary. Thus, the quantity S in eqn (3.5) denotes the light energy flux. It can be seen from eqn (3.3) that the dimensionality of S is $\text{erg}\,\text{cm}^{-2}\,\text{s}^{-1}$.

In the case of a plane monochromatic wave eqn (3.5) turns out to be trivial because $S(t,z) = S(t,0)$ and $W = \text{constant}$. However, for a non-monochromatic wave and, specifically, for an optical pulse, the derivative $\partial W/\partial t$ can be non-zero because the pulse crosses the parallelepiped boundaries at different moments of time.

Thus, eqn (3.3) expresses the energy conservation law for a plane optical wave in vacuum. This law can now be generalized to an arbitrary optical field interacting with charged particles.

We shall begin with Maxwell's equations

$$\text{rot}\vec{E} = -\frac{1}{c}\frac{\partial \vec{H}}{\partial t}, \quad \text{rot}\vec{H} = \frac{1}{c}\frac{\partial \vec{E}}{\partial t} + \frac{4\pi}{c}\vec{j}, \tag{3.7}$$

where $\vec{j}$ is the density of the current produced by moving charges. Multiplying the first equation by $\vec{H}$ and the second by $\vec{E}$, and subtracting the second scalar product from the first, we obtain

$$\frac{1}{c}\left(\vec{E}\frac{\partial\vec{E}}{\partial t}+\vec{H}\frac{\partial\vec{H}}{\partial t}\right)+\frac{4\pi}{c}\vec{j}\vec{E}=\vec{E}\operatorname{rot}\vec{H}-\vec{H}\operatorname{rot}\vec{E}. \tag{3.8}$$

Using the vector identity

$$\operatorname{div}[\vec{E},\vec{H}]=\vec{H}\operatorname{rot}\vec{E}-\vec{E}\operatorname{rot}\vec{H}, \tag{3.9}$$

we can rewrite eqn (3.8) as

$$\frac{\partial w}{\partial t}+\vec{j}\vec{E}=-\operatorname{div}\vec{S}, \tag{3.10}$$

where

$$w=\frac{1}{8\pi}(E^2+H^2) \tag{3.11}$$

is the energy density of the field, and

$$\vec{S}=\frac{c}{4\pi}[\vec{E},\vec{H}] \tag{3.12}$$

is the vector of energy flow, or the *Poynting vector*.

Integrating eqn (3.10) over the volume V, we obtain

$$\frac{\partial W}{\partial t}+\int_V \vec{j}\vec{E}\,dV=-\int_V \operatorname{div}\vec{S}\,dV, \tag{3.13}$$

where the total energy of the electromagnetic field, W, within the volume V is defined by eqn (3.6). Assuming the charged particles to be point charges, we can replace the integral on the left-hand side of eqn (3.13) by the sum involving the individual particles:

$$\int_V \vec{j}\vec{E}\,dV=\sum_i e_i\vec{v}_i\vec{E}, \tag{3.14}$$

where $\vec{v}_i$ is the velocity of a charge e_i and the sum is taken over all the charges in the volume V. The integral on the right-hand side of eqn (3.13) can be transformed applying Gauss's theorem,

$$\int_V \operatorname{div}\vec{S}\,dV=\int_\Sigma \vec{S}\vec{l}_{\rm n}\,d\sigma, \tag{3.15}$$

where $\vec{l}_{\rm n}$ is the unit outward normal to the element $d\sigma$ of the surface enclosing the volume V (Fig. 3.2).

Using the equation of motion of the charged particle, $m_i\dot{\vec{v}}_i=e_i\vec{E}$, one can easily demonstrate that

$$\sum_i e_i\vec{v}_i\vec{E}=\frac{dK}{dt}, \tag{3.16}$$

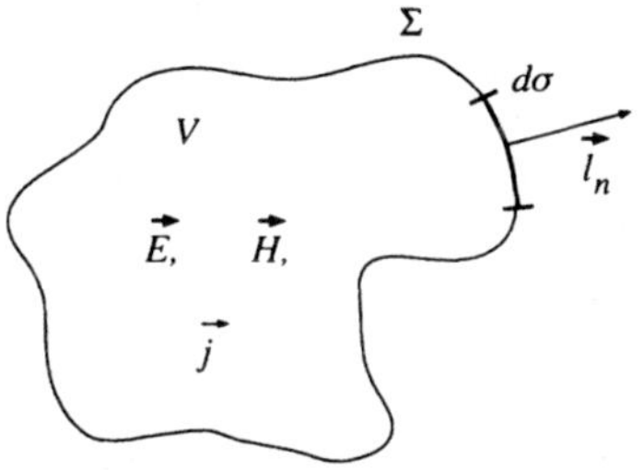

Fig. 3.2 The derivation of the energy conservation law for a system of charged particles in an electromagnetic field.

where

$$K = \sum_i \tfrac{1}{2} m_i v_i^2 \tag{3.17}$$

is the kinetic energy of the system of charges. Thus, we obtain the equation

$$\frac{\partial}{\partial t}(W + K) = -\int_\Sigma \vec{S}\vec{l}_n \, d\sigma, \tag{3.18}$$

which shows that the combined energy of the electromagnetic field and the charged particles within some volume varies at a rate equal to that of the energy flow across the boundary surface of the volume.

Using eqn (2.37a), we can write the following relation for the plane monochromatic wave:

$$\vec{S} = \frac{c}{4\pi}[\vec{E}, \vec{H}] = \frac{c}{4\pi}\left[\vec{E}, [\vec{n}, \vec{E}]\right] = \frac{c}{4\pi}E^2\vec{n} = \frac{c}{4\pi}H^2\vec{n},$$

which, combined with eqn (3.11), yields

$$\vec{S} = cw\vec{n}, \quad S = cw. \tag{3.19}$$

Here $\vec{n}$ is the unit vector oriented along the wave propagation direction. Thus, the optical energy is transferred in the direction of propagation of the plane wave. The energy is transferred by the light wave propagating in vacuum at the velocity c.

3.1.1 The intensity of light

Consider a plane harmonic optical wave having the following components:

$$E_x = A\cos(\omega t - kz), \quad H_y = A\cos(\omega t - kz). \tag{3.20}$$

In this case the modulus of the Poynting vector is

$$S = \frac{c}{4\pi}A^2\cos^2(\omega t - kz) = \frac{c}{8\pi}A^2 + \frac{c}{8\pi}A^2\cos[2(\omega t - kz)]. \tag{3.21}$$

Equation (3.21) shows that the energy flow comprises two components, one of which is constant while the other oscillates in time and space. The light wave frequencies are known to be equal to about 10^{15} Hz. Since no measuring instrument can register such rapid temporal variations of the energy flow, only the flow of energy averaged over the oscillation period can be experimentally measured:

$$I = \tilde{S} = \frac{1}{T}\int_0^T S\,dt, \quad T = 2\pi/\omega. \tag{3.22}$$

This quantity is called the *intensity* of the optical wave. It can be expressed through the electric field strength E of the wave as

$$I = \frac{c}{4\pi}\tilde{E^2}, \tag{3.23}$$

where

$$\tilde{E^2} = \frac{1}{T}\int_0^T E^2 dt. \tag{3.24}$$

Using this definition, we can derive the relations between the intensity I, the real amplitude A, and the complex amplitude $\mathcal{E}$ of the wave. Assuming that

$$E = A\cos(\omega t - kz) \tag{3.25}$$

or

$$E = \tfrac{1}{2}\mathcal{E}\exp[i(\omega t - kz)] + \text{c. c.}, \tag{3.26}$$

we obtain

$$I = \frac{c}{8\pi}A^2 \tag{3.27}$$

or

$$I = \frac{c}{8\pi}|\mathcal{E}|^2. \tag{3.28}$$

These formulae are valid for a plane monochromatic optical wave propagating in vacuum. The corresponding relations for a wave propagating in a medium will be given in Part IV of this book.

The dimensionality of the intensity, and of the energy flux, is $\text{erg}\,\text{cm}^{-2}\,\text{s}^{-1}$. Together with the Gaussian unit, the off-system unit of intensity, $\text{W}\,\text{cm}^{-2}=10^7$ $\text{erg}\,\text{cm}^{-2}\,\text{s}^{-1}$, is widely used in optics.

3.2 Optical beams and pulses: energy, power, and intensity

Unlike a plane wave, the real light beam has a finite transverse size. The beam can be described by a quasi-plane wave (see Chapter 1), i.e. by a wave the amplitude of which depends on the coordinates within the plane perpendicular to the propagation direction:

$$E = A(x, y)\cos(\omega t - kz). \tag{3.29}$$

Similarly, the field of an optical pulse can be described by a quasi-plane and quasi-harmonic wave, the amplitude of which depends not only on the coordinates but also on time:

$$E = A(x, y, t)\cos(\omega t - kz). \tag{3.30}$$

The energy parameters of light beams and pulses are the light intensity, the total power of the beam, and the total energy of the pulse. Inserting eqns (3.29) and (3.30) into eqn (3.23), we obtain expressions for the light intensity for the models of (i) a quasi-plane wave and (ii) quasi-plane and quasi-harmonic waves. Note that the light intensity in the beam depends on the coordinates within the plane of the cross-section:

$$I(x, y) = \frac{c}{8\pi}A^2(x, y). \tag{3.31}$$

The intensity of the pulse also depends on time:

$$I(x, y, t) = \frac{c}{8\pi}A^2(x, y, t). \tag{3.32}$$

The total power of the optical beam is defined as an integral of the intensity over the beam cross-section:

$$P = \iint\limits_{-\infty}^{\infty} I(x, y)dx\,dy. \tag{3.33}$$

Similarly, the total energy of the pulse is

$$W = \iiint\limits_{-\infty}^{\infty} I(x, y, t)dx\,dy\,dt. \tag{3.34}$$

For example, an axisymmetric Gaussian beam with the intensity distribution

$$I(r) = I_0 \exp(-r^2/r_0^2), \tag{3.35}$$

where r is the distance of the given point from the beam axis, I_0 is the intensity at the beam axis, and r_0 is the radius of the beam, has the total power

$$P = \int_0^{\infty} I(r)2\pi r\,dr = \pi r_0^2 I_0. \tag{3.36}$$

To measure the energy characteristics of light, we can use its thermal effects, as well as the photoeffect. Using instruments that ensure a sufficiently high temporal and spatial resolution, we can measure the transverse intensity distributions of continuous, $I(x, y)$, and pulsed, $I(x, y, t)$, radiations.

For evaluation purposes it is convenient to use such parameters as the effective light intensity, I_{eff}, the effective power of the pulse, P_{eff}, and the effective electric field strength of the light wave, E_{eff}. The effective intensity of continuous radiation is defined as the ratio of the beam power to the cross-sectional area:

$$I_{\text{eff}} = P/S. \tag{3.37}$$

For the light pulse,

$$P_{\text{eff}} = W/\tau, \quad I_{\text{eff}} = W/(\tau S), \tag{3.38}$$

where W is the energy of the pulse and τ is its duration. The effective electric field strength is defined as

$$E_{\text{eff}} = \sqrt{8\pi I_{\text{eff}}/c}. \tag{3.39}$$

Now we shall consider a few examples.

3.2.1 Sunlight

The intensity of sunlight near the Earth's surface is about $I = 0.1\ \text{W cm}^{-2}$. Knowing the radius of the Earth's orbit ($R = 150 \times 10^6$ km) we can calculate the total power of the Sun's radiation to be equal to $P = 4\pi R^2 I = 3 \times 10^{26}$ W.

3.2.2 A helium–neon laser

This is a continuous wave (CW) gaseous laser emitting monochromatic light at a wavelength of $\lambda = 0.63\ \mu$m. The typical radiation power is $P = 10^{-2}$ W. For a beam radius of $r = 0.2$ cm its effective intensity is

$$I_{\text{eff}} = P/(\pi r^2) = 0.1\ \text{W cm}^{-2}. \tag{3.40}$$

This rather low intensity is close to the intensity of sunlight at the Earth's surface. The corresponding electric field strength is also relatively low. According to eqns (3.39) and (3.40)

$$E_{\text{eff}} = 3 \times 10^{-2}\ \text{SGSE} = 10\ \text{V cm}^{-1}. \tag{3.41}$$

3.2.3 A carbon dioxide laser

This laser generates infrared radiation at a wavelength of $\lambda = 10.6\ \mu$m. In the CW regime the generated power is usually $P = 10$–100 W. In the pulse regime the laser generates pulses with energy $W = 0.1$ J and effective power $P_{\text{eff}} = 10^6$ W for a pulse duration $\tau = 10^{-7}$ s.

3.2.4 A YAG : Nd^{3+} laser

A solid-state laser using an yttrium–aluminium garnet crystal activated by neodymium ions emits radiation in the near infrared region at a wavelength of $\lambda = 1.06\ \mu$m. In the CW regime a laser with an active element 6–8 cm long and about 1 cm in diameter, excited by a discharge lamp, produces the power

$P = 1$–10 W. In the free-running pulse regime the laser generates pulses with a duration of $\tau = 10^{-3}$ s, an energy $W = 1$ J, and an effective radiation power $P_{\text{eff}} = 10^3$ W. In the Q-switched regime the laser generates a "giant" pulse with duration $\tau = 10^{-8}$ s, and power $P_{\text{eff}} = 10^7$ W.

3.3 The spatial and temporal compression of light energy

Equations (3.38) and (3.39) show that, for a given pulse energy, the power, intensity, and optical field strength depend on the cross-section of the beam and the pulse duration. It is interesting to explore the ultimate cases of optical energy compression in time and space.

A high monochromaticity and directionality of laser radiation make it possible to concentrate the laser energy on very small spatial and temporal scales. For example, if we focus a YAG : Nd^{3+} laser radiation with power $P = 10$ MW on a spot the radius of which is $r = 0.01$ cm, we will obtain the light intensity $I = P/\pi r^2 = 3 \times 10^{10}$ W cm^{-2} and an effective optical field strength of $E = 5$ MV cm^{-1}.

In the visible range the limiting diameter of the focal spot is of the order of the optical wavelength (see Part III),

$$d_{\text{min}} \approx \lambda \approx 10^{-4}\,\text{cm}, \tag{3.42}$$

and the limiting pulse duration is determined by the optical vibration period

$$\tau_{\text{min}} \approx T = 2\pi/\omega \approx 10^{-14}\,\text{s}. \tag{3.43}$$

By going down to this scale the intensity can be raised by eight to ten orders of magnitude while the total light energy remains the same.

The ultimate values of the optical energy defined by eqns (3.42) and (3.43) have been already attained. Some interesting possibilities resulting from energy compression are discussed in the monograph by Akhmanov, Vysloukh, and Chirkin (1991). Here we shall only mention that the concentration of radiation produced by ultra-high-power laser systems results in enormous magnitudes of intensity and field strength. For instance, a laser arrangement used in thermonuclear fusion research generates pulses with an energy $W = 10^4$–10^5 J. For a duration of $\tau = 10^{-9}$ s, such a pulse has the power $P = 10^{13}$–10^{14} W. By focusing the beam, an intensity of $I = 10^{17}$–10^{19} W cm^{-2} and an optical field strength of $E = 10^{10}$ V cm^{-1} can be realized. The latter value exceeds the strength of intra-atomic electric fields,

$$E_{\text{a}} = e/a_0^2 = 5 \times 10^9\,\text{V cm}^{-1}, \tag{3.44}$$

where $e = 4.8 \times 10^{-10}$ SGSE is the electron charge and $a_0 = 0.5 \times 10^{-8}$ cm is the first Bohr radius.

3.4 Basic concepts of photometry

Conventional (non-laser) sources of light are described by photometric quantities, such as radiant flux and intensity, radiant exitance, radiance and surface irradiance. In this section we shall define all these quantities and consider examples of their application).[1]

3.4.1 Radiant flux

Radiant flux is defined as the ratio of the energy dW, carried by the light across an arbitrary surface, to the time interval dt:

$$\Phi = dW/dt.$$

This quantity has units of $\mathrm{erg\,s^{-1}}$.

3.4.2 Radiant intensity

The radiant intensity is defined as the ratio of the radiant flux $d\Phi$ to the solid angle $d\Omega$ in which the radiation spreads:

$$I = d\Phi/d\Omega.$$

The dimensionality of this quantity is $\mathrm{erg\,s^{-1}\,sr^{-1}}$.

3.4.3 Irradiance

The irradiance of the surface is defined as the ratio of the radiant flux $d\Phi$ to the unit area dS of the illuminated surface:

$$\varepsilon = d\Phi/dS.$$

The dimensionality of this quantity is $\mathrm{erg\,s^{-1}\,cm^{-2}}$.

3.4.4 Radiant exitance (emittance) of the surface

The radiant exitance (emittance) of the surface is defined as the ratio of the radiant flux $d\Phi$ to the unit area dS_e of the radiating surface:

$$R = d\Phi/dS_e.$$

The dimensionality of radiant emittance is $\mathrm{erg\,s^{-1}\,cm^{-2}}$.

3.4.5 Radiance of a light source

The radiance of a light source is defined as the ratio of the radiant intensity dI to the area of the projection $dS_\perp$ of the radiating element of the surface on the plane, perpendicular to the direction of observation:

[1] We use energy units for the photometric quantities. Units based on the visual effect of light (candela and so on) differ from energy units by a factor depending on the wavelength — see, for example, Born and Wolf (1968).

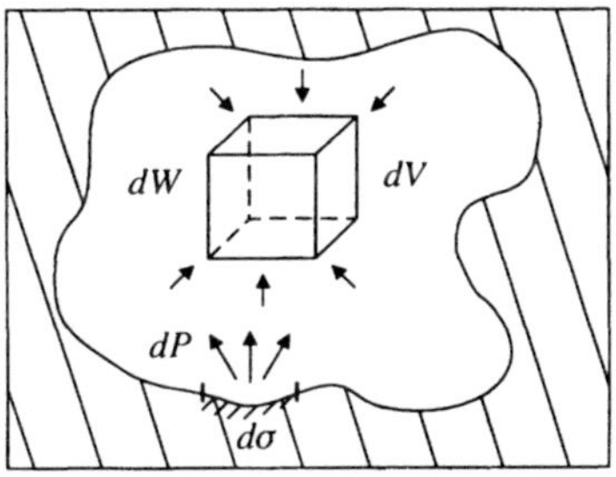

Fig. 3.3 The calculation of the energy characteristics of thermal radiation.

$$B = dI/dS_\perp,$$

where

$$dS_\perp = dS_0 \cos\theta,$$

in which dS_0 is the square of the surface element of the source, and θ is the angle between the normal to the radiating surface and the direction to the point of observation. The dimensionality of radiance is $\mathrm{erg\,s^{-1}\,sr^{-1}\,cm^{-2}}$.

Now let us illustrate the usage of photometric quantities by some examples.

3.4.6 Emissivity of a body and thermal radiation energy

As an example, let us consider the energy characteristics of the thermal radiation. Suppose that inside a heated body there is a cavity filled with the equilibrium thermal radiation (Fig. 3.3). Disregarding for a moment the spectral composition of radiation, we shall find the relationship between the radiant emittance of the cavity walls and the energy density of radiation in the cavity.

The following notation will be used: $d\sigma$ is the square of the cavity surface element, dP is the power emitted by the surface element $d\sigma$ in all directions, dV is the cavity volume element, dW is the energy of radiation in the volume dV,

$$\rho = dP/d\sigma \tag{3.45}$$

is the emissivity of the cavity surface, and

$$u = dW/dV \tag{3.46}$$

is the energy density of radiation.

Let us represent the power of radiation emitted by the surface element $d\sigma$ in the direction of the volume dV in the form:

$$dP(d\sigma, \theta, d\Omega) = B(\theta)\cos\theta\, d\sigma\, d\Omega, \tag{3.47}$$

where $B(\theta)$ is the radiance, θ is the angle between the normal to the surface element $d\sigma$ and the straight line connecting $d\sigma$ and dV, and $d\Omega$ is the solid

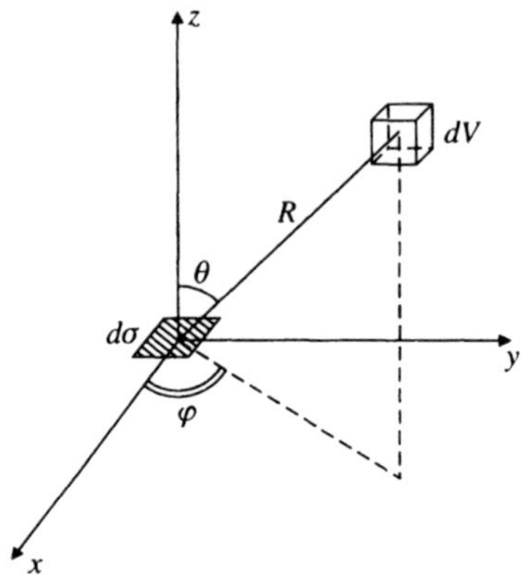

Fig. 3.4 The calculation of the energy density of thermal radiation.

angle subtending a volume dV at the centre of the surface element $d\sigma$ (Fig. 3.4). Equation (3.47) can be considered as the definition of radiance $B(\theta)$.

Figure 3.4 shows the spherical coordinates R, θ, and φ of the elementary volume dV with respect to the surface element $d\sigma$. The volume dV is subtended at the centre of the surface element $d\sigma$ by the solid angle

$$d\Omega = \frac{dS}{R^2} = \sin\theta \, d\theta \, d\varphi. \tag{3.48}$$

On the other hand, the surface element $d\sigma$ is subtended at the centre of the volume dV by the solid angle

$$do = \frac{d\sigma_\perp}{R^2} = \frac{\cos\theta \, d\sigma}{R^2}. \tag{3.49}$$

Using eqns (3.47) and (3.48), we can write the total radiation power dP emitted omnidirectionally by the surface element $d\sigma$ as

$$dP = d\sigma \int_0^{2\pi} d\varphi \int_0^{\pi/2} d\theta B(\theta) \cos\theta \, \sin\theta. \tag{3.50}$$

For the specific case of the equilibrium thermal radiation, the dependence $B(\theta)$ can be adequately described by the Lambert law expressing the angular independence of radiance:

$$B(\theta) = B = \text{constant}, \tag{3.51}$$

Substituting eqn (3.51) into eqn (3.50), we obtain

$$dP = \pi B \, d\sigma. \tag{3.52}$$

Assuming that the radiation reaches the volume dV omnidirectionally, we can write

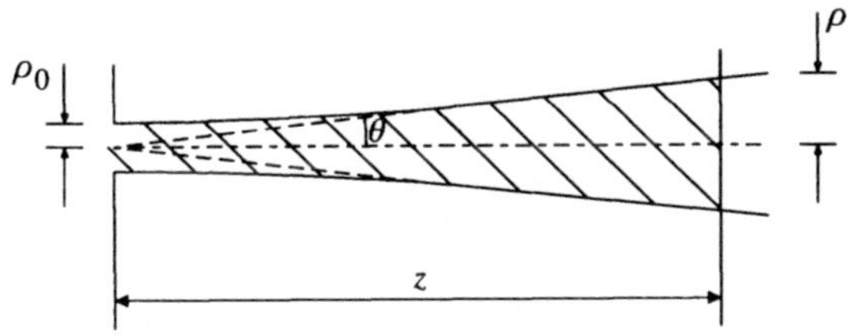

Fig. 3.5 The calculation of the radiance of laser radiation.

$$dW(dV, dt, d\sigma) = \frac{do}{4\pi} dW(dV, dt), \tag{3.53}$$

where $dW(dV, dt, d\sigma)$ is the energy flowing into the volume dV from the surface element $d\sigma$ during the time interval dt and $dW(dV, dt)$ is the total energy flowing omnidirectionally into the volume during the time interval dt. The quantities $dW(dV, dt, d\sigma)$ and $dP(d\sigma, \theta, d\Omega)$ are interrelated by the formula

$$dW(dV, dt, d\sigma) = dP(d\sigma, \theta, d\Omega)dt. \tag{3.54}$$

The energy density of the radiation is

$$u = \frac{dW}{dV} = \frac{dW(dV, dt)}{dV}, \tag{3.55}$$

where

$$dV = cdS\, dt, \tag{3.56}$$

and c is the velocity of light. From eqns (3.45)–(3.49) and (3.51)–(3.56) we finally obtain

$$\rho = \frac{c}{4} u. \tag{3.57}$$

This formula expresses the relation between the emissivity ρ of the cavity surface and the energy density u for equilibrium thermal radiation in the cavity of the heated body. Later we shall use this result to estimate the heating of the body by radiation (see Chapter 9).

3.4.7 Comparative characteristics of laser radiation and sunlight

As applied to a laser beam, the radiance B can be defined as the intensity I related to the solid angle Ω:

$$B = I/\Omega.$$

Let us introduce the divergence angle of the beam θ (see Fig. 3.5). Then

$$\Omega = \pi\theta^2,$$

assuming that $\theta \lll 1$. Writing the intensity I as

$$I = P/\pi\rho_0^2,$$

where P is the power of the opical beam and ρ_0 is its radius, we obtain

$$B = \frac{P}{\pi^2\rho_0^2\theta^2}.$$

The divergence of the laser beam is determined by the following formula (see Part III)

$$\theta = \frac{\lambda}{2\pi\rho_0},$$

where λ is the radiation wavelength. Consequently,

$$B = 4P/\lambda^2.$$

For example, for a He–Ne laser with $\lambda = 630$ nm and $P = 1$ mW, we have $B = 10^6$ W cm^{-2} sr^{-1}.

The radiance of the Sun can be calculated by the formula

$$B = P/(4\pi R)^2,$$

where P is the total power of the Sun's radiation and R is the radius of the Sun. Taking $P = 3.83 \times 10^{26}$ W and $R = 6.96 \times 10^{10}$ cm, we obtain $B = 5 \times 10^2$ W cm^{-2} sr^{-1}. Thus, the radiance of the laser exceeds that of the Sun by several orders of magnitude.

The superiority of the laser is even greater in terms of the spectral radiance B_λ, defined as the ratio of the radiance B to the spectral bandwidth $\Delta\lambda$ of the light:

$$B_\lambda = B/\Delta\lambda.$$

The bandwidth for a narrow-band He–Ne laser is $\Delta\lambda = 6 \times 10^{-10}$ nm (which corresponds to $\Delta\nu = 0.5$ kHz), yielding $B_\lambda = 1.5 \times 10^{15}$ W cm^{-2} sr^{-1} nm^{-1}. For the Sun, $\Delta\lambda = 400$ nm and $B_\lambda = 1.25$ W cm^{-2} sr^{-1} nm^{-1}.

Bibliography

Akhmanov, S. A., Vysloukh, V. A., and Chirkin A. S. (1991). *Optics of femtosecond laser pulses.* AIP, New York.

Born, M. and Wolf, E. (1968). *Principles of optics.* Pergamon Press, Oxford.

4

The pressure of light

The momentum of an optical wave. Measurement of the pressure of light and Lebedev's experiments. The pressure exerted by light on the classical atom. The angular momentum of an optical wave. The photon. Optical cooling.

The notions of the momentum and angular momentum of an optical wave are introduced. The relation between energy and the momentum of light is established. The problems of the measurement and application of light pressure are discussed.

4.1 The momentum of an optical wave

As follows from Maxwell's theory, an optical wave carries not only energy, but also *momentum*. In a plane monochromatic optical wave the momentum density (momentum per unit volume) is

$$G = I/c^2, \tag{4.1}$$

where I is the intensity and c is the velocity of light. Therefore, in accordance with the laws of mechanics, all bodies that absorb, refract, or reflect light experience the action of forces referred to as the forces of *light pressure*.

The light pressure on the body that totally absorbs the light is

$$\mathcal{P} = I/c = \tilde{w}, \tag{4.2}$$

where $\tilde{w}$ is the energy density (energy per unit volume) of the optical field, averaged over the optical vibration period. If the body partially reflects the light, with the reflection coefficient being R, then in accordance with the law of momentum conservation,

$$\mathcal{P} = (1 + R)I/c. \tag{4.2a}$$

In particular, the light pressure exerted on a totally reflecting body for which $R = 1$ is twice that exerted on a totally absorbing body for which $R = 0$. The pressure of light on the mirror is thus similar to the pressure of gas on the wall of a containing vessel. It is known that the gas pressure is caused by the molecules colliding with the wall and transferring their momentum to the wall as a result of the impact.

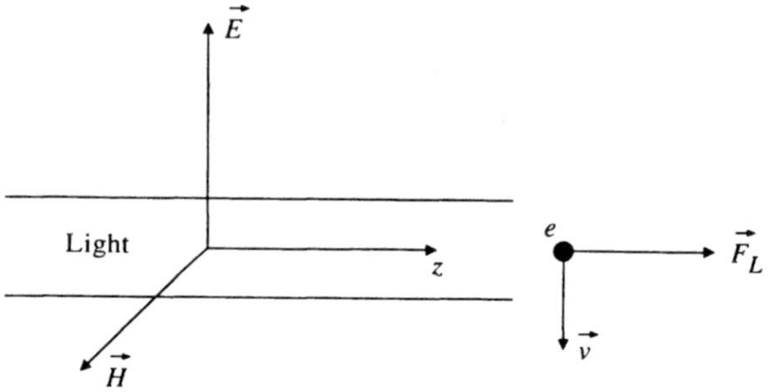

Fig. 4.1 The mechanism of light pressure.

The forces of light pressure are usually very small. However, in a laser beam they can be strong enough to control the motion of microparticles (Ashkin 1972). The pressure exerted by light on the atom is able to slow down its thermal motion (the "optical cooling" effect). In principle, it is also possible to build a propulsion engine using a powerful beam of light as a jet.

The existence of light pressure seemed quite natural to the proponents of the corpuscular theory of light. Indeed, the moving corpuscles were expected to transfer their momenta to the interfering bodies. As far back as 1619, Kepler explained the bending of comet tails by the pressure of the solar radiation, assuming the sunlight to be a flow of corpuscles. In terms of wave theory, the appearance of constant non-zero forces in a rapidly alternating wave field does not seem obvious. Nevertheless, soon after the establishment of the electromagnetic nature of light, Maxwell calculated the magnitude of the light pressure exerted on the surface of absorbing and reflecting bodies.

The mechanism of light pressure can be described as follows. Suppose that the surface of a certain body is irradiated by a normally falling plane monochromatic optical wave of "E_x, H_y" type, i.e. by a wave the electric field vector of which oscillates along the x-axis and the magnetic field vector along the y-axis (Fig. 4.1). Then the elementary electric charges of the body, excited by the electric field of the wave, will oscillate along the x-axis. Simultaneously, the magnetic field of the wave will exert a Lorentz force on the charges:

$$\vec{F}_{\mathrm{L}} = \frac{e}{c}[\vec{v}, \vec{H}], \tag{4.3}$$

which is directed along the z-axis, i.e. along the wave propagation direction and at right angles to the surface of the body. This is the elementary force of light pressure.

Let us calculate the light pressure, assuming that the light is totally absorbed by the body. We shall make use of the Maxwell equations

$$\mathrm{rot}\vec{E} = -\frac{1}{c}\frac{\partial \vec{H}}{\partial t}, \quad \mathrm{rot}\vec{H} = \frac{1}{c}\frac{\partial \vec{E}}{\partial t} + \frac{4\pi}{c}\vec{j}. \tag{4.4}$$

where $\vec{j}$ is the current density,

$$\vec{j} = ne\vec{v}, \tag{4.5}$$

e and $\vec{v}$ are the charge and velocity of the electron, and n is the number of electrons per unit volume. According to eqn (4.3) the electrons located in the unit volume of the body experience a force

$$f = nF_{\mathrm{L}} = n\frac{e}{c}vH,$$

directed along the z-axis. Using eqn (4.5) we can write this force in the form

$$f = \frac{1}{c}jH. \tag{4.6}$$

Then the sought-after light pressure is equal to

$$\mathcal{P} = \int_0^{\infty} \tilde{f}\, dz, \tag{4.7}$$

where the tilde denotes averaging over the optical vibration period.

For a plane monochromatic optical wave with the field components

$$E_x = A\cos(\omega t - kz), \quad H_y = A\cos(\omega t - kz) \tag{4.8}$$

we obtain from eqn (4.4):

$$\frac{\partial E_x}{\partial z} = -\frac{1}{c}\frac{\partial H_y}{\partial t}, \quad \frac{\partial H_y}{\partial z} = -\frac{1}{c}\frac{\partial E_x}{\partial t} - \frac{4\pi}{c}j_x. \tag{4.9}$$

Multiplying the first of eqn (4.9) by E_x and the second by H_y and summing up, we obtain

$$\frac{\partial w}{\partial z} = -\frac{1}{4\pi c}\frac{\partial}{\partial t}(E_x H_y) - f, \tag{4.10}$$

where

$$w = \frac{1}{8\pi}(E_x^2 + H_y^2) \tag{4.11}$$

is the energy density (energy per unit volume) of the optical field, and the force f is defined by the formula

$$f = \frac{1}{c}j_x H_y. \tag{4.12}$$

Now let us average eqn (4.10) over the optical vibration period $T = 2\pi/\omega$. Taking into account that

$$\left\langle \frac{\partial}{\partial t}E_x H_y \right\rangle = 0,$$

we obtain

$$\frac{\partial \tilde{w}}{\partial z} = -\tilde{f}. \tag{4.13}$$

Finally, substituting eqn (4.13) into eqn (4.7) we obtain the following expression for the light pressure:

$$\mathcal{P} = \tilde{w}, \tag{4.14}$$

which coincides with eqn (4.2). Writing the light pressure in the form $\mathcal{P} = F/S$, where F is the force and S is the optical beam cross-sectional area, and using the Newtonian law $F = dp/dt$, we arrive at eqn (4.1). The momentum density G of the light wave is defined by the formula $dp = G\,dV$, where $dV = Sc\,dt$ and c is the velocity of light. This result was obtained by Maxwell in 1873.

In 1879 Bartoli calculated the light pressure from thermodynamic considerations. According to the second law of thermodynamics, thermal energy cannot be transferred from a cold body to a hot body without any work being done. Bartoli indicated that any body emits thermal radiation, and, in principle, one could build a machine that could transfer the thermal radiation energy from a cold body to a hot body. Let us suppose that we have a hollow mirror cylinder in which a mirror piston can move. Then we could draw the radiant energy of the cold body into the cylinder by partially pulling the piston from the cylinder. Thereafter we could transfer the radiant energy to the hot body by placing the cylinder against it and pushing the piston into the cylinder. This procedure would be similar to filling a suction-pump with water, which can then be ejected to a considerable height. Although such a thermal machine is extremely difficult to realize practically, it is not forbidden in principle. Therefore, Bartoli assumed that the second law of thermodynamics would not be violated only if the moving piston overcame a certain force caused by the pressure of light. Maxwell and Bartoli thus came to the identical conclusion that light exerts pressure on the bodies that it is falling upon.

4.2 Measurement of the pressure of light and Lebedev's experiments

The estimates made using eqn (4.2a) show that the pressure of light is very small. For example, the pressure of sunlight of intensity $I = 0.1\ \mathrm{W\,cm^{-2}}$ on a mirror ($R = 1$) is $\mathcal{P} = 2I/c \approx 10^{-4}\ \mathrm{dyne\,cm^{-2}}$. That is why the registration of light pressure is a rather difficult experimental problem. The light pressure is often concealed by various accompanying effects, the main one being radiant heating of the probe.

The first correct measurement of light pressure was performed by the Russian physicist Peter N. Lebedev in 1898. In his experiments the light of an arc lamp was directed with the help of an optical system on to a tiny piece of platinum foil (Fig. 4.2). The foil was suspended on an elastic thread within an evacuated vessel. The angular deflection of the foil piece under the action of light could be measured and the energy of the incident optical beam found from calorimetric measurements. The major complication of the experiment was associated with the masking of the light pressure by radiometric forces and by convection effects.

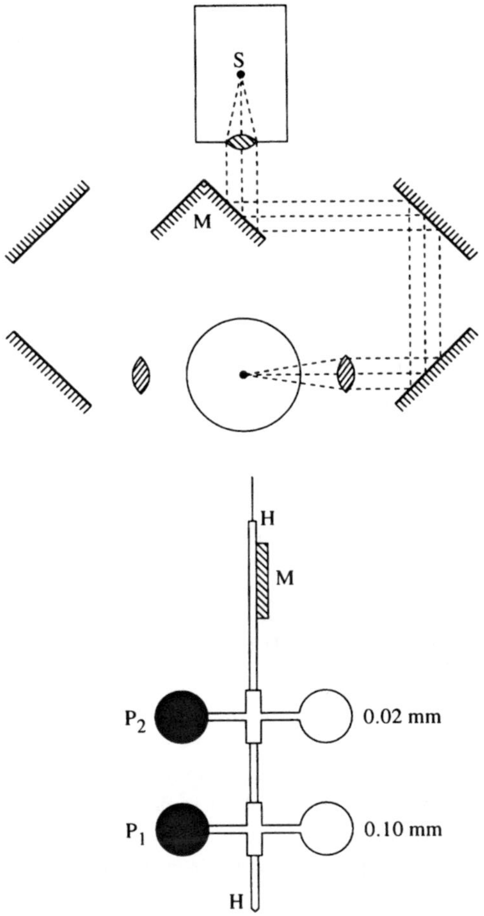

Fig. 4.2 A sketch of Lebedev's experiment on the measurement of light pressure.

The radiometric forces appear when gas molecules that have collided with the heated surface of the foil acquire additional velocity and thus produce an extra pressure on that surface. To exclude the radiometric forces, Lebedev used the thinnest foil available (about 0.02 mm thick) to ensure rapid equalization of temperatures on both sides of the foil via thermal conduction. The radiometric forces can be further reduced by the evacuation of the vessel. To get rid of the convection, Lebedev cooled his device and used a movable mirror system, which allowed him to direct the light on to both surfaces of the suspended foil. The results obtained have been formulated by Lebedev as follows:

"1. The incident beam of light exerts pressure both on the absorbing and reflecting surfaces; these ponderomotive forces are not associated with already known convective or radiometric forces caused by heating.

2. The forces of light pressure are proportional to the energy of the incident beam and independent of the colour (of the wavelength).

3. The observed forces of light pressure are equal, within the limits of experimental uncertainty, to the Maxwell–Bartoli pressure forces of radiant energy.

Thus, the existence of the light pressure has been proved experimentally."

4.3 The pressure exerted by light on the classical atom

The pressure of light can be used to control the motion of atoms and molecules. Let us evaluate the pressure exerted by light on the atom using the classical model of the atom.

According to this model an electron is confined near the nucleus by an electrostatic force. When the electron is displaced by a distance x with respect to the nucleus, the quasi-elastic Coulomb force $F_C = -\alpha x$ appears, tending to return the electron back to the equilibrium position.

The equation of motion of the electron has the form

$$\ddot{x} + \Gamma\dot{x} + \omega_0^2 x = \frac{1}{m}F(t), \tag{4.15}$$

where $\omega_0 = \sqrt{\alpha/m}$ is the frequency of natural vibrations of the electron in the atom, m is the mass of the electron, Γ is the parameter describing the damping of the vibrations, and $F(t)$ is the external force. Equation (4.15) is well known from mechanics. It should be noted that in optics the oscillator is also one of the most important models.

We will calculate the light pressure on the atom in two stages. First, we shall analyse the oscillator behaviour in the electric field of the optical wave, and then we shall calculate the Lorentz force arising due to the magnetic field of the wave.

Let the atom be irradiated by a harmonic optical wave with the components E_x, H_y:

$$E_x = \tfrac{1}{2}A\exp(i\omega t) + \text{c.c.}, \tag{4.16}$$

$$H_y = \tfrac{1}{2}A\exp(i\omega t) + \text{c.c.} \tag{4.17}$$

Neglecting for a moment the action of the magnetic field, we set

$$F = eE_x. \tag{4.18}$$

Then eqn (4.15) takes the form

$$\ddot{x} + \Gamma\dot{x} + \omega_0^2 x = \frac{e}{m}E_x. \tag{4.19}$$

The solution of eqn (4.19) will be sought in the form

$$x(t) = \tfrac{1}{2}\tilde{x}\exp(i\omega t) + \text{c.c.} \tag{4.20}$$

Inserting eqns (4.16) and (4.20) into eqn (4.19), we find that

$$\tilde{x} = \frac{e}{m} \frac{A}{\omega_0^2 - \omega^2 + i\omega\Gamma}. \tag{4.21}$$

Hence, the motion of the electron driven by the electric field of the wave is described by eqns (4.20) and (4.21).

Now let us calculate the Lorentz force produced by the magnetic field of the wave. Putting $\vec{v} = \vec{x}_0 \dot{x}$, and $\vec{H} = \vec{y}_0 H_y$, we obtain

$$\vec{F}_{\mathrm{L}} = \frac{e}{c}[\vec{v}, \vec{H}] = \vec{z}_0 \frac{e}{c} \dot{x} H_y, \tag{4.22}$$

where $\vec{x}_0$, $\vec{y}_0$, and $\vec{z}_0$ are the unit vectors directed along the x-, y-, and z-axes of the Cartesian frame of reference and $\dot{x} = dx/dt$. The force of light pressure is the Lorentz force averaged over the optical vibration period:

$$F_{\mathrm{p}} = \bar{F}_{\mathrm{L}}. \tag{4.23}$$

Inserting eqns (4.17) and (4.20) into eqn (4.22) and averaging over the period $T = 2\pi/\omega$, we obtain

$$\vec{F}_{\mathrm{p}} = \vec{z}_0 \frac{e}{c} \tfrac{1}{4} i\omega(\tilde{x}A^* - \tilde{x}^* A) \tag{4.24}$$

or, taking into account eqn (4.21),

$$\vec{F}_{\mathrm{p}} = \vec{z}_0 \frac{e^2}{2mc} |A|^2 \frac{\omega^2 \Gamma}{(\omega_0^2 - \omega^2)^2 + \omega^2 \Gamma^2}. \tag{4.25}$$

Thus, eqn (4.25) defines the force exerted by light on the classical atom. The obtained result can be conveniently presented in the form

$$\vec{F}_{\mathrm{p}} = \vec{z}_0 \mathcal{P} \sigma, \tag{4.26}$$

where $\vec{z}_0$ is the unit vector oriented along the propagation direction of the optical wave, $\mathcal{P} = I/c$ is the pressure of light, $I = c|A|^2/8\pi$ is the intensity of light, and

$$\sigma = \frac{4\pi e^2}{mc} \frac{\omega^2 \Gamma}{(\omega_0^2 - \omega^2)^2 + \omega^2 \Gamma^2} \tag{4.27}$$

is the cross-section of absorption of light by the atom (see Chapter 7). Note that the light pressure acting on the atom has a resonant character: it reaches a maximum when the frequency of light coincides with the frequency of natural vibrations of the electron in the atom, i.e. when $\omega = \omega_0$. In this case

$$\sigma = \sigma_{\max} = 4\pi e^2/(mc\Gamma). \tag{4.28}$$

Let us make a numerical estimate. The radiation decay time of an optical oscillator can be evaluated as $\tau = 10^{-8}$ s (see Chapter 5). Consequently, $\Gamma = 1/\tau =$

10^8 s^{-1} and calculation by eqn (4.28) gives $\sigma_{\max} = 10^{-9}$ cm^2. The obtained values are close to the parameters of the real systems. For example, the sodium atom has $\tau = 1.6 \times 10^{-8}$ s, and $\sigma = 1.6 \times 10^{-9}$ cm^2 ($3s(^2S_{1/2}) \to 3p(^2P_{3/2})$ transition).

4.4 The angular momentum of an optical wave

Under the action of the electric field of a circularly polarized optical beam, the charged particles of the matter start to rotate around the axis of the beam. This means that the light possesses an angular momentum. Let us calculate the angular momentum of a circularly polarized optical wave by considering the interaction of the wave with the classical oscillator.

Let the electric field of the optical wave be

$$\vec{E} = \vec{x}_0 E_x + \vec{y}_0 E_y, \tag{4.29}$$

where

$$E_x = \tfrac{1}{2} iA \exp(i\omega t) + \text{c.c.}, \quad E_y = \tfrac{1}{2} A \exp(i\omega t) + \text{c.c.} \tag{4.30}$$

The motion equation for an elementary optical oscillator can be written as

$$\ddot{\vec{r}} + \Gamma \dot{\vec{r}} + \omega_0^2 \vec{r} = \frac{e}{m} \vec{E}, \tag{4.31}$$

where e and m are the charge and mass of the electron, ω_0 is the frequency of its natural vibrations, and Γ is the damping constant. The solution of eqn (4.31) will be sought in the form

$$\vec{r} = \vec{x}_0 x + \vec{y}_0 y, \tag{4.32}$$

where

$$x(t) = \tfrac{1}{2} ia \exp(i\omega t) + \text{c.c.}, \quad y(t) = \tfrac{1}{2} a \exp(i\omega t) + \text{c.c.} \tag{4.33}$$

Substituting eqns (4.29), (4.30), (4.32), and (4.33) into eqn (4.31), we obtain

$$a = \frac{e}{m} \frac{A}{\omega_0^2 - \omega^2 + i\omega\Gamma}. \tag{4.34}$$

Equations (4.32)–(4.34) show that in the field of a circularly polarized optical wave the electron rotates around the axis of the beam.

The power transmitted to the electron is

$$dW/dt = e\vec{E}\dot{\vec{r}} = e(E_x \dot{x} + E_y \dot{y}). \tag{4.35}$$

Simultaneously, the light transmits the angular momentum to the electron. The rate of this process is

$$d\vec{N}/dt = e[\vec{r}, \vec{E}] = \vec{z}_0 e(xE_y - yE_x). \tag{4.36}$$

After substituting eqns (4.30) and (4.33) into eqns (4.35) and (4.36) and averaging over the period of optical vibrations, we obtain

$$\langle d\vec{N}/dt\rangle = \vec{z}_0(1/\omega)\langle dW/dt\rangle; \tag{4.37}$$

whence we conclude that in a circularly polarized optical wave the angular momentum $\vec{N}$ and the energy W averaged over the period of optical vibrations are interrelated:

$$\tilde{\vec{N}} = \vec{z}_0\tilde{W}/\omega, \tag{4.38}$$

where ω is the frequency of light and $\vec{z}_0$ is the unit vector oriented along the wave propagation direction.

Equation (4.38) yields the sought-after magnitude of the angular momentum of a circularly polarized optical wave. Let us make a numerical evaluation. An optical pulse with energy $W = 1$ J at a wavelength of $\lambda = 0.5$ μm, having circular polarization, carries the angular momentum

$$N = W/\omega = W\lambda/2\pi c = 2.5\times 10^{-9}\,\text{erg s}.$$

The angular momentum of light can be found experimentally by measuring the moment of force that a beam of polarized light exerts on a plate of anisotropic crystal while passing through it. Specifically, a quarter-wave plate transforms linear polarized light to circularly polarized light and vice versa (see Chapters 2 and 21). It is easy to see that in this transformation the angular momentum of light is totally transmitted to the plate because the linear polarized light possesses no angular momentum. Consequently, the magnitude of the moment of force acting on the plate will be

$$M = P/\omega = P\lambda/2\pi c, \tag{4.39}$$

where P is the power of the light. For example, if $P = 100$ W and $\lambda = 0.5$ μm, then $M = 2.5\times 10^{-7}$ dyne cm. This estimate shows that the measurement of the angular momentum of light is rather a complex experimental problem. Successful experiments were performed with quartz plates when the light, having crossed them, changed the rotation direction of the vector $\vec{E}$ to the opposite sense.

4.4.1 Measurement of the angular momentum of light

The idea that a circularly polarized light should possess an angular momentum was first formulated by A. I. Sadovsky in 1897. Simultaneously, he proposed a procedure to measure the angular momentum of light by passing an optical beam through an anisotropic crystal plate. In his publication, Sadovsky wrote that "...any apparatus transforming a plane polarized light into a circularly polarized light must tend to rotate." In 1935–36 Holborn in England and Beth in the USA proved experimentally that circularly polarized light carries an angular momentum. In Beth's experiment a circularly polarized wave passed through a half-wave quartz plate twice. The plate could turn around the optical beam axis.

The plate was observed to acquire an angular momentum proportional to the energy of the light that passed through. This subtle experiment not only upheld Sadovsky's hypothesis but enabled a determination of Planck's constant, with an uncertainty of about 10%.

4.5 The photon

The physical investigations to be described later (see Chapter 9) led to the idea that in some cases light behaves as a flow of particles, so-called *photons*. The photon is meant to be an elementary indivisible portion of light with an energy determined by Planck's formula

$$W = \hbar\omega, \tag{4.40}$$

where ω is the frequency of light and

$$\hbar = 1.05 \times 10^{-27}\,\text{erg s} \tag{4.41}$$

is Planck's constant. The relations between the energy, momentum, and angular momentum of light, which we derived earlier, make it possible to calculate the momentum and angular momentum of the photon.

The energy density (energy per unit volume) in an optical wave can be presented in the form

$$dW/dV = I/c, \tag{4.42}$$

where I is the intensity and c is the velocity of light (see Chapter 3). It follows from eqn (4.1) defining the momentum density of light (momentum per unit volume), $G = dp/dV$, and eqn (4.42) that the energy W and the momentum p of light are related:

$$W = pc. \tag{4.43}$$

Inserting eqn (4.43) into eqn (4.40) we obtain the following expression for the photon momentum:

$$p = \hbar k, \tag{4.44}$$

where $k = \omega/c$ is the wave number. From the expression

$$W = N\omega, \tag{4.45}$$

relating the energy and angular momentum of circularly polarized light, we shall obtain the angular momentum of the photon. Inserting eqn (4.45) into eqn (4.40) we obtain

$$N = \hbar. \tag{4.46}$$

The concept of the photon can be also introduced in the theory of relativity. In this theory the momentum p and the energy W of a material point are defined as

$$\vec{p} = \frac{m\vec{v}}{\sqrt{1 - v^2/c^2}}, \quad W = \frac{mc^2}{\sqrt{1 - v^2/c^2}}, \tag{4.47}$$

where m is the mass of the point, $\vec{v}$ is its velocity, and c is the velocity of light. It follows from eqns (4.47) that

$$\vec{p}c^2 = \vec{v}W, \quad W^2 = p^2c^2 + m^2c^4. \tag{4.48}$$

Setting $m = 0$, we obtain

$$W = pc, \quad v = c. \tag{4.49}$$

Thus, the theory of relativity predicts the existence of particles with zero mass moving at the velocity of light in any inertial frame of reference. Such particles, possessing energy and momentum related by eqn (4.43), can be called photons.

4.5.1 Photon jet engine

Let us estimate a thrust of an engine employing a powerful light beam as a jet. According to Newton's law a force is equal to the derivative of the momentum with respect to time: $F = \dot{p}$. Differentiating eqn (4.43) and taking into account that the derivative of energy with respect to time is power, we obtain

$$F = P/c, \tag{4.50}$$

where P is the power of the optical beam and c is the velocity of light. Equation (4.50) shows that a substantial thrust would require an enormous light power. Indeed, taking $F = 1$ N, we obtain $P = 300$ MW. In the theory of relativity the mass of a body is a measure of its rest energy: $W_0 = mc^2$. Imagine an ideal photon jet engine, in which the fuel is totally converted into light. In this case the thrust equals $F = \mu c$, where $\mu = \dot{m}$ is the rate of mass (fuel) consumption. For example, setting $\mu = 1\ \mathrm{g\,s^{-1}}$, we obtain $F = 3 \times 10^{10}$ dyne $=$ 30 tonnes. Thus, a photon jet engine with a large thrust is possible in principle.

4.6 Optical cooling

The momentum transmitted to the atom in the process of light absorption is able to slow down its thermal motion. This effect is used for optical cooling of matter.

The principle of optical cooling is shown in Fig. 4.3. A laser beam the frequency of which corresponds to the atomic transition frequency is directed towards a bunch of atoms moving in the opposite direction (Fig. 4.3(a)). The atoms absorb the laser light and re-emit it in all directions (Fig. 4.3(b)). The total momentum of the light re-emitted by the atom is equal to zero, because all directions of photon emission are equally probable. On the other hand, the total momentum of the light absorbed by the atom is non-zero, because all of the photons composing the laser beam have the same propagation direction. As a result, the atoms decelerate and their temperature drops.

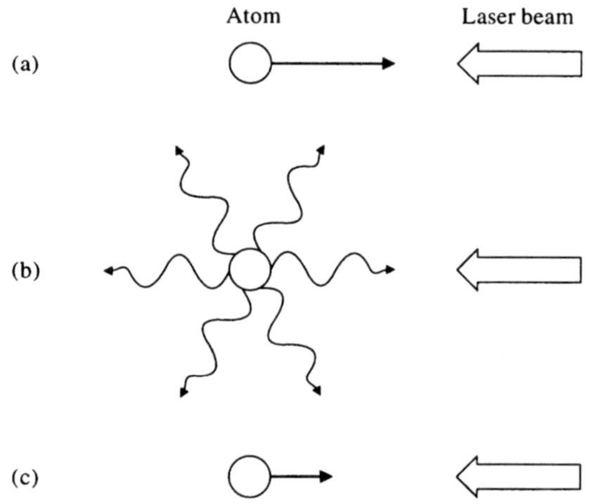

Fig. 4.3 The principle of optical cooling.

Experiments show that optical cooling makes it possible to obtain ultra-low temperatures of about 10^{-5} K. In such a state, atoms demonstrate quite unusual properties. For example, they can form a quasi-crystalline structure on the macroscopic scale (Diedrich *et al.* 1987). Speedy progress in optical methods of localization and cooling of atoms allows precise spectroscopic measurements, and assists in the development of most accurate optical frequency standards and in research into the fundamental properties of individual atoms and ions.

Bibliography

Ashkin, A. (1972). The pressure of laser light. *Scientific American,* **226**, 63–71.

Beth, B. A. (1935). *Physical Review,* **48**, 471; (1936). ibid., **50**, 115–125.

Diedrich, F., Peik, E., Chen, J., Quint, W., and Walther, H. (1987). *Physical Review Letters,* **59**, 2931–2934.

Holbourn, A. H. S. (1936). *Nature,* **137**, 31.

Lebedev, P. N. (1901). *Annalen der Physik,* **6**, 433–458.

Sadovsky, A. I. (1897). *Jurnal Russkogo Fiziko-Himicheskogo Obschestva (Fizika),* **29**, Section 1, (2), 82. (In Russian).

Part II

Light emission and optical oscillators

5

Classical radiation physics

Hertz's experiment. The atom as an elementary source of light. The classical model of the atom. Emission from a charge and Emission conditions. Calculation of the vortical component of the field. A rigorous solution of the problem of dipole emission. Harmonic oscillations of a dipole. The total power of dipole radiation. The decay of dipole oscillations due to emission of radiation.

How and why does optical radiation emerge? What is the source of electromagnetic waves? Only quantum theory can give comprehensive answers to these questions. Moreover, it is the investigation of radiation, that has led to the discovery of the quantum laws of nature. The very notion of a "quantum" was first introduced by Planck in connection with research into emission from heated bodies. Nevertheless, we shall start to discuss radiation physics using the classical model developed by Maxwell, Lorentz, and Hertz. The advantage of this model lies in its simplicity and clearness. It provides the possibility of explaining the emission of light in terms of the electrodynamics of macroscopic bodies. It should also be noted that many of the conclusions of classical theory find qualitative, and sometimes quantitative, confirmation in the quantum theory of radiation.

5.1 Hertz's experiment

The emission of electromagnetic waves was first demonstrated using a device known as the *Hertzian vibrator*. It consists of a pair of metal rods with spheres at their ends, located along a common line at a small distance from each other (Fig. 5.1). The rods are mounted on isolating supports and separated by a discharge gap. When a high voltage generated by a powerful induction coil is supplied to the rods, a high-frequency spark discharge takes place and the vibrator emits an electromagnetic wave. This wave can be detected with the help of a similar receiving vibrator (Fig. 5.2).

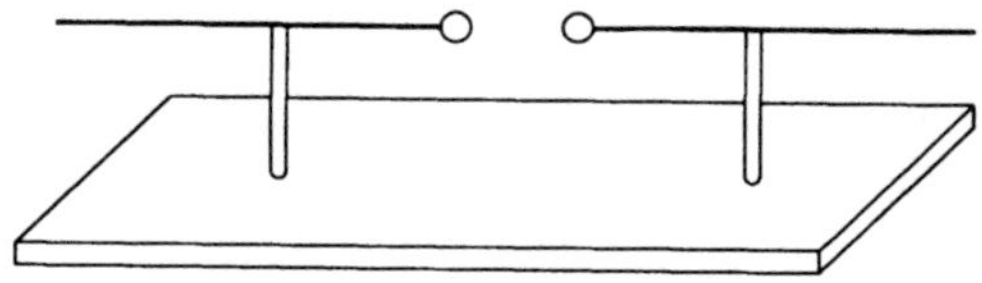

Fig. 5.1 The Hertzian vibrator.

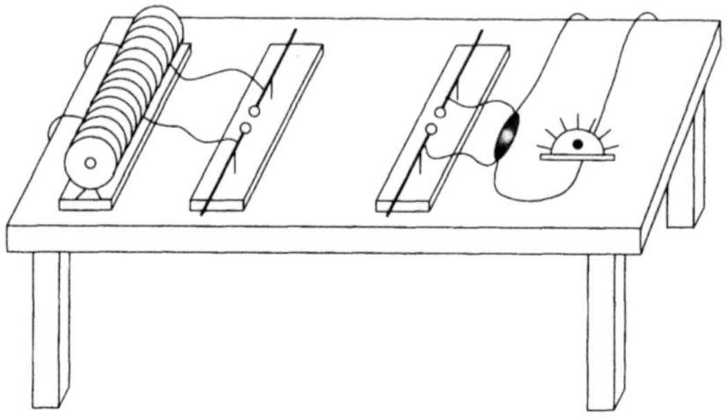

Fig. 5.2 The layout of the Hertz experiment.

5.2 The atom as an elementary source of light

The radiation frequency of the Hertzian vibrator equals 10–100 MHz. The frequency of visible light is several orders of magnitude higher ($10^{14} - 10^{15}$ Hz). Here the question arises: Does an optical frequency version of the Hertzian vibrator exist? If such a vibrator existed, it could emit light. Since the radiation frequency grows as the size of the vibrator diminishes, we can suppose that an optical oscillator should be extremely small. The idea has been proposed that an atom or molecule could perform as an elementary optical oscillator.

5.3 The classical model of the atom

The classical model of the atom is a pair of opposite electric charges bonded by an elastic Coulomb force. What structure could such a system have? One of the feasible arrangements is shown in Fig. 5.3. The positive point electric charge, a "nucleus", is located at the centre of the atom, while the negative electric charge, an "electron", is uniformly spread over the sphere enclosing the nucleus. The system is in equilibrium if the centre of the sphere coincides with the nucleus (Fig. 5.3(a)). If, however, the centre of the sphere is displaced relative to the nucleus, the Coulomb force appears, which is proportional to the displacement and strives to bring the atom back to its equilibrium state (Fig. 5.3(b)). Thus, the charges interact like bodies connected by a spring (Fig. 5.3(c)).

The motion of the electron relative to the nucleus is described by the following equation:

$$\ddot{x} + \frac{1}{\tau}\dot{x} + \omega_0^2 x = \frac{e}{m}E, \tag{5.1}$$

where e and m are the charge and mass of the electron, E is the external electric field, x is the displacement of the electron from the equilibrium position, and ω_0 and τ are the frequency and decay time of the free vibrations of the electron. The dot above the letter denotes time differentiation. An important characteristic of the atom is its *dipole moment*

$$p(t) = e\,x(t). \tag{5.1a}$$

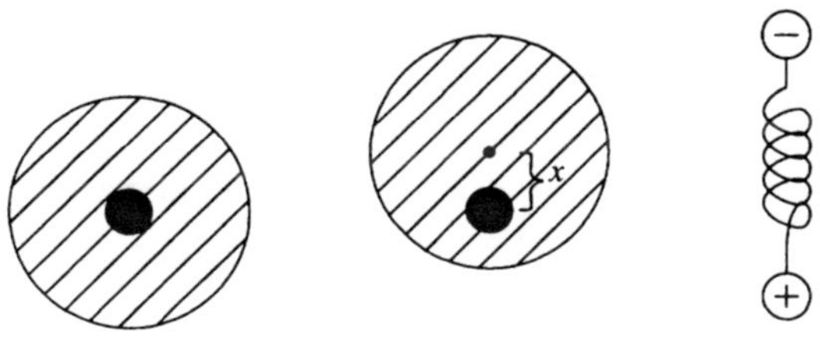

Fig. 5.3 The classical model of the atom.

Hence, according to the classical model, the atom is a system of bonded charges, in which the electron can oscillate relative to the nucleus. Now let us discuss how the oscillating electron generates an electromagnetic wave.

5.4 Emission from a charge and emission conditions

Let us clarify under what conditions a moving charge can induce a vortical field. Without resorting to Maxwell's equations, we shall consider an elementary pattern of lines of force.

5.4.1 A resting or uniformly moving charge does not emit

It is well known that the lines of electric force of a resting point charge are straight lines. If the charge moves uniformly and rectilinearly, the lines of electric force are also straight. They originate from the point of instantaneous location of the charge. Since the lines of electric force do not deflect, no vortical electromagnetic field, required to produce radiation, can form (see Chapter 1).

If the velocity of the charge motion is much less than the velocity of light, then the electric field is described by the formula

$$\vec{E} = \vec{n}\frac{q}{r^2}, \tag{5.2}$$

where $\vec{r}$ is the radius vector drawn from the point of instantaneous location of the charge to the observation point and $\vec{n} = \vec{r}/r$ is the unit vector directed along $\vec{r}$. As the velocity of the charge motion approaches the velocity of light c, the electric field of the moving charge becomes equal to

$$\vec{E} = \vec{n}\frac{q}{r^2}\frac{(1-\beta^2)}{(1-\beta^2\sin^2\theta)^{3/2}}, \tag{5.3}$$

where $\beta = v/c$ and θ is the angle between the vectors $\vec{r}$ and $\vec{v}$. The patterns of lines of electric force produced by a charge moving rectilinearly and uniformly are shown in Fig. 5.4.

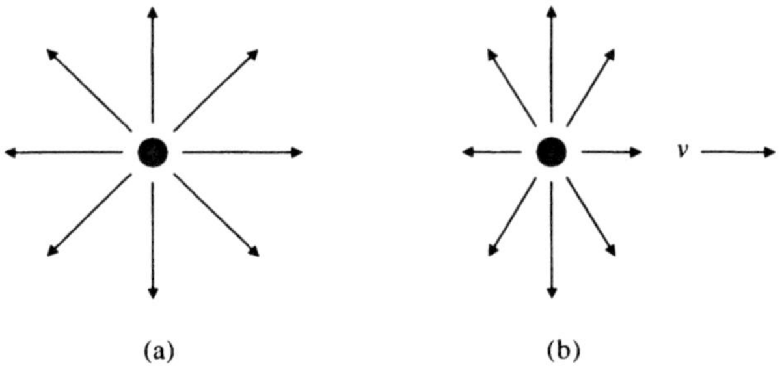

Fig. 5.4 The pattern of lines of electric force for a point charge moving uniformly and rectilinearly: (a) $v \ll c$; (b) $v \approx c$.

5.4.2 Acceleration of the charge generates a vortical electric field

Now we shall consider the accelerated motion of a point electric charge. Suppose that the charge q, being initially at rest, accelerates during a time interval Δt and then moves uniformly. The sequence of motion states of the charge is as follows:

$-\infty < t < 0$	the charge rests at the point $x = 0$,
$0 \le t \le \Delta t$	the charge moves along the x-axis with constant acceleration a,
$\Delta t < t < \infty$	the charge moves uniformly with velocity $v = a\,\Delta t$.

Assuming the charge motion velocity to be much less than the velocity of light, we consider the time moment $t \gg \Delta t$. The pattern of the lines of electric force at this moment is shown in Fig. 5.5. Beyond the sphere of radius $r = ct$ the centre of which is located at the origin of coordinates, the electric field coincides with that of a charge resting at the origin of coordinates. This is because the disturbance of the lines of electric force caused by the displacement of the charge propagates in space at the velocity of light c and, consequently, cannot get outside the sphere of radius ct by time t. Inside the sphere of smaller radius $r' = c(t-\Delta t)$, the centre of which is located at the point of instantaneous location of the charge $x \approx vt$, the electric field coincides with the field of a resting charge located at the point x. Therefore, the lines of electric force must *deflect* in the space between the two spheres described (Fig. 5.5).

Such a deflection attests to the emergence of a transverse (vortical) component of the electric field produced by the charge. It is this component that causes emission of the radiation. The deflection in the lines of electric force propagates in space at the velocity of light c. According to Maxwell's equations, the vortical electric field gives rise to the magnetic field. We can thus conclude that these travelling deflections in the lines of electric force constitute an electromagnetic pulse propagating in the form of a spherical wave.

It should be noted that the appearance of deflections in the lines of electric force—that is, a radiation—is associated with the *acceleration of the charge*, since the lines of electric force produced by a charge moving uniformly are rectilinear.

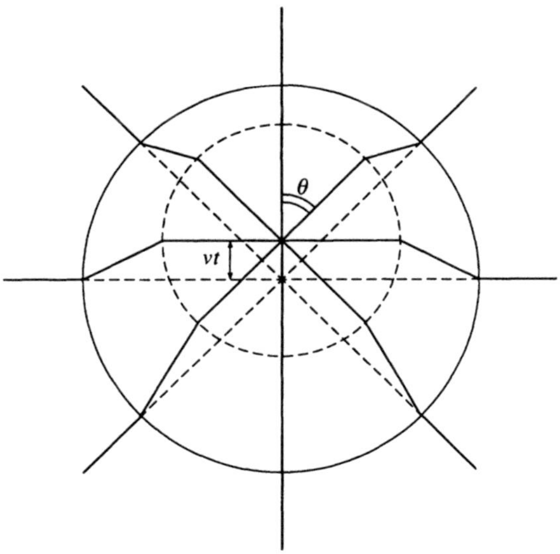

Fig. 5.5 The pattern of lines of electric force for an accelerating charge.

We can thus conclude that electromagnetic radiation is caused by the acceleration of electric charges.

5.5 Calculation of the vortical component of the field

The model of the charge motion considered above allows not only a qualitative explanation of the mechanism of electromagnetic radiation, but also a quantitative calculation of the vortical electromagnetic field appearing due to the charge acceleration.

Let us consider one of the lines of electric force at a time $t \gg \Delta t$ (Fig. 5.6). Draw the radius vector $\vec{r}$ from the initial charge position ($x = 0$) to the point of observation. The displacement of the charge during time t can be approximately written as $x = vt$, since for $t \gg \Delta t$ the inequality $a(\Delta t)^2/2 \ll vt$ holds. Let $v_\parallel$ and $v_\perp$ denote the longitudinal and transverse components of the charge velocity with respect to the direction of the radius vector $\vec{r}$. Similarly, we shall introduce the longitudinal $E_\parallel$ and transverse $E_\perp$ components of the electric field. As can be seen from Fig. 5.6, the longitudinal and transverse components of the field in the region of deflection of the line of the electric force are related as follows:

$$\frac{E_\perp}{E_\parallel} = \frac{v_\perp t}{c\Delta t}. \tag{5.4}$$

Using the formulae

$$e_\parallel = q/r^2, \quad r = ct, \quad v = a\Delta t, \quad v_\perp t = vt\sin\theta, \tag{5.5}$$

we obtain

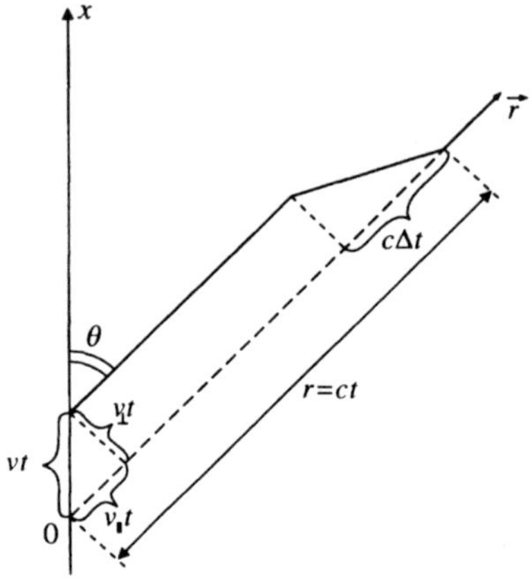

Fig. 5.6 The calculation of the transverse component of the electric field of an accelerating charge.

$$E_\perp = \frac{a\,q\sin\theta}{c^2 r}. \tag{5.6}$$

Thus, we have calculated the transverse component of the electric field associated with the charge acceleration. In eqn (5.6) a is the acceleration of the charge q, r is the distance between the charge and the observation point, θ is the angle between the direction of acceleration and the radius vector drawn from the charge location to the observation point, and c is the velocity of light.

By virtue of Maxwell's equations, the vortical electric field gives rise to a magnetic field. It can be shown that at a remote observation point the magnetic field vector is perpendicular to the transverse electric field vector and to the radius vector $\vec{r}$, and is equal in magnitude to the electric field (as it is in the case of the plane optical wave described in Chapter 1):

$$H_\perp = E_\perp. \tag{5.7}$$

Using eqns (5.6) and (5.7), we can calculate the energy flux density of the electromagnetic radiation emitted from an accelerating charge (see Chapter 3):

$$S = \frac{c}{4\pi}E_\perp H_\perp = \frac{(a\,q\sin\theta)^2}{4\pi c^3 r^2}. \tag{5.8}$$

The direction of the vector $\vec{S}$ coincides with that of the vector $\vec{r}$, i.e. the energy flows away from the charge. Equation (5.8) shows that the energy flux density varies inversely with the square of the distance between the charge and the observation point, with the total energy flux remaining constant. The most intensive radiation propagates at right angles to the charge acceleration direction: $S_{\max} = S(\theta = \pi/2)$. It can be seen from eqns (5.6) and (5.7) that the strength of the electromagnetic field produced by the charge varies as the reciprocal of the distance r from the charge, which is indicative of a spherical wave (see Chapter 1).

Applying the obtained results to the specific case of the vibrational motion of a charge, we can conclude that an oscillating charge (or a dipole) emits a spherical electromagnetic wave the frequency of which coincides with the oscillation frequency of the dipole.

5.6 A rigorous solution of the problem of dipole emission

Bearing in mind the classical model of the atom, let us consider the radiation of the pair of the electric charges, $+q$ and $-q$, bonded by an elastic force. Such a system will be called a *dipole*. The main characteristic of the dipole is its dipole moment, defined as

$$\vec{p} = \sum_i q_i \vec{r}_i = q(\vec{r}_1 - \vec{r}_2), \tag{5.9}$$

where $\vec{r}_1$ and $\vec{r}_2$ are the radius vectors of the charges. The dynamics of the dipole moment of an atom are described by eqns (5.1) and (5.1a). Using vector notation, we can write

$$\vec{p}(t) = \vec{x}_0\, p(t), \qquad p(t) = q\, x(t), \tag{5.10}$$

where $\vec{x}_0$ is the unit vector directed from one charge to the other.

The rigorous solution to the problem of dipole emission can be obtained by solving Maxwell's equation. Taking into account the alternating current arising due to the motion of the charges, we rewrite Maxwell's equations in the form

$$\begin{aligned} \operatorname{rot} \vec{E} &= -\frac{1}{c}\frac{\partial \vec{H}}{\partial t}, & \operatorname{div} \vec{E} &= 4\pi\rho, \\ \operatorname{rot} \vec{H} &= \frac{1}{c}\frac{\partial \vec{E}}{\partial t} + \frac{4\pi}{c}\vec{j}, & \operatorname{div} \vec{H} &= 0. \end{aligned} \tag{5.11}$$

Then the solution takes the form (see, e.g., Jackson 1962)

$$\begin{aligned} \vec{E} &= \frac{1}{c^2 r}\left[\vec{n}, [\vec{n}, \ddot{\vec{p}}(t - r/c)]\right], \\ \vec{H} &= -\frac{1}{c^2 r}\left[\vec{n}, \ddot{\vec{p}}(t - r/c)\right], \end{aligned} \tag{5.12}$$

where $\vec{r}$ is the radius vector drawn from the dipole to the observation point, $\vec{n} = \vec{r}/r$ is the unit vector in the direction of $\vec{r}$, and c is the velocity of light; the dot above the letter denotes time differentiation. The solution (5.12) is valid for the so-called far-field zone, i.e. for the spatial region removed from the dipole by a distance much larger than the size of the dipole and the radiation wavelength.

The structure of the dipole radiation field is illustrated in Fig. 5.7. Note that, in fact, it coincides with the structure of the field which we found earlier by analysing the pattern of the lines of electric force.

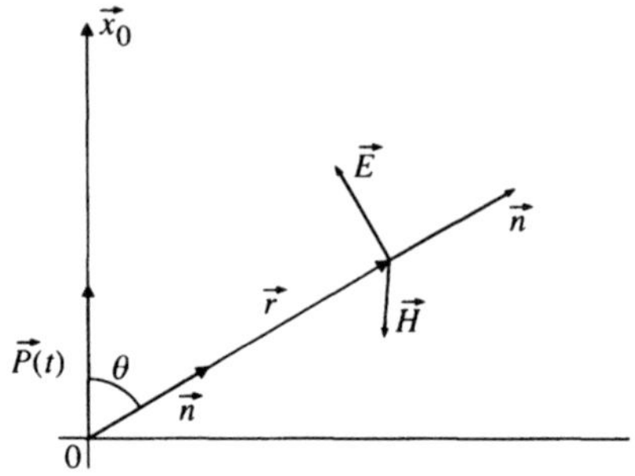

Fig. 5.7 Electromagnetic dipole radiation.

5.7 Harmonic oscillations of a dipole

Let us calculate the energy characteristics of dipole radiation, supposing that the dipole makes harmonic oscillations:

$$x(t) = X \cos \omega t. \tag{5.13}$$

Using eqns (5.10), (5.12), and (5.13), we obtain

$$\vec{E} = \vec{e} A \cos \omega(t - r/c), \qquad \vec{H} = \vec{h} A \cos \omega(t - r/c), \tag{5.14}$$

where

$$A = \frac{\omega^2 q X}{c^2 r} \sin \theta, \quad \vec{e} = -[\vec{n}, \vec{h}], \quad \vec{h} = \frac{[\vec{n}, \vec{x}_0]}{\sin \theta}, \tag{5.15}$$

in which θ is the angle between the vectors $\vec{n}$ and $\vec{x}_0$. Equations (5.14) and (5.15) show that the dipole radiation is linearly polarized. The vector $\vec{E}$ lies within the plane of the vectors $\vec{n}$ and $\vec{x}_0$, while the vector $\vec{H}$ is perpendicular to this plane (Fig. 5.7). The energy flux vector is

$$\vec{S} = \frac{c}{4\pi}[\vec{E}, \vec{H}] = \frac{cA^2}{4\pi}[\vec{e}, \vec{h}] \cos^2 \omega(t - r/c). \tag{5.16}$$

Taking into account eqns (5.15), we can write

$$[\vec{e}, \vec{h}] = \left[\vec{h}, [\vec{n}, \vec{h}]\right] = \vec{n} h^2 = \vec{n}. \tag{5.17}$$

The radiation intensity is as follows:

$$I = \langle S \rangle = \frac{cA^2}{8\pi} = \frac{q^2 X^2 \omega^4}{8\pi c^3 r^2} \sin^2 \theta. \tag{5.18}$$

The directivity pattern of dipole radiation is shown in Fig. 5.8.

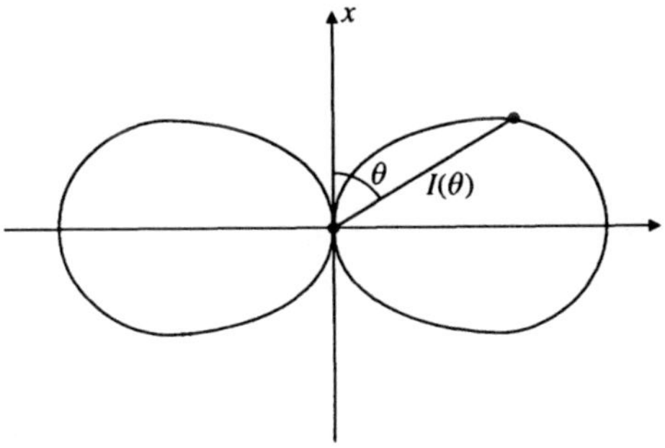

Fig. 5.8 The directivity pattern of dipole radiation.

5.8 The total power of dipole radiation

Using eqn (5.18), it is easy to find the total power of dipole radiation, i.e. the total power of the radiation emitted by a dipole in all directions. This can be done by integrating the radiation intensity over the spherical surface enclosing the dipole.

Let us draw a sphere of radius r around the dipole and introduce the angles θ and φ of the spherical coordinate system, as shown in Fig. 5.9. Making the angular increments $d\theta$ and $d\varphi$, we can obtain the surface element on the sphere:

$$d\sigma = r^2 \sin\theta \, d\theta \, d\varphi. \tag{5.19}$$

The total power of dipole radiation is expressed as

$$P = \int I \, d\sigma = \int_0^{2\pi} d\varphi \int_0^{\pi} I r^2 \sin\theta \, d\theta. \tag{5.20}$$

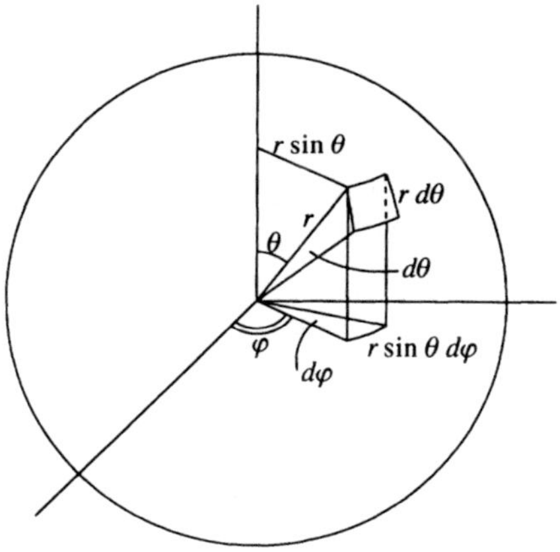

Fig. 5.9 The calculation of dipole radiation power.

Substituting eqn (5.18) into eqn (5.20) and integrating, we finally obtain

$$P = \frac{q^2 X^2 \omega^4}{3c^3}. \tag{5.21}$$

where q is the dipole charge, X and ω are the amplitude and frequency of the oscillations, c is the velocity of light, and P is the radiation power. Note that the radiation power depends strongly on the dipole oscillation frequency.

5.9 The decay of dipole oscillations due to emission of radiation

The emission of electromagnetic waves results in a loss of energy by the dipole and attenuation of the free oscillations (*radiation decay*). The characteristic time of radiation decay can be estimated as the ratio of the initial energy of an oscillator to the power of radiation:

$$\tau = W/P. \tag{5.22}$$

It is known from mechanics that the energy of an harmonic oscillator can be written in the form

$$W = m\omega^2 X^2/2, \tag{5.23}$$

where m is the oscillator mass, and X and ω are the amplitude and frequency of the oscillations. Substituting eqns (5.21) and (5.23) into eqn (5.22), we can estimate the radiation decay time:

$$\tau = \frac{3mc^3}{2q^2\omega^2}. \tag{5.24}$$

It is interesting to evaluate the decay time of electron oscillations in the atom at optical frequencies. Assuming $m = 9.1 \times 10^{-28}$ g, $q = e = 4.8 \times 10^{-10}$ CGSE, $c = 3 \times 10^{10}$ cm s^{-1}, and $\omega = 4 \times 10^{15}$ s^{-1}, we obtain

$$\tau = 10^{-8} \text{ s}. \tag{5.25}$$

Thus, according to our evaluation, the radiation decay time for the classical atom is about 10^{-8} s. Experimental investigations confirm this result.

According to eqn (5.1), the decay of atomic free oscillations can be approximately described by the formula

$$x(t) = x_0 e^{-t/2\tau} \cos \omega_0 t \tag{5.26}$$

or

$$x(t) = X \cos \omega_0 t, \tag{5.27}$$

where the oscillation amplitude,

$$X = x_0 e^{-t/2\tau}, \tag{5.28}$$

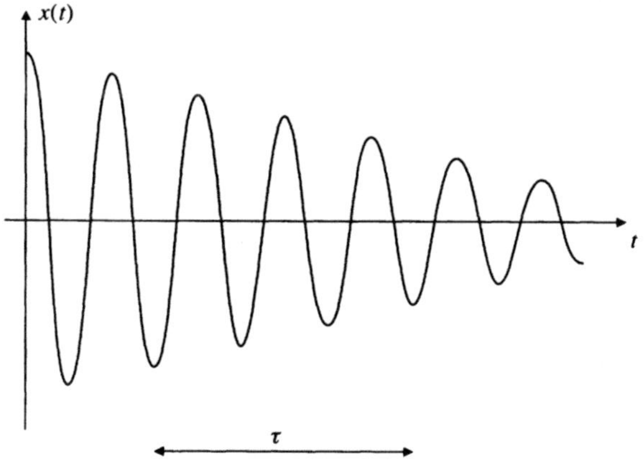

Fig. 5.10 The decay of free oscillations of an atomic oscillator.

decreases with time (Fig. 5.10). Substituting eqn (5.28) into eqn (5.23), we can find the law of oscillator energy decay

$$W = W(t) = W_0 e^{-t/\tau}, \tag{5.29}$$

where

$$W_0 = m\omega_0^2 x_0^2/2 = W(t=0) \tag{5.30}$$

is the initial energy of the oscillator.

In comparison with the period of optical oscillations ($T_0 \approx 10$ fs), the oscillation amplitude declines very slowly ($\tau \approx 10^{-8}$ s). The radiation field of an individual atom can thus be approximated as

$$E(t) = a_0 e^{-\frac{t-t_0}{2\tau}} \cos[\omega(t-t_0) - \psi], \tag{5.31}$$

where t_0 is the moment at which the oscillations are excited (atomic oscillations can be caused, for example, by a collision with another atom or with an electron), a_0 and ψ are the initial amplitude and phase of the electric field oscillations at the observation point, and ω and τ are the frequency and decay time of the free electron oscillations. The radiation field of the individual atom varies with time according to a law resembling that shown in Fig. 5.10. At a fixed point in space, the radiation field has the form of a *modulated oscillation.*

Bibliography

Born, M. and Wolf, E. (1968). *Principles of optics.* Pergamon Press, Oxford.

Crawford, F. S. (1965). *Waves. Berkeley course of physics,* Vol. 3. McGraw-Hill, New York.

Drude, P. (1912). *Lehrbuch der Optik.* Verlag von S. Hirzel, Leipzig.

Jackson, J. D. (1962). *Classical electrodynamics.* John Wiley, New York.

6

Emission from an oscillator ensemble

From the dynamics of an oscillator to the statistics of an ensemble. Non-laser sources of light: intensity, polarization, and radiation directivity pattern. Radiation statistics of independent oscillators. The radiation spectrum. Mechanisms of spectral line broadening. Natural broadening. Doppler broadening. Collision-induced broadening and the shift of a spectral line. Dicke's effect: collision-induced narrowing of a Doppler spectral line in a dense gas.

6.1 From the dynamics of an oscillator to the statistics of an ensemble

While describing an emission from real sources of light, we have to take into account the contributions of a great number of individual oscillators; that is, atoms or molecules. An effective way to do this is to replace summation by averaging and to use the rules of statistics for the calculation of averages.

6.2 Non-laser sources of light: intensity, polarization, and radiation directivity pattern

Consider the radiation of an oscillator ensemble at a fixed point in space (Fig. 6.1). The electric field strength at this point can be described in the form

$$\vec{E} = \sum_{i=1}^{N} \vec{E}_i, \tag{6.1}$$

where N is the number of oscillators and

$$\vec{E}_i = \vec{e}_i \, a_i \cos(\omega t - \varphi_i) \tag{6.2}$$

is the radiation field of an individual oscillator. Equation (6.1) can be rewritten as

$$\vec{E} = N \langle \vec{E}_i \rangle, \tag{6.3}$$

where

$$\langle \vec{E}_i \rangle = \frac{1}{N} \sum_{i=1}^{N} \vec{E}_i \tag{6.4}$$

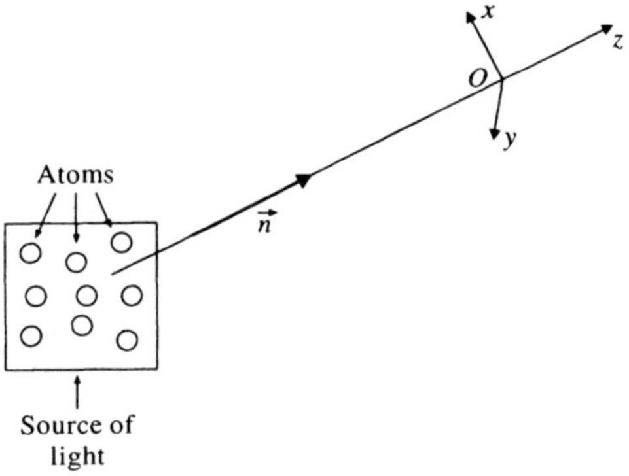

Fig. 6.1 Calculation of the emission characteristics of an oscillator ensemble.

is the value of the field averaged over the oscillator ensemble. Let us calculate this value assuming the parameters $\vec{e}_i$, a_i, and φ_i to be independent random variables. We obtain

$$\langle \vec{E}_i \rangle = \langle \vec{e}_i \rangle \langle a_i \rangle \langle \cos(\omega t - \varphi_i) \rangle = 0, \tag{6.5}$$

as the phases of the oscillators are distributed uniformly (Fig. 6.2)

$$w(\varphi_i) = 1/2\pi \qquad (-\pi \le \varphi_i \le \pi) \tag{6.6}$$

and, consequently,

$$\begin{aligned} \langle \sin \varphi_i \rangle &= \int_{-\pi}^{\pi} \sin \varphi_i \, w(\varphi_i) \, d\varphi_i = 0, \\ \langle \cos \varphi_i \rangle &= \int_{-\pi}^{\pi} \cos \varphi_i \, w(\varphi_i) \, d\varphi_i = 0, \\ \langle \cos(\omega t - \varphi_i) \rangle &= \cos \omega t \langle \cos \varphi_i \rangle + \sin \omega t \langle \sin \varphi_i \rangle = 0. \end{aligned} \tag{6.7}$$

Thus, the average strength of the optical field generated by the oscillator ensemble turns to be equal to zero due to the chaotic phase distribution. For the same

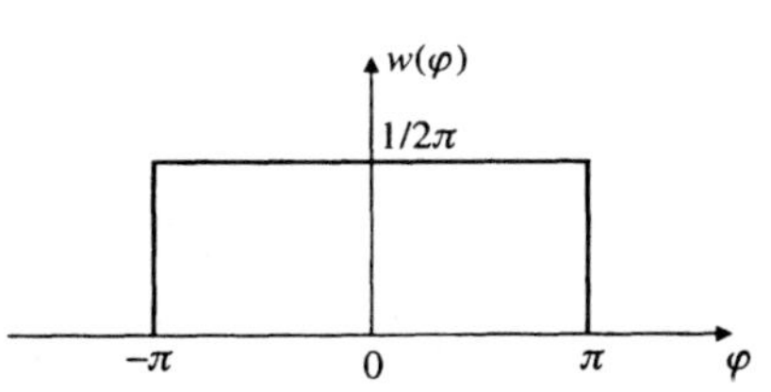

Fig. 6.2 The probability distribution of an oscillation phase.

reason, the total radiation intensity

$$I = \frac{c}{4\pi}\langle E^2 \rangle \tag{6.8}$$

proves to be equal to the sum of the average radiation intensities produced by individual oscillators:

$$I = N\langle I_i \rangle, \tag{6.9}$$

where

$$\langle I_i \rangle = \frac{c}{4\pi}\langle E_i^2 \rangle, \qquad \langle E_i^2 \rangle = \frac{1}{N}\sum_{i=1}^{N} E_i^2. \tag{6.10}$$

Indeed,

$$E^2 = \sum_{i=1}^{N} \vec{E}_i \sum_{j=1}^{N} \vec{E}_j = \sum_{i=1}^{N} E_i^2 + \sum_{i=1}^{N} \sum_{j\neq i=1}^{N} \vec{E}_i\vec{E}_j$$
$$= N\langle E_i^2 \rangle + (N^2 - N)\langle \vec{E}_i\vec{E}_j \rangle = N\langle E_i^2 \rangle. \tag{6.11}$$

In deriving eqn (6.11) we used eqns (6.1) and (6.5) together with the condition of independence of oscillators, expressed as

$$\langle \vec{E}_i\vec{E}_j \rangle = \langle \vec{E}_i \rangle \langle \vec{E}_j \rangle \qquad (i \neq j). \tag{6.12}$$

It follows from eqn (6.9) that the radiation power of the oscillator ensemble is equal to the sum of the average radiation powers of individual oscillators:

$$P = N\langle P_i \rangle. \tag{6.13}$$

The radiation directivity pattern is isotropic owing to the random orientation of the dipole moments.

6.2.1 Polarization of radiation

Let us project the vector $\vec{E}$ defined by eqns (6.1) and (6.2) on to the x- and y-axes located in the wavefront plane of the optical wave (Fig. 6.1). Using Euler's formula

$$\cos(\omega t - \varphi_i) = \tfrac{1}{2}\mathrm{e}^{i(\omega t - \varphi_i)} + \mathrm{c.\,c.}, \tag{6.14}$$

we obtain

$$E_x = \tfrac{1}{2}\mathcal{E}_x \mathrm{e}^{i\omega t} + \mathrm{c.\,c.}, \qquad E_y = \tfrac{1}{2}\mathcal{E}_y \mathrm{e}^{i\omega t} + \mathrm{c.\,c.}, \tag{6.15}$$

where

$$\mathcal{E}_x = \sum_{i=1}^{N} \mathcal{E}_{xi}, \qquad \mathcal{E}_y = \sum_{i=1}^{N} \mathcal{E}_{yi},$$
$$\mathcal{E}_{xi} = \cos\alpha_i a_i \mathrm{e}^{-i\varphi_i}, \quad \mathcal{E}_{yi} = \sin\alpha_i a_i \mathrm{e}^{-i\varphi_i}, \tag{6.16}$$

and α_i is the angle between the vector $\vec{e}_i$ and the x-axis (Fig. 6.3).

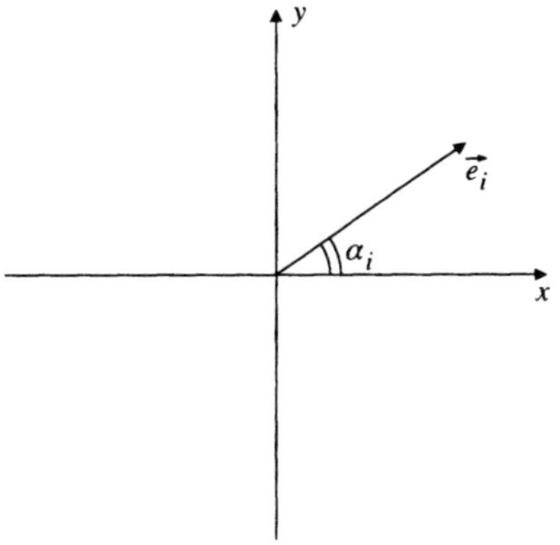

Fig. 6.3 An analysis of the polarization state of the radiation from an oscillator ensemble.

Now let us calculate the coherence matrix (see Chapter 2):

$$\hat{\mathcal{J}} = \begin{pmatrix} \langle \mathcal{E}_x \mathcal{E}_x^* \rangle & \langle \mathcal{E}_x \mathcal{E}_y^* \rangle \\ \langle \mathcal{E}_y \mathcal{E}_x^* \rangle & \langle \mathcal{E}_y \mathcal{E}_y^* \rangle \end{pmatrix}. \tag{6.17}$$

Inserting eqn (6.16) into eqn (6.17) and using a procedure similar to that described earlier (see eqn (6.11)), we find that

$$\hat{\mathcal{J}} = \begin{pmatrix} \frac{1}{2} N \langle a_i^2 \rangle & 0 \\ 0 & \frac{1}{2} N \langle a_i^2 \rangle \end{pmatrix}. \tag{6.18}$$

Equation (6.18) is derived under the assumption that the angle α_i is a random parameter with probability density

$$w(\alpha_i) = 1/2\pi \qquad (-\pi \leq \alpha_i \leq \pi). \tag{6.19}$$

The expression (6.18) obtained for the coherence matrix shows that the radiation of the oscillator ensemble has *natural polarization* (see Chapter 2). Such a polarization state is obviously caused by the random spatial orientation of the dipole moments of individual oscillators.

6.3 Radiation statistics of independent oscillators

Let us write the radiation field generated by an oscillator ensemble in the form

$$\vec{E} = \vec{x}_0 E_x + \vec{y}_0 E_y, \tag{6.20}$$

where

$$E_x = \sum_{i=1}^{N} \cos\alpha_i a_i \cos(\omega t - \varphi_i), \quad E_y = \sum_{i=1}^{N} \sin\alpha_i a_i \cos(\omega t - \varphi_i). \tag{6.21}$$

Consider one of the field components, e.g. E_x. It can be written in the form

$$E_x = a\cos\omega t + b\sin\omega t, \tag{6.22}$$

where the parameters a and b are called the *quadrature components* of the vibration E_x and are defined as

$$a = \sum_{i=1}^{N} \cos\alpha_i a_i \cos\varphi_i, \quad b = \sum_{i=1}^{N} \cos\alpha_i a_i \sin\varphi_i. \tag{6.23}$$

The field E_x is also characterized by amplitude, phase, and intensity. The amplitude A and phase φ of the vibration E_x are defined by the formula

$$E_x = A\cos(\omega t - \varphi) \qquad (A \geq 0). \tag{6.24}$$

The *intensity* I is the quantity

$$I = \langle E_x^2 \rangle = A^2/2, \tag{6.25}$$

where "$\langle\,\rangle$" denotes an averaging over the optical vibration period $T = 2\pi/\omega$, i.e.

$$I = \frac{1}{T}\int_t^{t+T} E_x^2\,dt. \tag{6.26}$$

It follows from eqns (6.22) and (6.24) that

$$a = A\cos\varphi, \qquad b = A\sin\varphi. \tag{6.27}$$

If we suppose that the parameters α_i, a_i, and φ_i in eqns (6.21) and (6.23) are random, then it can easily be shown that the parameters E_x, a, b, A, φ, and I in eqns (6.22), (6.23), (6.25), and (6.27) are also random. Let us find the probability distributions for these values.

Individual oscillators of a non-laser source of light emit independently of one another. This suggests that the parameters α_i, a_i, and φ_i are statistically independent. Turning to eqns (6.21) and (6.23), we see that the optical field strength E_x, as well as the quadrature components a and b, are the sums of a great number of independent random values. Then, according to the central limit theorem of probability theory, E_x, a, and b have a Gaussian probability distribution:

$$w(E_x) = \frac{1}{\sigma\sqrt{2\pi}} \mathrm{e}^{-E_x^2/2\sigma^2}, \tag{6.28}$$

$$w(a) = \frac{1}{\sigma\sqrt{2\pi}} e^{-a^2/2\sigma^2}, \qquad w(b) = \frac{1}{\sigma\sqrt{2\pi}} e^{-b^2/2\sigma^2}. \tag{6.29}$$

Here we use the notation $\langle E_x^2 \rangle = \sigma^2$ and take into account that

$$\langle E_x \rangle = \langle a \rangle = \langle b \rangle = 0, \quad \langle a^2 \rangle = \langle b^2 \rangle = \langle E_x^2 \rangle, \quad \langle ab \rangle = 0. \tag{6.30}$$

Assuming the quadrature components to be statistically independent, we can find the two-dimensional probability density

$$w(a, b) = w(a)\, w(b). \tag{6.31}$$

Inserting eqn (6.29) into eqn (6.31), we obtain

$$w(a, b) = \frac{1}{2\pi\sigma^2} e^{-(a^2+b^2)/2\sigma^2}. \tag{6.32}$$

Now we can easily find the probability distributions $w(A)$, $w(\varphi)$, and $w(I)$. Using the rule for substituting variables in the probability distribution and eqns (6.25), (6.27), and (6.32), we obtain

$$w(A, \varphi) = \frac{A}{2\pi\sigma^2} \exp(-A^2/2\sigma^2), \tag{6.33}$$

$$w(A) = \int_{-\pi}^{\pi} w(A, \varphi)\, d\varphi = \frac{A}{\sigma^2} e^{-A^2/2\sigma^2}, \tag{6.34}$$

$$w(\varphi) = \int_0^{\infty} w(A, \varphi)\, dA = \frac{1}{2\pi}, \tag{6.35}$$

$$w(I) = (1/\overline{I}) \exp(-I/\overline{I}), \qquad \overline{I} = \sigma^2, \tag{6.36}$$

where

$$\overline{I} = \int_0^{\infty} I\, w(I)\, dI \tag{6.37}$$

is the average intensity.

The probability distribution (6.34) is referred to as a *Rayleigh distribution*.

To summarize, we shall note that the basic characteristics of non-laser (e.g. thermal) radiation have the following probability distributions: the optical field strength and its quadrature components have a Gaussian distribution, while the amplitude, phase, and intensity of the field have Rayleigh, rectangular (uniform), and exponential distributions, respectively. The shapes of these distributions are shown in Fig. 6.4.

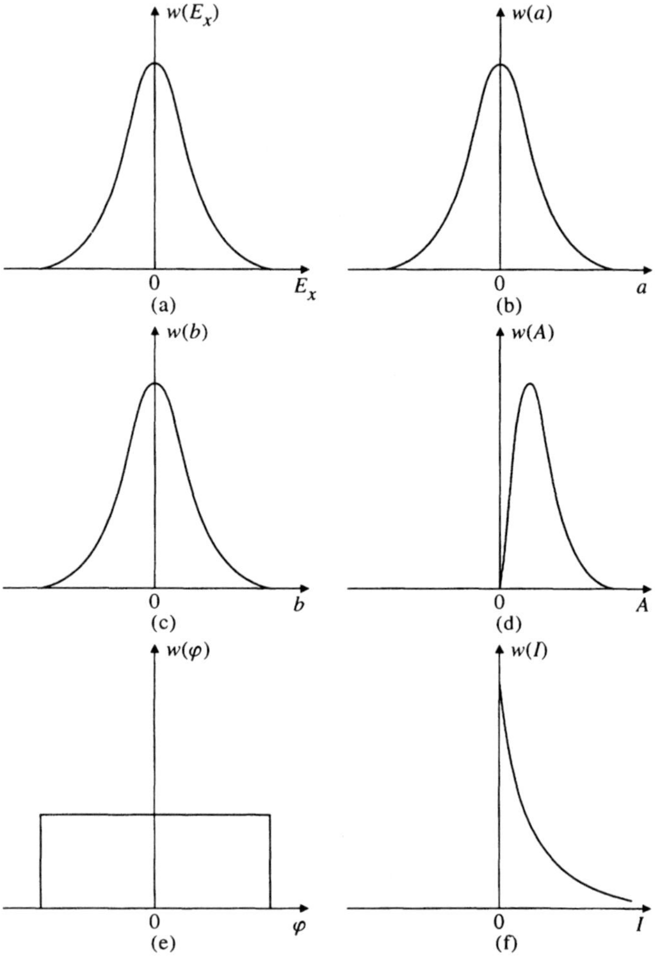

Fig. 6.4 Probability distributions characterizing the radiation of a non-laser source of light.

6.4 The radiation spectrum

It is natural to consider the emission from an ensemble of independent oscillators as a steady-state stochastic process. Then the radiation spectrum is determined by the Wiener–Khinchin theorem

$$S(\omega) = \frac{1}{\pi} \int_{-\infty}^{\infty} B(\tau) \mathrm{e}^{-i\omega\tau} d\tau, \tag{6.38}$$

where

$$B(\tau) = \langle E_x(t) E_x(t+\tau) \rangle \tag{6.39}$$

is the correlation function of the optical field.

Let us write $E_x(t)$ as the sum of the fields generated by individual oscillators:

$$E_x(t) = \sum_{j=1}^{N} E_{xj}(t). \tag{6.40}$$

Then

$$E_x(t)E_x(t+\tau) = N\langle E_{xj}(t)E_{xj}(t+\tau)\rangle, \tag{6.41}$$

where N is the total number of oscillators, and the angular brackets denote averaging over the oscillator ensemble. While deriving eqn (6.41), we allowed for the independence of oscillators and substituted statistical averaging for summation. The essence of this operation can be expressed by the formula

$$\frac{1}{N}\sum_{j=1}^{N} E_{xj}(t)E_{xj}(t+\tau) = \langle E_{xj}(t)E_{xj}(t+\tau)\rangle \equiv B_0(\tau). \tag{6.42}$$

Inserting eqns (6.41) and (6.42) into eqns (6.38) and (6.39), we obtain

$$B(\tau) = NB_0(\tau), \tag{6.43}$$

and

$$S(\omega) = NS_0(\omega), \tag{6.44}$$

where

$$S_0(\omega) = \frac{1}{\pi}\int_{-\infty}^{\infty} B_0(\tau)\mathrm{e}^{-i\omega\tau}\,d\tau \tag{6.45}$$

or, since the correlation function is even,

$$S_0(\omega) = \frac{1}{\pi}\int_{0}^{\infty} B_0(\tau)\left(\mathrm{e}^{i\omega\tau} + \mathrm{e}^{-i\omega\tau}\right)\,d\tau. \tag{6.45a}$$

Thus, the calculation of the radiation spectrum is reduced to calculation of the correlation function $B_0(\tau)$ defined by eqn (6.42).

6.5 Mechanisms of spectral line broadening

As seen from eqns (6.38), (6.42), and (6.43), the radiation spectrum is determined by the time variation of the optical field strength $E_{xj}(t)$. The form of this function, in turn, is determined by physical factors such as the oscillation decay due to emission of radiation, the thermal motion of oscillators, dephasing collisions, the dispersion of oscillator frequencies, etc. Each of these factors determines its specific mechanism of spectral line broadening: natural, Doppler, collision-induced, inhomogeneous broadening, etc. Let us consider some of these mechanisms.

6.6 Natural broadening

This most fundamental mechanism of spectral line broadening is caused by the decay of dipole oscillations due to the emission of radiation.

Suppose that the oscillator emits a sequence of bursts (pulses) of the standard form $F(t)$ at random time moments t_{pj}. Then we can write

$$E_{xj}(t) = \sum_{p=1}^{n} F(t - t_{pj}), \tag{6.46}$$

where n is the number of bursts emitted during a time interval T. The process described by eqn (6.46) is referred to as a *random pulse sequence.* Now we shall calculate the correlation function of this process.

Proceeding from the physical picture of the emission by an ensemble of independent oscillators excited, for example, via the collision mechanism, we can make some assumptions about the statistical characteristics of the random process $E_{xj}(t)$. Specifically, we shall assume that the individual pulses occur independently of one another, that the random moments of time t_{pj} have a uniform probability distribution over the interval $0 \le t \le T$, and that the number of pulses n is an independent random quantity obeying the Poisson distribution. In addition, we shall suppose that the duration of an individual pulse is much shorter than the observation time T and that the average value of the function $F(t)$ is equal to zero. Then

$$B_0(\tau) = \overline{n}\langle F(t - t_{pj}) F(t + \tau - t_{pj})\rangle, \tag{6.47}$$

or

$$B_0(\tau) = \frac{\overline{n}}{T}\int_{-\infty}^{\infty} F(\theta)F(\theta + \tau)\, d\theta, \tag{6.48}$$

where $\overline{n}$ is the average number of pulses emitted by the oscillator during the time interval T. Inserting eqn (6.48) into eqn (6.45), we obtain

$$\begin{aligned} S_0(\omega) &= \frac{\Omega}{\pi}\iint_{-\infty}^{\infty} F(\theta)\, F(\theta + \tau)\mathrm{e}^{-i\omega\tau}\, d\theta\, d\tau \\ &= \frac{\Omega}{\pi}\int_{-\infty}^{\infty} d\theta\, F(\theta)\mathrm{e}^{i\omega\theta}\int_{-\infty}^{\infty} F(\theta + \tau)\mathrm{e}^{-i\omega(\theta+\tau)} d\tau = \frac{\Omega}{\pi}|F(\omega)|^2, \end{aligned} \tag{6.49}$$

where $\Omega = \overline{n}/T$ is the average pulse repetition frequency and

$$F(\omega) = \int_{-\infty}^{\infty} F(t)\mathrm{e}^{-i\omega t}\, dt \tag{6.50}$$

is the spectral amplitude determined by the shape of the individual pulse.

Thus, we can conclude that the spectrum of a random pulse sequence is governed by the spectral density of the individual pulse. One can easily show

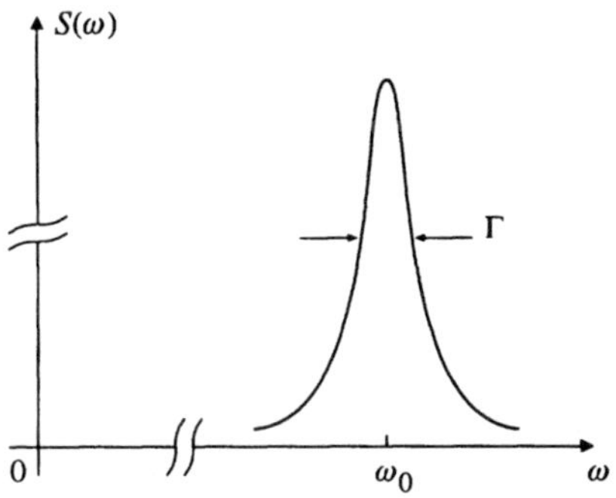

Fig. 6.5 The natural broadening of a spectral line.

that this is associated with the fact that a random moment of time, at which the pulse occurs, has no effect on the spectral density of the pulse.

In particular, assuming that

$$F(t) = a(t)\cos\omega_0 t, \tag{6.51}$$

where

$$a(t) = a_0 \begin{cases} 0, & t < 0, \\ \exp(-t/2\tau), & t \geq 0, \end{cases} \tag{6.52}$$

ω_0 is the intrinsic frequency, and τ is the oscillation decay time due to emission of radiation (see Chapter 5), we obtain

$$S_0(\omega) = \frac{\Omega}{\pi}\frac{a_0^2/4}{\Gamma^2/4 + (\omega-\omega_0)^2}. \tag{6.53}$$

We have thus shown that the natural spectrum has a Lorentzian shape with central frequency ω_0 and width $\Gamma = 1/\tau$ (Fig. 6.5). The radiation power is proportional to the number of oscillators, N, and to the average pulse frequency Ω of the individual oscillator. The typical value of the natural broadening is equal to

$$\Delta\omega_e = 1/\tau = 10^8 \text{ Hz}. \tag{6.54}$$

In practice, however, such narrow spectral lines are rare. As a rule, the natural broadening is augmented by other stronger broadening mechanisms, one of which is the Doppler mechanism.

6.7 Doppler broadening

Doppler broadening of a spectral line is caused by chaotic thermal motion of atoms or molecules. It is most typical of rarified luminous gases. Specifically, the radiation emitted laterally by the gas-discharge tube of a He–Ne laser has a

Doppler spectrum. Let us calculate the shape and bandwidth of a Doppler spectral line. As before, we shall describe the emission from the oscillator ensemble by eqn (6.40). Taking into account the thermal motion of the particles and neglecting (for simplicity) the oscillation decay due to the emission of radiation, we can write

$$E_{xj} = a_0 \cos(\omega_0 t - k_0 z_j), \tag{6.55}$$

where $k_0 = \omega_0/c$ is the wave number of the optical wave, and z_j is the oscillator coordinate with respect to the z-axis directed from the source of light to the observation point (see Fig. 6.1).

In the absence of collisions,

$$z_j = z_j(t) = z_{0j} + v_{zj}t, \tag{6.56}$$

where t is time, z_{0j} is the initial coordinate, and v_{zj} is the velocity component of a given molecule along the z-axis. Inserting eqn (6.56) into eqn (6.55), we obtain

$$E_{xj}(t) = a_0 \cos(\omega_j t - \varphi_j), \tag{6.57}$$

where

$$\varphi_j = k_0 z_{0j}, \qquad \omega_j = \omega_0 - k_0 v_{zj}. \tag{6.58}$$

To calculate the correlation function defined by eqn (6.42), we shall suppose that the random quantities z_{0j} and v_{zj} are statistically independent and that the phase φ_j is distributed uniformly over the periodicity interval $-\pi \leq \varphi_j \leq \pi$, i.e. $w(\varphi_j) = 1/2\pi$. It is easy to show that in this case

$$B_0(\tau) = \frac{a_0^2}{2}\langle \cos \omega_j \tau \rangle = \frac{a_0^2}{4}\langle \mathrm{e}^{i\omega_j \tau}\rangle + \mathrm{c.\,c.} \tag{6.59}$$

Inserting eqn (6.58) into eqn (6.59), we obtain

$$B_0(\tau) = \frac{a_0^2}{4}\mathrm{e}^{i\omega_0 \tau}\langle \mathrm{e}^{-ik_0 v_{zj}\tau}\rangle + \mathrm{c.\,c.} \tag{6.60}$$

The Cartesian component of the velocity of a gas molecule is distributed according to the Gaussian probability function

$$w(v_{zj}) = \frac{1}{\sigma_v\sqrt{2\pi}}\mathrm{e}^{-v_{zj}^2/2\sigma_v^2}, \tag{6.61}$$

where

$$\sigma_v^2 = kT/m \tag{6.62}$$

is the variance of thermal velocity, k is the Boltzmann constant, T is the absolute temperature of the gas, and m is the mass of the molecule. Consequently,

$$\langle \mathrm{e}^{-ik_0 v_{zj}\tau}\rangle = \mathrm{e}^{-\frac{1}{2}k_0^2\sigma_v^2\tau^2} \tag{6.63}$$

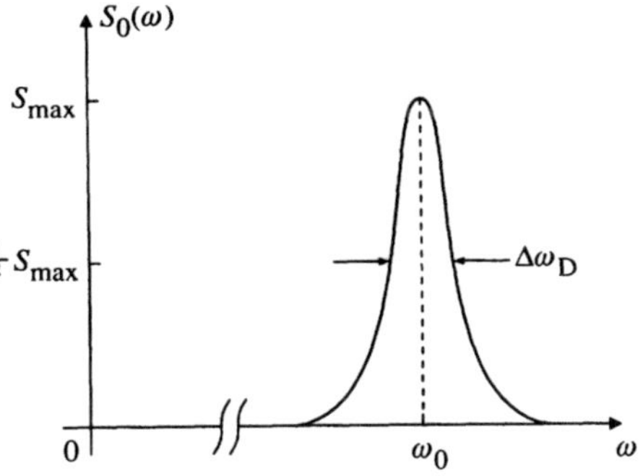

Fig. 6.6 A Doppler spectral line.

and

$$B_0(\tau) = \frac{a_0^2}{4} e^{i\omega_0\tau - \frac{1}{2}k_0^2\sigma_v^2\tau^2} + \text{c. c.} \tag{6.64}$$

Finally, inserting eqn (6.64) into eqn (6.45), we obtain (in the range of positive frequencies):

$$S_0(\omega) = \frac{a_0^2}{2} \frac{1}{\sigma_\omega\sqrt{2\pi}} e^{-(\omega-\omega_0)^2/2\sigma_\omega^2}, \qquad \sigma_\omega = k_0\sigma_v. \tag{6.65}$$

The function $S_0(\omega)$ is illustrated in Fig. 6.6.

Thus, the Doppler spectral line has a Gaussian form with centre frequency ω_0 and a full width at half maximum (FWHM) of

$$\Delta\omega_D = \omega_0\sqrt{\frac{8kT}{mc^2}\ln 2}. \tag{6.66}$$

Let us estimate the Doppler bandwidth of the He–Ne laser spectral line. Setting $\lambda = 6328$ Å, $m = 3 \times 10^{-23}$ g (the mass of a neon atom), $T = 300$ K, $k = 1.38 \times 10^{-16}$ erg K^{-1}, and $c = 3 \times 10^{10}$ cm s^{-1}, we obtain, from eqn (6.66):

$$\Delta f_D = 10^9 \text{ Hz} \quad \text{or} \quad \Delta\nu_D = 0.04 \text{ cm}^{-1} \quad \text{or} \quad \Delta\lambda_D = 0.016\,\text{Å}. \tag{6.67}$$

Note that the Doppler broadening exceeds the natural broadening of the spectral line by about an order of magnitude.

6.8 Collision-induced broadening and the shift of a spectral line

When a light-emitting oscillator collides with another particle, it experiences an external force causing modulation of the oscillator frequency and dephasing of the oscillation. The collisions, rapidly following one another, lead to random phase modulation of the oscillations, which, in turn, results in a spectral line broadening.

Let us evaluate the collision-induced broadening. We shall write the oscillator radiation field in the form of random phase-modulated vibration:

$$E_{xj}(t) = a_0 \cos[\omega_0 t + \varphi_j(t)]. \tag{6.68}$$

Inserting eqn (6.68) into eqn (6.42), we obtain

$$B_0(\tau) = \frac{a_0^2}{4} e^{i\omega_0\tau} \langle e^{i\Delta\varphi_j(t,\tau)} \rangle + \text{c. c.}, \tag{6.69}$$

where

$$\Delta\varphi_j(t,\tau) = \varphi_j(t+\tau) - \varphi_j(t) \tag{6.70}$$

is the phase shift accumulated over the time interval τ.

Suppose that $\varphi_j(t)$ describes a steady-state stochastic process. Then the averaged quantity appearing in eqn (6.69) is independent of t and can be denoted as

$$\langle e^{i\Delta\varphi_j(t,\tau)} \rangle = b(\tau). \tag{6.71}$$

Consider the infinitesimal time interval $d\tau$ and calculate the differential of the function $b(\tau)$. Using eqns (6.70) and (6.71), we obtain

$$db(\tau) = \langle e^{i\Delta\varphi_j(t,\tau)} (e^{i\varepsilon} - 1) \rangle, \tag{6.72}$$

where

$$\varepsilon = \Delta\varphi_j(t, \tau + d\tau) - \Delta\varphi_j(t,\tau) \tag{6.73}$$

is the phase shift accumulated over time $d\tau$.

Suppose also that the phase shift induced by the collision does not depend on the value of oscillation phase before the collision. Then we can consider the values $\Delta\varphi_j$ and ε as statistically independent and write

$$db(\tau) = b(\tau)\langle e^{i\varepsilon} - 1 \rangle. \tag{6.74}$$

Statistical averaging in eqn (6.74) can be replaced by averaging over the parameters characterizing the collision process. Let us perform averaging over the principal parameter R, which is referred to as an impact parameter and defined as a minimum distance between the centres of mass of the colliding particles.

Let N be the gas density (the number of particles per unit volume) and $\langle v \rangle$ the average relative velocity of the particles. Then the number of collisions that the oscillator will be subjected to during time interval $d\tau$ equals

$$dn = 2\pi N \langle v \rangle R \, dR \, d\tau \tag{6.75}$$

and the impact parameter will vary within the range from R to $R + dR$. Let $\varepsilon(R)$ denote the collision-induced phase shift characterized by the impact parameter R. Then we can write

$$\langle e^{i\varepsilon} - 1\rangle = 2\pi N\langle v\rangle d\tau \int_0^\infty \left[e^{i\varepsilon(R)} - 1\right] Rdr \tag{6.76}$$

or

$$\langle e^{i\varepsilon} - 1\rangle = -N\langle v\rangle d\tau(\sigma_1 - i\sigma_2), \tag{6.77}$$

where

$$\sigma_1 = 2\pi \int_0^\infty [1 - \cos\varepsilon(R)]R\,dR, \qquad \sigma_2 = 2\pi \int_0^\infty \sin\varepsilon(R)R\,dR. \tag{6.78}$$

Substitution of eqn (6.77) into eqn (6.74) leads to the following differential equation for the function $b(\tau)$:

$$db(\tau) = -b(\tau)N\langle v\rangle(\sigma_1 - i\sigma_2)d\tau, \tag{6.79}$$

whence it follows that

$$b(\tau) = e^{-N\langle v\rangle(\sigma_1 - i\sigma_2)\tau}. \tag{6.80}$$

Now it is easy to calculate the radiation spectrum. Using eqns (6.45), (6.69), (6.71), and (6.80) and taking into account that the correlation function is even, we obtain (for positive frequency values)

$$S_0(\omega) = \frac{a_0^2}{2\pi}\frac{N\langle v\rangle\sigma_1}{(\omega - \omega_0 - N\langle v\rangle\sigma_2)^2 + (N\langle v\rangle\sigma_1)^2}. \tag{6.81}$$

Equation (6.81) indicates that the radiation spectrum generated by elastically colliding oscillators has a Lorentzian shape and a bandwidth of

$$\Delta\omega_{\text{col}} = 2N\langle v\rangle\sigma_1. \tag{6.82}$$

The collision-induced shift of the spectral line centre is

$$\delta\omega = N\langle v\rangle\sigma_2. \tag{6.83}$$

Note that both the bandwidth and the shift of the spectral line are proportional to the gas density N and to the average relative velocity of the particles $\langle v\rangle$.

6.9 Dicke's effect: collision-induced narrowing of a Doppler spectral line in a dense gas

In the previous section we have calculated the broadening of a Doppler spectral line under the assumption that the thermal velocities of molecules do not vary with time. This approximation is valid for low-density gases in which molecules travel freely over large distances, because the collisions between them are relatively rare. As the gas density increases, the more frequent collisions between molecules restrict the free motion of oscillators. In very dense gases, as well as in condensed media, the oscillators are virtually localized. It is thus natural that a change in the character of oscillator motion should cause a modification in the shape and bandwidth of the spectral line. In this section we shall consider how the Doppler spectrum varies with an increase in the gas density.

Generalizing eqn (6.56) to the case in which the velocity of the molecule varies with time, we can write

$$z_j = z_j(t) = z_{0j} + \int_0^t v_{zj}(\theta)\, d\theta. \tag{6.84}$$

Inserting eqn (6.84) into eqn (6.55), we obtain

$$E_{xj} = a_0 \cos[\omega_0 t - \varphi_j(t)], \tag{6.85}$$

where

$$\varphi_j(t) = k_0 z_{0j} + k_0 \int_0^t v_{zj}(\theta)\, d\theta. \tag{6.86}$$

Using eqn (6.42), we find that

$$B_0(\tau) = \frac{a_0^2}{4} \mathrm{e}^{i\omega_0\tau} \langle \mathrm{e}^{-i\Delta\varphi_j(t,\tau)} \rangle + \text{c. c.}, \tag{6.87}$$

where

$$\Delta\varphi_j(t,\tau) = \varphi_j(t+\tau) - \varphi(t) = k_0 \int_t^{t+\tau} v_{zj}(\theta)\, d\theta. \tag{6.88}$$

It is natural to consider the motion of the molecule with thermal velocity $v_{zj}(t)$ as a steady-state stochastic process. Then the averaged quantity appearing in eqn (6.87) is independent of t and can be denoted by

$$\langle \mathrm{e}^{-i\Delta\varphi_j(t,\tau)} \rangle = b(\tau). \tag{6.89}$$

The stochastic process $v_{zj}(t)$ has a zero average value and a Gaussian probability distribution. It is well known that linear transformations do not change the statistics (probability distribution) of the Gaussian noise. Consequently, the stochastic process $\Delta\varphi_j(t,\tau)$, which is related to the process $v_{zj}(t)$ by a linear transformation (6.88), is also the Gaussian process with zero average value. Thus,

$$b(\tau) = \exp\left[-\tfrac{1}{2}\langle \Delta\varphi_j^2(t,\tau)\rangle\right]. \tag{6.90}$$

Inserting eqn (6.88) into eqn (6.90), we obtain

$$\langle \Delta\varphi_j^2(t,\tau)\rangle = k_0^2 \int_0^\tau \int_0^\tau \langle v_{zj}(\theta_1) v_{zj}(\theta_2)\rangle\, d\theta_1\, d\theta_2 \tag{6.91}$$

or

$$\langle \Delta\varphi_j^2(t,\tau)\rangle = 2k_0^2 \int_0^\tau (\tau - \theta) B_v(\theta)\, d\theta, \tag{6.92}$$

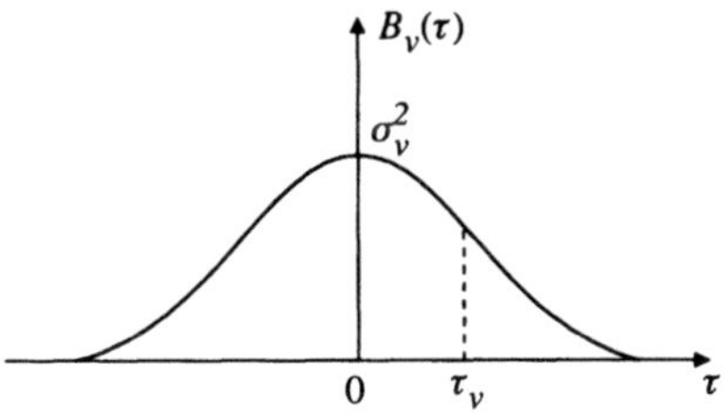

Fig. 6.7 The correlation function of the thermal velocities of gas molecules.

where

$$B_v = \langle v_{zj}(t) v_{zj}(t+\theta) \rangle \tag{6.93}$$

is the correlation function of the thermal velocity of the molecules. This function is characterized by two basic parameters, the variance $\sigma_v^2 = kT/m$ and the correlation time τ_v of the thermal velocity. The meaning of these parameters is illustrated in Fig. 6.7. Using this figure and eqns (6.90) and (6.92), one can easily show that

$$b(\tau) = \begin{cases} \exp\left[-\frac{1}{2}(k_0 \sigma_v \tau)^2\right], & t \ll \tau_v, \\ \exp\left[-(\tau - \tau_0) D\right], & \tau \gg \tau_v, \end{cases} \tag{6.94}$$

where

$$D = \int_0^\infty B_v(\tau) d\tau = \sigma_v^2 \tau_v, \qquad \tau_0 = D^{-1} \int_0^\infty \tau B_v(\tau) d\tau. \tag{6.95}$$

The parameter D is referred to as a correlation constant or a diffusion coefficient. Note that eqn (6.95) can be regarded as a definition of the correlation time τ_v.

Let us analyse the obtained results. In the limiting case of an extremely rarefied gas, when molecular collisions are absent, we can set $\tau_v = \infty$. As can be seen from eqns (6.87) and (6.94), the quantity $B_0(\tau)$ is determined in this case by eqn (6.64) and the radiation spectrum is therefore described by eqn (6.65). Thus, the adopted model correctly describes the spectral parameters of the radiation emitted by rarefied gases.

For the opposite limiting case $\tau_v \to 0$ (dense gas) we can obtain from eqns (6.87) and (6.95):

$$S_0(\omega) = \frac{a_0^2}{2\pi} \frac{D e^{\tau_0 D}}{(\omega - \omega_0)^2 + D^2}. \tag{6.96}$$

This formula indicates that the radiation spectrum has a Lorentzian shape and a bandwidth of

$$\Delta\omega = 2D = 2\sigma_v^2 \tau_v. \tag{6.97}$$

The correlation time of the thermal velocity—that is, the mean time between molecular collisions—decreases as the gas density N increases according to the following formula:

$$\tau_v^{-1} = N \langle v \rangle S, \tag{6.98}$$

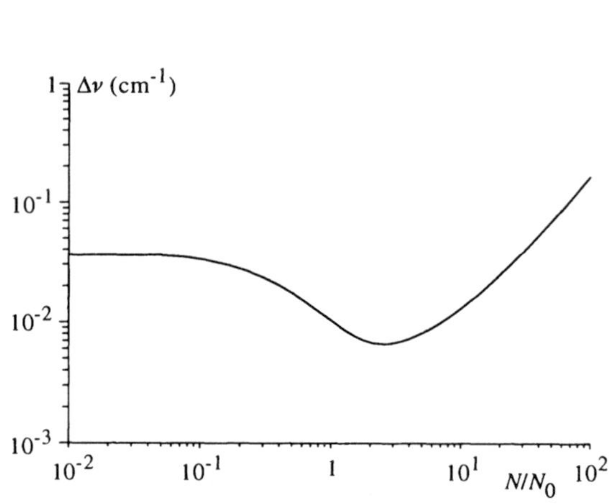

Fig. 6.8 Dicke's effect.

where $\langle v \rangle$ is the average relative thermal velocity and S is the gas-kinetic cross-section of the molecule. Hence, the spectral emission line narrows as the gas density increases. This phenomenon is called *collisional narrowing of the spectral line*, or *Dicke's effect*, after the scientist who was first to study the effect (Dicke 1953).

To calculate the radiation spectrum for an arbitrary gas density, we should apply a certain model for the thermal velocity correlation function. Suppose, for example, that

$$B_v(\tau) = \sigma_v^2 e^{-|\tau|/\tau_v}. \tag{6.99}$$

Then we can obtain from eqns (6.90) and (6.92) that:

$$b(\tau) = \exp\left[-\beta(y - 1 + e^{-y})\right], \tag{6.100}$$

where

$$\beta = (k_0 \sigma_v \tau_v)^2, \qquad y = \tau/\tau_v. \tag{6.101}$$

The radiation spectrum is governed by eqns (6.45a), (6.87), (6.89), and (6.100).

Figure 6.8 shows the bandwidth of the spontaneous Raman scattering in molecular hydrogen as a function of the gas pressure. One can see the spectral line narrowing due to Dicke's effect in the range of pressures from 0.1 to 3.0 atm. The broadening of the spectral line at higher pressures is caused by the collisional mechanism.

Bibliography

Dicke, R. H. (1953). *Physical Review*, **89**, 472–473.

Wittke, J. P. and Dicke, R. H. (1956). *Physical Review*, **103**, 620–631.

7

A linear oscillator in an optical field

Absorption of light by an oscillator and by an oscillator ensemble. The Bouguer law. Absorption spectroscopy. Decrease in the phase velocity of light propagating in a medium and the index of refraction. An inhomogeneous ensemble of oscillators. Homogeneous and inhomogeneous widths of spectra. Excitation by a short pulse—energy relaxation and dephasing. Principles of frequency- and time-domain spectroscopy.

In this Chapter we shall apply the model of an harmonic oscillator to consider the basic effects of the interaction of light with matter: (i) absorption of light and (ii) the decrease in the phase velocity of light propagating in a medium. We shall also examine the interaction of light with an inhomogeneous ensemble of oscillators, in which the individual oscillators have different intrinsic frequencies of vibration. Finally, we shall discuss optical methods of investigation of matter and general principles of frequency- and time-domain spectroscopy.

7.1 Absorption of light by an oscillator and by an oscillator ensemble

One basic effect resulting from the interaction of light with matter is absorption of light. A propagating optical wave loses its energy by exciting elementary optical oscillators (atoms or molecules) of the medium. Consequently, the energy of the medium increases while the energy of the optical wave decreases, i.e. the medium absorbs the light.

The main features of this phenomenon can be described in terms of the simple classical model treating the medium as an ensemble of harmonic oscillators. We shall begin our consideration with an estimate of the power transmitted by the optical wave to an individual oscillator.

7.1.1 Oscillator excitation by a harmonic field

The excitation of an oscillator by the optical field is described by the equation

$$\ddot{x} + \Gamma\dot{x} + \omega_0^2 x = \frac{e}{m}E. \tag{7.1}$$

where e and m are the charge and mass of the oscillator, ω_0 and Γ are the intrinsic frequency and damping coefficient of oscillations, x is the coordinate, and E is the electric field strength of the optical wave (Fig. 7.1).

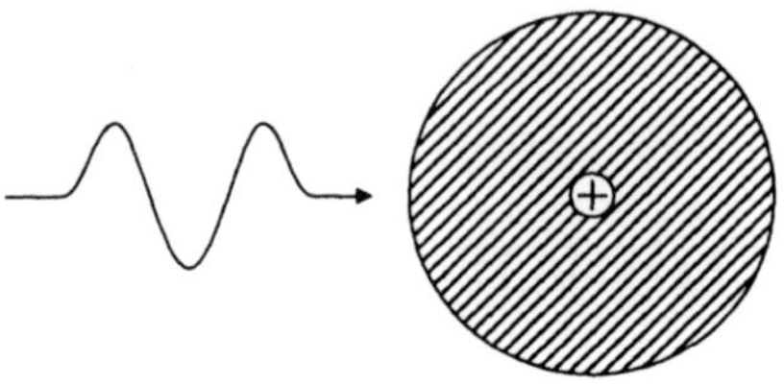

Fig. 7.1 Oscillator excitation by an optical field.

Let us write the optical field in the form

$$E = \tfrac{1}{2}\mathcal{E}\exp(i\omega t) + \text{c. c.}, \tag{7.2}$$

where ω is the frequency and $\mathcal{E}$ is the complex amplitude of the field. Inserting eqn (7.2) into eqn (7.1), we obtain the equation of forced oscillations:

$$\ddot{x} + \Gamma\dot{x} + \omega_0^2 x = \frac{e}{m}\tfrac{1}{2}\mathcal{E}\exp(i\omega t) + \text{c. c.} \tag{7.3}$$

We shall seek the solution to eqn (7.3) in the form

$$x = \tfrac{1}{2}\tilde{x}\exp(i\omega t) + \text{c. c.}, \tag{7.4}$$

where $\tilde{x}$ is the complex amplitude of oscillations. Inserting eqn (7.4) into eqn (7.3), we find that

$$\tilde{x} = \frac{e}{m}\frac{1}{\omega_0^2 - \omega^2 + i\omega\Gamma}\mathcal{E}. \tag{7.5}$$

Equations (7.4) and (7.5) describe the steady-state vibrations of the oscillator in the field of a monochromatic optical wave.

7.1.2 Induced absorption

Let us calculate the power P transmitted by light to the oscillator. By definition,

$$P = dA/dt, \tag{7.6}$$

where

$$dA = \vec{F}d\vec{r} = eE\,dx \tag{7.7}$$

is the elementary work that the optical field performs on the oscillator. Inserting eqn (7.7) into eqn (7.6), we obtain

$$P = eE\dot{x}. \tag{7.8}$$

According to eqns (7.2), (7.4), and (7.8),

$$P = \frac{e}{4}i\omega\left(\mathcal{E}^*\tilde{x} + \mathcal{E}\tilde{x}e^{2i\omega t}\right) + \text{c. c.},$$

i.e. the instantaneous power P oscillates at the frequency 2ω. The power averaged over the optical vibration period is

$$\langle P \rangle \equiv P_0 = \frac{e}{4} i\omega \mathcal{E}^* \tilde{x} + \text{c. c.} \tag{7.9}$$

Inserting eqn (7.5) into eqn (7.9), we obtain

$$P_0 = \frac{e^2}{2m} \frac{\omega^2 \Gamma}{(\omega_0^2 - \omega^2)^2 + \omega^2 \Gamma^2} |\mathcal{E}|^2$$

or

$$P_0 = \frac{4\pi e^2}{mc} \frac{\omega^2 \Gamma}{(\omega_0^2 - \omega^2)^2 + \omega^2 \Gamma^2} I, \tag{7.10}$$

where

$$I = c|\mathcal{E}|^2 / 8\pi \tag{7.11}$$

is the optical wave intensity.

One can see from eqn (7.10) that $P_0 > 0$. This means that the light is absorbed by transmitting its power to the oscillator. To characterize the value of absorption, it is helpful to introduce the cross-section σ, defined by the formula

$$P_0 = \sigma I. \tag{7.12}$$

Then

$$\sigma(\omega) = \frac{4\pi e^2}{mc} \frac{\omega^2 \Gamma}{(\omega_0^2 - \omega^2)^2 + \omega^2 \Gamma^2}. \tag{7.13}$$

7.1.3 Resonance curve

At frequencies near the resonance $|\omega - \omega_0| \ll \omega_0$, eqn (7.13) takes the form

$$\sigma(\omega) = \frac{\pi e^2}{mc} \frac{\Gamma}{(\omega_0 - \omega)^2 + \Gamma^2/4}. \tag{7.14}$$

Thus, the dependence of the absorption cross-section on the frequency of light has a Lorentzian form (Fig. 7.2).

The absorption cross-section reaches a maximum at the frequency $\omega = \omega_0$, coinciding with the intrinsic frequency of the oscillator. The FWHM of the distribution $\sigma(\omega)$ is equal to the damping coefficient of the oscillation:

$$\Delta\omega = \Gamma.$$

The maximum absorption cross-section,

$$\sigma_{\max} = \frac{4\pi e^2}{mc\Gamma}, \tag{7.14a}$$

is inversely proportional to the parameter Γ. In physical terms, it is this parameter Γ, defining the amount of loss, that restricts the amplitude of the steady-state forced oscillations (see eqn (7.5)). Assuming that the oscillation damping

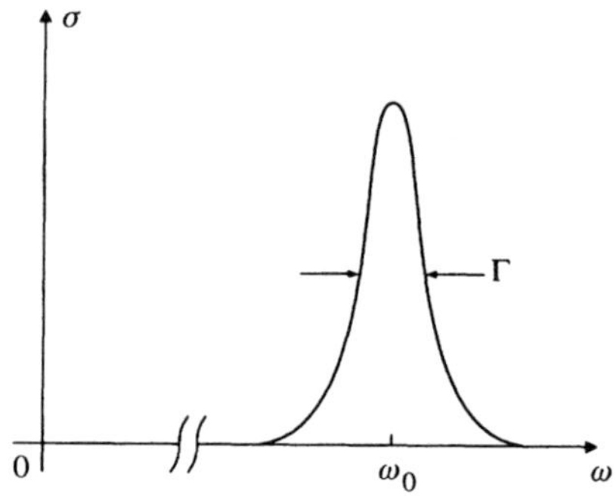

Fig. 7.2 The frequency dependence of the absorption cross-section.

is caused by emission of radiation, and making use of eqn (5.24), we can evaluate the parameter Γ as

$$\Gamma = \frac{1}{\tau} = \frac{2e^2\omega^2}{3mc^3}. \tag{7.14b}$$

Inserting eqn (7.14b) into eqn (7.14a), we obtain

$$\sigma_{\max} = \frac{3}{2\pi}\lambda^2, \tag{7.14c}$$

where λ is the wavelength of the optical wave. Therefore, the cross-section of light resonant absorption by an atom has the order of the square of the optical wavelength. For example, if $\lambda = 0.5\ \mu$m, then $\sigma_{\max} \approx 10^{-9}$ cm^2.

7.2 The Bouguer law

Consider the light absorption by a model medium representing an ensemble of harmonic oscillators. Let a plane monochromatic optical wave travel over the section of length dz in a medium of density N. Here N is the number of elementary oscillators (atoms) in a unit volume (Fig. 7.3). Denoting the optical beam cross-section area by S, we can obtain the total number of oscillators irradiated by the optical wave:

$$dN = NS\,dz. \tag{7.15}$$

Suppose that every oscillator absorbs the power P_0 determined by eqns (7.12) and (7.13). Then the total variation of the power of light over the given section of the medium will be

$$dP = -P_0\,dN. \tag{7.16}$$

Correspondingly, the variation of light intensity will be $dI = S^{-1}dP$ or, in view of eqns (7.15) and (7.16),

$$dI = -P_0 N\,dz. \tag{7.17}$$

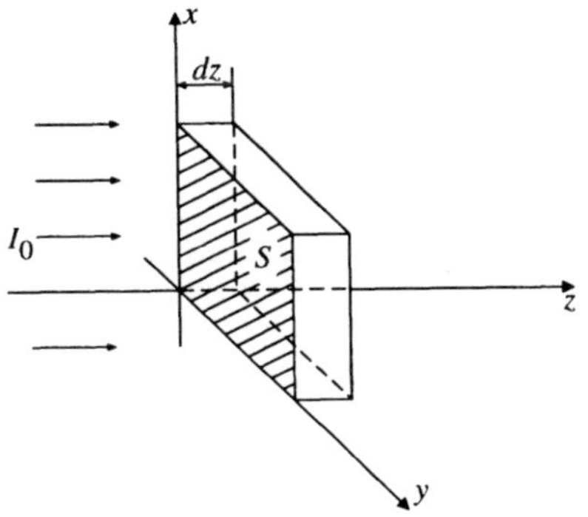

Fig. 7.3 The derivation of the Bouguer law.

Inserting eqn (7.12) into eqn (7.17), we obtain the equation for the light intensity, $dI = -I\sigma N\,dz$, the solution of which is

$$I(z) = I_0 \exp(-\delta z), \tag{7.18}$$

where I_0 is the intensity of the incident wave, and

$$\delta = \sigma N \tag{7.19}$$

is the absorption coefficient.

The law of light absorption in the medium described by eqn (7.18) is called the Bouguer law. According to this law, the light intensity decreases exponentially with the distance travelled by the light in the medium (Fig. 7.4).

As follows from eqns (7.14) and (7.19), the absorption coefficient can be written in the following form:

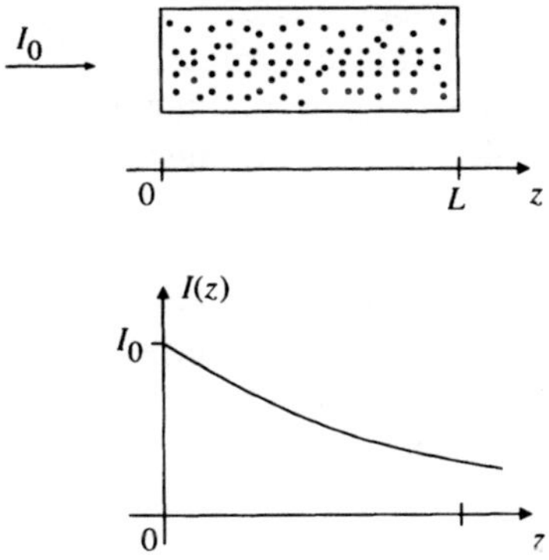

Fig. 7.4 The exponential damping of light in a medium (the Bouguer law).

Fig. 7.5 An experimental set-up for optical absorption spectroscopy.

$$\delta(\omega) = \frac{\pi N e^2}{mc} \frac{\Gamma}{(\omega_0 - \omega)^2 + \Gamma^2/4}. \tag{7.20}$$

One can see from eqn (7.20) that the absorption coefficient δ does not depend either on the incident light intensity or on the size of the irradiated sample, being a characteristic of the absorbing material itself. The absorption coefficient may vary drastically for different materials and physical conditions. For example, optical fibres have $\delta = 10^{-4} - 10^{-6}$ cm^{-1}, while metals and semiconductors have $\delta = 10^3 - 10^6$ cm^{-1}. Along with the index of refraction, the light absorption coefficient represents one of the most important optical characteristics of the substance.

We have derived the Bouguer law from the simplest model of the medium as an ensemble of stationary non-interacting classical oscillators. Experimental investigations show, however, that this law is valid for a wide class of objects: liquid, solid, and gaseous media, as well as mixtures of different substances, and over a very broad range of light intensity. Thus, the Bouguer law is quite universal. Deviations from this law become appreciable only for very powerful (laser) optical beams. These deviations are caused by a non-linear response of matter to strong optical fields (see Chapters 8, 22, and 23), by quantum effects etc.

7.3 Absorption spectroscopy

The dependence of the absorption coefficient of the substance, δ, on the frequency of light ω is called an *absorption spectrum*. By measuring the absorption spectrum, we can determine experimentally the characteristics of elementary optical oscillators, such as intrinsic frequencies and damping coefficients. These characteristics, in turn, provide important information about the structure, properties, and interaction of atoms and molecules, making it possible to determine the composition and thermodynamic state of the medium. The collection and analysis of this information forms the subject of optical spectroscopy.

An ideal device for absorption spectrum measurement is a narrow-band frequency-tunable laser. The block diagram of an absorption measurement arrangement is shown in Fig. 7.5. A laser beam, having passed through the medium under investigation, is directed on to a measuring instrument to record the intensity (power, energy) of the light. Variations of the output signal due to laser wavelength tuning yield the frequency dependence of the absorption coefficient $\delta(\omega)$.

Figure 7.6 illustrates schematically the absorption spectroscopy technique using a broadband light source (e.g. an arc lamp). The spectrum is analysed

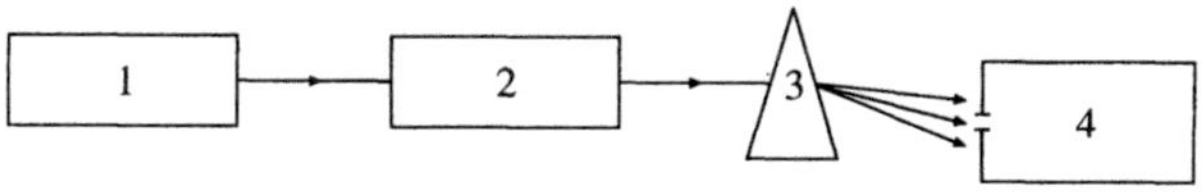

Fig. 7.6 The "broadband" version of absorption spectroscopy. 1, Broadband source of light (e.g. an arc lamp); 2, sample under investigation; 3, spectral device (a prism, diffraction grating, etalon); 4, measuring equipment.

by a spectral device, such as a prism or diffraction grating. As an example, Fig. 7.7 shows an experimental arrangement in which the absorption spectrum of a $KMnO_4$ water solution is observed. The white light beam produced by an arc lamp passes a cell containing the $KMnO_4$ water solution, crosses a glass prism, and falls on to the screen. A dark band is then observed in the yellow–green region of the spectrum on the screen. The appearance of this band is caused by the absorption of yellow–green light by the solution. A similar experiment can be carried out using sodium vapour. Putting a sodium burner in the way of a light beam, we can observe the light absorption in the yellow region of the spectrum.

7.4 Decrease in the phase velocity of light propagating in a medium and the index of refraction

In the previous section we considered the effect of absorption of light in a medium. Now let us consider the process of re-radiation of light by an ensemble of oscillators. This will ensure an insight into the mechanism of yet another important effect accompanying the interaction of light with matter; that is, the decrease in the phase velocity of light propagating in a medium.

Consider a plane monochromatic optical wave travelling in a transparent medium. It is well known that the medium generally reduces the phase velocity of the propagating light. The velocity of light travelling through the medium is equal to

$$v = c/n, \tag{7.21}$$

where c is the velocity of light in vacuum and n is the real positive value exceeding unity and usually called the *index of refraction*. In this section we shall derive

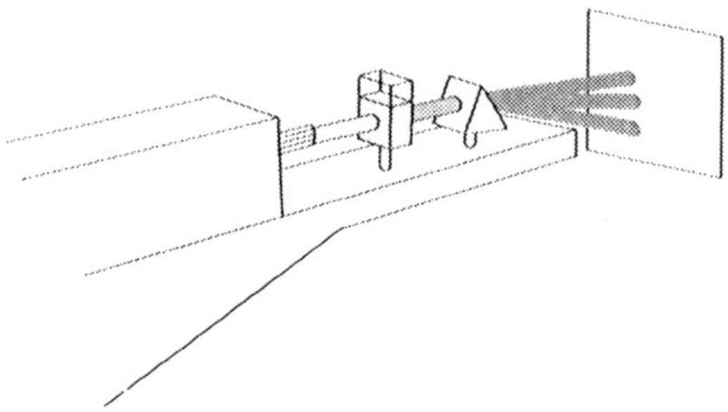

Fig. 7.7 Observation of the absorption spectrum of a $KMnO_4$–water solution.

the expression for n, starting from the model of the medium as an ensemble of oscillators.

Suppose first that the index of refraction of the medium is known. Then the propagation of light in the medium can be described as

$$E = \tfrac{1}{2}\mathcal{E}\mathrm{e}^{i(\omega t - kz)} + \text{c. c.}, \tag{7.22}$$

where ω is the frequency, $\mathcal{E}$ is the complex amplitude of the wave, and

$$k = \omega n/c \tag{7.23}$$

is the wave number for the light in the medium. Introducing the wave number of optical radiation in vacuum

$$k_0 = \omega/c, \tag{7.24}$$

we can rewrite eqn (7.23) in the form

$$k = k_0 + (n-1)k_0. \tag{7.25}$$

One can see from eqns (7.22) and (7.25) that the optical wave acquires in the medium an additional phase shift proportional to the thickness of the medium. For example, for the medium layer of thickness Δz the additional phase shift is

$$\Delta\varphi = (n-1)k_0\Delta z. \tag{7.26}$$

Then the optical field leaving the medium is

$$E = \tfrac{1}{2}\mathcal{E}\mathrm{e}^{i(\omega t - k\Delta z)} + \text{c. c.} = \tfrac{1}{2}\mathcal{E}\mathrm{e}^{i(\omega t - k_0\Delta z) - i\Delta\varphi} + \text{c. c.} \tag{7.27}$$

Suppose that the thickness Δz of the medium layer is so small (much less than the optical wavelength) that the phase shift $\Delta\varphi$ satisfies the inequality

$$\Delta\varphi \ll 1. \tag{7.28}$$

Then, using the formula $\exp(-i\Delta\varphi) \approx 1 - i\Delta\varphi$, we obtain from eqn (7.27),

$$E = E_{\text{in}} + E_{\text{ad}}, \tag{7.29}$$

where

$$E_{\text{in}} = \tfrac{1}{2}\mathcal{E}\mathrm{e}^{i(\omega t - k_0\Delta z)} + \text{c. c.},$$

$$E_{\text{ad}} = \tfrac{1}{2}\mathcal{E}_g\mathrm{ad}^{i(\omega t - k_0\Delta z)} + \text{c. c.}, \qquad \mathcal{E}_{\text{ad}} = -i\Delta\varphi\mathcal{E}.$$

The quantity E_{in} in eqn (7.29) can be treated as the field of the initial incident optical wave crossing the medium layer without any distortion. The second quantity, E_{ad}, describes an additional weak field the phase of which is shifted

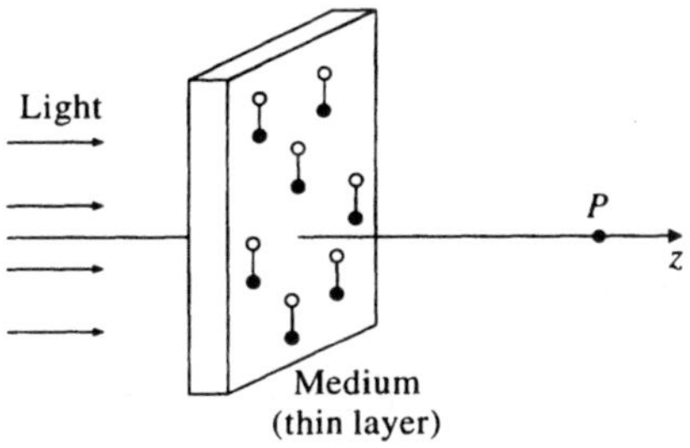

Fig. 7.8 A calculation of the index of refraction for a model medium. A thin layer of the medium is irradiated by a plane monochromatic optical wave. P is the observation point.

with respect to that of the incident field by $\pi/2$. Considering eqn (7.26), we can describe the amplitude of the additional field in the form

$$\mathcal{E}_{\text{ad}} = -ik_0 \Delta z (n-1)\mathcal{E}. \tag{7.30}$$

Note that in the calculation made above the additional field arises due to the decrease in phase velocity of the light travelling through the medium. On the other hand, this field cannot be anything other than the field *re-radiated* by the elementary oscillators (atoms) of the medium. Let us calculate this re-radiated field by employing the notion of an emitting oscillator making forced oscillations in the field of the incident optical wave.

Figure 7.8 shows a thin layer of a medium irradiated by a plane monochromatic optical wave. Let Δz be the thickness of the layer and let N be the number of atoms per unit volume. Suppose that the medium is so rare that we can neglect the interaction between atoms.

Neglecting the oscillation decay also, we can write the equation of forced oscillations for the oscillator in the incident optical field,

$$E = \tfrac{1}{2}\mathcal{E}\mathrm{e}^{i\omega t} + \text{c. c.}, \tag{7.31}$$

as

$$\ddot{x} + \omega_0^2 x = \frac{e}{m}E, \tag{7.32}$$

where e and m are the charge and mass of the electron, and ω_0 is the intrinsic frequency of the oscillator. The solution to eqn (7.32) can be sought in the form

$$x = \tfrac{1}{2}\tilde{x}\mathrm{e}^{i\omega t} + \text{c. c.} \tag{7.33}$$

Inserting eqns (7.31) and (7.33) into eqn (7.32), we obtain

$$\tilde{x} = \frac{e}{m}\frac{1}{\omega_0^2 - \omega^2}\mathcal{E}. \tag{7.34}$$

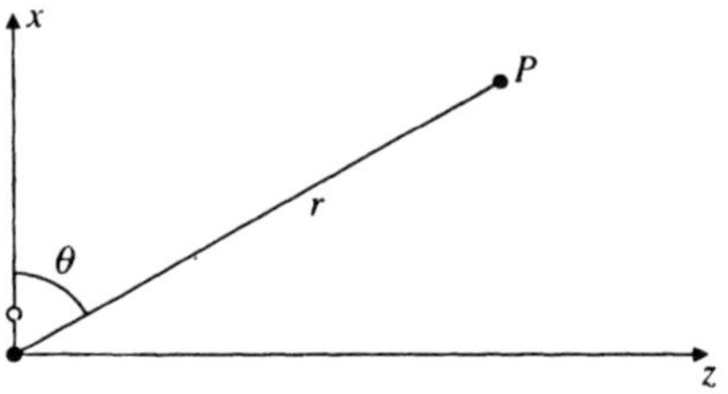

Fig. 7.9 The calculation of the secondary emission field.

Equation (7.34) defines the relation between the complex amplitudes $\tilde{x}$ of dipole vibrations and the incident wave field $\mathcal{E}$.

The oscillating dipole itself turns into a source of secondary radiation. Using the results of Chapter 5, we can represent the secondary radiation field in the form

$$E_0 = \tfrac{1}{2}\mathcal{E}_0 e^{i\omega(t-r/c)} + \text{c. c.}, \tag{7.35}$$

where

$$\mathcal{E}_0 = \frac{e\omega^2 x}{c^2 r}\sin\theta, \tag{7.36}$$

r is the distance from the dipole to the observation point, and θ is the angle between the dipole axis and the direction to the observation point (Fig. 7.9). Hereinafter we shall restrict our consideration to the angles $\theta \approx \pi/2$, for which $\sin\theta \approx 1$. Then eqn (7.36) yields the amplitude of the secondary wave in the form

$$\mathcal{E}_0 = \frac{e\omega^2 x}{c^2 r}. \tag{7.37}$$

Now we turn to the summation of radiations emitted by the oscillators contained within the thin layer of the medium under consideration. Let us choose the observation point P at a sufficiently long distance from the layer (Fig. 7.8). The radiation of an individual atom at the observation point is described by eqns (7.35) and (7.37). While making the summation, we should take into account the fact that different oscillators are located at different distances r from the observation point and, consequently, the light waves emitted by them reach the observation point with different phases.

Suppose that the layer thickness is much less than the optical wavelength: $\Delta z \ll \lambda$. Then the layer thickness can be disregarded by assuming all the oscillators to be located within one plane. Let us introduce the x- and y-axes in this plane, and direct the z-axis at right angle to the plane through the observation point P (Fig. 7.10(a)).

Figure 7.10(b) shows a thin ring of the radius ρ and the width $d\rho$, located in the plane of oscillators so that the z-axis coincides with the axis of the ring. All of the oscillators located within the ring are equidistant from the point P and, consequently, the optical waves emitted by these oscillators have the same phase

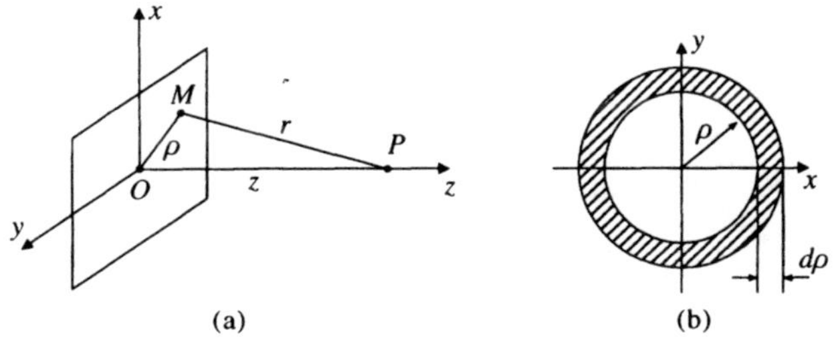

Fig. 7.10 The calculation of the radiation field produced by an ensemble of oscillators located on a plane.

at the point P and can be added up directly. Therefore, the resulting optical field is

$$dE = E_0\, dN, \tag{7.38}$$

where E_0 is determined by eqn (7.35) and describes the field of an individual oscillator, and

$$dN = N\Delta z 2\pi\rho\, d\rho \tag{7.39}$$

is the number of oscillators located within the ring. To find the total optical field produced by all the oscillators of the plane at the point P, we shall integrate eqn (7.38) over ρ and obtain

$$E_{\mathrm{ad}} = N\Delta z 2\pi \int_0^\infty E_0\rho\, d\rho. \tag{7.40}$$

It can be seen from Fig. 7.10 that $r^2 = z^2 + \rho^2$. Since we are concerned with the case $z \gg \rho$, we can write a fairly accurate approximation:

$$r = z + \rho^2/2z. \tag{7.41}$$

Using eqns (7.35), (7.40), and (7.41), and the designation (7.24), we obtain

$$E_{\mathrm{ad}} = \tfrac{1}{2}\mathcal{E}_g \mathrm{ad}^{i(i\omega t - k_0 z)} + \mathrm{c.\,c.}, \tag{7.42}$$

where

$$\mathcal{E}_{\mathrm{ad}} = 2\pi N\Delta z\mathcal{E}_0\mathcal{J} \tag{7.43}$$

and

$$\mathcal{J} = \int_0^\infty \exp(-ik_0\rho^2/2z)\rho\, d\rho. \tag{7.44}$$

The integral (7.44) can be evaluated as

$$\mathcal{J} = \lim_{\alpha\to 0}\int_0^\infty \exp(-\alpha\rho^2 - ik_0\rho^2/2z)\rho\, d\rho = \frac{z}{ik_0}. \tag{7.45}$$

Inserting eqns (7.37) and (7.45) into eqn (7.43), and neglecting the difference between r and z, we obtain

$$\mathcal{E}_{\text{ad}} = -ik_0 \Delta z 2\pi N e x$$

or, considering eqn (7.34),

$$\mathcal{E}_{\text{ad}} = -ik_0 \Delta z \frac{2\pi N e^2}{m} \frac{1}{\omega_0^2 - \omega^2} \mathcal{E}. \tag{7.46}$$

Thus, the expressions for the complex amplitude of the additional field arising due to the interaction of light with matter are obtained using two different approaches. The phenomenological derivation of eqn (7.30) is based on the idea of a decelerated phase velocity of light propagating in a medium characterized by the index of refraction n. Equation (7.46) is derived using the model in which the medium is treated as an ensemble of harmonic oscillators. Equating the expressions (7.30) and (7.46), we obtain the formula

$$n = 1 + \frac{2\pi N e^2}{m} \frac{1}{\omega_0^2 - \omega^2}, \tag{7.47}$$

which defines the index of refraction n through the microscopic parameters of the medium (the charge e, mass m, and intrinsic frequency ω_0 of the elementary oscillator), the number of atoms N per unit volume, and the optical wave frequency ω. According to the approximations adopted here, eqn (7.47) is valid for rarefied transparent media. The calculation made above shows that the mechanism underlying the decrease in phase velocity of light is associated with the interaction, or interference (see Chapter 11), of the secondary optical waves emitted by the atoms of the medium.

7.5 An inhomogeneous ensemble of oscillators

Consider an ensemble in which individual oscillators have different intrinsic frequencies of vibration. Such an ensemble is called *inhomogeneous*. In real systems the inhomogeneity can be caused by the difference in quantum states of atoms or molecules, by inhomogeneity of the isotopic or chemical composition of the medium, by chaotic thermal motion of particles, etc. We shall look into the characteristics of the interaction of the inhomogeneous ensemble with the optical field.

Let subscript "J" assign numbers to different resonant frequencies, i.e. types of oscillators in a non-homogeneous ensemble. The main characteristics of the ensemble are the sets of intrinsic frequencies, $\{\omega_J\}$, and statistical weights, $\{p_J\}$, of the oscillators.

The oscillator motion driven by the optical field $E(t)$ can be described by the following equation:

$$\ddot{x}_J + \Gamma \dot{x}_J + \omega_J^2 x_J = \frac{e}{m} E. \tag{7.48}$$

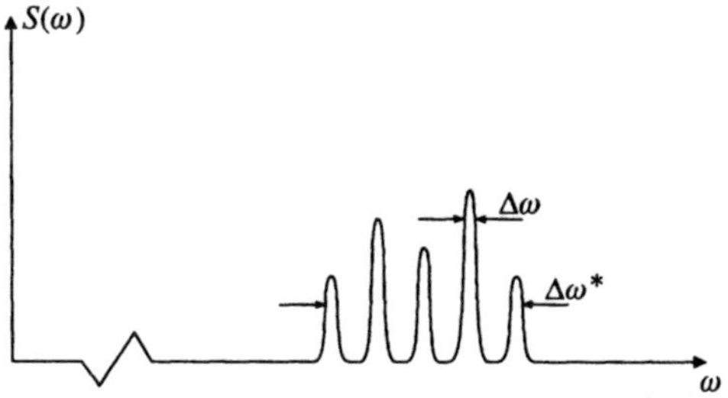

Fig. 7.11 The vibration spectrum of an inhomogeneous ensemble of oscillators.

The inverse action of the oscillators on the optical wave (emission, absorption, or scattering of light) is governed by the entire dipole moment of the medium, i.e. the *polarization* (see Part IV), which in this specific case is equal to

$$P = Ne\langle x\rangle, \qquad \langle x\rangle = \sum_J p_J x_J, \tag{7.49}$$

where N is the mean number of particles per unit volume of the medium.

7.6 Homogeneous and inhomogeneous widths of spectra

Figure 7.11 shows a typical spectrum of optical emission from an inhomogeneous ensemble. A similar structure is also observed, for example, in absorption spectra or Raman scattering spectra in molecular gases at moderate pressures.

The oscillation spectrum generated by an inhomogeneous ensemble can be characterized by two basic parameters: the homogeneous width $\Delta\omega$ and the inhomogeneous width $\Delta\omega^*$. The homogeneous width describes the oscillation spectrum of an individual oscillator, whereas the *inhomogeneous width* is a measure of the dispersion of the oscillator frequencies. Usually, the latter is much wider than the former (Fig. 7.11):

$$\Delta\omega^* \gg \Delta\omega. \tag{7.50}$$

Specifically, such a relation exists between the Doppler (inhomogeneous) width and the natural (homogeneous) width of the spectrum of a gas.

For the model described by eqns (7.48) and (7.49), we can write

$$\Delta\omega = \Gamma, \qquad \Delta\omega^* = \sqrt{\langle\tilde{\omega}_J^2\rangle}, \tag{7.51}$$

where

$$\langle\tilde{\omega}_J^2\rangle = \sum_J p_J \tilde{\omega}_J^2 \tag{7.52}$$

is the variance of resonant frequencies and

$$\tilde{\omega}_J = \omega_J - \overline{\omega} \tag{7.53}$$

are the oscillator frequency deviations from the mean frequency

$$\overline{\omega} = \sum_J p_J \omega_J. \tag{7.54}$$

Thus, the homogeneously broadened line is a spectral line resulting from a superposition of spectral lines of individual oscillators having the same intrinsic frequency of vibration ω_0. Conversely, the inhomogeneously broadened line is a spectral line formed by a superposition of spectral lines of oscillators having different intrinsic frequencies of vibration $\omega_J \neq \omega_0$. In the former case the spectral line width of an individual oscillator equals that of the whole ensemble, whereas in the latter case the spectral line of the ensemble is usually much broader than that of an individual oscillator. The natural and collision-induced broadening are examples of homogeneous broadening, which is characteristic both of the spectrum of each individual particle and of the spectrum of the whole ensemble. On the other hand, examples of non-homogeneous broadening include Doppler broadening, isotopic broadening, rotational broadening (in molecular gases), etc. These kinds of broadening are inherent only in the spectrum of the oscillator ensemble and are not observed for an individual oscillator.

7.7 Excitation by a short pulse—energy relaxation and dephasing

Suppose that an oscillator ensemble has been excited by a short optical pulse. Then its return to the equilibrium state can be described by two different times; that is, the *energy relaxation time* T_1 and the *phase relaxation time* T_2. According to eqns (7.48) and (7.49), the optical polarization of the medium (a dipole moment per unit volume) can decay not only via energy loss, but also due to the relative phase variation (dephasing) of oscillators. Generally, the following inequality holds

$$T_1 \geq T_2. \tag{7.55}$$

In particular, the strong inequality is possible:

$$T_1 \gg T_2, \tag{7.56}$$

meaning that the optical polarization decay (free induction decay) is caused entirely by the dephasing.

In terms of the model described by eqns (7.48) and (7.49), the energy relaxation time is

$$T_1 = 1/\Gamma = 1/\Delta\omega. \tag{7.57}$$

As to dephasing, it obviously proceeds more quickly when the dispersion of individual oscillator frequencies is higher. Using the notion of the inhomogeneous width of the spectrum, we can therefore write

$$T_2 \sim 1/\Delta\omega^*. \tag{7.58}$$

7.8 Principles of frequency- and time-domain spectroscopy

The oscillator ensemble described by eqns (7.48) and (7.49) is a linear system. As for any linear system, it can be characterized by the frequency–response function (in optics, called the linear optical susceptibility), $\chi(\omega)$, and the pulse–response function (the influence function or Green function), $h(t)$. The frequency–response function characterizes the response of the system to the harmonic (sinusoidal) excitation, while the influence function describes its reaction to a very short exciting pulse. Generally, these two characteristics of the linear system are interrelated via the Fourier transformation:

$$h(t) = \frac{1}{2\pi}\int_{-\infty}^{\infty} \chi(\omega)e^{i\omega t}d\omega, \tag{7.59}$$

$$\chi(\omega) = \int_{-\infty}^{\infty} h(t)e^{-i\omega t}dt, \tag{7.60}$$

According to the *causality principle*, a system cannot respond to an external action before the action starts. Consequently, the influence function can be nonzero only for the positive values of its argument, whereas for the negative values of time it should be equal to zero:

$$h(t < 0) = 0. \tag{7.61}$$

Therefore, eqn (7.60) can be rewritten in the form

$$\chi(\omega) = \int_{0}^{\infty} h(t)e^{-i\omega t}dt. \tag{7.62}$$

The spectroscopic methods developed and applied in optics make it possible to measure both the susceptibility dispersion $\chi(\omega)$ and the influence function $h(t)$ of the medium. The methods of stationary or frequency-domain spectroscopy that allow the characteristics of the medium (absorption coefficient, index of refraction, and scattering intensity) to be measured as the function of the radiation frequency ω provide information about the susceptibility dispersion $\chi(\omega)$. An ideal device for frequency-domain spectroscopy is a narrow-band frequency-tunable laser. Transient or time-domain methods based on the application of lasers generating ultrashort optical pulses enable direct measurements of the influence function $h(t)$ of the medium. As the functions $\chi(\omega)$ and $h(t)$ are related by Fourier transformation, both groups of methods give, in principle, the equivalent spectroscopic information. In practice, time-domain spectroscopy is used, as a rule, for the investigation of narrow spectral lines.

As an example, Fig. 7.12(a) shows the spectrum of vibrations of hydrogen molecules measured by a narrow-band frequency-tunable laser (Owyoung 1978). Figure 7.12(b) illustrates the pulse-response function (dephasing) of excited vibrations of hydrogen molecules, measured using ultrashort laser pulses (Dyakov *et al.* 1983).

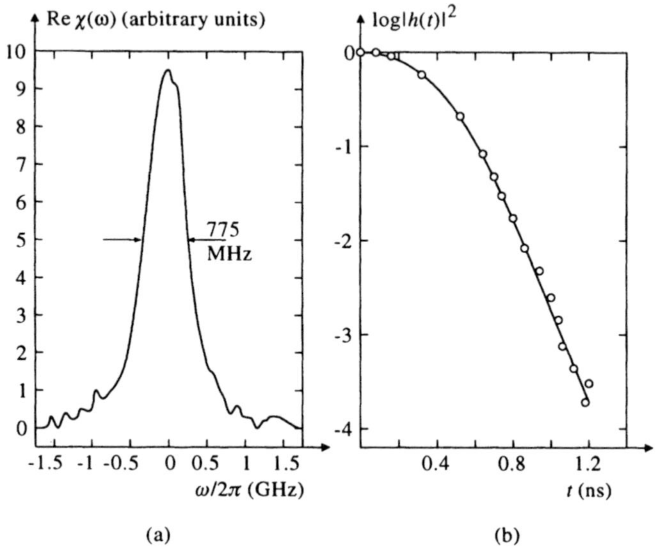

Fig. 7.12 (a) The vibration spectrum of molecular hydrogen, measured using a narrow-band frequency-tunable laser (from Owyoung 1978). The spectrum is obtained at room temperature and a pressure of 200 Torr. The spectral line bandwidth at this pressure (775 MHz) is close to the Doppler bandwidth. The mean frequency of molecular vibration of hydrogen is $\nu_0 = 4155 \text{ cm}^{-1}$. (b) The pulse–response (dephasing) function of vibrationally excited hydrogen molecules, measured using ultra-short laser pulses (from Dyakov *et al.* 1983). The pulse–response function is obtained at room temperature and a pressure of 100 Torr. Molecular vibration dephasing at this pressure is caused primarily by Doppler dephasing. The dephasing time (about 1 ns) is inversely proportional to the spectral line bandwidth.

7.8.1 Frequency susceptibility and pulse–response function of an inhomogeneous ensemble of oscillators

For the sake of illustration, let us calculate the functions $\chi(\omega)$ and $h(t)$ for the inhomogeneous system described by eqns (7.48) and (7.49).

The following relations are generally well satisfied in optics:

$$\overline{\omega} \gg \Delta\omega^* \geq \Gamma, \tag{7.63}$$

i.e. the mean oscillator frequency exceeds essentially both the homogeneous and inhomogeneous broadenings. Then, instead of eqn (7.48), it is reasonable to use the tapered equation for the vibration amplitude.

Suppose that the optical field corresponds to a narrow-band (quasi-harmonic) process and the central frequency of the field is tuned to the centre of the inhomogeneously broadened band:

$$E(t) = \tfrac{1}{2}\mathcal{E}(t)\mathrm{e}^{i\overline{\omega}t} + \text{c. c.} \tag{7.64}$$

Inserting eqn (7.64) into eqn (7.48), we shall seek the solution in the form

$$x_J(t) = \tfrac{1}{2}\tilde{x}_J(t)\mathrm{e}^{i\overline{\omega}t} + \text{c. c.} \tag{7.65}$$

In the approximation of slowly varying amplitudes, signifying that $\mathcal{E}(t)$ and $\tilde{x}_J(t)$ vary only slightly during the optical vibration period, we obtain

$$\dot{\tilde{x}}_J + \frac{\Gamma}{2}\tilde{x}_J = i\overline{\omega}_J\tilde{x}_J + A(t), \tag{7.66}$$

where

$$A(t) = \gamma\mathcal{E}(t), \qquad \gamma = \frac{e}{2im\overline{\omega}}. \tag{7.67}$$

Inserting eqn (7.65) into eqn (7.49), we obtain

$$P = \tfrac{1}{2}\mathcal{P}\mathrm{e}^{i\overline{\omega}t} + \text{c. c.} \tag{7.68}$$

where

$$\mathcal{P} = Ne\langle\tilde{x}\rangle, \qquad \langle\tilde{x}\rangle = \sum_J p_J\tilde{x}_J. \tag{7.69}$$

Thus, to calculate the optical polarization of the medium, we have to find the mean amplitude of vibration $\langle\tilde{x}\rangle$ determined by eqns (7.66) and (7.69). Let us find the solution to eqn (7.66). Using the spectral expansion technique, we shall represent the functions $A(t)$ and $\tilde{x}_J(t)$ as the Fourier integrals:

$$A(t) = \frac{1}{2\pi}\int_{-\infty}^{\infty} A(\omega)\mathrm{e}^{i\omega t}d\omega, \tag{7.70}$$

$$\tilde{x}_J(t) = \frac{1}{2\pi}\int_{-\infty}^{\infty} \tilde{x}_J(\omega)\mathrm{e}^{i\omega t}d\omega, \tag{7.71}$$

Inserting eqns (7.70) and (7.71) into eqn (7.66), we find relation between the spectral amplitudes:

$$\tilde{x}_J(\omega) = \chi_J(\omega)A(\omega), \tag{7.72}$$

where

$$\chi_J(\omega) = \frac{1}{i\omega - i\overline{\omega}_J + \Gamma/2} \tag{7.73}$$

is the spectral-response function (linear optical susceptibility) for a given oscillator.

Inserting eqn (7.72) into eqn (7.71) and making use of the formula for inverse Fourier transformation,

$$A(\omega) = \int_{-\infty}^{\infty} A(t)\mathrm{e}^{-i\omega t} dt, \tag{7.74}$$

which follows from eqn (7.70), we obtain

$$\tilde{x}_J(t) = \int_0^{\infty} A(t-\theta)h_J(\theta)\, d\theta, \tag{7.75}$$

where

$$h_J(t) = \frac{1}{2\pi}\int_{-\infty}^{\infty} \chi_J(\omega)\mathrm{e}^{i\omega t} d\omega \tag{7.76}$$

is the oscillator influence function corresponding to the frequency–response function $\chi_J(\omega)$. Finally, inserting eqn (7.73) into eqn (7.76), we find the explicit expression for $h_J(t)$:

$$h_J(t) = \begin{cases} \exp(i\tilde{\omega}_J t - \frac{1}{2}\Gamma t), & t \geq 0, \\ 0, & t < 0. \end{cases} \tag{7.77}$$

Note that, according to eqn (7.77),

$$h_J(0) = 1, \qquad h_J(\infty) = 0. \tag{7.78}$$

Equation (7.77) allows eqn (7.75) to be rewritten in the form

$$x_J(t) = \int_{-\infty}^{\infty} A(t-\theta)h_J(\theta)\, d\theta. \tag{7.79}$$

The solution of eqn (7.66) is thus given by eqns (7.79) and (7.77). The integral (7.79) is called the *Duhamel integral.* Equation (7.79) suggests that the "response" of the linear system $\tilde{x}_J(t)$ is the linear functional delayed with respect to the "input signal" $A(t)$. Using eqn (7.79), we can easily define the physical meaning of the influence function $h_J(t)$. Indeed, if

$$A(t) = \text{constant} \cdot \delta(t), \tag{7.80}$$

where $\delta(t)$ is the delta function, we obtain

$$\tilde{x}_J(t) = \text{constant} \cdot h_J(t). \tag{7.81}$$

Thus, the influence function characterizes the system response to a very short excitation pulse.

Having inserted eqn (7.79) into eqn (7.66), one can easily show that the influence function satisfies the following equation and the initial conditions

$$\dot{h}_J + \frac{\Gamma}{2} h_J = i\tilde{\omega}_J h_J, \qquad h_J(0) = 1. \tag{7.82}$$

Thus, the influence function can be found by Fourier transformation of the frequency–response function or by solving the homogeneous equation describing free oscillations of the linear system. Evidently, both approaches should give the same result.

Using eqns (7.79) and (7.69), we can write the mean amplitude of vibration $\langle\tilde{x}\rangle$ as the Duhamel integral

$$\langle\tilde{x}\rangle = \int_0^\infty A(t-\theta)h(\theta)\,d\theta, \tag{7.83}$$

where

$$h(t) = \sum_J p_J h_J(t) \tag{7.84}$$

is the influence function for the oscillator ensemble. It follows from the probability normalization condition $\sum_J p_J = 1$ and eqn (7.82) that, generally, $h(0) = 1$. Additionally, eqns (7.84) and (7.77) lead to eqn (7.61).

Thus, the influence function of the inhomogeneous oscillator ensemble is described by eqns (7.84) and (7.77). The frequency susceptibility of the medium is determined by eqn (7.62).

7.8.2 Example: Doppler dephasing

Consider the excitation of an atomic or molecular gas by a short optical pulse. The electrical field strength of the pulse can be written as

$$E(t) = \tfrac{1}{2}\mathcal{E}(t)\mathrm{e}^{i(\omega t - kz)} + \text{c. c.} \tag{7.85}$$

Since the velocity of light is not infinite, the elementary oscillators at different points z of the medium start to oscillate at different moments in time. After the passage of the optical pulse, all oscillators will oscillate with the same frequency, but with different phases. The spatial variation of the phase will be governed by the wavelength of the exciting radiation. Thus, the optical pulse excites a wave of coherent vibrations. Then, the chaotic thermal motion of particles will lead to a decay of the coherently driven vibrations (Fig. 7.13) and, consequently, to relaxation of the optical polarization of the medium (a dipole moment per unit volume). This mechanism of the optical polarization decay (free induction decay) is called *Doppler dephasing*.

Let us derive the expressions for the susceptibility function of the medium $\chi(\omega)$ and the influence function $h(t)$ corresponding to Doppler dephasing. To do this, we shall use the formula for the Doppler frequency shift,

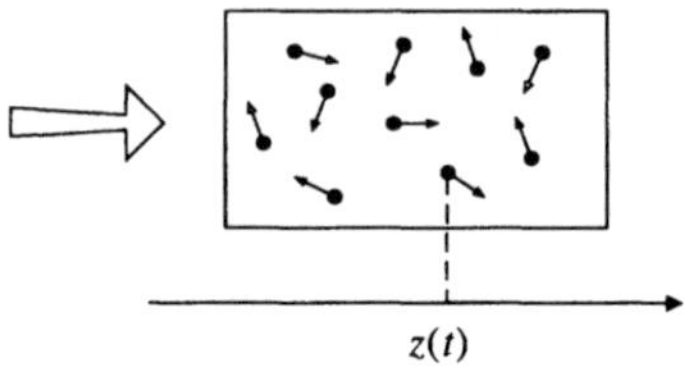

Fig. 7.13 The Doppler dephasing mechanism.

$$\tilde{\omega} = kv_z, \tag{7.86}$$

and the probability density function for the thermal velocity,

$$w(v_z) = \frac{1}{\sigma_v\sqrt{2\pi}} e^{-v_z^2/2\sigma_v^2}. \tag{7.87}$$

Here k is the wave number of the optical wave,

$$\sigma_v^2 = k_B T/M \tag{7.88}$$

is the thermal velocity dispersion, k_B is the Boltzmann constant, T is the absolute temperature of the gas, and M is the mass of the oscillator (atom or molecule).

Neglecting the damping of oscillations, we can write the influence function of the individual oscillator in the form

$$h(v_z) = \exp(ikv_z t), \tag{7.89}$$

where v_z is the velocity of the oscillator in the direction of the z-axis. The influence function of the whole oscillator ensemble can be written as

$$h = \langle h(v_z) \rangle = \int_{-\infty}^{\infty} h(v_z) w(v_z) dv_z. \tag{7.90}$$

Inserting eqns (7.87) and (7.89) into eqn (7.90), we find that

$$h(t) = \exp\left(-\tfrac{1}{2}\sigma^2 t^2\right) \tag{7.91}$$

and

$$\chi(\omega) = \int_{-\infty}^{\infty} h(t) e^{-i\omega t} dt = \frac{\sqrt{2\pi}}{\sigma} \exp(-\omega^2/2\sigma^2), \tag{7.92}$$

where

$$\sigma = k\sigma_v. \tag{7.93}$$

Thus, the problem is solved. The obtained functions $\chi(\omega)$ and $h(t)$ are plotted in Fig. 7.14 (compare with Fig. 7.12). Note that the results obtained can be generalized to allow for the collisions changing the thermal velocities, quantum

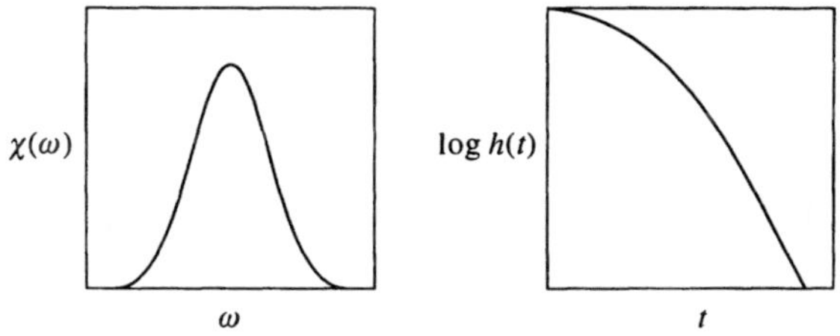

Fig. 7.14 The characteristics of Doppler dephasing.

states, and, consequently, the intrinsic frequencies of the oscillators. The relevant theory was developed by Dyakov *et al.* (1983) and Kolomoitsev and Nikitin (1991) (see also Chapter 6).

Bibliography

Dyakov, Yu. E., Krikunov, S. A., Magnitslii, S. A., Nikitin, S. Yu., and Tunkin, V. G. (1983). *Jurnal Eksperimentalnoi i Teoreticheskoi Fiziki*, **84**, 2013–2025. (In Russian).

Feynman, R. P., Leighton, R. B., and Sands, M. (1963). *Feynman lectures on physics, part 3: waves, quanta, radiation.* Addison-Wesley, Reading, Massachusetts.

Kolomoitsev, D. V. and Nikitin, S. Yu. (1991). *Proceedings SPIE*, **1402**, 11–30.

Owyoung, A. (1978). *Optics Letters*, **2**, 91–93.

8

A non-linear oscillator in an optical field

A model of a non-linear oscillator. The perturbation method. An oscillator with quadratic non-linearity. Optical second-harmonic generation. An oscillator with cubic non-linearity. The dependence of the oscillation frequency on amplitude. Oscillation phase variations caused by an external field. Self-action effects of light: self-focusing of optical beams and self-modulation of pulses. Optical third-harmonic generation. Non-linear resonance and hysteresis. Optical bistability. Parametric generation of light. Parametric resonance. Raman scattering of light. A model of non-linearly bonded oscillators. A non-homogeneous ensemble of non-linear oscillators. "Photon echo".

The phenomena associated with a *non-linear response* of matter to the optical field rank high in modern optics. Among such phenomena are optical harmonic generation, e.g. doubling and tripling of the light frequency, parametric processes, (specifically, doubling of the optical wavelength), stimulated Raman scattering of light, self-focusing of optical beams, self-modulation of pulses, etc. All of these are connected by the notion of "non-linear optics" and characterized by a strong dependence on the intensity of light; that is, they become appreciable only in high-powered laser beams.

In this Chapter we shall consider the simple picture of non-linear optics phenomena using a *non-linear oscillator model.* A more detailed discussion of this issue is given in Part IV.

8.1 A model of a non-linear oscillator

In the general case, the motion of a one-dimensional optical oscillator can be described by the equation

$$\ddot{x} + \Gamma\dot{x} + \frac{1}{m}\frac{\partial U}{\partial x} = \frac{e}{m}E, \tag{8.1}$$

where e and m are the charge and mass of the oscillator, Γ is the damping factor, $U = U(x)$ is the potential energy of the bonded charge, describing the internal force field of the oscillator (atom, molecule), and E is the electric field of the optical wave.

Let us determine the coordinate x and energy U with reference to the values corresponding to the equilibrium position of the oscillator, given that for the equilibrium point $x = 0$, $U = 0$, $U^{(1)}(0) = 0$. At this point the potential energy of the oscillator has an extremum (a minimum) and the derivative $U^{(1)} = \partial U/\partial x$ is therefore equal to zero. Expanding the function $U(x)$ into a power series in x, we obtain

$$U(x) = \frac{x^2}{2!}U^{(2)}(0) + \frac{x^3}{3!}U^{(3)}(0) + \frac{x^4}{4!}U^{(4)}(0) + \ldots$$

or

$$U(x) = ax^2 + bx^3 + cx^4 + \ldots . \tag{8.2}$$

In the case of small oscillations we can limit the expansion (8.2) to the quadratic term

$$U(x) = ax^2. \tag{8.3}$$

This approximation corresponds to the harmonic oscillator and to the linear oscillation equation

$$\ddot{x} + \Gamma\dot{x} + \omega_0^2 x = \frac{e}{m}E, \tag{8.4}$$

where ω_0 is the natural frequency of small oscillations. If the oscillation amplitude becomes large, one should take into account in eqn (8.2) the terms proportional to x^3, x^4, etc. In this case the oscillation equation (8.1) becomes non-linear and the oscillator motion acquires new qualitative features. It can be seen from eqn (8.1) that the non-linear effects are to be expected primarily in strong optical fields.

A specific kind of non-linearity is determined by the type of the oscillator. If the function $U(x)$ is even, i.e. the system has a *centre of symmetry*, we have in the lowest non-linear approximation,

$$U(x) = ax^2 + cx^4, \tag{8.5}$$

and the motion equation contains the cubic term

$$\ddot{x} + \Gamma\dot{x} + \omega_0^2 x - \beta x^3 = \frac{e}{m}E. \tag{8.6}$$

When the first non-linear correction is taken into account in systems *without a centre of symmetry*, we obtain

$$U(x) = ax^2 + bx^3 \tag{8.7}$$

and, accordingly, the quadratic non-linear oscillation equation

$$\ddot{x} + \Gamma\dot{x} + \omega_0^2 x + \gamma x^2 = \frac{e}{m}E. \tag{8.8}$$

In eqns (8.6) and (8.8), β and γ are the non-linearity parameters.

Isotropic media (gases, liquids, and glasses) possess cubic optical non-linearity, while anisotropic crystals (see Part IV) possess quadratic non-linearity.

The potential energies of the harmonic oscillator, symmetric non-linear oscillator, and asymmetric oscillator are plotted in Fig. 8.1.

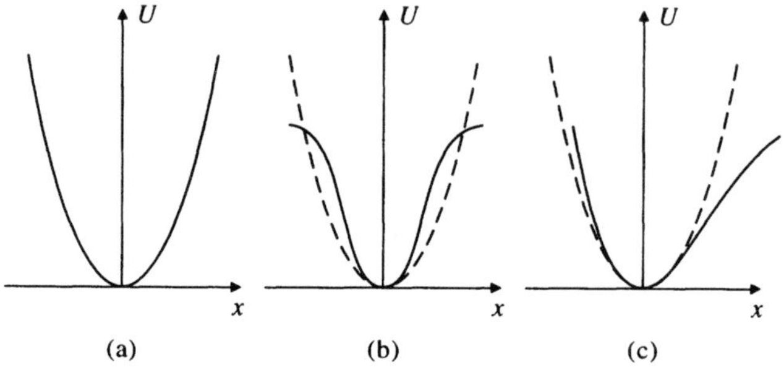

Fig. 8.1 The potential energy of oscillators: (a) harmonic; (b) non-linear symmetric; (c) asymmetric. In (b) and (c) the dashed line shows the energy in the parabolic (harmonic) approximation.

8.2 The perturbation method

As a rule, precise analytical solutions of non-linear differential equations either do not exist at all or are too complicated for easy application. Consequently, various *approximate* methods are used to analyse non-linear systems, one of which is the *perturbation method.*

The main idea of the method is to consider the motion of the system in a *weakly non-linear regime,* i.e. under conditions in which the non-linearity reveals itself as a *small perturbation.* For example, in the case of the oscillator, one can examine oscillations the amplitudes of which are sufficiently low. It is clear from physical reasoning that the oscillations taking place near the potential energy minimum should not differ appreciably from the harmonic oscillations (see Fig. 8.1). This allows the non-linear distortion of the oscillations to be treated as a small correction for which an approximate *linear equation* can be obtained.

8.3 An oscillator with quadratic non-linearity and optical second-harmonic generation

We shall begin to discuss the non-linear phenomena in optics by considering the effect of light frequency doubling in the crystal; that is, *optical second-harmonic generation.* This effect consists of the appearance of a doubled frequency in a non-linear crystal under the action of high-power laser radiation (Fig. 8.2).

The generation of the second harmonic of neodymium laser radiation in a KDP crystal can be demonstrated in a class experiment. The original infrared radiation ("pumping") has a wavelength of $\lambda_p = 1.06\ \mu$m. When the laser pulse is

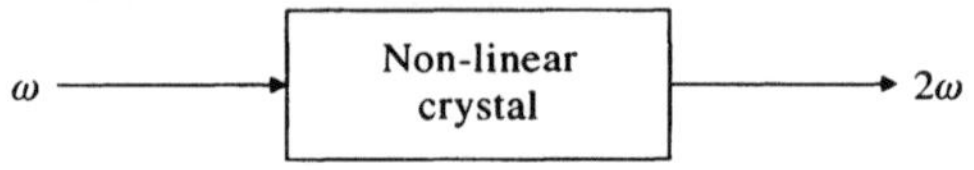

Fig. 8.2 Optical second-harmonic generation.

focused on to the crystal, a bright green flash appears on its exit face (a "second harmonic"). The wavelength of the generated radiation, $\lambda_h = 0.53\ \mu m$, proves to be precisely half the wavelength of the original laser radiation.

The mechanism of second-harmonic generation can be made clear using a model of a quadratic non-linear oscillator the equation of motion of which has the form of eqn (8.8). If in the optical field E the oscillation frequency is equal to ω then, according to eqn (8.8), the *non-linear term* γx^2 describes the oscillations at the *doubled* frequency 2ω. It is these oscillations that cause the second-harmonic radiation.

8.4 An oscillator with cubic non-linearity—the dependence of the oscillation frequency on amplitude

An important property of the non-linear oscillator is the dependence of its oscillation frequency on amplitude:

$$\omega = \omega(A). \tag{8.9}$$

This property is intrinsic to non-linear oscillators of different natures, e.g. to a mathematical pendulum, and is referred to as *non-isochronism*. In optics this property brings about some peculiar self-action effects in light waves, such as self-focusing and self-defocusing of light beams, self-modulation of pulses, etc.

As an example, let us consider the non-isochronic oscillations of a cubic non-linear oscillator without damping. Its free oscillations are described by the equation

$$\ddot{x} + \omega_0^2 x - \beta x^3 = 0. \tag{8.10}$$

The method of small perturbations is applicable when the non-linear term in this equation is much less than the linear one. This happens when the inequality

$$\beta x^2 \ll \omega_0^2 \tag{8.11}$$

holds, i.e. when the oscillation amplitude is sufficiently low.

Suppose that under the conditions of weak non-linearity the oscillations are nearly harmonic, although their frequency ω differs from ω_0, the frequency of infinitesimal oscillations. Denoting the oscillation amplitude by A, we seek an approximate solution of eqn (8.10) in the form

$$x(t) = x_1(t) + x_2(t), \tag{8.12}$$

where

$$x_1(t) = A\cos\omega t \tag{8.13}$$

and

$$|x_2| \ll |x_1|. \tag{8.14}$$

Let us derive the equation for x_2. Substituting eqn (8.12) into eqn (8.10), we obtain

$$\ddot{x}_1 + \ddot{x}_2 + \omega_0^2(x_1 + x_2) - \beta(x_1 + x_2)^3 = 0.$$

The condition (8.14) allows one to linearize this equation with respect to x_2, keeping only the linear terms in x_2:

$$\ddot{x}_1 + \ddot{x}_2 + \omega_0^2(x_1 + x_2) - \beta(x_1^3 + 3x_1^2 x_2) = 0. \tag{8.15}$$

Using eqn (8.13), we shall rewrite this equation as

$$\ddot{x}_2 + (\omega_0^2 - 3\beta x_1^2)x_2 = (\omega^2 - \omega_0^2)x_1 + \beta x_1^3$$

or, taking into account eqns (8.11), (8.12), and (8.14), as

$$\ddot{x}_2 + \omega_0^2 x_2 = (\omega^2 - \omega_0^2)x_1 + \beta x_1^3. \tag{8.16}$$

Now we shall insert eqn (8.13) into eqn (8.16) and, using the well-known trigonometric formula

$$\cos^3\alpha = \tfrac{1}{4}(3\cos\alpha + \cos 3\alpha),$$

obtain the equation

$$\ddot{x}_2 + \omega_0^2 x_2 = (\omega^2 - \omega_0^2 + \tfrac{3}{4}\beta A^2)x_1 + \tfrac{1}{4}\beta A^3 \cos 3\omega t. \tag{8.17}$$

The time variation of the small component x_2 is determined by the linear equation (8.17). To ensure that the x_2 component contains no oscillations at the frequency ω of the fundamental oscillation of x_1, the coefficient of x_1 in eqn (8.17) is to be set equal to zero. This leads to the relation

$$\omega^2 - \omega_0^2 + \tfrac{3}{4}\beta A^2 = 0, \tag{8.18}$$

and the equation for x_2 takes the form

$$\ddot{x}_2 + \omega_0^2 x_2 = \tfrac{1}{4}\beta A^3 \cos 3\omega t. \tag{8.19}$$

The relation (8.18) defines the sought-for dependence of the oscillation frequency ω of the non-linear oscillator on amplitude A. Taking into account eqn (8.11), this dependence can be approximated in the form

$$\omega = \omega_0 - \kappa A^2, \tag{8.20}$$

where

$$\kappa = 3\beta/8\omega_0. \tag{8.21}$$

The dependence $\omega(A)$ is illustrated in Fig. 8.3. Now we shall discuss how non-isochronic oscillations manifest themselves in optics.

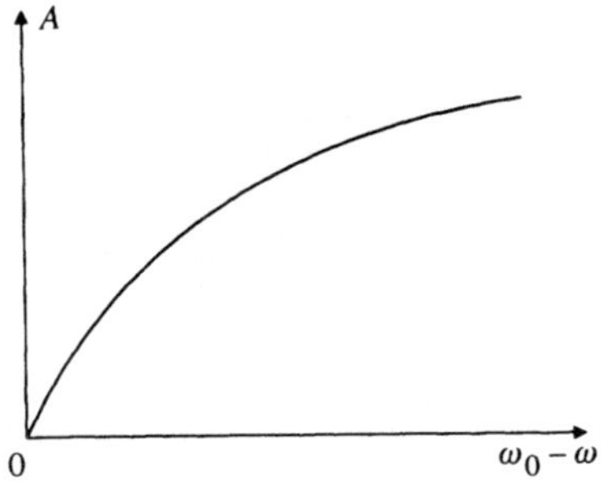

Fig. 8.3 The non-isochronism of a non-linear oscillator: the dependence of the oscillation frequency ω on the oscillation amplitude A. ω_0 is the frequency of infinitesimal (linear) oscillations.

8.5 Oscillation phase variations caused by an external field

In optics, the non-isochronism of a non-linear optical oscillator manifests itself in the dependence of the frequency of the oscillator, excited by the light field, on the excitation intensity:

$$\omega = \omega(I). \tag{8.22}$$

This dependence causes the appearance of an additional phase difference $\varphi = \int \omega(t)\, dt$ proportional to the light intensity:

$$\varphi = \varphi(I). \tag{8.23}$$

Thus, the possibility arises of controlling the oscillation phase by use of the external field ("phasing"). This circumstance drastically distinguishes the non-linear oscillator from the classical harmonic oscillator.

The non-linear phasing of oscillators brings about the effects of self-action of powerful modulated light waves, e.g. the self-focusing of beams and self-modulation of pulses.

8.6 Self-action effects of light: self-focusing of optical beams and self-modulation of pulses

In a spatially limited light beam the light intensity depends on the transverse coordinate:

$$I = I(\vec{r}). \tag{8.24}$$

By the mechanism of non-linear phasing, eqn (8.23), this dependence is transferred to the oscillation phase

$$\varphi = \varphi(\vec{r}) \tag{8.25}$$

of the oscillators and, consequently, to the phase of secondary light waves emitted by these oscillators. This results in a bending of the wavefront of the intense light beam propagating in the medium, which, in particular, can lead to a change in the

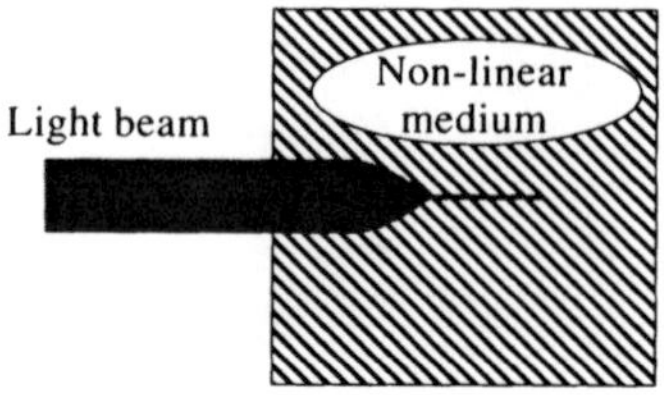

Fig. 8.4 The self-focusing of a high-powered light beam in a non-linear medium.

cross-section of the beam, i.e. to phenomena such as *self-focusing* and *defocusing* of light (Fig. 8.4).

Certainly, the non-linear phase difference may depend not only on the spatial coordinates, but also on time. In a short optical pulse the radiation intensity varies rapidly with time:

$$I = I(t). \tag{8.26}$$

According to eqns (8.22), (8.23), and (8.26), the frequency and phase of oscillations excited by the pulse, as well as the frequency and phase of the secondary radiation emitted by the oscillators, turn out to be modulated in time:

$$\omega = \omega(t), \qquad \varphi = \varphi(t). \tag{8.27}$$

Thus, the intense light pulse experiences *self-modulation* in a non-linear medium. In some cases this can lead to a considerable broadening of the frequency spectrum of the pulse. A similar effect is observed, for example, when a high-powered short laser pulse propagates in an optical fibre (Fig. 8.5).

8.7 Optical third-harmonic generation

Suppose that the cubic non-linear oscillator is in the field of a monochromatic light wave the frequency ω of which is close to the natural frequency of the oscillator, ω_0. According to eqn (8.6), the resonance field will excite forced oscillations which, by virtue of eqns (8.12), (8.13), and (8.19), will contain a strong spectral component at the frequency of the exciting light wave ω ("pumping"), and a

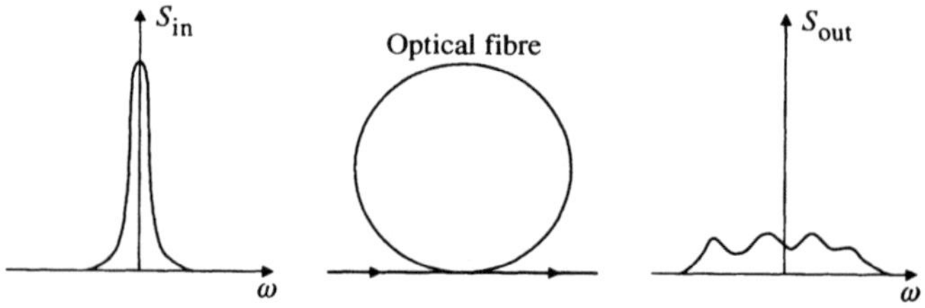

Fig. 8.5 Broadening of the spectrum of a high-powered light pulse in an optical fibre due to self-modulation.

weak spectral component at the frequency 3ω. The frequencies ω and 3ω will also be present in the secondary emission spectrum of the oscillator, indicating that the non-linear oscillator generates the *optical third harmonic* in the field of the light wave. It can be anticipated that harmonic generation will be most efficient when powerful laser radiation is used for pumping.

Experiments confirm these conclusions. Optical third-harmonic generation has been observed in various gases pumped by laser radiation. The highest efficiency of the conversion process was achieved in experiments with alkali metal vapours. With a Nd:YAG laser pumping 30 ps 300 MW pulses into rubidium vapour, third-harmonic generation at a wavelength of 354.7 nm has been accomplished with an energy conversion efficiency of 10% (Bloom *et al.* 1975). It should be also mentioned that a utilization of higher-order optical non-linearities enables the generation of coherent radiation in the vacuum ultra-violet (VUV) and soft X-ray ranges. For example, a coherent emission of radiation at a wavelength of 38 nm was obtained by generating the seventh harmonic in He (Reintjes 1987). A conversion efficiency of 10^{-6} was attained in this case.

8.8 Non-linear resonance and hysteresis—optical bistability

Suppose that the cubic non-linear oscillator described by the motion equation

$$\ddot{x} + \Gamma\dot{x} + \omega_0^2 x - \beta x^3 = \frac{e}{m}E, \tag{8.28}$$

is excited by the monochromatic light field

$$E = \tfrac{1}{2}\mathcal{E}\mathrm{e}^{i\omega t} + \text{c. c.} \tag{8.29}$$

Let us calculate the dependence of the oscillation amplitude A of the oscillator on the frequency ω of the field.

Applying the perturbation method to the solution of this problem, we shall consider a weakly non-linear oscillation regime, i.e. we shall assume the amplitude A to be small enough for condition (8.11) to be satisfied. First, we calculate the amplitude of forced oscillations in the linear approximation. In this case the motion equation is

$$\ddot{x} + \Gamma\dot{x} + \omega_0^2 x = \frac{e}{m}E. \tag{8.30}$$

The solution of the problem (8.29), (8.30) is well known (see Chapter 7). It has the form

$$x = \tfrac{1}{2}\tilde{x}\mathrm{e}^{i\omega t} + \text{c. c.}, \tag{8.31}$$

with the real oscillation amplitude $A = |\tilde{x}|$ in the frequency region near the resonance ($\omega \approx \omega_0$) being defined by the formula

$$4\omega_0^2 A^2 = \frac{B^2}{(\omega_0 - \omega)^2 + \Gamma^2/4}, \qquad B = \frac{e}{m}|\mathcal{E}|. \tag{8.32}$$

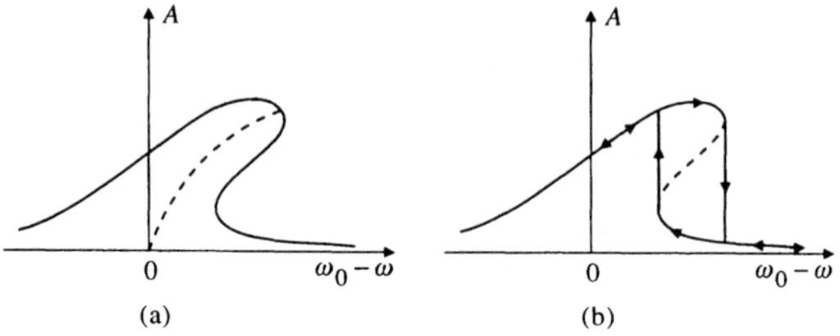

Fig. 8.6 (a) Non-linear resonance. (b) Hysteresis.

Thus, the resonance curve of the oscillator $A(\omega)$ is given in the linear approximation by eqn (8.32). Now we shall modify this formula, allowing for non-isochronism of the non-linear oscillator. According to eqn (8.20), this can be done by substituting ω_0 for $\omega_0 - \kappa A^2$, where κ is the oscillator non-linearity parameter defined by eqn (8.21). Finally, we obtain the relation

$$4\omega_0^2 A^2 = \frac{B^2}{(\omega_0 - \omega - \kappa A^2)^2 + \Gamma^2/4}. \tag{8.33}$$

This formula defines the sought-for dependence $A(\omega)$. However, it is rather difficult to derive this dependence explicitly, because one needs to solve a cubic equation in A^2. Instead of this, we shall write a simple equation for the reciprocal function $\omega = \omega(A)$:

$$\omega_0 - \omega = \kappa A^2 \pm \sqrt{\left(\frac{B}{2\omega_0 A}\right)^2 + \frac{\Gamma^2}{4}}. \tag{8.34}$$

By assuming different values of A, we can calculate from this formula the corresponding values of ω and plot the resonance curve $A = A(\omega)$ (see the beak-shaped curve in Fig. 8.6(a)). The dashed line corresponds to $\Gamma = B = 0$ (the so-called "skeleton" line) and describes the free non-isochronic oscillations of the non-linear oscillator (cf. Figs 8.3 and 8.6(a)).

The distinctive characteristic of the non-linear resonance is the existence of a frequency range in which different oscillation amplitudes are possible. Depending on the past history, that is, on the frequency of the external force (above and below the resonance) in the preceding moments in time, the amplitude settles at a certain value (Fig. 8.6(b)). The indicated property of the non-linear oscillator is called *hysteresis*. This can form the basis for the operation of bistable optical elements that are indispensable for the development of an all-light computer.

One of the possible schemes for a bistable optical element is shown in Fig. 8.7(a). The element consists of an optical resonator formed by mirrors M_1 and M_2 and filled up by a non-linear medium. When one channels a light beam into this system, the output light intensity I_{out} may have one of the two possible val-

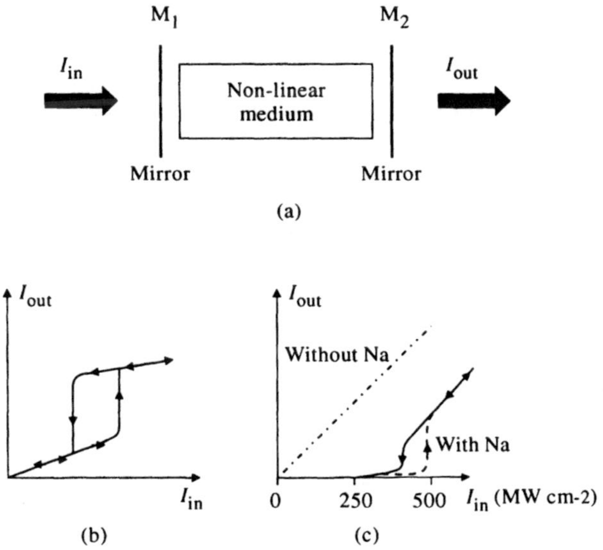

Fig. 8.7 Optical bistability: (a) a schematic of a bistable element; (b) a bistability characteristic; (c) optical bistability in sodium vapour (from McCall *et al.* 1975).

ues for an input beam intensity I_{in} lying within a certain range of its variation (Fig. 8.7(b)). In other words, the given system possesses two stable states (with a high and low optical transmittance) which can be used as logical "0" and "1". Clearly, the bistable element is principally non-linear because the output intensity I_{out} cannot be obtained by multiplying the input intensity I_{in} by a constant coefficient.

Optical bistability of this type was first observed in 1974 by McCall and coworkers, who used in their experiments the resonance non-linearity of sodium vapour (Fig. 8.7(c)).

8.9 Parametric generation of light and parametric resonance

One of the significant effects of non-linear optics is the parametric generation of light. This effect is observed in anisotropic crystals with quadratic non-linearity (KDP, lithium niobate, and others) when high-powered laser radiation of frequency ω_p ("pumping") generates in the crystal a pair of light waves with frequencies ω_1 and ω_2, the sum of which is equal to the pump frequency:

$$\omega_1 + \omega_2 = \omega_p \tag{8.35}$$

(Fig. 8.8). In particular, in the *degenerated regime* the parametric generator of light produces an *optical sub-harmonic* the frequency of which is half of the pump frequency:

$$\omega = \tfrac{1}{2}\omega_p. \tag{8.36}$$

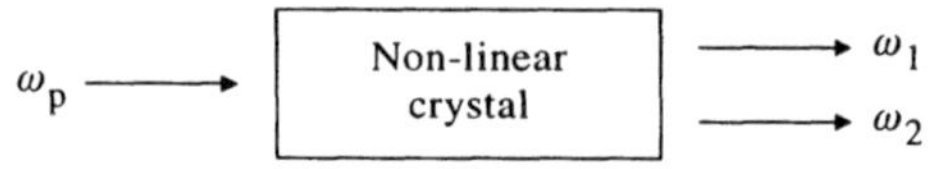

Fig. 8.8 The parametric generation of light.

As the crystal is being rotated with respect to the beam, the frequencies of the generated light waves vary, allowing continuous tuning of the radiation frequency. For example, when an ADP crystal is pumped by radiation with a wavelength of 347 nm, continuous tuning of the radiation wavelength can be carried through the 450–1000 nm range. In the process, the rotating of the crystal through one angular degree from the orientation of the degenerated regime results in a frequency retuning of the order of 1000 cm^{-1} (Fig. 8.9).

At the present time, parametric generators of light serve as sources of coherent radiation that can be tuned in the visible and infrared ranges.

The mechanism of parametric generation of light can be made clear using the model of a *parametric oscillator*, i.e. an oscillator the parameter (frequency) of which varies with time in the preset manner. The motion of such an oscillator is described by the equation

$$\ddot{x} + \omega^2(t)x = 0. \tag{8.37}$$

The fundamental physical effect in such a system is a *parametric resonance*. This phenomenon consists of an abrupt increase in the oscillation amplitude when the frequency ω_p of the parameter variations (pump frequency) coincides with the doubled natural frequency ω_0 of the oscillator:

$$\omega_p = 2\omega_0. \tag{8.38}$$

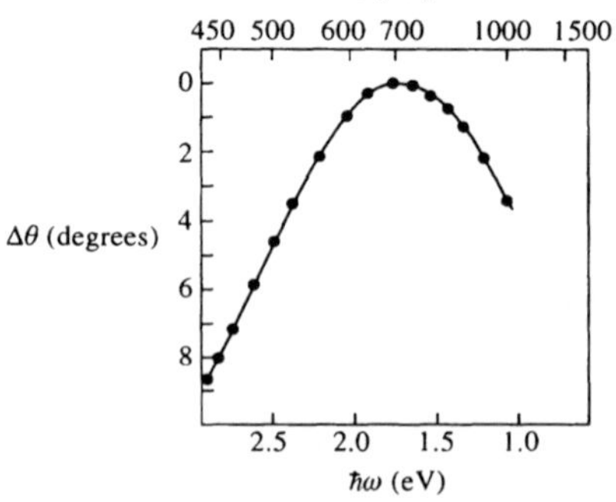

Fig. 8.9 The tuning curve of a parametric oscillator utilizing a KDP crystal. The pump radiation wavelength is 347 nm. The crystal turning angle, $\Delta\theta$, is measured relative to the orientation of the degenerate generation regime (from Madge, D. and Mahr, H. (1967). *Physical Review Letters*, **18**, 905).

Expressed otherwise, the oscillations of the pump frequency are efficiently converted under the conditions of parametric resonance into oscillations the frequency of which is half as high as that of pumping:

$$\omega_0 = \tfrac{1}{2}\omega_{\mathrm{p}}. \tag{8.39}$$

In other words, a process similar to the degenerated parametric generation of light takes place [cf. eqns (8.39) and (8.36)].

Let us demonstrate that, provided that the conditions of the parametric resonance (eqn (8.38)) are satisfied, the oscillation amplitude of the oscillator starts to grow exponentially, i.e. the *parametric instability* arises.

Assuming that

$$\omega^2(t) = \omega_0^2(1 + h\cos 2\omega_0 t), \tag{8.40}$$

where h is the degree of modulation, we obtain from eqn (8.37)

$$\ddot{x} + \omega_0^2(1 + h\cos 2\omega_0 t)x = 0. \tag{8.41}$$

Hereafter, we suppose that $h \ll 1$.

We try a solution of eqn (8.41) in the form

$$x(t) = a(t)\cos\omega_0 t + b(t)\sin\omega_0 t, \tag{8.42}$$

where $a(t)$ and $b(t)$ are slowly varying amplitudes. Inserting eqn (8.42) into eqn (8.41) and neglecting the quantities $\ddot{a}$, and $\ddot{b}$, we obtain

$$-2\dot{a}\sin\omega_0 t + 2\dot{b}\cos\omega_0 t + \omega_0 h(a\cos\omega_0 t + b\sin\omega_0 t)\cos 2\omega_0 t = 0.$$

Making use of the formulae

$$\cos\omega_0 t\cos 2\omega_0 t = \tfrac{1}{2}(\cos\omega_0 t + \cos 3\omega_0 t),$$

$$\sin\omega_0 t\cos 2\omega_0 t = \tfrac{1}{2}(-\sin\omega_0 t + \sin 3\omega_0 t),$$

and omitting the "non-resonant" terms oscillating at the frequency $3\omega_0$, we obtain

$$-(2\dot{a} + \omega_0 hb/2)\sin\omega_0 t + (2\dot{b} + \omega_0 ha/2)\cos\omega_0 t = 0.$$

Equating to zero the coefficients of $\sin\omega_0 t$ and $\cos\omega_0 t$ gives the system of equations

$$2\dot{a} + \omega_0 hb/2 = 0, \qquad 2\dot{b} + \omega_0 ha/2 = 0. \tag{8.43}$$

The solution of this system takes the form

$$a(t) = c_1 e^{\gamma t} + c_2 e^{-\gamma t}, \qquad b(t) = -c_1 e^{\gamma t} + c_2 e^{-\gamma t}, \tag{8.44}$$

where

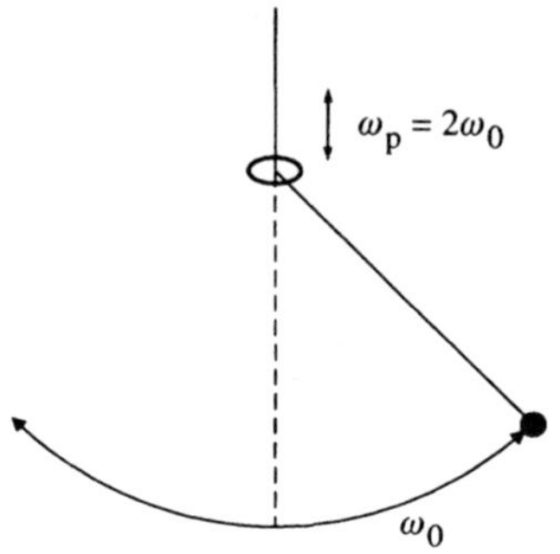

Fig. 8.10 A demonstration of parametric resonance by the example of a mathematical pendulum.

$$\gamma = \omega_0 h/4 \tag{8.45}$$

is the *parametric gain increment*; c_1 and c_2 are constants determined from the initial conditions. Specifically, if

$$x(t=0) = x_0, \qquad \dot{x}(t=0) = 0, \tag{8.46}$$

then $c_1 = c_2 = x_o/2$ and

$$x(t) = x_0(\cosh \gamma t \cos \omega_0 t - \sinh \gamma t \sin \omega_0 t).$$

At sufficiently long times ($\gamma t \gg 1$) the approximate formula

$$x(t) = A(t)\cos(\omega_0 t + \pi/4),$$

is valid, where

$$A(t) = \frac{x_0}{\sqrt{2}} e^{\gamma t}.$$

Hence, in the conditions of parametric resonance, the oscillation amplitude of the oscillator grows exponentially with time. The parametric gain increment is determined in this case by eqn (8.45).

The build-up of parametric oscillation can be easily demonstrated using a simple mechanical system, e.g. a pendulum or a swing (Fig. 8.10).

8.10 Raman scattering of light—a model of non-linearly bonded oscillators

In 1928, G. S. Landsberg and L. I. Mandelstam in Russia, and C. V. Raman and K. S. Krishnan in India, independently discovered the phenomenon that later came to be called the *Raman scattering of light*. On irradiating a quartz crystal

(a)

(b)

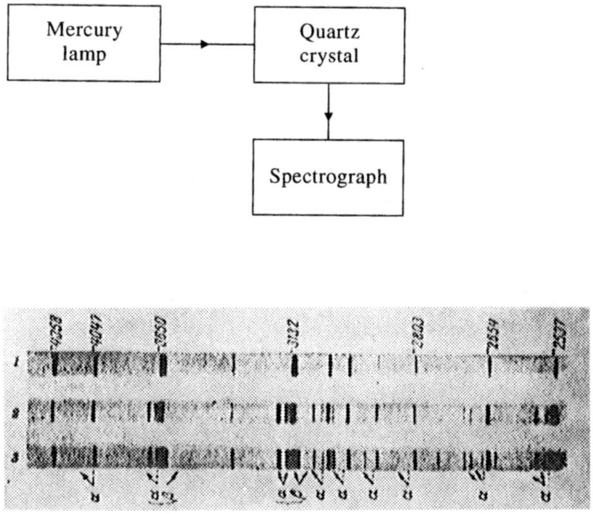

Fig. 8.11 The observation of Raman scattering of light in quartz (from Landsberg and Mandelstam 1928). (a) A schematic of the experiment. (b) The spectrum of light scattered by quartz: 1, the radiation spectrum of a mercury arc lamp (given for comparison); 2 and 3, spectra of scattered light recorded for 20°C and 210°C (the exposure time is 105 hours); α refers to the red satellites; β refers to the violet satellites.

with a mercury vapour lamp, Landsberg and Mandelstam observed that in the scattered light each spectral line of the incident radiation was accompanied by a group of new spectral lines ("satellites").

The scheme of the experiment carried out by Landsberg and Mandelstam, as well as the spectrograms obtained by them, are shown in Fig. 8.11. It should be pointed out that the new spectral lines were extremely weak and that their photographic recording required an exposure of about 100 hours.

The spectrograms of the light scattered by the quartz crystal showed 72 new lines (not present in the mercury spectrum), which were classified into five groups, so that in each group the frequency difference between a satellite and the fundamental line is constant in absolute value. Touching upon the mechanism of the phenomenon, Landsberg and Mandelstam proposed the idea that the appearance of the new spectral lines in the scattered light is caused by a *modulation* of the incident radiation by the natural vibrations of the quartz lattice.

In 1962, Woodbury and Ng discovered that the ruby laser containing a cell with nitrobenzene in its resonator emitted, in addition to the usual radiation at a wavelength of 694.3 nm, an intensive radiation at a wavelength of 767.0 nm. When other liquids were placed between the mirrors of the laser resonator, the additional emission was observed at different frequencies. In all cases, the new frequencies were shifted with respect to that of the ruby radiation by a value equal to the vibration frequency of the molecules of the liquid. The correct explanation of this effect was given by Hellworth, who interpreted it as the *stimulated Raman scattering of light.*

The stimulated Raman scattering (SRS) of light belongs to the class of non-linear optics effects in which a high-powered optical wave induces elementary excitations in a medium (optical and acoustical phonons, polaritons, thermal waves, etc.) and is coherently scattered by them. SRS involves a coherent excitation of optical phonons; this process is a forced analogue of the spontaneous Raman scattering that occurs when light is scattered by chaotic thermal vibrations in a medium.

SRS can be described in terms of classical theory by using a *model of non-linearly bonded oscillators.* Let us consider an isotropic medium, e.g. a liquid. Let y denote the normal vibration coordinate of the atoms in a molecule and x the normal vibration coordinate of the electrons. In the linear approximation, the atomic vibrations and the electron vibrations, which determine the optical polarization of the medium, proceed independent of each other. The situation changes when the non-linear terms in the expression for the potential energy of the molecule are taken into account. With an allowance made for one of the lower terms, the potential energy is written as

$$U(x,y) = \tfrac{1}{2}\alpha x^2 + \tfrac{1}{2}\beta y^2 + \tfrac{1}{2}\gamma x^2 y. \tag{8.47}$$

Here α and β are the elastic constants of intramolecular bonds, and γ describes the non-linear interaction of the electrons and the nuclei. It is this interaction, the mutual influence of electron and nuclear vibrations, that underlies SRS.

In accordance with eqn (8.47), the equations of motion for the electrons and the nuclei take the form

$$\ddot{x} + \omega_0^2 x = \frac{e}{m}E - \frac{\gamma}{m}xy, \tag{8.48}$$

$$\ddot{y} + \Omega^2 y = -\frac{\gamma}{2M}x^2. \tag{8.49}$$

Here M and m are the reduced masses of the atomic and electron oscillators, $\omega_0 = \sqrt{\alpha/m}$ is the natural frequency of the electron vibrations, $\Omega = \sqrt{\beta/M}$ is the natural frequency of nuclear vibrations in the molecule, e is the electron charge, and E is the electric field of light waves. For the real molecules the vibration frequencies are of the following orders of magnitude: $\omega_0 \simeq 10^{15}$ Hz, $\Omega \simeq 10^{13}$ Hz.

In the case of small perturbations the non-linear terms in eqns (8.48) and (8.49) can be regarded as additional forces acting on the electron and molecular oscillators.

Let us consider a molecule located in an optical field of frequency ω. According to eqn (8.48), the light excites electron oscillations of the same frequency. Suppose, then, that nuclear vibrations take place concurrently in the molecule ("molecular vibrations") at frequency Ω (these could be, for example, thermal vibrations). Then, $x \sim \cos\omega t$, $y \sim \cos\Omega t$, and $xy \sim \cos(\omega - \Omega)t + \cos(\omega + \Omega)t$.

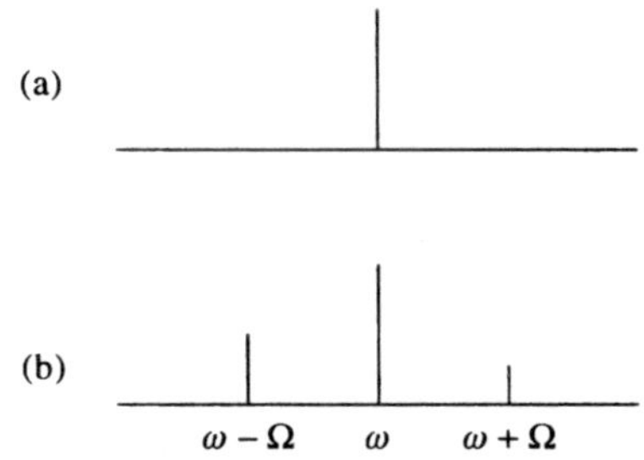

Fig. 8.12 The Raman scattering of light. Spectra of (a) the exciting radiation and (b) the scattered radiation. Ω is the frequency of molecular oscillations of the medium.

Therefore, such coupling of electron and nuclear motions in the molecule results in an effective force acting on the electron oscillator, which is proportional to the product of the oscillation coordinates xy and contains oscillations of frequencies $\omega - \Omega$ and $\omega + \Omega$. In other words, the nuclear vibrations *modulate* the electron vibrations in the molecule. As a result, the molecule re-emits light not only at the frequency ω of the external light wave, but also at the new "Raman" frequencies $\omega \pm \Omega$, where Ω is the natural frequency of the molecular oscillations of the medium. This provides an explanation of the Raman scattering of light (see Fig. 8.12).

Let us now examine eqn (8.49), describing the motion of the nuclei in the molecule. It can easily be seen that when the electron coordinate x contains oscillations of frequency ω, excited by the initial light wave ("pumping"), together with oscillations of frequency $\omega - \Omega$, excited by the scattered radiation ("the Stokes component"), the quantity x^2 contains oscillations of frequency Ω, which is the frequency of natural molecular oscillations. Thus, as a result of the Raman scattering of light, a *resonance force* appears, exciting the molecular oscillations. Provided that this force is strong enough, a *positive feedback* arises in the system: the scattering of light enhances the molecular oscillations and the molecular oscillations in turn amplify the scattering of light. As a result, the system becomes self-excited and passes into a regime of *stimulated* scattering, in which the scattering intensity increases abruptly, by many orders of magnitude, to become comparable with the intensity of laser radiation.

It is easy to show that a pair of light waves (pumping and the Stokes component of SRS) not only trigger resonance molecular oscillations but also *match their phases* in a large volume of the medium. In other words, SRS produces a wave of *coherent* molecular oscillations. This wave, induced by the light field, interacts effectively with that field, leading to the amplification of radiation at the Stokes frequency. Thus, the non-linearity results in the appearance of quite new effects in the system of classical oscillators; namely, the phasing of oscillations of individual oscillators due to their interaction via the radiation field, and the amplification of light. In this respect, the ensemble of non-linear classical oscillators resembles that of quantum oscillators (see below), ensuring coherence and amplification of light in the laser.

In practice, stimulated Raman scattering can give rise to instability in a high-powered light wave, e.g. in the process of propagation in a fibre waveguide or in a fusion plasma. Stimulated scattering finds extensive application in the control of laser radiation parameters; specifically, in frequency conversion, pulse shortening (compression), and improvement in spatial coherence. For example, Raman oscillators utilizing compressed hydrogen and liquid nitrogen accomplish the Raman conversion of laser radiation with a quantum efficiency of up to 90%. In the case of stimulated backward scattering, the efficient compression of laser pulses is possible with a simultaneous increase in their peak power. This technique makes it possible to compress pulses by a factor of 10–20; the use of several compression cascades raises this factor to 100–1000.

Stimulated Raman scattering of light can be used for optical phase conjugation. Another possibility is a wavefront correction. In the latter case we deal with the conversion of energy from partially coherent laser radiation into the energy of a totally coherent light beam of another (Stokes) frequency. This method can be called "the SRS correction of the wavefront". Experiments show that this technique can substantially (by a factor of 100 or more) reduce the angular divergence of radiation. In this case, the efficiency of energy conversion is usually of the order of 50%.

8.11 A non-homogeneous ensemble of non-linear oscillators and "Photon echo"

Let us consider an ensemble of non-linear oscillators which have differing natural frequencies. Such an ensemble is said to be non-homogeneous. Here we shall discuss a peculiar optical phenomenon emerging in a non-linear, non-homogeneous ensemble, the phenomenon of the *photon echo.*

This phenomenon consists of radiation, by a medium, of a light pulse some time after the medium was excited by a pair of short laser pulses separated in time (Fig. 8.13).

The photon echo was predicted in 1963 by Kopvillem and Nagibarov, and later experimentally discovered in ruby by Kurnit, Abella, and Hartmann (1964). Since the echo signals are emitted by the medium some time after the application of excitation pulses, some authors suggest utilizing the photon echo phenomenon in information recording devices.

In order to interpret the mechanism of echo emergence, we shall consider the motion of a cubic non-linear oscillator excited by a pair of short light pulses. Ignoring oscillation damping, we can write the motion equation in the form

$$\ddot{x} + \omega_0^2 x - \gamma x^3 = F(t). \tag{8.50}$$

Assuming the oscillation amplitude to be small enough, we can apply the perturbation method. The approximate solution of eqn (8.50) can be represented as

$$x = x_{\rm L} + x_{\rm NL}, \tag{8.51}$$

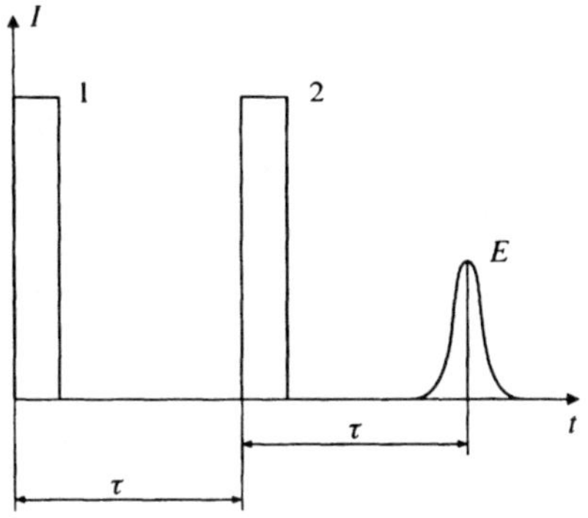

Fig. 8.13 Photon echo: a non-linear medium excited successively by a pair of short laser pulses (1 and 2) spontaneously emits an echo signal (E) after a time interval equal to the delay between the pulses.

where x_{L} is the solution of the linear equation

$$\ddot{x}_{\mathrm{L}} + \omega_0^2 x_{\mathrm{L}} = F(t), \tag{8.52}$$

and x_{NL} is a small, additional non-linear term. Inserting eqn (8.51) into eqn (8.50), we obtain the approximate linear equation for x_{NL}:

$$\ddot{x}_{\mathrm{NL}} + \omega_0^2 x_{\mathrm{NL}} = \gamma x_{\mathrm{L}}^3. \tag{8.53}$$

Let us now assume that the oscillator is excited by two short pulses following one after the other with a time interval τ. In this case,

$$F(t) = f_1(t) + f_2(t - \tau), \tag{8.54}$$

where the functions $f_1(t)$ and $f_2(t)$ describe the pulse shapes. In the limit of infinitesimally short pulses we can assume that $f(t) \sim \delta(t)$, where $\delta(t)$ is the delta function.

By virtue of the linearity of eqn (8.52) the solution of eqns (8.52) and (8.54) can be represented as a sum of the responses to each individual pulse. For example, in the case of infinitesimally short pulses we can use the well-known expression for the Green function for eqn (8.52), to obtain

$$x_{\mathrm{L}}(t) = a_1 \sin \omega_0 t + a_2 \sin \omega_0 (t - \tau), \tag{8.55}$$

where a_1 and a_2 are constants proportional to the energies of the excitation pulses. Equation (8.53) is non-linear in x_{L}, so that the oscillation phases featuring in eqn (8.55) will appear in eqn (8.53) in various combinations. Omitting the terms with frequency $3\omega_0$, we obtain

$$x_{\mathrm{L}}^3 = \tfrac{3}{4}(a_1^3 + 2a_1 a_2^2) \sin \omega_0 t + \tfrac{3}{4}(a_2^3 + 2a_1^2 a_2) \sin \omega_0 (t - \tau)$$

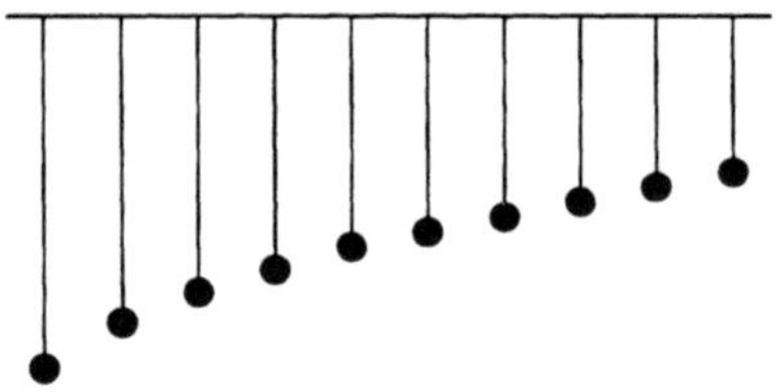

Fig. 8.14 The Chebotayev pendulum: a mechanical model for the demonstration of oscillator echo.

$$+\tfrac{3}{4}a_1 a_2^2 \sin \omega_0(t - 2\tau) + \tfrac{3}{4}a_1^2 a_2 \sin \omega_0(t + \tau). \quad (8.56)$$

There are certain moments of time at which the oscillation phases, eqn (8.56), have strictly definite values irrespective of the values of natural frequency of the non-linear oscillator. These are the moments

$$t = 0, \qquad t = \tau, \qquad t = 2\tau. \quad (8.57)$$

Let us consider the non-homogeneous ensemble of oscillators with different natural frequencies. One can conclude that if the ensemble is excited by two short light pulses, a phasing (synchronization) of weak non-linear oscillation components x_{NL} takes place at the moments of time determined by eqns (8.57), for all oscillators belonging to the ensemble. This leads to the formation of pulses of macroscopic optical polarization (a total dipole moment) of the medium and to the emission of light pulses at the appropriate moments of time. The last of these pulses emitted at the moment $t = 2\tau$ is what is called the photon echo (Fig. 8.13).

This signal is also referred to as the *primary echo*. Its existence can be demonstrated using a model of mechanical oscillators (Fig. 8.14).

Similarly, it can be shown that, on excitation by three short light pulses sent at the moments $t = 0$, $t = \tau$, and $t = \tau + T$, the medium emits, at the moment $t = T + 2\tau$, a light pulse called a *signal of stimulated echo* (Fig. 8.15). The stimulated echo can be utilized in information recording devices. Indeed, with the times τ and T being independent, one can assume that the first two pulses record the information in the oscillator system where it is *stored* for time T, and that the third pulse *reads* the information out by producing an echo signal.

In 1979, the photon echo was experimentally studied in $LaF_3 : Pr^{3+}$ crystals (see Chen *et al.* 1979, Morsink *et al.* 1979) in which the time interval between the second and third excitation pulses can be extended up to several minutes (and longer) by imposing an external magnetic field. This effect is referred to as a *long-lived photon echo*. It was suggested that this effect could be used as a basis for an optical storage device.

In conclusion, we shall point out once again the most significant (from the standpoint of optics) properties of non-linear oscillators. These are the sensitivity

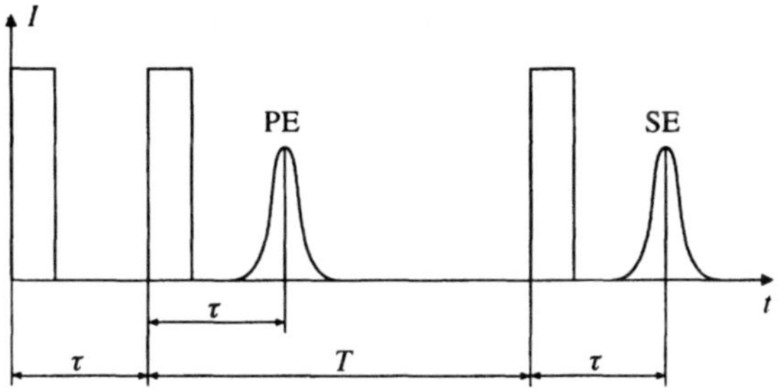

Fig. 8.15 The excitation of the signals of the primary (PE) and stimulated (SE) echo.

of an oscillation phase to the intensity of the exciting light wave, the re-emission of light at optical harmonics and Raman frequencies, and the possibility of the amplification of the light wave interacting with the oscillator ensemble. It is these properties of non-linear elementary oscillators that underlie the diverse effects and phenomena of non-linear optics.

Bibliography

Akhmanov, S. A. and Khokhlov, R. V. (1972). *Problems of nonlinear optics.* Gordon and Breach, New York.

Bloembergen, N. (1965). *Nonlinear optics.* W. A. Benjamin, New York.

Bloom, D. M., Bekkers, G. W., Young, J. F., and Harris, S. E. (1975). *Applied Physics Letters*, **26**, 687–689.

Chebotayev, V. P. and Dubetsky, B. Ya. (1983). *Applied Physics*, **B31**, 45–52.

Chen, Y. C., Chiang, K., and Hartmann, S. R. (1979). *Optics Communications*, **29**, 181–185.

Franken, P. A., Hill, A. E., Peters, and C. W., Weinreich, G. (1961). *Physical Review Letters*, **7**, 118–119.

Gibbs, H. M. (1985). *Optical bistability: controlling light with light.* Academic Press, New York.

Kopvillem, W. H. and Nagibarov, V. R. (1963). *Fizika Metallov i Metallovedenie*, **15**, 313–315. (In Russian.)

Kurnit, N. A., Abella, I. D., and Hartmann, S. R. (1964). *Physical Review Letters*, **13**, 567–568.

Landsberg, G. S. and Mandelstam, L. I. (1928). *Zeitschrift für Physik*, **50**, 769–780.

McCall, S. L., Gibbs, H. M., Churchill, G. G., and Venkateson, T. N. C. (1975). *Bulletin of the American Physical Society*, **20**, 636.

Morsink, J. B. W. and Wiersma, D. A. (1979). *Chemical Physics Letters*, **65**, 105–108.

Raman, C. V. and Krishnan, K. S. (1928). *Nature*, **121**, 501–502.

Reintjes, G. R. (1987). *Nonlinear optical parametric processes in liquids and gases.* Academic Press, New York.

Shen, Y. R. (1984). *Principles of nonlinear optics.* John Wiley, New York.
Woodbury, E. J. and Ng, W. K. (1962). *Proceedings of the Institute of Radio Engineers,* **50**, 2367.
Zeldovich, B. Ya., Pilipetsky, N. F., and Shkunov, V. V. (1985). *Principles of phase conjugation.* Springer-Verlag, Berlin.

9

Thermal radiation

The main experimental facts and methods of investigation of thermal radiation. Emissivity and absorptivity of physical bodies. Equilibrium thermal radiation. Kirchhoff's law. Spectral density of thermal radiation. Thermodynamics of equilibrium thermal radiation. The Rayleigh–Jeans formula. The "ultraviolet catastrophe". Planck's formula. The necessity of a quantum-mechanical approach. Thermal radiation laws (Stefan–Boltzmann law, Wien's displacement law).

In this Chapter we present the foundations of thermal radiation theory, together with the results of experimental investigations. It is shown how thermal radiation research led to the idea of a quantum.

Thermal radiation is the light emitted by a heated body. The Sun, a candle flame, an incandescent lamp, an electric arc, lightning, a laser spark, and a human body emitting infrared radiation are all examples of thermal radiation sources. Thermal radiation substantially affects the thermal balance in many physical systems. Specifically, it governs the mean temperature of the Earth's surface. The theory of this kind of radiation is very important for physics, because thermal radiation is universal and is not associated with any specific substance, or even with atomic structure. Moreover, the origin of thermal radiation is due to the fundamental thermodynamics of light and matter, and the properties of thermal radiation clearly indicate that atoms emit and absorb light at a given frequency not in arbitrary quantities, but only in strictly definite portions (quanta).

9.1 The main experimental facts and methods of investigation of thermal radiation

It is known from experience that strongly heated bodies emit light. The brightness of the emission increases as the temperature grows, and the colour of the emission changes from red to white. We can notice such phenomena by observing, for example, the heating of a piece of metal in a forging furnace or the filament of an incandescent lamp.

A characteristic feature of thermal radiation is a broad continuous frequency spectrum. One can easily see it in the experiment shown schematically in Fig. 9.1. The radiation of an arc lamp is directed on to a glass prism, which resolves it into a spectrum. The spectrum is observed on a screen in a darkroom. The spectrum appears as a bright rainbow band containing red, yellow, green, and blue colours, which change continuously. One can appraise the distribution of radiation energy over the spectrum by shifting a thermocouple along the rainbow band and

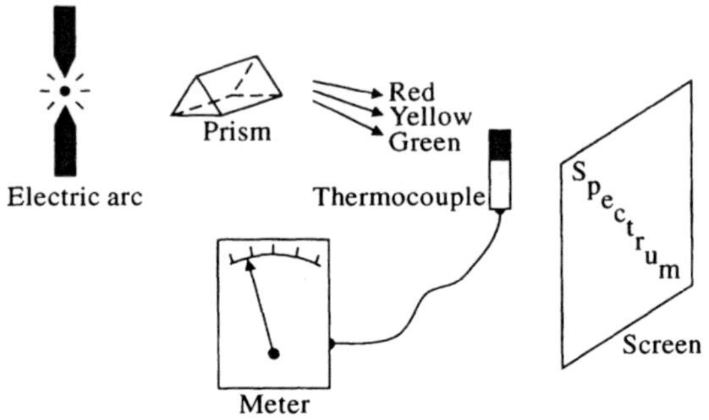

Fig. 9.1 Observation of the radiation spectrum of an electric arc.

measuring the output signal by means of a meter. Such measurements show that the radiation intensity in the visible range grows steadily as the wavelength increases. In fact, the maximum intensity of emission from the arc lamp lies in the infrared region. The measurement results can be plotted as a dependence of the radiation intensity on wavelength or light frequency.

Figure 9.2 shows emission spectra of heated bodies as measured by Lummer and Pringsheim (1899). The graphs clearly demonstrate that the increase in temperature of the emitting body is accompanied by a growth in the total power of thermal radiation and a shift of the spectral density maximum to the shorter wavelength region. At the end of the 19th century, the explication of the thermal radiation spectrum offered a serious challenge to physicists. Investigations of this problem have led to the advent of quantum notions.

9.2 Emissivity and absorptivity of physical bodies

Different physical bodies absorb light differently. Black bodies are known to absorb light strongly, whereas white ones absorb only a small fraction of the incident radiation. For example, the water in a metal cask gets warm in the sunlight much faster when the cask is painted black.

To define the absorption capability of a body we shall introduce the quantity

$$\alpha = \frac{W_{\text{abs}}}{W_{\text{inc}}}. \tag{9.1}$$

Here W_{inc} is the energy incident on the body and W_{abs} is the energy absorbed by the body. We shall call the quantity α the absorptivity of the body. The absorptivity can be assessed from the heating of the body by light.

For opaque bodies, the sum of the energies of absorbed and reflected light is equal to the energy of the incident light (Fig. 9.3):

$$W_{\text{inc}} = W_{\text{ref}} + W_{\text{abs}}.$$

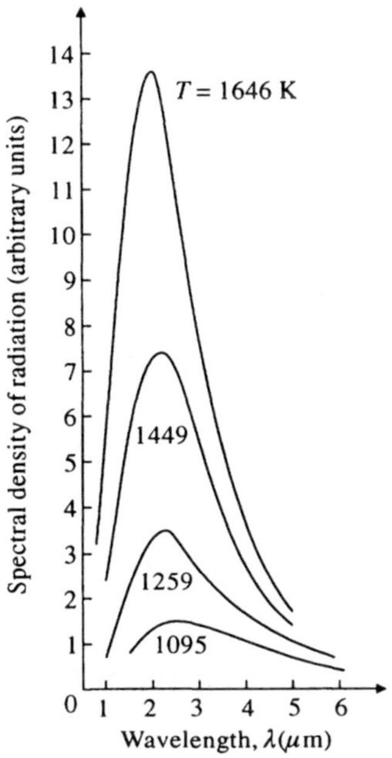

Fig. 9.2 Emission spectra of heated bodies, measured by Lummer and Pringsheim.

Consequently, the dimensionless value α lies within the limits

$$0 \leq \alpha \leq 1.$$

When $\alpha = 0$, the body is called *perfectly white*, and when $\alpha = 1$, the body is called *perfectly black*. A piece of chalk is a good example of a white body. A piece of coal or soot are good examples of black bodies.

Experiments show that the absorption capability of a body depends on its temperature T and the frequency of the incident light, $\nu = \omega/2\pi$. Therefore, any

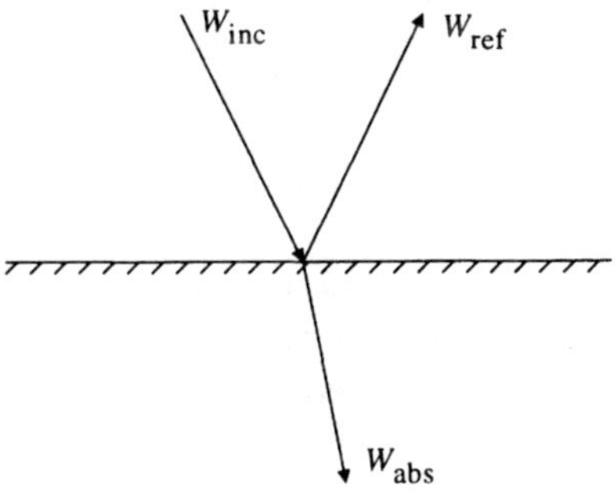

Fig. 9.3 The reflection and absorption of light by an opaque body.

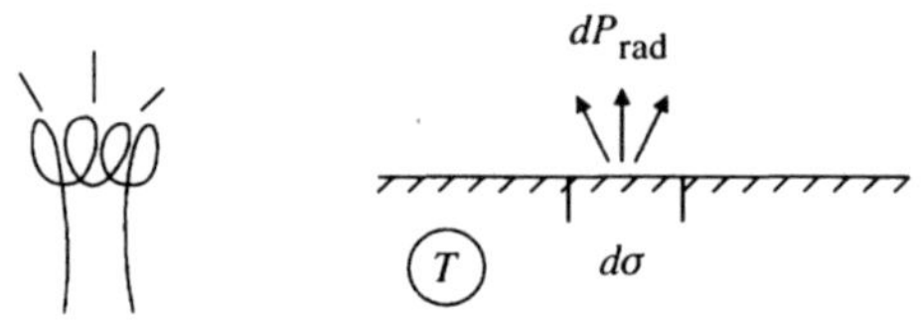

Fig. 9.4 The definition of the emissivity of a body.

body can be characterized by a function

$$\alpha = \alpha(\nu, T)$$

(absorptivity). Needless to say, different bodies have different absorptivities.

9.2.1 Emission of light

Let us consider a heated body; for example, the filament of an incandescent lamp. Suppose that the area element $d\sigma$ of the body surface emits in the frequency range from ν to $\nu + d\nu$ radiation of power $dP_{\rm rad}$ (see Fig. 9.4). It is clear that the quantity $dP_{\rm rad}$ is proportional to the quantities $d\sigma$ and $d\nu$. Consequently, we can write

$$dP_{\rm rad} = \varepsilon \, d\nu \, d\sigma. \tag{9.2}$$

The coefficient in eqn (9.2) is referred to as the emissivity of the body. Experiments show that the emissivity, just like the absorptivity, depends on the body temperature T and on the frequency of light ν, i.e.

$$\varepsilon = \varepsilon(\nu, T).$$

9.3 Equilibrium thermal radiation

Suppose that a certain body heated to temperature T has an internal cavity (see Fig. 9.5). The cavity is filled up by the thermal radiation emitted by the cavity walls. Under steady-state conditions, when the body temperature remains constant, the characteristics of the thermal radiation within the cavity will also have constant and quite definite values; that is, those ensuring a balance between the processes of light emission and absorption by the cavity walls. In other words, under steady-state conditions thermal equilibrium sets in between the cavity walls and the radiation within the cavity. In this case the thermal radiation within the cavity is called *equilibrium thermal radiation*.

We can study the equilibrium radiation experimentally by making a pinhole (channel) joining the cavity and the outer space. The equilibrium thermal radiation will leave the cavity through the channel, so that its characteristics can be measured (Fig. 9.5(b)).

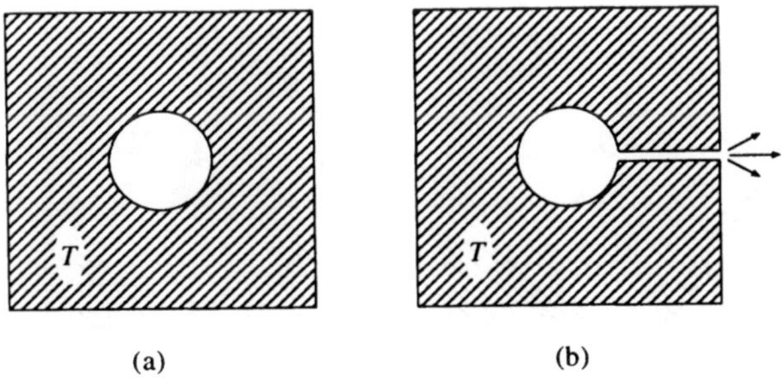

Fig. 9.5 The concept of equilibrium thermal radiation. (a) Equilibrium thermal radiation is the radiation that is in equilibrium with a heated body. It can be, for example, the radiation filling a cavity within a heated body. (b) Measurement of the characteristics of equilibrium thermal radiation.

9.4 Kirchhoff's law

In 1859, Kirchhoff established the law that under thermal equilibrium conditions the ratio of the emissivity to the absorptivity is the same for all bodies:

$$\frac{\varepsilon(\nu, T)}{\alpha(\nu, T)} = \rho(\nu, T) = \text{inv}. \tag{9.3}$$

Here $\rho(\nu, T)$ is the universal function of frequency ν and temperature T, which is called the *emissivity of a perfectly black body*. The meaning of this term is clear from the definition of a perfectly black body, for which $\alpha(\nu, T) = 1$ and, consequently, $\varepsilon(\nu, T) = \rho(\nu, T)$. This law, derived originally by Kirchhoff using thermodynamic considerations, was later confirmed experimentally.

Kirchhoff's law can be derived as follows. Suppose that a certain body, for example a piece of chalk, is put into a cavity filled with equilibrium thermal radiation (see Fig. 9.6(a)). Then, the power of the radiation absorbed by the body will obviously be equal to the power of the radiation emitted by the body

$$dP_{\text{abs}} = dP_{\text{em}}, \tag{9.4}$$

because otherwise the body temperature would vary and the equilibrium would be violated. By definition,

$$dP_{\text{abs}} = \alpha\, dP_{\text{inc}}, \tag{9.5}$$

$$dP_{\text{em}} = \varepsilon\, d\nu\, d\sigma, \tag{9.6}$$

where ε and α are the emissivity and absorptivity of the body, and dP_{inc} is the power of the radiation falling on to an area element $d\sigma$ of the body surface and having a frequency within the spectral interval from ν to $\nu + d\nu$. It follows from eqns (9.4)–(9.6) that

$$\alpha\, dP_{\text{inc}} = \varepsilon\, d\nu\, d\sigma. \tag{9.7}$$

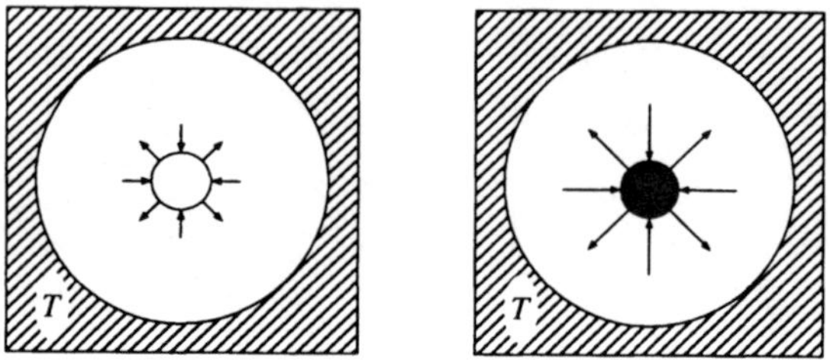

(a) (b)

Fig. 9.6 A derivation of the Kirchhoff law. The arrows show the radiation flows absorbed and emitted by the bodies.

Now suppose that we replace the probe body by another body of the same size and shape; for example, by a piece of coal (Fig. 9.6(b)). As the probe bodies are assumed to be small compared to the thermostat and thus cannot affect the characteristics of the thermal radiation within the cavity, the area element $d\sigma$ of the new body will be irradiated by radiation of the same power $dP_{\rm inc}$ in the same spectral bandwidth $d\nu$. As before, eqns (9.4)–(9.7), remain valid, although the coefficients α and ε now have different values. Thus, all of the quantities in eqn (9.7), with the exception of α and ε, are invariant. Then, the ratio of these quantities should be also invariant:

$$\varepsilon/\alpha = \text{inv}.$$

This equality constitutes Kirchhoff's law.

In the experimental demonstration of Kirchhoff's law we can use a white ceramic plate decorated with a sharp dark pattern, e.g. a flower (see Fig. 9.7). Having been heated to a temperature about 1000°C in a furnace, the plate is pulled out by a steel wire tied to it. When observed in a darkroom, the pattern shines brightly against the background of a weak glow coming from the unpainted parts of the plate. However, as the ceramic plate cools down, the bright shining pattern grows dimmer until the dark flower again appears against the white background. This experiment demonstrates that the parts of the body surface possessing higher absorptivity also have higher emissivity, i.e. $\varepsilon \sim \alpha$, in accordance with Kirchhoff's law.

The following circumstance is also essential. When observing the plate inside the heated furnace through a small opening, we cannot discern any pattern at all (Fig. 9.7). In fact, the plate itself is barely perceptible against the background of the furnace's glowing walls. The equilibrium thermal radiation escaping from the furnace is uniform and carries no image. This means that strongly absorbing parts of the body emit light just as strongly, i.e. the emissivity of the body is proportional to its absorptivity—in full compliance with Kirchhoff's law.

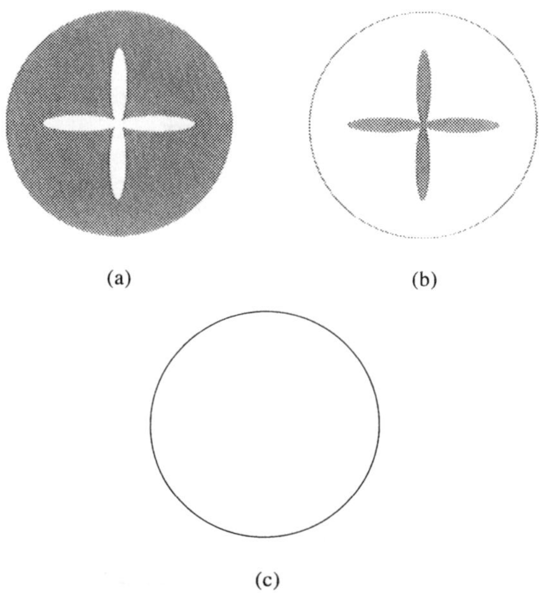

Fig. 9.7 A demonstration of the relation between the emissivity and absorptivity of a body. (a) The luminescence of a heated ceramic plate in a darkroom. (b) A view of the plate at room temperature. (c) A view of the plate in the furnace.

9.5 The spectral density of thermal radiation

The equilibrium thermal radiation at temperature T is characterized by the spectral density $u = u(\omega, T)$, defined as the equilibrium radiation energy per unit volume and per infinitesimal frequency interval $d\omega$ near the frequency ω. The energy of the thermal radiation in volume V within the frequency band from ω to $\omega + d\omega$ is

$$dW = V\, u(\omega, T)\, d\omega \qquad [\mathrm{erg}]. \tag{9.8}$$

This formula can be regarded as the definition of the function $u(\omega, T)$.

The function $u(\omega, T)$ is also referred to as the *spectral density of black body emission*, because this function is directly related to the universal Kirchhoff function $\rho(\nu, T)$ signifying the emissivity of the perfectly black body.

Lummer and Pringsheim studied the spectral density of equilibrium thermal radiation experimentally using a device a schematic of which is shown in Fig. 9.8. The principal element of the device is an iron cavity with a hole to extract radiation. The cavity is heated by means of a special furnace with double walls. The gas heated by the flame flows between the cavity wall and the internal wall of the furnace, then through the space between the double walls of the furnace and, finally, escapes through the flue. The temperature is measured by mercury thermometers or by a thermocouple.

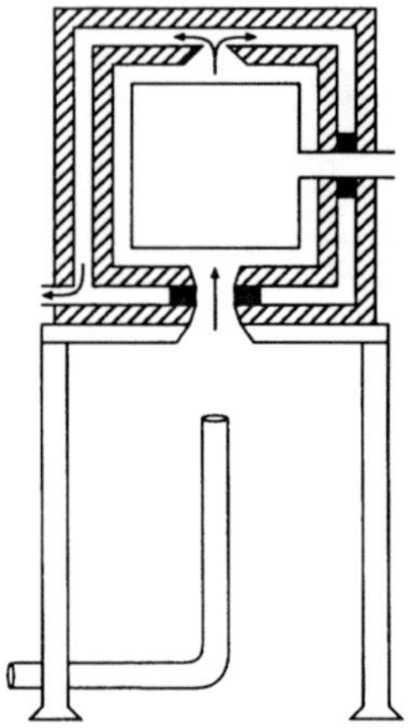

Fig. 9.8 A schematic of the Lummer and Pringsheim device.

The power of the thermal radiation issuing from the cavity was measured by a bolometer. In this device, the radiation to be measured falls on to one of the two blackened platinum wires. By absorbing the radiation, the wire becomes hot and its ohmic resistance grows. The change in resistance is then measured by comparison with the resistance of the other wire.

All possible precautions were taken to make sure that the registered power was indeed radiated only by the heated cavity. The measurement results were corrected to allow for the possible absorption of the radiation by the air on the way from the cavity to the sensor. A spectroscopic prism was used for spectral resolution of the radiation.

The results of this research were published in 1899–1900 and are shown in Fig. 9.2. Let us now turn to the theoretical calculation of the spectral density of the equilibrium thermal radiation $u(\omega, T)$.

9.6 The thermodynamics of equilibrium thermal radiation

The function $u(\omega, T)$ can be calculated on the basis of the *thermodynamical law of energy equipartition*. According to this law, the same mean energy $kT/2$ falls on each degree of freedom of the system in thermal equilibrium. Here T is the absolute temperature of the system and k is the Boltzmann constant. To apply this law to thermal radiation, we need to calculate the number of degrees of freedom for an electromagnetic field contained inside a closed cavity.

The main idea of the calculation is to treat the electromagnetic field within the cavity as a *set of standing waves*. In fact, we need to perform spectral expansion of the spatial structure of the equilibrium radiation into simple elementary structures; that is, sinusoidal standing waves.

The following two conditions are essential. First, the spatial structure of the equilibrium thermal radiation must be steady-state and, second, the energy of the radiation incident on the cavity walls must be equal to the energy of the

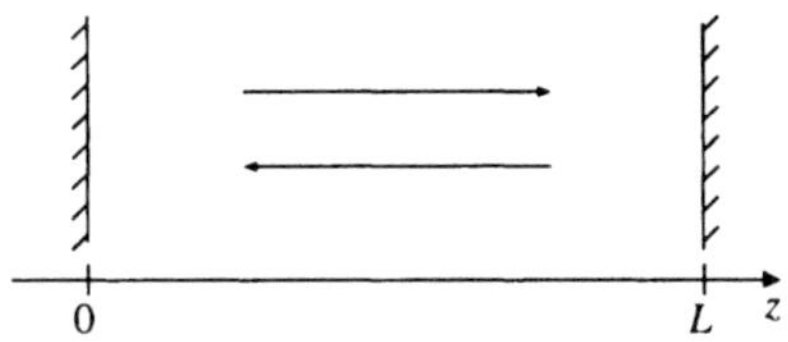

Fig. 9.9 An optical resonator as an elementary system allowing one to calculate the number of the optical field's degrees of freedom.

radiation emitted by the walls. The simplest system that meets both of these requirements is a pair of parallel mirrors, i.e. an *optical resonator* (Fig. 9.9). It is known that the optical field in such a resonator can be regarded as a discrete numerable set of standing waves ("field oscillators"). The number of different standing waves is the sought-after number of the field's degrees of freedom.

Let us consider an optical field confined between two parallel plane mirrors. Let L be the distance between the mirrors and z the axis perpendicular to the mirror surfaces. The optical field in the plane resonator satisfies the one-dimensional wave equation

$$\frac{\partial^2 E}{\partial z^2} - \frac{1}{c^2}\frac{\partial^2 E}{\partial t^2} = 0. \tag{9.9}$$

Assuming the mirrors to be ideal conductors, we can write the boundary conditions in the form

$$E(0,t) = E(L,t) = 0. \tag{9.10}$$

The solution to eqns (9.9) and (9.10) is

$$E(z,t) = A(t)\sin k_z z. \tag{9.11}$$

Inserting eqn (9.11) into eqn (9.9), we obtain an equation for the amplitude $A(t)$:

$$\frac{\partial^2 A}{\partial t^2} + c^2 k_z^2 A = 0. \tag{9.12}$$

The amplitude of the field is thus governed by the harmonic oscillator equation. The solution to this equation is

$$A(t) = A_0 \cos(\omega t + \varphi), \qquad \omega = ck_z. \tag{9.13}$$

On the other hand, inserting eqn (9.11) into eqn (9.10), we obtain

$$\sin k_z L = 0, \tag{9.14}$$

whence

$$k_z L = m_z \pi, \qquad m_z = 1, 2, 3, \ldots\ . \tag{9.15}$$

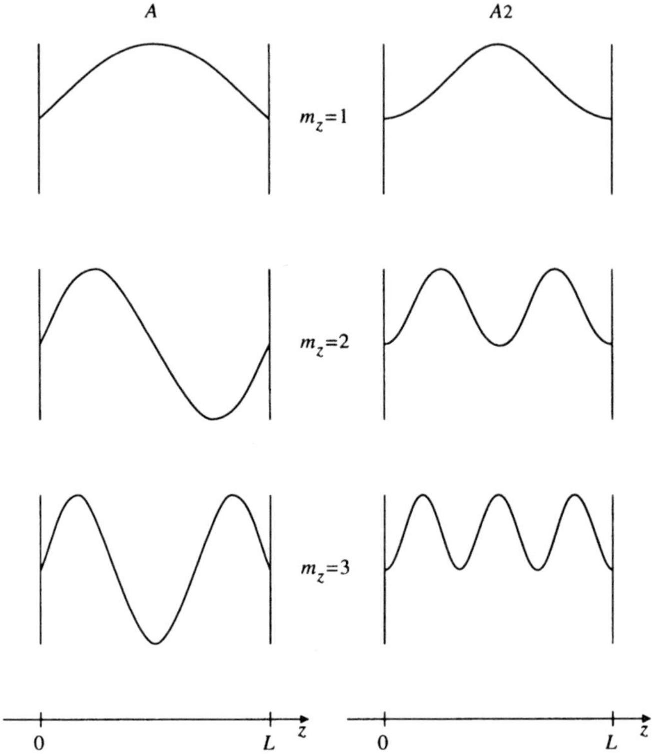

Fig. 9.10 Examples of the spatial structures of field oscillators in an optical resonator.

Correspondingly,

$$\omega = ck_z = m_z \frac{\pi c}{L}, \qquad m_z = 1, 2, 3, \ldots \; . \tag{9.16}$$

Thus, the optical field in the resonator has the structure of a discrete set of waves. Each standing wave characterized by a definite spatial distribution of light intensity can be called a *field oscillator*. A number of the most simple structures are shown in Fig. 9.10. Importantly, the field oscillators in the resonator form a numerable set, i.e. they can be numbered. As different oscillators are independent and their amplitudes are random, the total number of oscillators is the number of the *field's degrees of freedom*.

According to eqn (9.15) each oscillator occupies, in the space of wave numbers k_z, a cell the size of which is $\Delta k = \pi/L$ (Fig. 9.11), where L is the resonator length.

Generalizing to the case of three spatial coordinates, we shall consider the optical field within a cubic mirror cavity (Fig. 9.12). Let L be the length of the cube edge, let $V = L^3$ be the volume of the cube, and let the coordinate axes be

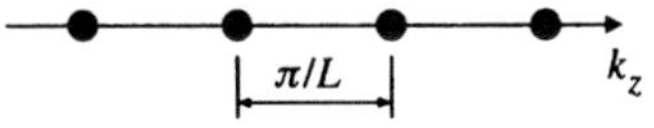

Fig. 9.11 Field oscillators in wave number space.

directed along the cube edges.

Each of the Cartesian components of the field, E_x, E_y, and E_z, obeys the wave equation

$$\Delta E - \frac{1}{c^2}\frac{\partial^2 E}{\partial t^2} = 0, \tag{9.17}$$

where

$$\Delta = \frac{\partial^2}{\partial x^2} + \frac{\partial^2}{\partial y^2} + \frac{\partial^2}{\partial z^2}$$

is the Laplacian operator. We shall seek a solution to eqn (9.17) in the form

$$E(x, y, z, t) = A_0 \sin(k_x x)\sin(k_y y)\sin(k_z z)\cos(\omega t + \varphi). \tag{9.18}$$

Substitution of eqn (9.18) into eqn (9.17) leads to the equation

$$k_x^2 + k_y^2 + k_z^2 \equiv k^2 = \omega^2/c^2, \tag{9.19}$$

relating the modulus of the wave number k and the frequency ω of the optical wave (the so-called "dispersion relation"). The boundary conditions lead to the following relations (cf. eqn (9.15)):

$$k_x L = m_x \pi, \qquad k_y L = m_y \pi, \qquad k_z L = m_z \pi, \tag{9.20}$$

where m_x, m_y, and m_z are integers, i.e.

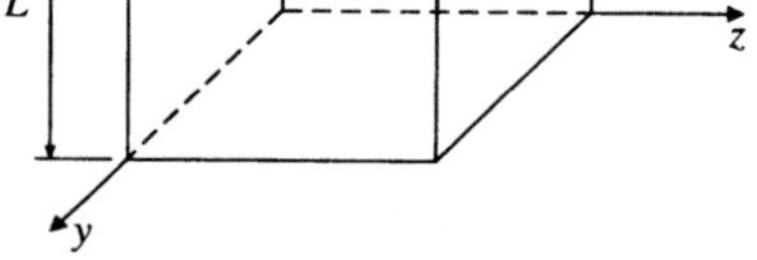

Fig. 9.12 Calculating the degrees of freedom of the optical field filling up a three-dimensional cavity.

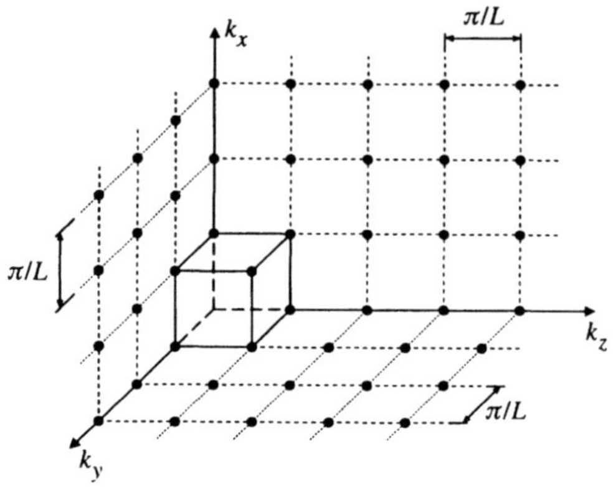

Fig. 9.13 Field oscillators shown in wave number space (three dimensions).

$$m_x, m_y, m_z = 1, 2, 3, \dots \,. \tag{9.21}$$

The quantities k_x, k_y, and k_z can be regarded as the Cartesian components of the wave vector $\vec{k}$ of the optical wave.

Thus, each oscillator of the three-dimensional field can be characterized by a triad of integers, m_x, m_y, and m_z. Figure 9.13 illustrates the field oscillators in the k-space. One can see that the k-space is subdivided into elementary cubes, each of which corresponds to an individual field oscillator. The volume of each elementary cube is

$$v_k = (\pi/L)^3. \tag{9.22}$$

It is now easy to calculate the total number of the field's degrees of freedom. Consider the optical field occupying the frequency range from 0 to ω. According to the dispersion relation (9.19), the wave numbers of the field move into the range from 0 to $k = \omega/c$. In the k-space this region turns into a sphere of radius k and volume $4\pi k^3/3$. According to eqn (9.18), the change in the signs of the quantities k_x, k_y, and k_z to the opposite leads only to alteration of the phase of the optical field, but does not change its spatial structure (see Fig. 9.10). Consequently, only the numbers k_x, k_y, and k_z with definite signs, for example,

$$k_x,\ k_y,\ k_z > 0. \tag{9.23}$$

correspond to the physically distinguishable configurations of the field; that is, to the field oscillators. The inequalities (9.23) describe the positive octant of the wave number space, into which only 1/8 of the sphere protrudes. The volume of this fraction of the sphere is

$$V_k = \frac{1}{8}\frac{4}{3}\pi k^3. \tag{9.24}$$

The sought-after number of field oscillators is apparently equal to the ratio of the volume V_k, which all field oscillators occupy in the k-space, to the volume v_k of the elementary cell corresponding to an individual oscillator. Denoting this number by Z', we can write

$$Z' = V_k/v_k. \tag{9.25}$$

Using eqns (9.19), (9.22), and (9.24), we can write the same number as

$$Z' = \frac{\omega^3 L^3}{6\pi^2 c^3}. \tag{9.26}$$

Thus, eqn (9.26) determines the number of field oscillators occupying the region of space of volume L^3 and the frequency range from 0 to ω. This formula yields the number of the field's degrees of freedom associated with all possible directions of propagation of optical waves. However, a wave propagating in a definite direction may have two independent polarization states. For example, a plane wave travelling along the z-axis is generally a superposition of waves of E_x, H_y type and E_y, H_x type (see Chapter 2). Consequently, the total number of degrees of freedom of the optical field is twice the number Z' and equals

$$Z = \frac{\omega^3 L^3}{3\pi^2 c^3}. \tag{9.27}$$

Now let us calculate the number of degrees of freedom falling within the spectral range from ω to $\omega + d\omega$. Taking the differential of both sides of eqn (9.27), we obtain

$$dZ = \frac{\omega^3 L^3}{\pi^2 c^3} d\omega. \tag{9.28}$$

Thus, the problem of calculation of the number of degrees of freedom of equilibrium thermal radiation is solved. Using this result, we can present the spectral density of equilibrium thermal radiation, i.e. the energy of radiation per unit volume of space and per spectral interval from ω to $\omega + d\omega$, in the form

$$u(\omega, T) = \frac{\omega^2}{\pi^2 c^3} \langle w \rangle, \tag{9.29}$$

where $\langle w \rangle$ is the mean energy corresponding to the individual field oscillator.

9.7 The Rayleigh–Jeans formula

According to the law of equipartition of energy, the mean energy of an oscillator in a thermal equilibrium state is

$$\langle w \rangle = kT, \tag{9.30}$$

where k is Boltzmann's constant and T is the absolute temperature of the system. For a mechanical oscillator the mean kinetic energy is equal to the mean potential

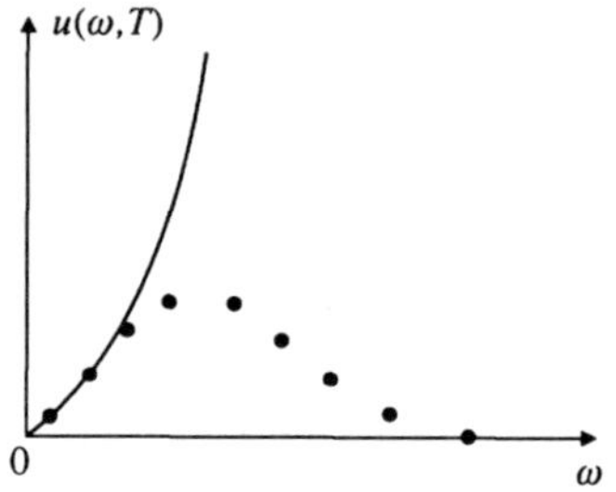

Fig. 9.14 The spectral density of equilibrium thermal radiation. •, Experimental data; —, calculated from the Rayleigh–Jeans formula.

energy, or to $kT/2$, while for the field oscillator the mean energy of the electric field is equal to the mean energy of the magnetic field, or to $kT/2$. In both cases the total mean energy of the oscillator is thus equal to kT.

Inserting eqn (9.30) into eqn (9.29), we obtain

$$u(\omega, T) = \frac{\omega^2}{\pi^2 c^3} kT. \tag{9.31}$$

This expression for the spectral density of equilibrium thermal radiation is known as the *Rayleigh–Jeans formula.*

Figure 9.14 shows the experimentally measured distribution $u(\omega, T)$ together with the theoretical curve calculated from the Rayleigh–Jeans formula. Since there is a qualitative difference between experimental and theoretical data, the Rayleigh–Jeans formula cannot be valid over the whole frequency range. At long wavelengths, however, the formula agrees well with the experiment. This fact, as well as the plain and obvious assumptions underlying the formula, allow us to assert that it must be valid at low frequencies. When found, a more general formula would describe the spectral density throughout the whole frequency range and turn into the Rayleigh–Jeans formula en route to the long-wavelength limit.

9.8 The "ultraviolet catastrophe"

The primary drawback of the Rayleigh–Jeans formula lies in its predicting an unlimited growth in the spectral density of radiation at high frequencies. This prediction is in sharp conflict with the experimental data, according to which the spectral density tends to zero when $\omega \to \infty$ (see Figs 9.2 and 9.14). Additionally, it follows from the Rayleigh–Jeans formula that the total (integral over the frequency) energy of the thermal radiation is equal to infinity,

$$\int_0^\infty u(\omega, T)\, d\omega = \infty, \tag{9.32}$$

which, of course, defies common sense. Since this circumstance reveals itself in the shortwave (ultraviolet) region of the spectrum, it has been named the *ultraviolet catastrophe*, or the *Rayleigh–Jeans paradox*.

9.9 Planck's formula

By the end of the year 1900, research into equilibrium thermal radiation had led to the following situation. Some precise experimental measurements of the spectral radiation density had been performed by Lummer and Pringsheim, and by Rubens, Kurlbaum, and coworkers (see Fig. 9.2). At that time the Rayleigh–Jeans formula (9.31) was known correctly to describe the low-frequency part of the curve $u(\omega, T)$. Also known at that time was the Wien formula

$$u(\omega, T) \sim \omega^3 \mathrm{e}^{-\gamma\omega/T}, \qquad \gamma = \text{constant}, \tag{9.33}$$

which agrees well with experiment in the range of medium and high frequencies. Thus, the asymptotic behaviour of the distribution $u(\omega, T)$ was known:

$$u(\omega, T) \sim \begin{cases} \omega^2 T, & \omega \to 0 \quad \text{(Rayleigh–Jeans law)} \\ \omega^3 \mathrm{e}^{-\gamma\omega/T}, & \omega \to \infty \quad \text{(Wien's law).} \end{cases} \tag{9.34}$$

In an attempt to construct an expression for the function $u(\omega, T)$ that would be applicable throughout the whole frequency range, Max Planck derived several formulae to meet the conditions (9.34). The elegance of one of the formulae amazed him. In modern notation, this formula takes the form

$$u(\omega, T) = \frac{1}{\pi^2 c^3} \frac{\hbar\omega^3}{\mathrm{e}^{\hbar\omega/kT} - 1} \tag{9.35}$$

and is known as *Planck's formula*. In this formula $\hbar$ is a constant, the numerical value of which is

$$\hbar = 1.05 \times 10^{-27}\,\text{erg s}. \tag{9.36}$$

This constant is called *Planck's constant.*

Very soon eqn (9.35) proved to be in excellent agreement with experiment. This raised questions about the justification for this formula and clarification of its physical sense. Max Planck began to solve the problem.

9.10 The necessity of a quantum-mechanical approach

Planck soon came to a discouraging conclusion. He discovered that eqn (9.35) can be derived under the assumption that the atoms constituting the walls of a heated cavity, filled with equilibrium thermal radiation, emit and absorb the light of a given frequency ω not in arbitrary amounts, but only in discrete portions ("quanta") having an energy

$$w_0 = \hbar\omega. \tag{9.37}$$

This assumption was in sharp contradiction with the ideas of classical physics which dominated at that time, according to which the energy of any physical

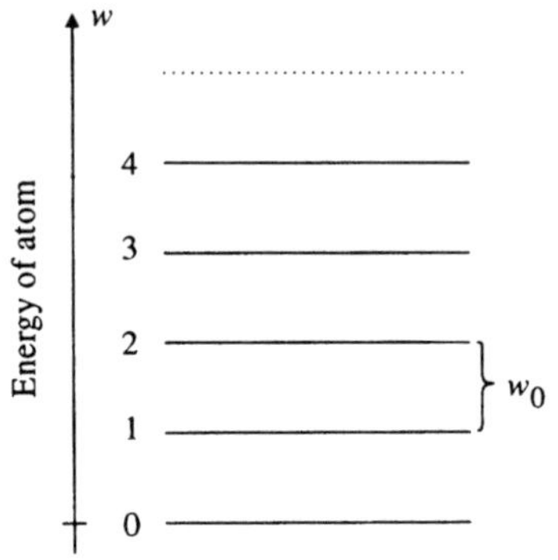

Fig. 9.15 Atomic energy levels in Planck's model.

system could vary continuously. One can easily see, however, that this assumption resolves the Rayleigh–Jeans paradox. Indeed, eqn (9.37) shows that the high-frequency radiation can be emitted only in sufficiently large portions, that under the condition $\hbar\omega \gg kT$ there would be no atoms in the cavity walls capable of emitting high-frequency quanta; the thermal energy of atoms having the order of kT would be too low to make the transition. This means that in the limiting case $\omega \to \infty$ the spectral density of radiation should tend to zero, in total agreement with the experimental data. On the other hand, at low frequencies, when $\hbar\omega \ll kT$, the discrete nature of emission and absorption does not play a significant role and the ideas of classical theory become valid.

Despite the bizarre impression that Planck's hypothesis initially made on the scientific community, strong arguments in its favour started to appear almost immediately. Further developments in physics have conclusively confirmed Planck's quantum ideas. The consequences of Planck's discovery have proved to be so far-reaching that it can rightfully be listed among the most important discoveries in the history of physics.

Let us discuss briefly how eqn (9.37) can be derived from eqn (9.35). Suppose that the atom interacting with radiation varies its energy not continuously, but in some discrete portions w_0. Then we can number the possible values of the energy of the atom, by writing

$$w_n = n w_0, \qquad n = 0, 1, 2, \dots . \tag{9.38}$$

We can also depict the energy levels in the form of a "staircase", as shown in Fig. 9.15.

It should be noted that the assumption made does not imply that the classical model has to be discarded if we pass to the limit $w_0 \to 0$.

The probability of finding the atom at the nth energy level is determined by the Boltzmann distribution

$$P(n) = C\,\mathrm{e}^{-n w_0/kT} \tag{9.39}$$

or

$$P(n) = C\,\mathrm{e}^{-nx}, \tag{9.40}$$

where the designation

$$x = w_0/kT \tag{9.41}$$

is introduced. In eqns (9.39)–(9.41), $n = 0, 1, 2, 3, \ldots$; w_0 is the energy element, T is the absolute temperature of the system, k is the Boltzmann constant, and C is a constant determined from the normalization condition

$$\sum_{n=0}^{\infty} P(n) = 1. \tag{9.42}$$

Substituting eqn (9.40) into eqn (9.42), we obtain

$$C^{-1} = \sum_{n=0}^{\infty} \mathrm{e}^{-nx} = \frac{1}{1 - \mathrm{e}^{-x}}, \tag{9.43}$$

whence

$$C = 1 - \mathrm{e}^{-x}. \tag{9.44}$$

In forming the summation in eqn (9.43), we used the known formula for the sum of the geometric progression

$$\sum_{n=0}^{\infty} q^n = \frac{1}{1-q}, \qquad q < 1. \tag{9.45}$$

The mean energy of the atom is determined by the expression

$$\langle w \rangle = \sum_{n=0}^{\infty} w_n\, P(n). \tag{9.46}$$

Inserting eqns (9.38) and (9.40) into eqn (9.46), we obtain

$$\langle w \rangle = C w_0 \sum_{n=0}^{\infty} n\mathrm{e}^{-nx}. \tag{9.47}$$

The sum of the series in eqn (9.47) can be calculated by differentiating eqn (9.43) with respect to the parameter

$$\sum_{n=0}^{\infty} n\mathrm{e}^{-nx} = -\frac{\partial}{\partial x} \sum_{n=0}^{\infty} \mathrm{e}^{-nx} = -\frac{\partial}{\partial x}\left(\frac{1}{1 - \mathrm{e}^{-x}}\right) = \frac{\mathrm{e}^{-x}}{(1 - \mathrm{e}^{-x})^2}. \tag{9.48}$$

Inserting eqns (9.48) and (9.44) into eqn (9.47), we find that

$$\langle w \rangle = \frac{w_0}{e^x - 1} \tag{9.49}$$

or, by virtue of eqn (9.41),

$$\langle w \rangle = \frac{w_0}{e^{w_0/kT} - 1}. \tag{9.50}$$

Thus, eqn (9.50) determines the mean energy of the atom in the Planck model of discrete energy levels. Note that according to this expression the mean thermal energy of an elementary oscillator (atom) of a medium is by no means equal to kT, as required by the law of equipartition of energy. The expression $\langle w \rangle = kT$ follows from eqn (9.50) only in the limiting case $w_0 \to 0$, i.e. when the Planck energy element w_0 tends to zero.

Planck modelled an atom of the medium by means of an harmonic oscillator. By virtue of the law of equipartition of energy, the mean energy of the oscillator (atom) of the substance in the thermal equilibrium state should be equal to the mean energy of the field oscillator

$$\langle w \rangle_{\text{substance oscillator}} = \langle w \rangle_{\text{field oscillator}}. \tag{9.51}$$

This serves as a justification for inserting eqn (9.50) into eqn (9.29) describing the spectral density of equilibrium thermal radiation. As a result we obtain

$$u(\omega, T) = \frac{\omega^2}{\pi^2 c^3} \frac{w_0}{e^{\omega_0/kT} - 1}. \tag{9.52}$$

In the limiting case $w_0 \to 0$ corresponding to the classical (continuous-energy) model of the atom, eqn (9.52) turns into the Rayleigh–Jeans formula (9.31). However, the comparison of eqn (9.52) with Planck's formula (9.35), which ideally describes the experiment, leads to the conclusion that the Planck energy element w_0 is not infinitesimal. It is a finite quantity proportional to the radiation frequency ω. If we make use of eqn (9.37) to define w_0, then eqn (9.52) turns into eqn (9.35).

Thus, as a result of studies on thermal radiation, it has been established that the atom possesses a *discrete set of energy levels*. This conclusion, confirmed by all subsequent investigations, has served a starting point for the development of quantum physics.

9.11 Thermal radiation laws

Before Planck's discovery, the two laws describing the properties of thermal radiation were known. These are the *Stefan–Boltzmann law* and *Wien's displacement law*. Here we shall derive these laws as consequences of Planck's formula (9.35).

9.11.1 Stefan–Boltzmann law

According to this law, the total power of thermal radiation grows in proportion to the fourth degree of the absolute temperature of the body. It follows from eqn (9.35) for the energy density (energy per unit volume) of the equilibrium thermal radiation that

$$\frac{dW}{dV} = \int_0^\infty u(\omega, T) dw = \frac{\hbar}{\pi^2 c^3} \int_0^\infty \frac{\omega^3 \, d\omega}{e^{\hbar\omega/kT} - 1}$$
$$= \frac{\hbar}{\pi^2 c^3} \left(\frac{kT}{\hbar}\right)^4 \int_0^\infty \frac{x^3 \, dx}{e^x - 1} = \text{constant} \cdot T^4. \quad (9.53)$$

Thus, Planck's formula confirms the Stefan–Boltzmann law.

Using the value of the integral

$$\int_0^\infty \frac{x^3 \, dx}{e^x - 1} = \pi^4/15, \quad (9.54)$$

we can write the Stefan–Boltzmann law in the form

$$\frac{dW}{dV} = aT^4, \quad (9.55)$$

where

$$a = \frac{\pi^2 k^4}{15 \hbar^3 c^3} = 7.55 \times 10^{-15} \, \text{erg} \, \text{cm}^{-3} \, \text{K}^{-4}. \quad (9.56)$$

For calculation of the constant a we used the known values of physical constants: the Boltzmann constant $k = 1.38 \times 10^{-16}$ erg K^{-1}, Planck's constant $\hbar = 1.05 \times 10^{-27}$ erg s, the velocity of light $c = 3 \times 10^{10}$ cm s^{-1}, and $\pi = 3.14$.

9.11.2 Wien's displacement law

According to this law, the maximum density of thermal radiation is observed at a wavelength λ_{max}, which is inversely proportional to the absolute temperature of the body:

$$\lambda_{\text{max}} = \frac{\text{constant}}{\text{T}}. \quad (9.57)$$

To derive this law, we shall express the spectral density of equilibrium thermal radiation in terms of the radiation wavelength λ. This can be done by using the relation between frequency and wavelength, $\omega = 2\pi c/\lambda$, and applying the rule for the probability density transformation

$$u(\omega, T)|d\omega| = u(\lambda, T)|d\lambda|. \quad (9.58)$$

Figure 9.16 clarifies the meaning of eqn (9.58).

It follows from eqn (9.58) that

$$u(\lambda, T) = u(\omega(\lambda), T) \left| \frac{d\omega(\lambda)}{d\lambda} \right|. \quad (9.59)$$

Inserting eqn (9.35) into eqn (9.59), we find that

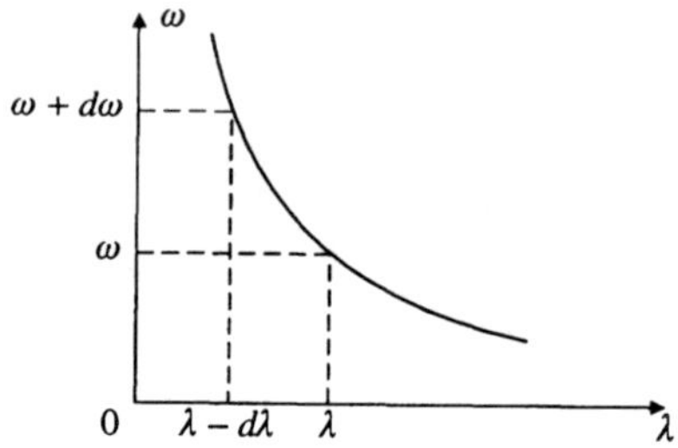

Fig. 9.16 The substitution of variables in the probability distribution.

$$u(\lambda, T) = \frac{8\pi ch}{\lambda^5} \frac{1}{e^{hc/\lambda kT} - 1}, \tag{9.60}$$

where $h = 2\pi\hbar$. A form of the distribution $u(\lambda, T)$ is shown in Fig. 9.17.

The sought-after wavelength can be found from the condition

$$\frac{\partial u(\lambda, T)}{\partial \lambda} = 0. \tag{9.61}$$

Substituting eqn (9.60) into eqn (9.61), we obtain

$$\frac{x e^x}{e^x - 1} = 5, \tag{9.62}$$

where

$$x = \frac{hc}{kT\lambda_{\mathrm{max}}}. \tag{9.63}$$

The solution to eqn (9.62) can be found numerically or graphically to yield

$$x = 4.965\,. \tag{9.64}$$

Consequently,

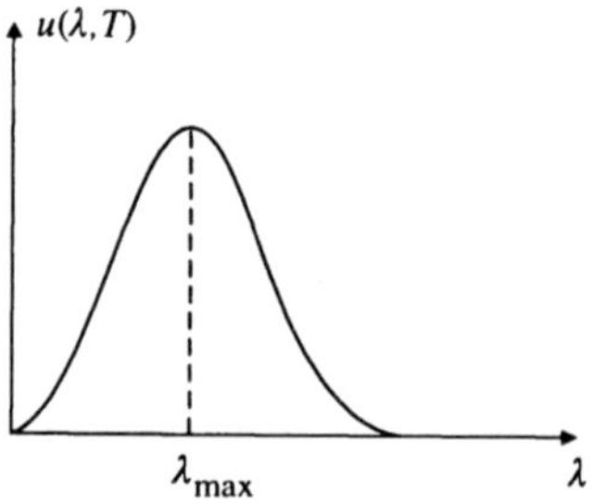

Fig. 9.17 The derivation of the Wien displacement law. The spectral density of equilibrium thermal radiation as a function of radiation wavelength.

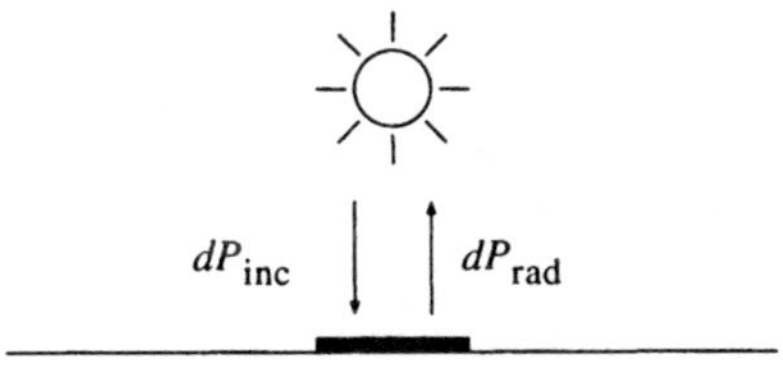

Fig. 9.18 The calculation of the limiting temperature of a black body heated by sunlight.

$$\lambda_{\max} = \frac{hc}{4.965kT} = \frac{0.29\,\mathrm{cm}}{T(\mathrm{K})}, \tag{9.65}$$

in accordance with Wien's displacement law.

9.12 Examples

Let us consider two examples of application of the thermal radiation laws.

1. *Evaluate the temperature of the Sun's surface, knowing that the maximum intensity of the solar radiation is observed at a wavelength of about 500 nm.*

Assuming the Sun to be an absolute black body and $\lambda_{\max} = 500$ nm, we obtain $T \approx 6000$ K from eqn (9.65). The Sun's surface temperature is thus about 6000 kelvins.

2. *Estimate the temperature to which sunlight can heat a black body on the Earth's surface.*

Suppose that the heating of the body is limited only by its own emission of thermal radiation. Then the thermal balance condition for a small element of the body surface $d\sigma$ (Fig. 9.18) can be obtained by equating the sunlight power dP_{inc} illuminating the given surface element and the power of its own thermal radiation dP_{rad}:

$$dP_{\mathrm{inc}} = dP_{\mathrm{rad}}. \tag{9.66}$$

The quantities entering eqn (9.66) can be written in the form

$$dP_{\mathrm{inc}} = I_{\mathrm{S}}\, d\sigma, \tag{9.67}$$

where I_{S} is the solar radiation intensity, and

$$dP_{\mathrm{rad}} = d\sigma \int_0^{\infty} \rho(\omega, T)\, d\omega, \tag{9.68}$$

where $\rho(\omega, T)$ is the emissivity of the perfectly black body. This quantity is related to the spectral density of the equilibrium thermal radiation $u(\omega, T)$ by the formula (see Chapter 3):

$$\rho(\omega, T) = \frac{c}{4} u(\omega, T), \tag{9.69}$$

where c is the velocity of light. Thus, using Planck's formula (9.35) and the Stefan–Boltzmann law,

$$\int_0^\infty u(\omega, T)\, d\omega = aT^4, \tag{9.70}$$

we obtain

$$\frac{dP_{\mathrm{rad}}}{d\sigma} = \kappa T^4, \tag{9.71}$$

where

$$\kappa = \frac{c}{4} a, \tag{9.72}$$

and the constant a is determined by eqn (9.56). The quantity dP_{rad} is called the *radiant exitance of the body*. Equation (9.71) expresses the Stefan–Boltzmann law in terms of this quantity.

Inserting eqn (9.56) into eqn (9.72), we obtain

$$\kappa = 5.66 \times 10^{-12}\,\mathrm{W\,cm^{-2}\,K^{-4}}. \tag{9.73}$$

and inserting eqns (9.67) and (9.71) into eqn (9.66), we find that

$$T = \sqrt[4]{I_{\mathrm{S}}/\kappa}. \tag{9.74}$$

The last formula determines the limiting temperature to which a black body can be heated by light. It is known that the solar radiation intensity near the Earth's surface is about

$$I_{\mathrm{S}} \approx 0.1\,\mathrm{W\,cm^{-2}}. \tag{9.75}$$

Substituting eqns (9.73) and (9.75) into eqn (9.74), we obtain the numerical estimate

$$T = 365\,\mathrm{K} = 92^\circ\mathrm{C}. \tag{9.76}$$

The experiment confirms the obtained result. Note that by concentrating sunlight using mirrors or lenses one can raise the body temperature considerably. This property is used in solar power devices.

Bibliography

Lummer, O. and Pringsheim, E. (1900). *Verhandlungen der Deutsche Physikalische Gesellschaft,* **2**, 163–180.

Planck, M. (1900). *Verhandlungen der Deutsche Physikalische Gesellschaft,* **2**, 202–204.

Trigg, G. L. (1971). *Crucial experiments in modern physics.* Van Nostrand Reinhold, New York.

10
The laser

Quantum properties of atoms and Bohr's postulates. Optical quanta. Spontaneous and stimulated emission in a quantum-mechanical system. The operational principle and design of a laser. The characteristics of laser radiation. The generation of ultra-short pulses.

It is hard to find a discovery that has affected the science and technology of the 20th century as much as the laser has. Superstrong optical fields and laser fusion, ultra-short optical pulses and ultralow temperatures, non-linear optics and laser spectroscopy, holography and optical communications, treatment of materials and the optical computer, laser medicine and environmental monitoring—this is only a concise list of laser applications, which continues to grow and to embrace useful new aspects. From the principal physical point of view, the laser demonstrates a new kind of radiation; that is, *stimulated emission* in a quantum system. The primary property of this radiation is its *coherence*. It implies that the laser radiation has a structure very similar to that of a plane monochromatic wave.

10.1 Quantum properties of atoms and Bohr's postulates

The investigation of thermal radiation, the discovery of energy quanta, and studies of the emission and absorption spectra of atoms, as well as other experimental and theoretical research performed at the end of the 19th century and the beginning of the 20th century, made it possible to establish the following two postulates that characterize the properties of atomic systems discovered by Niels Bohr in 1913.

1. Atoms may exist in steady states in which they neither emit nor absorb the light. The energies of these states form a discrete set W_1, $W_2, \ldots$ (energy levels).

2. The radiation emitted or absorbed by the atom when it transits from the energy level W_m to W_n is monochromatic and its frequency obeys the condition

$$W_m - W_n = \hbar\omega, \tag{10.1}$$

where $\hbar$ is Planck's constant.

These postulates sharply contradict the basic concepts of classical mechanics and electrodynamics, but prove to be in complete agreement with experimental

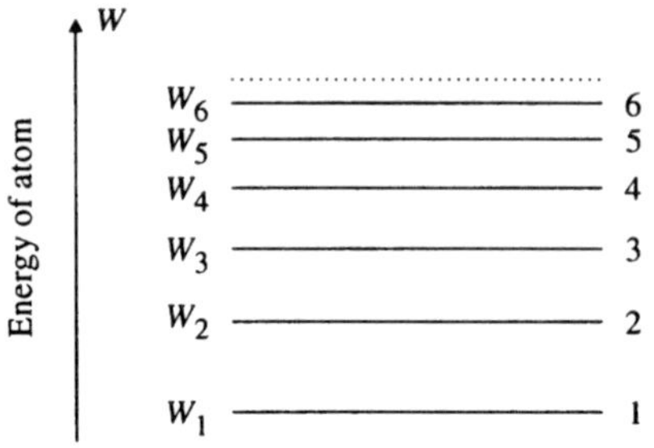

Fig. 10.1 Atomic energy levels.

findings. Bohr's postulates can be considered to be a brief and accurate description of the behaviour of atomic systems, established by the analysis and generalization of experimental data.

Figure 10.1 illustrates a scheme of atomic energy levels. Each level is characterized by its number n and its energy W_n.

10.2 Optical quanta

Even the Planck formula itself,

$$W = \hbar\omega, \tag{10.2}$$

seems to state that the concept of an energy quantum should be related to light because the only physical variable on the right-hand side of eqn (10.2), that is, the frequency ω, characterizes light and not the substance. However, quite a number of years had passed after Planck's discovery before the concept of a "quantum of light" was adopted.

In 1905, Einstein wrote in his paper "An heuristic point of view on the emission and transformation of light" that: "... It may happen that the theory of light operating with continuous spatial functions would lead to a contradiction with experiment when applied to the phenomena of emission and transformation of light". He went on to say that: "...According to the assumption made, the energy of an optical beam emerging from a point does not spread continuously in ever extending volume, but comprises a finite number of spatially localized indivisible energy quanta absorbed or generated only as a whole. Below I set forth a chain of arguments and facts that made me suggest this idea, which hopefully can be of advantage to other researchers".

In this paper, Einstein examines the density of thermal radiation and concludes that at high frequencies it fluctuates as the energy density of a gas of independent particles, each of which has the energy $\hbar\omega$. Einstein writes: "If a monochromatic radiation ... behaves as a discrete medium consisting of energy quanta $\hbar\omega$, then one can hypothesize that the laws of emission and transformation of light are such as if the light consists of the similar energy quanta".

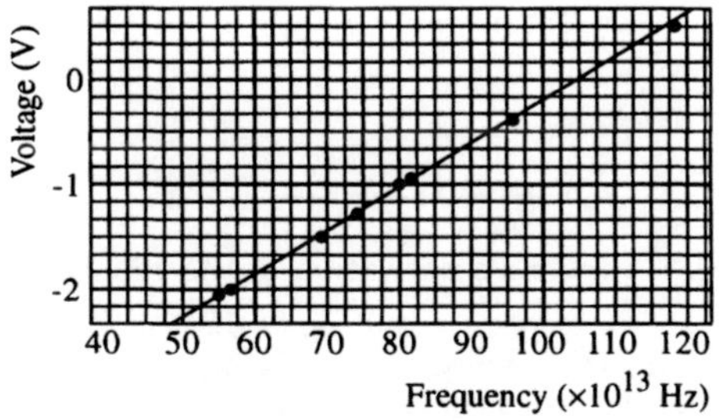

Fig. 10.2 The maximum energy of sodium photoelectrons (measured using a retarding voltage) as a function of the incident light frequency (from Millikan, R. A. (1916). *Physical Review*, **7**, 355).

10.2.1 The photoeffect

An important argument in favour of the hypothesis of optical quanta has been provided by investigations of the photoeffect, discovered by Hertz in 1887 while studying the propagation of electromagnetic waves. When Hertz screened a receiving vibrator with a metal sheet, in order to see a spark discharge better, he observed that the screen affects the conditions of discharge formation by obstructing the light coming from the spark of an emitting vibrator. Hertz examined this phenomenon and showed that when the screen was illuminated by an electric arc light, the ionization of air near the screen increased so that the discharge occurred at a lower voltage.

In 1888, Stoletov studied the photoeffect in more detail to establish that a metal plate irradiated by light issues a flux of negatively charged particles producing an electric current, the magnitude of which is proportional to the incident radiation intensity. In 1902, Lennard discovered that the energy of photoelectrons does not depend on the incident radiation intensity.

Using the hypothesis of optical quanta, Einstein obtained the following formula for the photoeffect:

$$W_{\max} = \hbar\omega - A, \qquad (10.3)$$

where $W_{\max}$ is the maximum energy of the photoelectrons and A is the work function, i.e. the energy required to remove an electron from the metal. Equation (10.3) predicts that the graph of the function $W_{\max}(\omega)$ is a straight line, the slope of which is independent of the substance and is governed by Planck's constant. The research carried out by Millikan in 1915 confirmed this conclusion (Fig. 10.2).

10.2.2 Compton's effect

In 1922, the US physicist A. H. Compton reported an increase in the wavelengths of X-rays due to scattering by free electrons. This effect proved experimentally that the optical quantum possesses the momentum

$$p = \hbar k, \qquad (10.4)$$

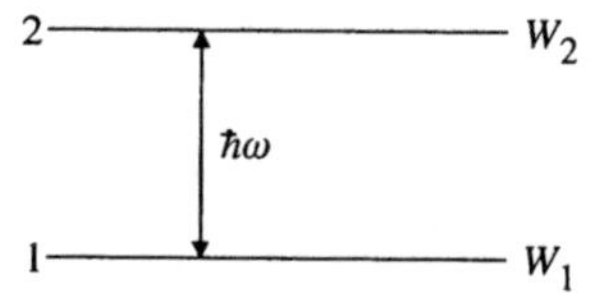

Fig. 10.3 Einstein's derivation of the Planck formula.

where $k = \omega/c$, ω is the frequency, and c is the velocity of light. Thereafter, optical quanta ceased to be hypothetical particles.

Thus, the *optical quantum*, or the *photon*, is an elementary indivisible portion of light, possessing an energy $W = \hbar\omega$ and a momentum $p = \hbar k$. Note that eqn (10.4) can be derived from eqn (10.2) and the formula

$$\mathcal{P} = I/c, \tag{10.5}$$

relating the energy flux and the flow of momentum in an optical wave (see Chapters 2 and 3). In eqn (10.5), I is the intensity of light and is $\mathcal{P}$ the pressure of light exerted on the absorbing body. The relation between the energy and the momentum of the light,

$$W = pc, \tag{10.6}$$

also follows from the Einstein theory of relativity where the given formula describes zero-mass particles moving at the velocity of light in any inertial frame of reference.

10.3 Spontaneous and stimulated emission in a quantum-mechanical system

A significant step towards the creation of a laser was taken by Einstein in 1916. In his article "On the quantum theory of radiation", Einstein considered the interaction of atoms with equilibrium thermal radiation in terms of the optical quanta hypothesis. Having introduced the notions of spontaneous and stimulated emission in a quantum-mechanical system, he found a surprisingly simple way to derive the Planck formula.

Consider a pair of atomic energy levels in a substance which is in equilibrium with the thermal radiation at temperature T. Let the subscript 1 be attributed to the lower energy level and the subscript 2 to upper level (Fig. 10.3). Suppose that N_1 is the number of atoms with energy W_1 and N_2 is the number of atoms with energy W_2. In the state of thermal equilibrium the level populations N_1 and N_2 obey the Boltzmann distribution:

$$N_i = \text{constant} \cdot \mathrm{e}^{-W_i/kT}. \tag{10.7}$$

The given pair of levels interacts with the radiation the frequency of which satisfies the Bohr condition:

$$W_2 - W_1 = \hbar\omega. \tag{10.8}$$

Following Einstein, we shall suppose that the next three kinds of radiation processes are possible.

1. *Spontaneous emission.* An optical quantum is emitted when the atom transits spontaneously from state "2" to state "1". The probability of this process is proportional to the number of atoms at the upper energy level and, consequently, we can write

$$P_{21}^{\text{spont}} = A_{21}N_2, \tag{10.9}$$

where A_{21} is the Einstein coefficient.

2. *Stimulated absorption.* The atom absorbs an optical quantum and transits from state "1" to state "2". The probability of this radiation-induced process is proportional to the spectral density of radiation at the frequency ω and to the number of atoms at the lower energy level. We can therefore write

$$P_{12}^{\text{stim}} = B_{12}N_1u(\omega, T). \tag{10.10}$$

The coefficient B_{12} in this formula is also called the Einstein coefficient.

3. *Stimulated emission.* The atom transition from state "2" to state "1" takes place under the action of the resonant optical quantum and is accompanied by the emission of the identical quantum. The probability of this process is

$$P_{21}^{\text{stim}} = B_{21}N_2u(\omega, T), \tag{10.11}$$

where B_{21} is the Einstein coefficient.

In the equilibrium state, the transitions accompanied by emission and absorption of light must apparently balance each other. This is possible if the radiation processes mentioned above obey the condition

$$P_{21}^{\text{spont}} + P_{21}^{\text{stim}} = P_{12}^{\text{stim}}. \tag{10.12}$$

Inserting eqns (10.9)–(10.11) into eqn (10.12), we obtain the equation

$$A_{21}N_2 + B_{21}N_2u(\omega, T) = B_{12}N_1u(\omega, T). \tag{10.13}$$

Taking into account eqns (10.7) and (10.8), we find from eqn (10.13) the following expression for the spectral density of equilibrium thermal radiation:

$$u(\omega, T) = \frac{A_{21}/B_{21}}{(B_{12}/B_{21})e^{\hbar\omega/kT} - 1}. \tag{10.14}$$

The ratios of the Einstein coefficients can be found by considering the limiting cases. Indeed, let the thermostat temperature tend to infinity:

$$T \to \infty. \tag{10.15}$$

Then the spectral density of the thermal radiation will grow infinitely, $u(\omega, T) \to \infty$, and the populations of the levels will tend to become equal, $N_1 \to N_2$. It follows from eqn (10.13) that, under these conditions,

$$B_{12} = B_{21}. \tag{10.16}$$

Now consider the low-frequency limit:

$$\omega \to 0. \tag{10.17}$$

In this case $\hbar\omega \ll kT$ and the quantum structure of radiation does not manifest itself. The classical Rayleigh–Jeans formula for the spectral density of thermal radiation is then valid (see Chapter 9):

$$u(\omega, T) = \frac{\omega^2}{\pi^2 c^3} kT. \tag{10.18}$$

On the other hand, it follows from eqns (10.14), (10.16), and (10.17) that

$$u(\omega, T) = \frac{A_{21}}{B_{21}} \frac{kT}{\hbar\omega}. \tag{10.19}$$

Comparing eqns (10.18) and (10.19), we find that

$$\frac{A_{21}}{B_{21}} = \frac{\hbar\omega^3}{\pi^2 c^3}. \tag{10.20}$$

Finally, inserting eqns (10.16) and (10.20) into eqn (10.14), we obtain

$$u(\omega, T) = \frac{1}{\pi^2 c^3} \frac{\hbar\omega^3}{e^{\hbar\omega/kT} - 1}. \tag{10.21}$$

Thus, Planck's formula is derived. The simplicity and clarity of the derivation favours Einstein's optical quanta hypothesis as well as his assumption about the existence in a quantum-mechanical system of three kinds of radiation processes; that is, *spontaneous emission*, *stimulated absorption*, and *stimulated emission* of light. Einstein's idea of describing the processes of emission and absorption of optical quanta by atoms in terms of probability theory also proves to be very fruitful.

10.4 The operational principle and design of a laser

The generation of light by a laser is based on the effect of stimulated emission in a quantum-mechanical system. In fact, the term "laser" is an abbreviation of "light amplification by stimulated emission of radiation". Let us discuss the

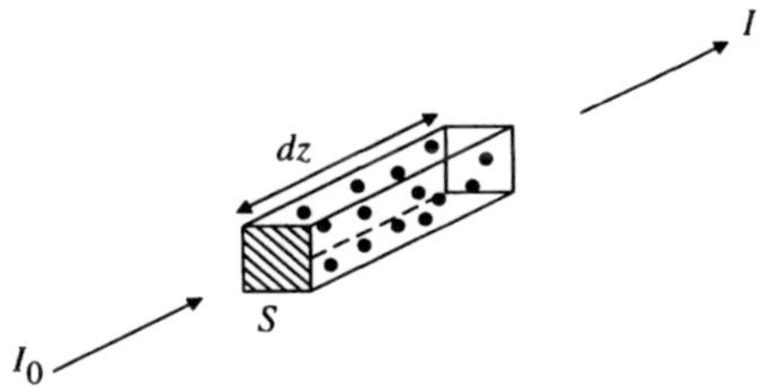

Fig. 10.4 A calculation of the stimulated emission gain.

conditions under which light amplification by stimulated emission of radiation can occur.

Using the Einstein model, we can calculate the change in the intensity of light after it passes through a layer of substance. Suppose that a plane optical wave of frequency ω, which is resonant with the quantum transition, falls on a layer of substance of depth dz (Fig. 10.4). Let us choose the area element S on the surface of the substance and consider the volume element $dV = S\,dz$ of the medium. Denoting the number of atoms per unit volume of the medium by N, we can write the number of atoms within the chosen volume element: $dN = N\,dV$. Of that number, $dN_1 = N_1 dV$ atoms populate the lower energy level and $dN_2 = N_2 dV$ atoms the upper energy level. N_i is the density of atoms (the number of atoms per unit volume) populating the energy level i.

According to the Einstein model, the power of spontaneous emission dP_{21}^{spont}, stimulated absorption dP_{12}^{stim}, and stimulated emission dP_{21}^{stim} for the given element of the medium can be written, respectively, as

$$\begin{aligned} dP_{21}^{\text{spont}} &= A_{21}\hbar\omega\, dN_2, \\ dP_{12}^{\text{stim}} &= B_{12}\hbar\omega\, u(\omega)\, dN_1, \\ dP_{21}^{\text{stim}} &= B_{21}\hbar\omega\, u(\omega)\, dN_2. \end{aligned} \tag{10.22}$$

Here $u(\omega)$ is the spectral density of incident radiation at frequency ω, $\hbar\omega$ is the energy of the quantum, A_{21} is the probability of spontaneous emission of an optical quantum by the atom per unit time, B_{12} is the probability of stimulated absorption of an optical quantum by the atom per unit time in the external radiation field of unit spectral density, and B_{21} is the probability of stimulated emission of an optical quantum by the atom per unit time in the external radiation field of unit spectral density. In fact, eqns (10.22) can be considered as a definition of the coefficients A_{21}, B_{12}, and B_{21}.

The total radiation power of the given element of the medium is equal to

$$dP = dP_{21}^{\text{spont}} + dP_{21}^{\text{stim}} - dP_{12}^{\text{stim}}. \tag{10.23}$$

Using, for simplicity, a rectangular approximation for the radiation spectrum (Fig. 10.5), it is easy to show that the intensity I and the spectral density u of

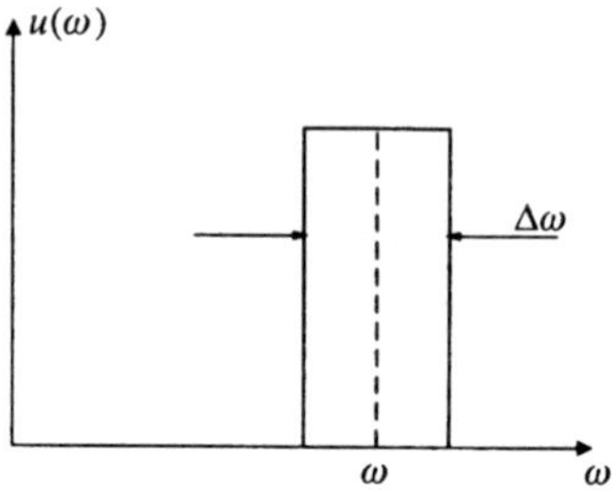

Fig. 10.5 A rectangular approximation of the spectrum of an optical wave.

radiation in a plane optical wave are related by formula

$$I = cu\Delta\omega, \tag{10.24}$$

where $\Delta\omega$ is the spectral bandwidth and c is the velocity of light. Using eqns (10.22)–(10.24), we can write the increment of light intensity $dI = dP/S$ as

$$dI = \left[A_{21}N_2 + (B_{21}N_2 - B_{12}N_1)\frac{I}{c\Delta\omega}\right]\hbar\omega\, dz. \tag{10.25}$$

Now suppose that the light intensity is so high that the spontaneous emission can be neglected. Then, from eqn (10.25), taking into account eqn (10.16) and denoting $B_{12} = B_{21} = B$, we obtain

$$dI = B(N_2 - N_1)\frac{I}{c\Delta\omega}\hbar\omega\, dz \tag{10.26}$$

or

$$dI = GI\, dz \tag{10.27}$$

where

$$G = B(N_2 - N_1)\frac{\hbar\omega}{c\Delta\omega}. \tag{10.28}$$

Equation (10.27) gives the light intensity as a function of the distance z traversed:

$$I(z) = I_0 e^{Gz}. \tag{10.29}$$

Thus, our calculation shows that the light intensity in the medium varies exponentially, with the sign of the exponent depending on the sign of the population difference $N_2 - N_1$.

If the medium is in a state of thermal equilibrium, then according to eqns (10.7) and (10.8), the ratio of the level populations is expressed by the formula

$$N_2/N_1 = \exp(-\hbar\omega/kT). \tag{10.30}$$

Let us estimate the value of the parameter $\hbar\omega/kT$ for the visible range of frequencies and for room temperature. Putting $T = 300$ K, $\omega = 2\pi c/\lambda$, $\lambda = 0.5 \times 10^{-4}$ cm, $c = 3\times10^{10}$ cm s^{-1}, $k = 1.38\times10^{-16}$ erg K^{-1}, and $\hbar = 1.05\times10^{-27}$ erg s, we obtain

$$\hbar\omega/kT \approx 10^2. \tag{10.31}$$

It follows from eqns (10.30) and (10.31) that the thermal population of the higher level can be neglected. Assuming $N_1 = N$ and $N_2 = 0$, we obtain from eqns (10.28) and (10.29)

$$I(z) = I_0 \exp(-|G|z), \tag{10.32}$$

where

$$|G| = NB\frac{\hbar\omega}{c\Delta\omega}. \tag{10.33}$$

Obviously, eqn (10.32) describes the conventional absorption of light by the medium (Bouguer's law). Note that expression (10.33) for the absorption coefficient, which can also be calculated from the classical model (see Chapter 7), allows us to establish the following relation between the parameters of the classical and quantum models:

$$G = \left(\frac{N_2 - N_1}{N}\right)\delta(\omega). \tag{10.34}$$

Here $\delta(\omega)$ is the classical absorption coefficient.

However, the quantum model, unlike the classical one, predicts the possibility of exponential light amplification by the medium. According to eqns (10.28) and (10.29), this requires the population of the higher level to exceed that of the lower one:

$$N_2 > N_1, \tag{10.35}$$

i.e. the inverse population is realized. It is this phenomenon that underlies laser generation.

The inverse population can be obtained by different ways. Most frequently, illumination by light, called optical pumping, is used. Other techniques are electric discharge, electric current, and chemical reactions.

To pass from *light amplification* conditions to those of *light generation*, a *feedback* is applied, as in any other generator. In lasers, the feedback is realized with the help of an *optical resonator*, which in the simplest case consists of a pair of parallel mirrors.

A basic scheme of a laser is shown in Fig. 10.6. It contains an active element, a resonator, and a pump source. Initially, the pump source, e.g. a powerful flashlamp, illuminates a working substance (the active element) of the laser and creates in it the inverse population. Then the "inverted" medium spontaneously begins to emit optical quanta. Under the action of spontaneous emission, the process of stimulated emission begins. Due to the inverse population, this process (being of an avalanche-type character) leads to exponential light amplification. The light going sideways abandons the active element without gaining

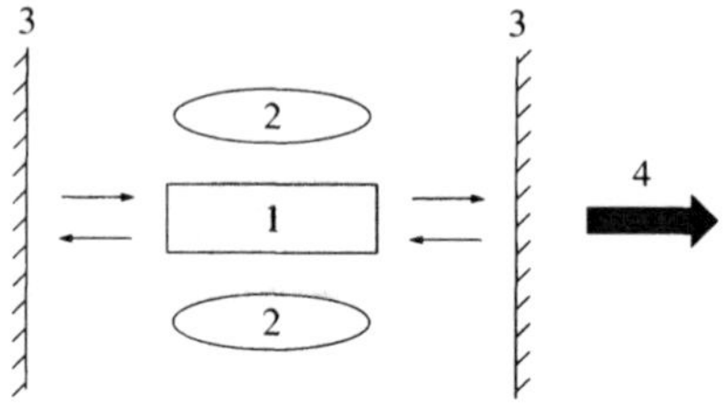

Fig. 10.6 A schematic diagram of a laser. 1, The active element; 2, the pump system; 3, the optical resonator; 4, the output beam.

substantial energy. At the same time, the light wave, travelling along the axis of the optical resonator and passing repeatedly through the active element, continuously gains energy. Due to partial transmission of the light through one of the resonator mirrors, the radiation comes out from the resonator in the form of a laser beam.

The concept of the laser was originally developed by Basov and Prokhorov in Russia and by Townes and Schawlow in the USA. The first ruby laser was built by Maiman in 1960.

10.4.1 Laser self-excitation conditions

Consider a model of a laser shown in Fig. 10.7. Suppose, for simplicity, that the laser resonator is filled completely by the active medium. Let λ be the wavelength of laser emission, G the optical gain in the inverted medium, L the resonator length, and R_1 and R_2 the reflection coefficients of the mirrors (in terms of the optical wave amplitude).

Laser self-excitation occurs when the light amplification by stimulated emission of radiation in the inverted medium exceeds the loss. Assuming that the loss is caused only by the output radiation, we can write the condition for laser generation in the form

$$R_1 R_2 \mathrm{e}^{GL} \geq 1. \tag{10.36}$$

Equation (10.36) is called the amplitude condition for laser self-excitation. Another condition, known as the *phase condition*, requires that the resonator length equals an integral number of laser radiation half-wavelengths :

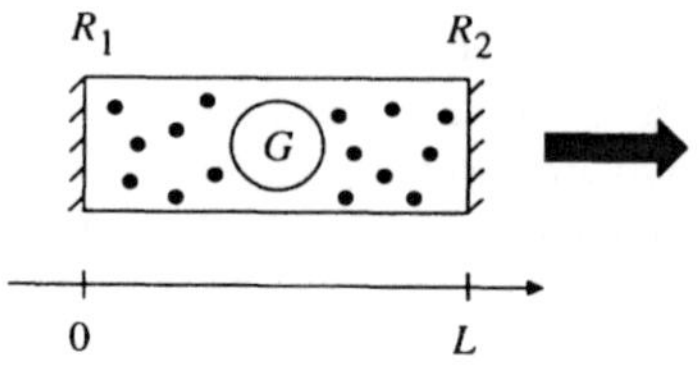

Fig. 10.7 A derivation of the laser self-excitation conditions.

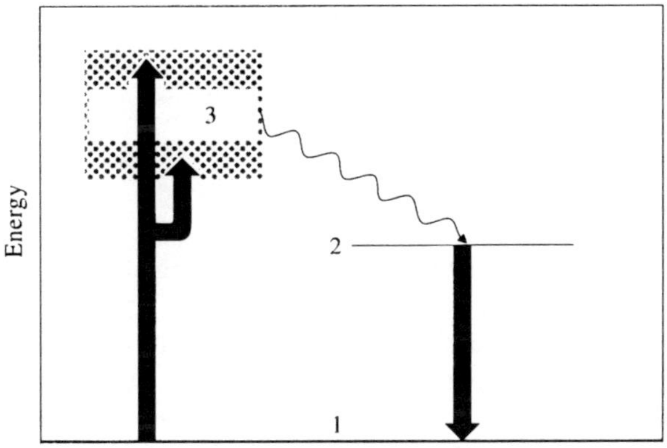

Fig. 10.8 Chromium energy levels in a ruby crystal and the operational principle of the ruby laser.

$$L = n\frac{\lambda}{2}, \qquad n = 1, 2, 3, \ldots \, . \tag{10.37}$$

As long as this condition is satisfied, the round-trip phase shift of an optical wave in the resonator is equal to a multiple of 2π, which ensures the optimum conditions for light amplification. Let us now consider some specific types of lasers.

10.4.2 The ruby laser

Ruby is a crystal of aluminium oxide, in which a fraction of the aluminium atoms are replaced by chromium atoms: the higher chromium concentration, the deeper is the colour of the crystal—therefore, ruby may be pink or red. The hue is caused by the chromium atoms which absorb the green and yellow light, transmitting only the red and blue light. In his first laser, Maiman used a pale-pink ruby crystal containing 0.05% of chromium. Figure 10.8 shows a scheme of the energy levels that chromium atoms have in a ruby crystal.

The ruby laser operates using the so-called three-level scheme. Initially, a powerful xenon flashlamp excites the chromium atoms to level 3, with a large spectral bandwidth. Then the atoms make a rapid non-radiative transition from level 3 to the narrow energy level 2. Level 2 is metastable, so that the lifetime of atoms at this level is about 1 ms. This allows an accumulation of chromium atoms at level 2. If the rate of atomic transfer from ground state 1 to excited state 3 is sufficiently high, then, in the short run, the population of level 2 may exceed that of level 1, i.e. population inversion will be attained. Eventually, the atoms transit from level 2 to level 1, emitting the radiation at a wavelength of 694.3 nm.

Figure 10.9 shows the basic structural elements of the Maiman laser; that is, a ruby crystal and a xenon flashlamp.

Note an important advantage of the three-level laser scheme. As seen in

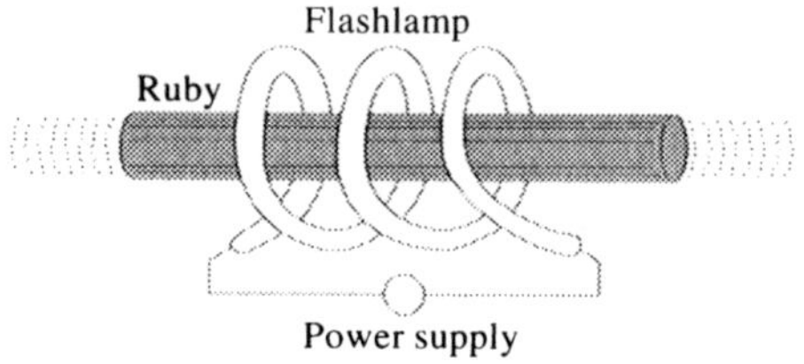

Fig. 10.9 Maiman's laser design.

Fig. 10.8, the frequency of the exciting light (optical pumping) does not coincide with the laser radiation frequency. On the contrary, in a two-level scheme, the pump frequency coincides with the laser radiation frequency and, consequently, population inversion cannot be achieved. The pump radiation quanta induce atomic transitions in both directions simultaneously, i.e. from the lower working level to the upper one and vice versa. As a result, even high-powered pumping attains only population equilization, but not population inversion.

As an active element, our laser uses a ruby rod 140 mm long and 10 mm in diameter. Optical pumping is effected by two xenon flashlamps. A 800 μF bank of capacitors charged to 3 kV supplies voltage to the lamps.

The physical processes taking place in the laser can be clarified using the phenomenon of luminescence (see Fig. 10.10). In a demonstration performed in a darkroom, a green beam from an argon laser is passed along the axis of a ruby crystal rod. An observer can see the red light luminescence emitted laterally from the ruby rod.

The luminescence process follows the pattern shown in Fig. 10.8, differing from the laser generation only by a lower pump power, which is insufficient to ensure population inversion between energy levels 1 and 2. Similarly, the Maiman laser failed to generate and produced only a luminescence when the flashlamp power was not high enough. Laser generation set in only when a very powerful flashlamp was used.

Let us evaluate, by way of an example, the flashlamp power needed for the ruby laser operation. Ruby is known to contain about 10^{19} chromium atoms per cm^3. To stimulate the generation, we need at least half of these atoms, i.e.

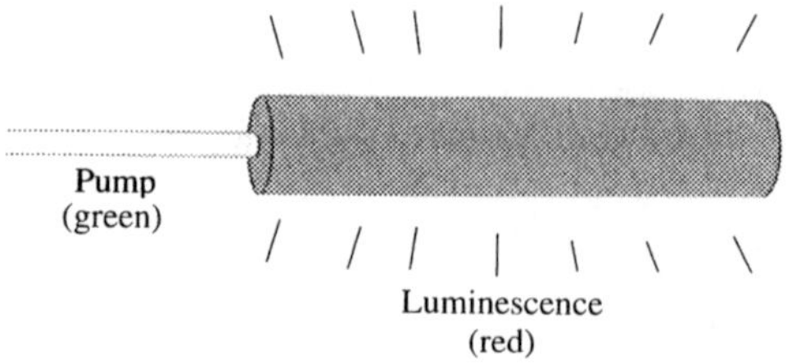

Fig. 10.10 The observation of luminescence from a ruby crystal irradiated by an argon laser.

5×10^{18} cm^{-3}, to be transferred to level 3 (see Fig. 10.8). Each atom requires an energy of $W_3 - W_1 = 4 \times 10^{-12}$ erg per atom, or 2×10^7 erg cm^{-3}, to accomplish the transition. The transition of atoms to state 2 should continue for a time no longer than the time $\Delta T \approx 2$ ms during which the atoms stay at level 2. If the atom transits to the excited state at a slower rate, then population inversion will not be attained, due to spontaneous emission and relaxation transitions to the ground state. Thus, a pump energy of about 2×10^7 erg has to be absorbed per millisecond within 1 cm^3 of the ruby crystal to produce laser generation. This means that the absorbed power should be about 2 kW per 1 cm^3 of the crystal, and the pump power about 20 kW for the total crystal volume of 10 cm^3. The flashlamp emits light which is close to white light, having a broad frequency spectrum. The energy bands of chromium in ruby utilize about 15% of the flashlamp light. Consequently, the total power of the flashlamp should be about 200 kW.

10.4.3 The helium–neon laser

This laser uses a mixture of helium and neon gases as a working medium. Neon is the active substance in the laser. Helium atoms play an auxiliary part by producing population inversion. The gaseous mixture is contained in a glass tube, in which a continuous electric discharge is sustained by a special device. The electrons gaining energy in the electric field of the discharge act as primary energy carriers.

The operation of the laser is shown schematically in Fig. 10.11. The helium atoms are excited in the electric discharge via collisions with electrons (Fig. 10.11(a)). When colliding with a neon atom, a helium atom transmits its energy to it. As a result, the helium atom comes back to the ground state while the neon atom transits to excited state 3 (Fig. 10.11(b)). The probability of collisional energy transfer is rather high due to the almost equal energies of the excited states of the helium and neon atoms. Thus, the atomic collisions bring a fraction of neon atoms up to excited state 3. Then a population inversion arises between this state and the lower-energy state 2, because the latter state, in which the neon atom energy is about 20 eV, is hardly populated. Finally, the neon atoms passing from state 3 to state 2 emit laser radiation at a wavelength of 632.8 nm.

A peculiar feature of the helium–neon laser is that the lower working level (level 2 in Fig. 10.11) is located at quite a high level and, consequently, scarcely populated. This promotes population inversion so that the laser can operate continuously, consuming a mere 50 W of energy (in the early models). The radiation power of the helium–neon laser is usually of the order of 10 mW.

10.4.4 The carbon dioxide laser

This laser is a member of the family of molecular lasers, in which the laser generation results from transitions between the vibrational energy levels of molecules. The CO_2 laser emits infrared radiation at a wavelength of 10.6 μm. Typically, this continuous-wave laser is highly efficient (15–20%) and its radiation power can reach 1 MW.

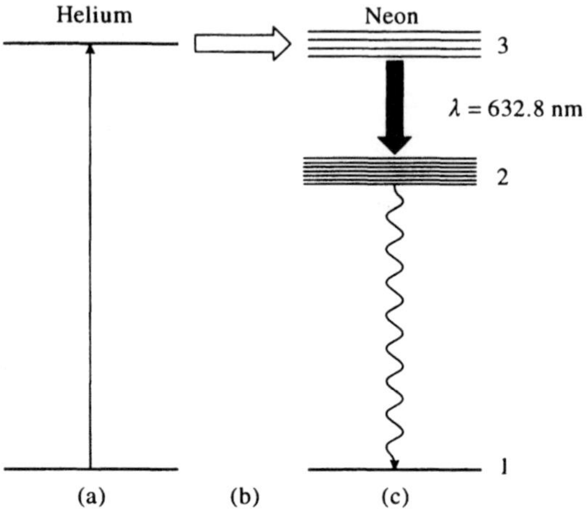

Fig. 10.11 The operational principle of the helium–neon laser.

Figure 10.12 illustrates the vibration types that are characteristic of the CO_2 molecule: symmetric, bending, and antisymmetric. The quantum energy levels corresponding to these vibrations are shown in Fig. 10.13.

The vibration of a CO_2 molecule is stimulated by electronic collisions in an electric discharge. The working substance of the CO_2 laser is a mixture of gaseous carbon dioxide with nitrogen and helium. Collisions of electrons, atoms, and molecules under the conditions of the electric discharge lead to effective population of the upper energy level and to rapid depletion of the lower energy level of the laser transition, which brings on population inversion and increases the output power of the laser.

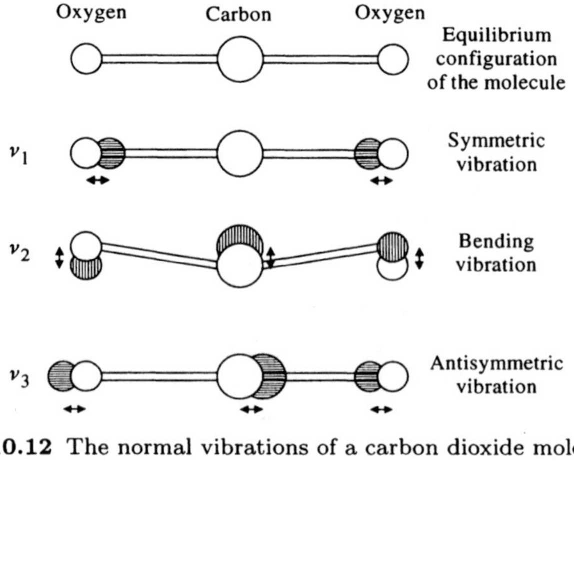

Fig. 10.12 The normal vibrations of a carbon dioxide molecule.

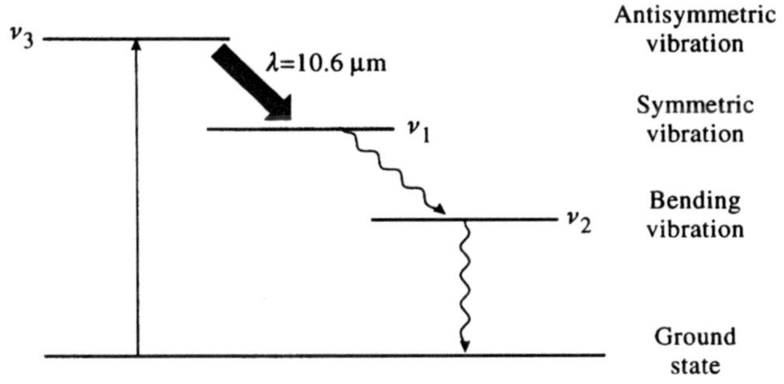

Fig. 10.13 The vibrational energy levels and operational principle of the carbon dioxide laser.

In one of the early models of the CO_2 laser, created by Patell, the working substance consisted of carbon dioxide and nitrogen gases at pressures of 3 Torr and helium at a pressure of 20 Torr. The laser generated continuous-wave radiation at a wavelength of 10.6 μm with an output power of 80 W power per metre of the discharge tube length. The operational efficiency exceeded 20%.

Typical CO_2 lasers have a length of about 2 m and generate continuous-wave radiation with an output power of about 150 W. Some of the latest models, providing up to 10 kW of output power, are succesfully used for the treatment of materials.

10.5 The characteristics of laser radiation

Laser radiation is characterized primarily by its wavelength, energy, pulse duration, CW (continuous-wave) power, spectral bandwidth, tuning range, and beam divergence. For different laser types, these parameters vary widely. For example, excimer lasers generate ultraviolet radiation in the 200–300 nm wavelength range, while CO_2 lasers operate in the infrared (10.6 μm). Dye lasers emit radiation that is tunable within a spectral band about 1000 cm^{-1} wide. Special models of helium–neon laser generate extremely narrow-band radiation, the bandwidth of which is about 100–1000 Hz. Angular divergence of the laser beam can be close to the diffraction limit. Modern laser techniques have made it possible to reach the fundamental limit of optical pulse duration. Pulse durations as short as 6 fs have been obtained, which is equivalent to only a few optical vibration periods. The output power of a continuous-wave CO_2 laser has been known to reach 1 MW. Superpowerful lasers developed for laser fusion applications can generate 1 ns optical pulses with an energy of 100 kJ and a power of 10 TW. By focusing such radiation, one can obtain light intensities of up to 10^{19} W cm^{-2} and electric field strengths up to 10^{10} V cm^{-1}, which exceed intra-atomic electric field strengths.

Such a wide range of laser characteristics offers considerable scope for practical applications, some of which were mentioned at the beginning of this chapter.

10.6 The generation of ultra-short pulses

Let us consider in more detail the spectral and temporal characteristics of laser radiation. The phase condition (10.37) for laser self-excitation shows that laser generation occurs only at discrete frequencies determined by the formula

$$\nu_n = n\,\Delta\nu, \quad \Delta\nu = 1/2L, \quad n = 1, 2, 3, \dots\ . \tag{10.38}$$

Since these frequencies depend only on the resonator length L, they can be called the *resonator eigenmodes*. The number M of excited eigenmodes equals the ratio of the amplification bandwidth $\Delta\nu_{\text{amp}}$ of the laser to the intermodal interval $\Delta\nu$; that is,

$$M = \Delta\nu_{\text{amp}}/\Delta\nu. \tag{10.39}$$

Note that according to eqn (10.34) the amplification bandwidth $\Delta\nu_{\text{amp}}$ coincides with the absorption bandwidth $\Delta\nu_{\text{abs}}$ of the active element of the laser in the absence of population inversion. Depending on the type of laser, this value varies over a wide range. For example, in a helium–neon laser the value of $\Delta\nu_{\text{amp}}$ is determined by the Doppler broadening of the spectral line, and $\Delta\nu_{\text{amp}} \approx 0.04\ \text{cm}^{-1}$ (see Chapter 6). For the ruby laser, as well as for the laser using a garnet crystal activated by neodymium ions, $\Delta\nu_{\text{amp}} = 2$–$4\ \text{cm}^{-1}$. For a neodymium–glass laser, $\Delta\nu_{\text{amp}} = 100\ \text{cm}^{-1}$, while for a dye laser, $\Delta\nu_{\text{amp}} = 1000\ \text{cm}^{-1}$. Assuming, for example, $L = 50$ cm, we obtain $\Delta\nu = 1/2L = 10^{-2}\ \text{cm}^{-1}$, while the number of generated modes $M \approx 4$ for a helium–neon laser, $M \approx 100$ for a ruby laser, and $M \approx 10^4$ for a neodymium–glass laser. Thus, as a rule, the laser emits *multiple-mode radiation*. A typical spectrum of multiple-mode laser radiation is shown in Fig. 10.14.

10.6.1 A single-mode laser

One of the basic problems in laser physics is to provide *monochromatic radiation*, i.e. the radiation the structure of which is close to that of the ideal harmonic wave. A *single-mode laser* emitting only one longitudinal mode is used to solve this problem.

A schematic diagram of the single-mode laser is shown in Fig. 10.15. The special feature of this laser is that its resonator contains an element performing frequency filtration of the radiation. The frequency filter consists of a dispersive element (prism, diffraction grating, or etalon) and a diaphragm. The dispersive element resolves the radiation frequency spectrum in space, while the diaphragm selects the desired frequency component (mode). Thus, the laser generates radiation at the frequency of a single longitudinal mode. By rotating the prism (or another dispersive element), the radiation frequency can be varied.

At present, the narrowest band is observed in the radiation from the helium–neon laser. Some special models generate radiation with a spectral line bandwidth of about 100 Hz. The quality factor of such radiation, that is, the ratio of the mean radiation frequency to its spectral bandwidth,

$$Q = \nu/\Delta\nu, \tag{10.40}$$

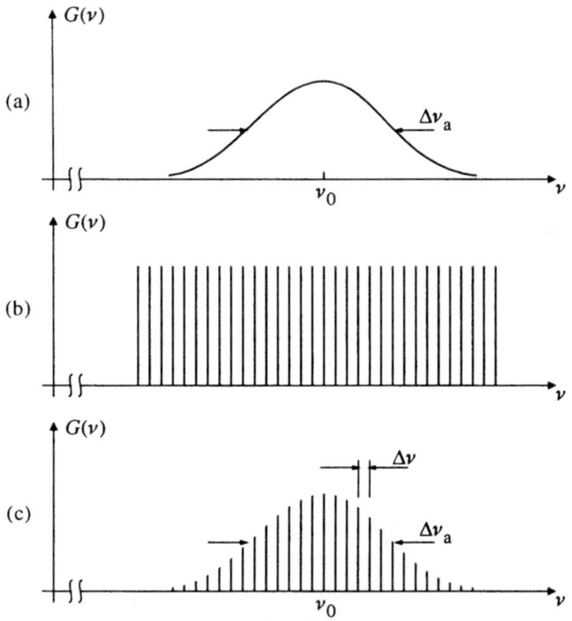

Fig. 10.14 The radiation spectrum of a multimode laser. (a) The amplification band spectrum. (b) The resonator eigenmode spectrum. (c) The laser emission spectrum.

amounts to 10^{13} (see Fig. 10.16). Note that this value exceeds the quality factor of radiation emitted by a non-laser source of light by many orders of magnitude.

10.6.2 The operation of a mode-locked laser—the generation of ultra-short optical pulses

The narrow-band single-mode laser generates an optical vibration of an almost sinusoidal form. The temporal behaviour of multimode laser radiation can vary depending on the amplitude and phase relations between the modes. This is illustrated in Fig. 10.17, which shows the temporal dependence of multimode laser radiation for the two opposite extreme cases of totally independent modes (a) and in mode-locked operation (b). In the former case the radiation resembles

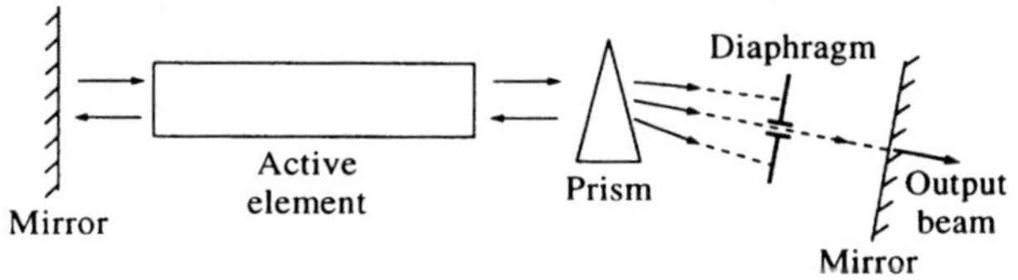

Fig. 10.15 A schematic diagram of a single-mode laser.

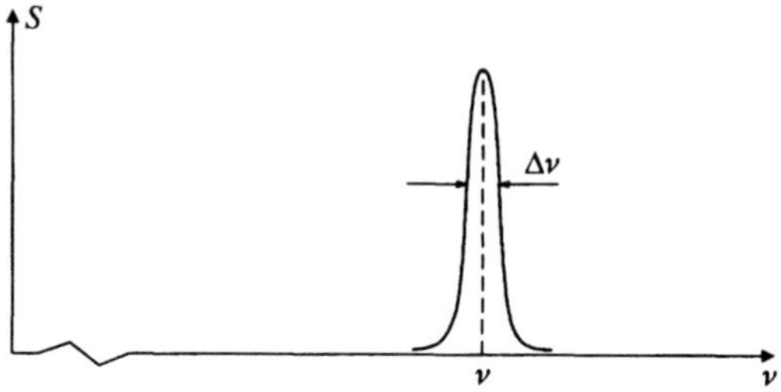

Fig. 10.16 The notion of a radiation quality factor.

a random optical noise, while in the latter case the laser generates a powerful short pulse typical of the so-called mode-locked regime. Let us consider mode-locked operation, which is of great practical interest, in detail. We shall prove that short light pulses can indeed be generated by a multimode laser and evaluate the limiting pulse duration.

According to eqn (10.38) the resonator eigenmode frequencies form an equidistant spectrum (see Fig. 10.14(b)). Let $\Delta\omega$ denote the intermodal interval, N the total number of modes, and ω_0 the carrier frequency. Suppose, for simplicity, that all of the modes have the same amplitude and phase. The optical field strength at a fixed point in space can then be represented in the form

$$E(t) = \sum_{n=1}^{N} A_0 \cos \omega_n t, \tag{10.41}$$

where

$$\omega_n = \omega_0 + n\Delta\omega, \qquad n = 1, 2, 3, \ldots, N. \tag{10.42}$$

To calculate the sum in eqn (10.41), we shall use the Eulerian formulae

$$\cos\varphi = \frac{e^{i\varphi} + e^{-i\varphi}}{2}, \qquad \sin\varphi = \frac{e^{i\varphi} - e^{-i\varphi}}{2i} \tag{10.43}$$

and the expression for the sum of a finite geometric progression,

$$S_N = a_1 \frac{1 - q^N}{1 - q}, \tag{10.44}$$

where a_1 is the first term of the progression, q is the progression ratio, and N is the number of terms in the progression. After some simple rearrangements, we obtain

$$E(t) = A(t) \cos \omega t, \tag{10.45}$$

where

$$\omega = \omega_0 + \left(\frac{N+1}{2}\right) \Delta\omega \tag{10.46}$$

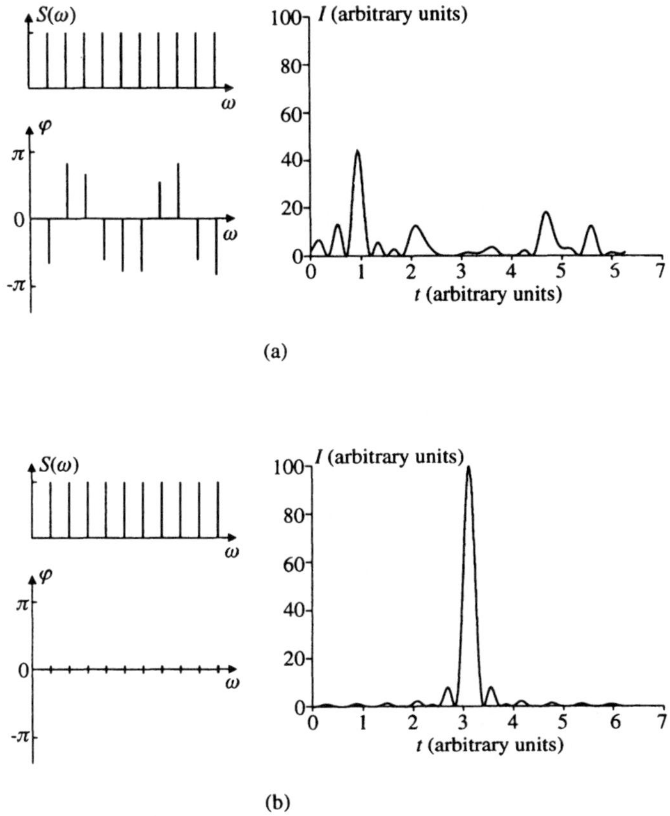

Fig. 10.17 The spectra (left) and temporal dependence (right) of laser radiation (a) in the case of independent modes and (b) in the mode-locked regime.

is the mean frequency and

$$A(t) = A_0 \frac{\sin(N\Delta\omega t/2)}{\sin(\Delta\omega t/2)} \tag{10.47}$$

is the amplitude of the field. The radiation intensity can be expressed as a function of time:

$$I(t) = I_0 \frac{\sin^2(N\Delta\omega t/2)}{\sin^2(\Delta\omega t/2)}, \tag{10.48}$$

where I_0 is the intensity corresponding to an individual mode. A typical form of dependence $I(t)$ for $N \gg 1$ is shown in Fig. 10.18. The laser radiation is seen as a succession of powerful short pulses. Using eqn (10.48), we can find the main parameters of the pulses. The duration of an individual pulse is inversely proportional to the total bandwidth of the laser radiation:

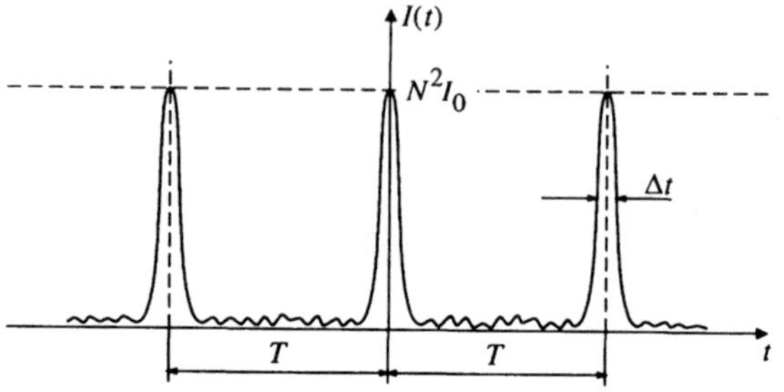

Fig. 10.18 The temporal dependence of radiation emitted by a mode-locked laser.

$$\Delta t = \frac{2\pi}{N\Delta\omega}. \tag{10.49}$$

The maximum radiation intensity is proportional to the square of the number of modes:

$$I_{\max} = I_0 N^2. \tag{10.50}$$

The pulse repetition period is governed by the value of the intermodal interval:

$$T = \frac{2\pi}{\Delta\omega}. \tag{10.51}$$

Mode locking can be made clear using the mechanical model shown in Fig. 10.19 ("Chebotaev's pendulums"). The demonstration device contains a set of pendulums, the lengths of which are chosen so that their oscillation frequencies form an equidistant sequence of the form of eqn (10.42). Having deflected all of the pendulums from equilibrium and simultaneously released them, we observe at first an oscillation dephasing followed by the reverse process of oscillation phasing, which is equivalent to mode locking.

Using eqn (10.38), we can present the pulse repetition period as

$$T = \frac{2L}{c}. \tag{10.52}$$

The formula shows that the pulse repetition period is equal to the round-trip time taken by the light in the resonator. This permits the conclusion that a short dense bunch of radiation is formed in the laser resonator under mode locking conditions. Every arrival of the bunch at the exit mirror of the resonator is accompanied by the escape of a fraction of radiation from the resonator forming a sequence of optical pulses with a repetition period of $T = 2L/c$. It becomes clear, therefore, that the modes can be locked by a *saturable absorber* (a bleachable filter) placed inside the laser resonator. By inhibiting the amplification of light of constant intensity, such an element becomes transparent only for the short powerful optical pulses. This technique for generating ultra-short laser pulses is

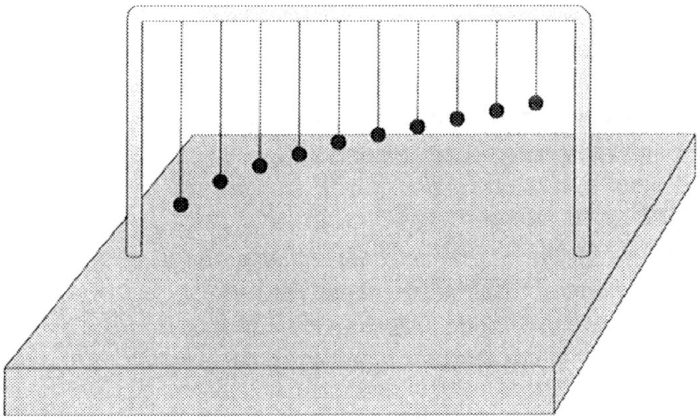

Fig. 10.19 A mechanical model for the demonstration of mode locking ("Chebotaev's pendulums").

widely used in practice and is called *passive mode locking*. The basic elements of a passively mode-locked laser are illustrated in Fig. 10.20. A thin cell filled with a special dye and placed into the resonator serves as a saturable absorber. In fact, it acts as a fast intensity modulator, cutting the pulse duration down to about 10 ps. There are a number of other mode locking methods.

10.6.3 Evaluations

Using the formula $\Delta\omega = 2\pi c\Delta\nu_{\text{amp}}$, where $\Delta\nu_{\text{amp}}$ is the laser amplification bandwidth expressed in cm^{-1}, we can rewrite eqn (10.49) in the form:

$$\Delta t = \frac{1}{c\Delta\nu_{\text{amp}}}. \tag{10.53}$$

Thus, the limiting pulse duration in the mode-locked regime is determined by the laser amplification bandwidth.

The neodymium–glass laser has a broad amplification bandwidth: $\Delta\nu_{\text{amp}} = 100\ \text{cm}^{-1}$. In this case, $\Delta t = 0.33 \times 10^{-12}\,\text{s} = 0.33\,\text{ps} = 330$ fs, where the following units of time are used: 1 picosecond = 1 ps = 10^{-12} s, 1 femtosecond = 1 fs = 10^{-15} s.

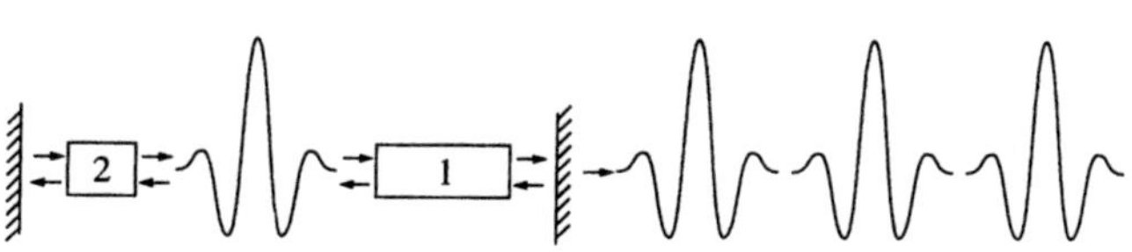

Fig. 10.20 The operation of a multimode passively mode-locked laser. 1, The active element; 2, the saturable absorber.

The dye laser has a broader amplification bandwidth: $\Delta\nu_{\mathrm{amp}} = 1000\ \mathrm{cm}^{-1}$. In this case the limiting pulse duration is $\Delta t = 33$ fs.

The fundamental limit of optical pulse duration is set by the optical vibration period, which is equal to about 1 fs. To date, optical pulses of 6 fs duration have been produced by methods of non-linear optics (see, e.g., Achmanov *et al.* 1991). This value is quite close to the ultimate limit.

Bibliography

Akhmanov, S. A., Vysloukh, V. A., and Chirkin, A. S. (1991). *Optics of femtosecond laser pulses.* AIP Press, New York.

Einstein, A. (1905). *Annalen der Physik,* **17**, 132–148.

Einstein, A. (1916). *Mitteilungen Physikalischen Gesellschaft,* No. 18, 47–62 (Zürich).

Maiman, T. H. (1960). *Nature,* **187**, 493–494.

Schavlow, A. L. and Townes, C. H. (1958). *Physical Review,* **112**, 1940–1949.

Part III

Interference, diffraction, and coherence

11

The interference of light

Interference phenomena in optics. Young's experiment. The Michelson interferometer. The interference of monochromatic waves. The interference of non-monochromatic light. The interference of a random optical wave. Multiple-beam interference. The Fabry–Perot interferometer. Eigenmodes and the quality factor of a laser resonator.

11.1 Interference phenomena in optics

Whenever the intensity of light resulting from the superposition of two optical beams on a screen turns out to be not equal to the sum of intensities of the beams, but varies periodically from point to point, forming a system of dark and bright bands, we say that the *interference of light* takes place. The principle of observation of interference is illustrated in Fig. 11.1. Two uniform light beams are superimposed on a screen to form a system of dark and bright bands (an *interference pattern*). The experiment can be conducted using laser beams. One may see that there are regions on the screen (dark bands) within which two optical beams completely quench each other. Briefly, the interference of light can be defined as a phenomenon in which

$$\text{light} + \text{light} = \text{darkness}.$$

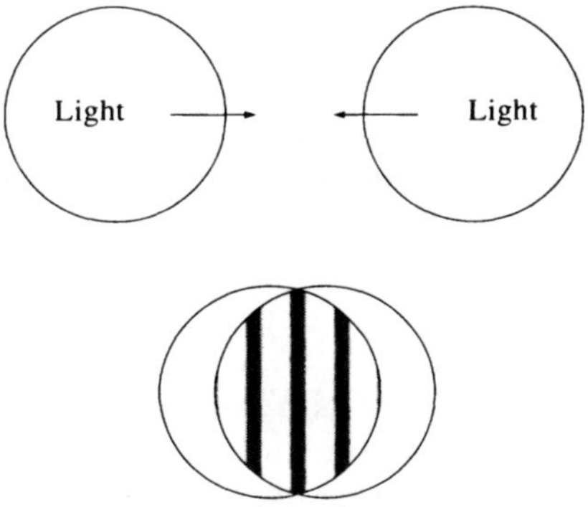

Fig. 11.1 The interference of light. The superposition of two uniform optical beams produces an interference pattern representing alternating dark and bright fringes.

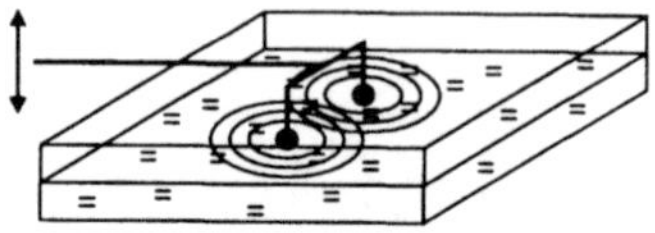

Fig. 11.2 The observation of the interference of waves on a water surface.

The interference of light is a subtle physical phenomenon which is rarely observed in everyday life. For example, it is well known from experience that when a room is lit by two lamps, the illumination at any point is equal to the sum of the illuminations created by each lamp individually. This phenomenon obeys the *law of intensity addition* and is observed for all independent sources of light: lamps, candles, stars, lasers, etc. There are, however, situations in which this law of intensity is violated. This may happen, for example, in the case of the superposition of two optical beams emanating from the same source, but traversing different optical paths. As experiments show, the superposition of optical beams under certain conditions can give rise to the formation of a system of alternating dark and bright bands called *interference fringes*.

11.1.1 Interference as a wave phenomenon

Assuming that light is a *wave*, we can explain the phenomenon of interference. When two identical spherical waves are superimposed, the vibrations reinforce each other at some points in space and annul each other at other points. This phenomenon is most easily demonstrated by means of waves on a liquid surface (Fig. 11.2).

A flat glass vessel is filled with water. Two vertical rods controlled by a special vibrating mechanism dip periodically into the water and excite surface waves. Superposition of the waves produces an interference pattern which can be observed on a screen by using stroboscopic illumination synchronized with the vibrating mechanism.

11.2 Young's experiment

Is it possible to observe interference in optics? The affirmative answer to this question was obtained by Young in 1801. A schematic diagram of Young's experiment is shown in Fig. 11.3. The light from source S passes through a pinhole in screen Q_1 to fall on two pinholes located a short distance apart on screen Q_2. The light passing through these pinholes falls onto screen Q_3, where the interference pattern is observed.

One can easily see that the idea of Young's experiment was borrowed from the experiment with waves on a water surface. If the light emanating from each pinhole is a wave, then the alternating dark and bright interference fringes have to be formed on screen Q_3.

The experiment confirmed this assumption. At first, Young let a sunlight ray pass into a darkroom and placed two screens in its way: a dark screen with two

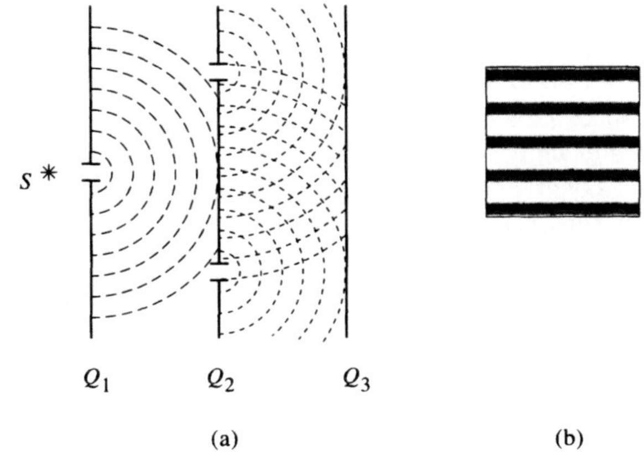

Fig. 11.3 Young's experiment. (a) A schematic diagram of the experiment. (b) A fragment of the interference pattern.

pinholes in it and a white screen at some distance behind the first one. Having observed on the white screen a bright central band and two adjoining darkish bands, he decided to repeat the experiment. As a source of light, Young used an alcohol burner flame into which he added a small amount of edible salt to obtain the bright yellow emission of sodium. With this source of light, Young observed a pattern of regularly alternating dark and bright bands.

Young's experiment was the first convincing demonstration that superposition of light may produce darkness. This direct observation of interference provided experimental evidence of the *wave nature of light.*

11.2.1 Problems with the interference theory

On the one hand, the theory of light interference should explain in detail the results of Young's experiment, as well as those of other interference experiments. On the other hand, it should explain why the superposition of optical beams emanating from independent sources does not bring about any interference, thus obeying the law of intensity addition. First, we can note that the latter fact is connected with the complicated random structure of the optical field generated by conventional, in other words, non-laser sources of light. The light from such sources is a superposition of a great number of elementary spherical waves emitted by independent oscillators (atoms). Its structure differs strongly from an ideal harmonic wave. That is why we shall use the notion of a random optical field and statistical methods while describing the optical field in the theory of interference. In the course of developing the theory of interference, we come to the important concept of the coherence of light.

11.2.2 Coherence of light

Historically, the idea of the coherence of light arose in connection with interference experiments. It was obvious, for example, that the appearance of an interference pattern in Young's interferometer depended on the kind of light source used to illuminate the screen with two pinholes. When a point source, that is, a pinhole in screen Q_1 (Fig. 11.3), is used, then interference takes place. But if the light comes from an extended source or a source covered with a matt plate, interference is absent. This *capability of light to produce an interference pattern* is called *coherence of light.*

It is clear that coherence is connected with the structure of light. We shall see later that light that has a structure close to that of a plane or spherical harmonic wave is coherent. In other words, the light of a *highly ordered structure* is coherent. On the contrary, the non-coherent light that fails to produce interference and obeys the law of intensity addition has the structure of a randomly modulated wave.

11.3 The Michelson interferometer

At the end of the 19th century, A. Michelson carried out a number of important optical experiments, the most famous of which were measurements of the velocity of light and an experiment on the interference of light. The latter experiment was performed using a special optical instrument created by Michelson and now known as the Michelson interferometer. Using this device, Michelson and Morley carried out a classic experiment to demonstrate that there is no difference between the velocities of light in two perpendicular directions. This result allowed them to proclaim the universal principle that light propagates with a *constant velocity* in any direction, irrespective of the motion of its source. Einstein used this principle to create the theory of relativity.

The Michelson interferometer is shown schematically in Fig. 11.4. A monochromatic light beam impinges on a light-splitting plate, which divides the beam into two parts of nearly equal intensity. One part passes through the plate, while the other is deflected perpendicular to the incident beam direction. Then both beams fall normally onto two mirrors and return to the beam-splitting plate. Once again, the plate partially reflects and partially transmits the incident light, producing an output beam composed of two beams that have traversed different arms of the interferometer. By moving one of the interferometer mirrors along the incident light beam, one can vary the optical path difference and observe a variation of the interference pattern.

The light-splitting plate is a plane-parallel glass plate the back surface of which is covered with a translucent silver film. The film is so thin that the plate partially reflects and partially transmits the light, thus serving as a semi-reflecting mirror.

11.4 The interference of monochromatic waves

Suppose that a light source sends to the Michelson interferometer a plane monochromatic wave of the form

$$E_0 = A_0 \cos(\omega t - kz). \tag{11.1}$$

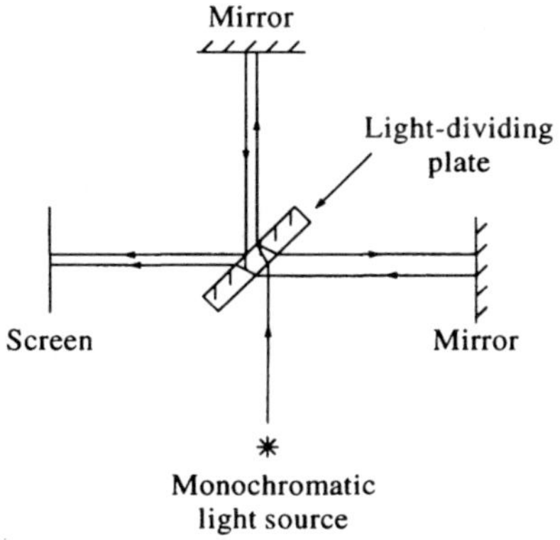

Fig. 11.4 A schematic diagram of the Michelson interferometer.

Let R and T denote the coefficients of reflection and transmission of light (in terms of intensity) by the light-splitting plate. Neglecting the absorption of light inside the plate, we can write

$$R + T = 1. \tag{11.2}$$

As the amplitude A of the plane monochromatic wave is related to its intensity I as (see Chapter 3)

$$I = \frac{cA^2}{8\pi}, \tag{11.3}$$

the amplitude coefficients of the reflection and transmission of light by the light-splitting plate are equal to $\sqrt{R}$ and $\sqrt{T}$ respectively.

As can be seen from Fig. 11.4, the input optical beam is split into two beams, each of which is then once transmitted and once reflected by the light-splitting plate. The output optical wave thus takes the form

$$E = \sqrt{RT} A_0 \left[\cos(\omega t - kl_1) + \cos(\omega t - kl_2)\right], \tag{11.4}$$

where l_1 and l_2 are the interferometer arm lengths. Equation (11.4) can also be written as

$$E = A\cos(\omega t - kl), \tag{11.5}$$

where

$$A = 2\sqrt{RT} A_0 \cos(\delta/2), \qquad l = (l_1 + l_2)/2. \tag{11.6}$$

The quantity δ signifies the phase difference of the waves that have traversed the different arms of the interferometer:

$$\delta = k\Delta, \qquad \Delta = l_1 - l_2, \qquad k = \frac{2\pi}{\lambda}. \tag{11.7}$$

Putting

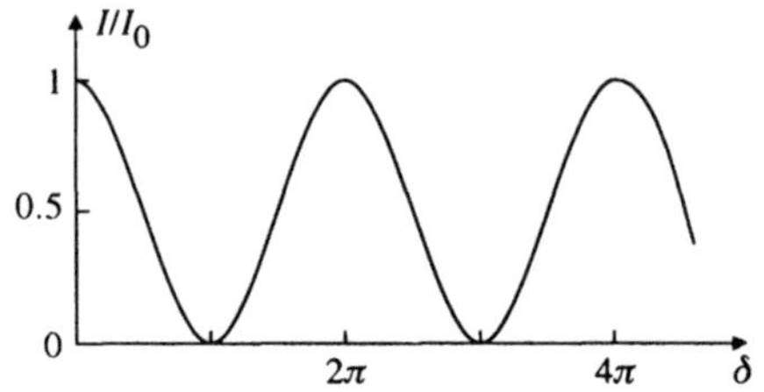

Fig. 11.5 The transmission of light by the Michelson interferometer as a function of the phase difference of the interfering waves.

$$R = T = 1/2 \tag{11.8}$$

and using eqns (11.3)–(11.7), we can write the intensity of light in the plane of observation as

$$I = \tfrac{1}{2} I_0 (1 + \cos\delta), \tag{11.9}$$

where $I_0 = cA_0^2/8\pi$ is the light wave intensity at the interferometer input. Usually, one of the mirrors of the Michelson interferometer is fixed, while the other is mounted on a carriage and can be moved towards or away from the dividing plate. If we displace the movable mirror by a distance Δz, then the path difference varies by $\Delta = 2\Delta z$ and the phase difference by $2\pi\Delta/\lambda$. Figure 11.5 shows the light intensity I in the plane of observation as a function of δ for the case of a plane monochromatic incident wave. It can be seen from the figure that the maximum intensity I of the output beam, observed when

$$\delta = 2m\pi, \qquad m = 0, 1, 2, \dots, \tag{11.10}$$

is equal to that of the incident wave, I_0. This means that, provided that condition (11.10) holds, the Michelson interferometer completely transmits the incident plane monochromatic wave. On the contrary, the minimum intensity of the output beam proves to be equal to zero when the condition

$$\delta = (2m + 1)\pi, \qquad m = 0, 1, 2, \dots \tag{11.11}$$

is satisfied. This means that the incident plane monochromatic wave is totally reflected back to the source. The Michelson interferometer can therefore serve as a filter, the transmission of which depends on the wavelength, or as a wavelength-selective reflector of light. The latter application of the Michelson interferometer is frequently used for mode selection in lasers.

In the case of a *diverging* incident beam, the path difference between the two waves depends on the angle of incidence (Fig. 11.6). The interference pattern observed with the Micheson interferometer consists of alternating dark and bright rings that are concentric with the output beam axis. Displacement of the movable mirror results in variation of the diameters of the rings.

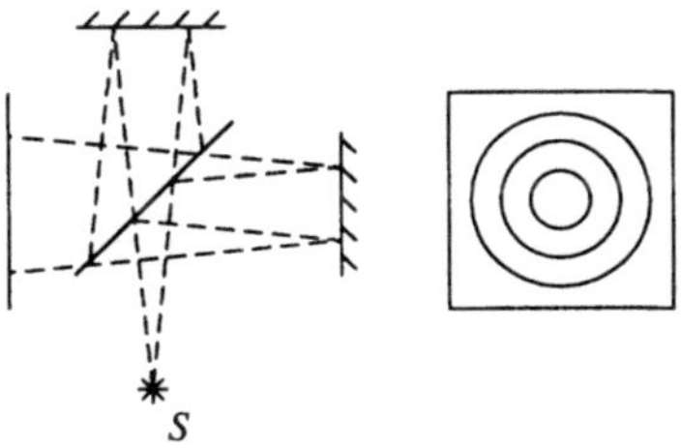

Fig. 11.6 Fringes observed with the Michelson interferometer, illuminated by a diverging beam of monochromatic light.

In the case of a parallel incident beam and slightly inclined interferometer mirrors, the interference pattern consists of parallel fringes which displace perpendicular to the fringes when the path difference Δ varies.

The Michelson interferometer can be used for the absolute measurement of the optical wavelength. For this purpose we need to count the number N of the maxima appearing on the observation screen after displacing the movable interferometer mirror through a known distance Δz. Then the measured wavelength is calculated from the following formula:

$$\lambda = 2\Delta z/N. \tag{11.12}$$

This technique was used, in particular, for the precise measurement of the laser wavelength.

11.4.1 The reference plane

Figure 11.7(a) shows a simplified diagram of the Michelson interferometer. Figure 11.7(b) shows an equivalent diagram in which the mirror M_1 is removed and the so-called "reference plane" R is introduced instead of the image of the mirror M_1 in the light-splitting plate. It can be seen from Fig. 11.7(b) that the operation of the Michelson interferometer can be adequately analysed by considering the interference of waves reflected by the reference plane R and by the mirror M_2. We thus have a problem of the interference of waves reflected from the surfaces of a plane-parallel plate.

11.4.2 Fringes of equal inclination

Let a plane monochromatic optical wave fall on to a plane-parallel plate, both surfaces of which partially reflect the light (Fig. 11.8). We shall calculate the optical path difference Δ for the beams reflected from the front and back surfaces of the plate. Let h be the thickness of the plate, α the angle of incidence, β the refraction angle, and n the refractive index of the plate. According to Snell's law,

$$\sin\alpha = n\sin\beta. \tag{11.13}$$

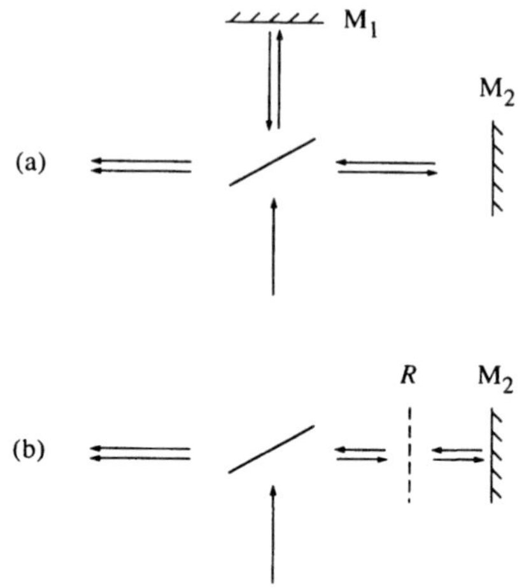

Fig. 11.7 The Michelson interferometer (a) and its equivalent (b). The dashed line shows the position of the reference plane.

Since the phase shift of the light wave travelling in the medium is proportional to the refractive index n (see Part IV), the sought-after path difference is

$$\Delta = 2ABn - AD. \tag{11.14}$$

It can be seen from Fig. 11.8 that

$$AD = AC\sin\alpha, \quad AC = 2AB\sin\beta, \quad AB = h/\cos\alpha. \tag{11.15}$$

Equations (11.13)–(11.15) yield

$$\Delta = 2hn\cos\beta. \tag{11.16}$$

Equations (11.13) and (11.16) show that the path difference of the beams reflected from the sides of the plane-parallel plate depends on the angle at which the light falls on the plate (the "angle of inclination"). The interference fringes

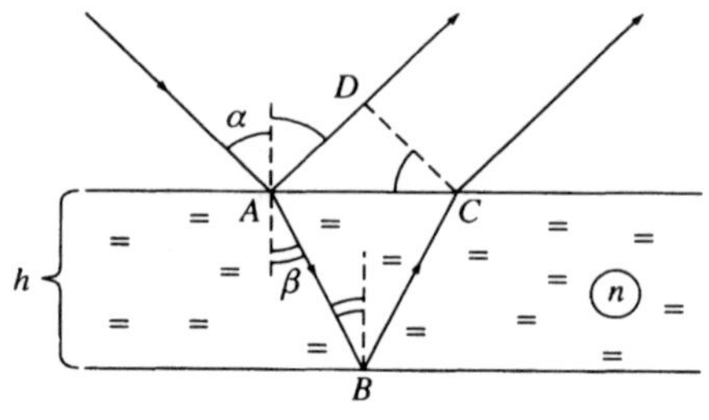

Fig. 11.8 The interference of light in a plane parallel plate.

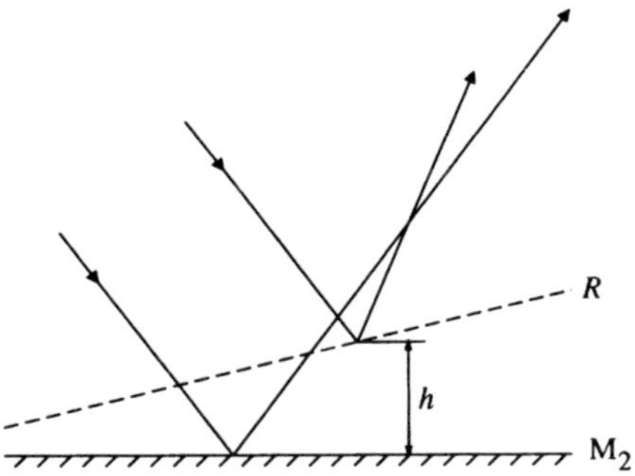

Fig. 11.9 The interference of light in a wedge-like plate.

formed by a diverging optical beam are therefore called *fringes of equal inclination*. In particular, an axially symmetric laser beam produces the interference pattern in the form of concentric rings. Such a pattern can be observed with the help of the Michelson interferometer, using a helium–neon laser as a source of light (Fig. 11.6).

11.4.3 Fringes of equal thickness

If one of the mirrors of the Michelson interferometer (Fig. 11.7(a)) is slightly inclined so that the planes of mirrors M_1 and M_2 are not mutually perpendicular, then the reference plane R, i.e. the plane occupying the place of the image of mirror M_1 in the split-dividing plate, and mirror M_2 are not parallel and form the faces of a wedge-like plate.

An illustration of the reflection of parallel optical rays from the wedge faces is given in Fig. 11.9. The path difference can be calculated approximately by eqn (11.16), where h is the thickness of the wedge at the reflection point. Since the value of h varies along the wedge-like plate, the resulting interference pattern is said to form *fringes of equal thickness.*

To demonstrate the operation of the Michelson interferometer, we use a helium–neon laser as a source of light. The interference pattern can be projected onto a screen or onto the ceiling of the lecture room. One of the mirrors of the demonstration interferometer is fixed, while the other is mounted on a carriage and can be moved along the optical beam over a distance of about 20 cm. The movable mirror is mounted on a special universal joint that can be rotated round two mutually perpendicular axes, lying in the vertical plane. With this arrangement we can demonstrate interference fringes of equal inclination and interference fringes of equal thickness. The linear size of the interference patterns can be increased by sending both interfering beams through a special reflecting plate and then through a diverging lens with a small local length, of 5–10 cm. A special feedback servo system is used for stabilization of the interference pattern.

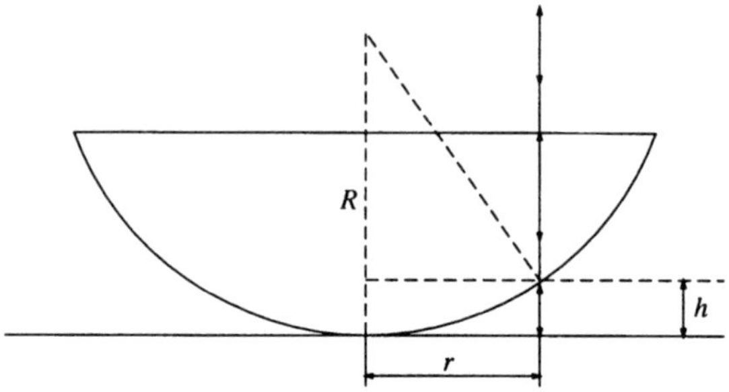

Fig. 11.10 The calculation of the radii of Newton's rings.

11.4.4 Newton's rings

The fringes of equal thickness can be observed with the help of a glass plate and a lens. If we press the convex surface of the lens to the plate and illuminate the system with a parallel beam of light, then the rays reflected by the plate and lens surface will interfere to produce a system of interference fringes in the form of concentric dark and bright rings ("Newton's rings"). Let us calculate the radii of the interference rings. The path difference of the interfering beams is $\Delta = 2h$, where h is distance from the lens surface to the plate. It can be seen from Fig. 11.10 that $h = R - \sqrt{R^2 - r^2}$, where R is the radius of curvature of the lens surface, and r is the distance from the point at which the lens touches the plate to the optical beam. Assuming $r \ll R$, we obtain the approximate expression $\Delta = r^2/R$. The radii of the bright rings are determined by the condition $\Delta = m\lambda$, where λ is the optical wavelength, and $m = 0, 1, 2, \ldots$. Thus we come to following expression for the radii of Newton's rings:

$$r_m = \sqrt{mR\lambda}, \qquad m = 0, 1, 2, \ldots . \tag{11.17}$$

This result can be verified experimentally.

11.4.5 Diagnostics interferometry

The interference pattern observed with the Michelson interferometer is very sensitive to such small perturbations as a mechanical displacement of the mirrors, and variation of the composition, temperature, and density of the medium through which the optical beams pass. This allows one to use the interferometer for various precise measurements. Specifically, mechanical displacements of the order of an optical wavelength, i.e. about 10^{-4} cm, can be measured. By using special techniques, the measurement precision can be raised by several more orders of magnitude. Such accurate measurements are made, for example, in seismology to register the motion of the Earth's crust.

11.5 The interference of non-monochromatic light

The characteristics of interference depend essentially on the spectral composition of light. We start the analysis of this dependence with a consideration of a simple case of biharmonic (two-frequency) wave interference.

Suppose that two plane monochromatic optical waves with frequencies ω_1 and ω_2 enter the Michelson interferometer. Assuming for simplicity that the wave amplitudes are equal, we shall write the input signal in the form

$$E_0 = A_0 \cos(\omega_1 t - k_1 z) + A_0 \cos(\omega_2 t - k_2 z). \tag{11.18}$$

Here k_1 and k_2 are the wave numbers related to the wavelengths as follows:

$$k_1 = 2\pi/\lambda_1, \qquad k_2 = 2\pi/\lambda_2. \tag{11.19}$$

Since the interferometer is a *linear system*, each of the monochromatic waves traverses it *independently* and gives rise to the interference pattern described by eqns (11.4)–(11.9). This allows us to write the output intensity I as a function of the path difference $\Delta = l_1 - l_2$ in the following form:

$$I = \tfrac{1}{2} I_0 (1 + \cos\delta_1) + \tfrac{1}{2} I_0 (1 + \cos\delta_2), \tag{11.20}$$

where

$$I_0 = cA_0^2/8\pi, \qquad \delta_1 = k_1\Delta, \qquad \delta_2 = k_2\Delta. \tag{11.21}$$

Equation (11.20) can also be presented as

$$I = I_0 + I_0 \cos k\Delta \cos\kappa\Delta, \tag{11.22}$$

where

$$k = (k_1 + k_2)/2, \qquad \kappa = (k_1 - k_2)/2. \tag{11.23}$$

Figure 11.11 shows the output intensity I as a function of the path difference Δ for the three cases in which plane monochromatic optical waves enter the Michelson interferometer: (a) one wave at frequency ω_1, (b) one wave at frequency ω_2, and (c) a pair of waves at frequencies ω_1 and ω_2. Distribution (c) is the sum of distributions (a) and (b). As seen from the figure, the *visibility* of the interference pattern depends on the path difference of the incoming non-monochromatic waves. The dependence is periodic for the two-frequency radiation: the visibility is the highest at $\Delta = 0$; as Δ grows, it decreases, then increases again and so on.

Let us determine the characteristic path difference Δ_0 at which the first smearing of the interference pattern takes place (Fig. 11.11(c)). It follows from eqn (11.22) that the condition determining Δ_0 is either $\cos\kappa\Delta = 0$ or $(k_1 - k_2)\Delta_0 = \pi$, whence

$$\Delta_0 = \pi/(k_1 - k_2) \tag{11.24}$$

or

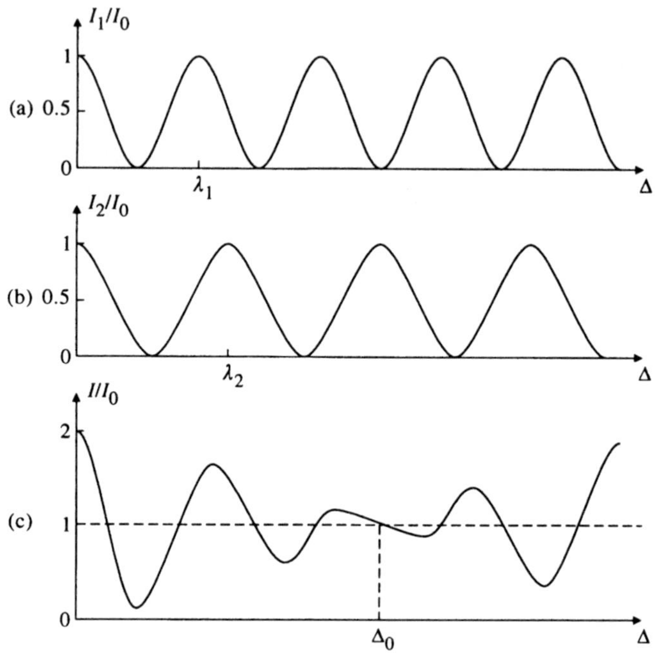

Fig. 11.11 The light intensity I at the exit of the Michelson interferometer as a function of the path difference of the beams, when the input signal is (a) a wave of frequency ω_1, (b) a wave of frequency ω_2, and (c) a pair of waves of frequencies ω_1 and ω_2.

$$\Delta_0 = \frac{1}{2\delta\nu}, \tag{11.25}$$

where $\delta\nu = \nu_1 - \nu_2$ is the spectral bandwidth of the radiation, and $\nu = 1/\lambda$ is the frequency expressed in cm^{-1}. If, for example, $\lambda_1 = 0.5\ \mu m$, and $\lambda_2 = 0.4\ \mu m$, then $\Delta_0 = 1\ \mu m$.

Thus, if non-monochromatic light enters the Michelson interferometer, then the period of the interference pattern is determined by the mean wavelength, whereas the visibility of the interference fringes depends on the path difference of the rays in the interferometer. Note that the characteristic path difference at which the first smearing of the interference pattern takes place is inversely proportional to the spectral bandwidth of the radiation.

11.5.1 The visibility of the interference pattern

To characterize the brightness contrast of the interference pattern, it is convenient to introduce a parameter called the *visibility*:

$$\gamma = \frac{I_{max} - I_{min}}{I_{max} + I_{min}}, \tag{11.26}$$

where $I_{\max}$ and $I_{\min}$ are the magnitudes of the light intensity at the neighbouring maximum and minimum of the interference pattern.

The parameter γ can be measured experimentally or calculated theoretically on the basis of a certain model of the optical wave. Generally, the dimensionless parameter γ lies in the range $0 \le \gamma \le 1$. The limiting case $\gamma = 1$ corresponds to a plane monochromatic wave that gives rise to an interference pattern with the highest contrast (see eqn (11.9)). Using eqn (11.22), we obtain, for two-frequency radiation, $I_{\max} = I_0(1 + \cos\kappa\Delta)$ and $I_{\min} = I_0(1 - \cos\kappa\Delta)$, whence

$$\gamma = \cos\kappa\Delta. \tag{11.27}$$

Thus, the visibility of the interference pattern formed by the non-monochromatic radiation depends on the path difference of the beams, $\gamma = \gamma(\Delta)$. Using eqn (11.27), we can rewrite eqn (11.22) in the form

$$I = I_0[1 + \gamma(\Delta)\cos k\Delta]. \tag{11.28}$$

11.6 The interference of a random optical wave

As was noted before (see Part II), the model of a random optical wave is the most adequate for the radiation emitted by a non-laser source of light. Let us consider the features of the interference of the *random* optical field in the Michelson interferometer. The electric field strength of an optical wave is generally a random vector function of spatial coordinates and time:

$$\vec{E} = \vec{E}(\vec{r}, t). \tag{11.29}$$

The description of such a field presents a complicated problem. To simplify it, we consider only one of the Cartesian components of the field $\vec{E}$ at a fixed point in space, $\vec{r}$ = constant. The field can then be described by a scalar function of time,

$$E = E(t), \tag{11.30}$$

representing a *random oscillation*. This model (11.30) is convenient for the description of the light field formed by a plane linearly polarized non-monochromatic wave. Using the model (11.30), we can calculate the interference pattern formed by the random optical field in the Michelson interferometer.

11.6.1 Characteristics of random optical oscillations

Suppose that the optical field strength $E = E(t)$ is a *steady-state stochastic process*. The multidimensional probability density is the most detailed statistical characteristic of an arbitrary stochastic process $x(t)$. There are, however, a few simpler characteristics of the process $x(t)$, such as the probability distribution $w(x)$, the mean value $\overline{x}$, the variance $\sigma^2 = \langle (x - \overline{x})^2 \rangle$, the correlation function $B(\tau) = \langle x(t)x(t+\tau) \rangle$, and the spectrum $G(\omega)$, which are frequently used in practice. Hereinafter, the angular brackets will denote statistical averaging which, for a steady-state process, is equivalent to time integration. The correlation function $B(\tau)$ and frequency spectrum $G(\omega)$ are generally real and even functions of their arguments. They are interrelated by the Fourier transformation (the Wiener–Khinchin theorem)

$$B(\tau) = \tfrac{1}{2}\int_{-\infty}^{\infty} G(\omega)\mathrm{e}^{-i\omega\tau}d\omega, \qquad G(\omega) = \frac{1}{\pi}\int_{-\infty}^{\infty} B(\tau)\mathrm{e}^{i\omega\tau}d\tau. \tag{11.31}$$

Additionally, the functions $w(x)$ and $G(\omega)$ are generally non-negative. The correlation function and the spectrum are characterized, in turn, by the correlation time τ_c and spectral bandwidth $\Delta\omega$, which define the widths of the distributions $B(\tau)$ and $G(\omega)$, respectively. Generally, the parameters τ_c and $\Delta\omega$ are inversely proportional to each other:

$$\tau_\mathrm{c} = \frac{2\pi}{\Delta\omega}. \tag{11.32}$$

To describe the optical field, we shall introduce the mean intensity

$$\overline{I} = \frac{c}{4\pi}\overline{E^2}, \tag{11.33}$$

the correlation function

$$B(\tau) = \langle E(t)E(t+\tau)\rangle, \tag{11.34}$$

and the spectrum $G(\omega)$. The mean intensity $\overline{I}$, the spectrm $G(\omega)$, and the spectral bandwidth $\Delta\omega$ of radiation can be measured experimentally.

The optical oscillation $E(t)$ can often be considered as a *narrow-band stochastic process,* i.e. a process the spectral bandwidth of which is much less than the mean frequency of oscillation:

$$\Delta\omega \ll \omega_0. \tag{11.35}$$

In this case it is worthwhile to introduce the complex amplitude of the field $\mathcal{E}(t)$, defining it by the formula

$$E(t) = \tfrac{1}{2}\mathcal{E}(t)\mathrm{e}^{i\omega_0 t} + \mathrm{c.\,c.} \tag{11.36}$$

Just like the field $E(t)$, the amplitude $\mathcal{E}(t)$ is a random function of time and, consequently, is characterized by a correlation function and a spectrum. Substituting eqn (11.36) into eqn (11.34) and taking into account that the stochastic process $E(t)$ is steady-state, we come to the stipulation

$$\langle \mathcal{E}(t)\mathcal{E}(t+\tau)\rangle = 0. \tag{11.37}$$

Therefore, the correlation function of the complex amplitude is introduced by the formula:

$$b(\tau) = \langle \mathcal{E}(t)\mathcal{E}^*(t+\tau)\rangle. \tag{11.38}$$

Assume $b(\tau)$ to be real. Then it follows from eqns (11.34) and (11.36)–(11.38) that

$$B(\tau) = \tfrac{1}{2}b(\tau)\cos\omega_0\tau. \tag{11.39}$$

The mean intensity of radiation is related to the complex amplitude of the field by

$$\overline{I} = \frac{c}{8\pi}\langle\mathcal{E}\mathcal{E}^*\rangle. \tag{11.40}$$

Together with the correlation function $B(\tau)$, it is also convenient to introduce a dimensionless function called the correlation coefficient:

$$R(\tau) = B(\tau)/B(0). \tag{11.41}$$

Similarly, the correlation coefficient of the complex amplitude is

$$r(t) = b(\tau)/b(0). \tag{11.42}$$

It follows from eqns (11.39), (11.41), and (11.42), that

$$R(\tau) = r(\tau)\cos\omega_0\tau. \tag{11.43}$$

11.6.2 The interference of a random optical wave in the Michelson interferometer

Suppose that a random optical field of the form (11.30) enters the Michelson interferomenter. The mean intensity of the field is

$$\overline{I} = \frac{c}{4\pi}\overline{E^2}. \tag{11.44}$$

In this case the resulting optical field at the interferometer exit can be represented in the form

$$E_{\mathrm{r}} = \sqrt{RT}\left[E(t) + E(t+\tau)\right], \tag{11.45}$$

where

$$\tau = \Delta/c, \tag{11.46}$$

$\Delta = l_1 - l_2$ is the path difference of the beams in the interferometer, c is the velocity of light, and R and T are the coefficients of light reflection and transmission by the light-dividing plate (in terms of the intensity of an optical wave). According to eqn (11.45), the field at the interferometer exit is proportional to the sum of the input fields, taken at two different moments in time. These moments are separated by a delay time τ proportional to the path difference of the beams in the interferometer.

Hereinafter, we shall assume condition (11.8) to be satisfied. Let us introduce the following designations:

$$E(t) = E, \qquad E(t+\tau) = E_\tau. \tag{11.47}$$

Then

$$E_r = \tfrac{1}{2}(E + E_\tau). \tag{11.48}$$

Since the field E is random, the field E_r is also random and can be characterized by the mean intensity

$$\overline{I_r} = \frac{c}{4\pi}\overline{E_r^2}. \tag{11.49}$$

Substituting eqn (11.48) into eqn (11.49) and taking into account that for the steady-state stochastic process

$$\overline{E^2} = \overline{E_\tau^2}, \qquad \overline{EE_\tau} = B(\tau), \qquad B(\tau)/B(0) = R(\tau), \tag{11.50}$$

we obtain

$$\overline{I_r} = \tfrac{1}{2}\overline{I}[1 + R(\tau)] \tag{11.51}$$

or, by virtue of eqn (11.43),

$$\overline{I_r} = \tfrac{1}{2}\overline{I}\left[1 + r(\tau)\cos\omega_0\tau\right]. \tag{11.52}$$

Thus, eqn (11.52) obtained using of the model of a random optical wave describes the mean intensity of light $\overline{I_r}$ at the exit of the Michelson interferometer via the mean intensity of the input beam $\overline{I}$, the mean frequency ω_0, of the input field, the delay time $\tau = \Delta/c$, depending on the path difference $\Delta = l_1 - l_2$ of the beams in the interferometer, and the correlation coefficient $r(\tau)$ of the complex amplitude of the field.

The visibility of the interference pattern can easily be calculated from eqn (11.52). Indeed, according to eqn (11.52), the maximum intensity of light in the interference pattern is

$$I_{\max} = \tfrac{1}{2}\overline{I}\left[1 + r(\tau)\right], \tag{11.53}$$

while the minimum intensity is

$$I_{\min} = \tfrac{1}{2}\overline{I}\left[1 - r(\tau)\right]. \tag{11.54}$$

It follows from eqns (11.26), (11.53), and (11.54), that

$$\gamma = r(\tau). \tag{11.55}$$

Thus, eqn (11.55) relates the visibility of the interference pattern γ defined by eqn (11.26) to the correlation coefficient of the complex amplitude of the field $r(\tau)$ defined by eqns (11.38) and (11.42). Since the visibility of the interference pattern can be measured experimentally with the help of the Michelson interferometer, eqn (11.55) provides a recipe for direct experimental measurements of the correlation coefficient $r(\tau)$.

Using the Wiener–Khinchin theorem (11.31) and specifying the model of the radiation spectrum $G(\omega)$, we can calculate the function $r(\tau)$. Let, for example,

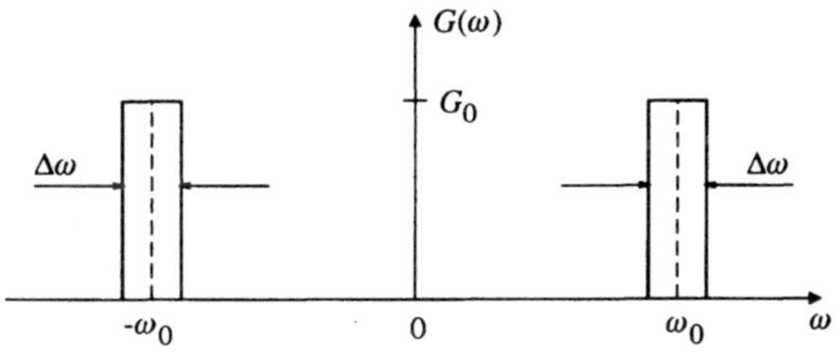

Fig. 11.12 A model of the radiation spectrum.

the spectrum be of rectangular form with a central frequency ω_0 and bandwidth $\Delta\omega$ (Fig. 11.12). Assuming

$$G(\omega) = \begin{cases} G_0, & |\omega \pm \omega_0| \leq \Delta\omega/2, \\ 0, & |\omega \pm \omega_0| > \Delta\omega/2 \end{cases} \tag{11.56}$$

and using eqn (11.31) for $B(\tau)$, we obtain

$$B(\tau) = G_0\Delta\omega\,\mathrm{sinc}(\Delta\omega\tau/2)\cos\omega_0\tau, \tag{11.57}$$

where

$$\mathrm{sinc}(x) \equiv \frac{\sin x}{x}. \tag{11.58}$$

Thus $B(0) = G_0\Delta\omega$, and for $R(\tau) = B(\tau)/B(0)$ we obtain the following expression for the correlation coefficient:

$$R(\tau) = \mathrm{sinc}(\Delta\omega\tau/2)\cos\omega_0\tau. \tag{11.59}$$

Finally, comparing eqns (11.59) and (11.43), we find that

$$r(\tau) = \mathrm{sinc}(\Delta\omega\tau/2). \tag{11.60}$$

Thus, the correlation coefficient (11.60) of the field amplitude corresponds to the model of the spectrum in eqn (11.56). Using eqns (11.60) and (11.52), we can calculate the light intensity distribution in the interference pattern. This distribution takes the form

$$\overline{I_r} = \tfrac{1}{2}\overline{I}\left[1 + \mathrm{sinc}(\Delta\tau/2)\cos\omega_0\tau\right]. \tag{11.61}$$

In the specific case of monochromatic radiation, when $\Delta\omega = 0$, we have

$$\mathrm{sinc}\,(\Delta\omega\tau/2) = 1,$$

and eqn (11.61) is reduced to eqn (11.9).

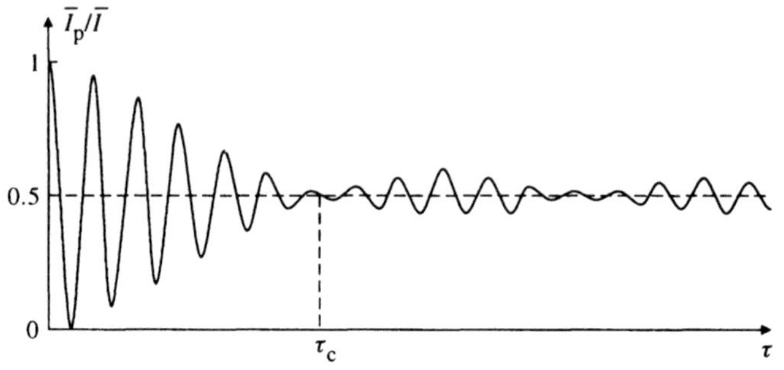

Fig. 11.13 The mean light intensity in the interference pattern calculated for the model of a narrow-band optical wave with a rectangular shape for the spectrum.

Figure 11.13 shows the graph of the function $\overline{I_r}(\tau)$, calculated from eqn (11.61) for a small, although finite, spectral bandwidth $\Delta\omega \ll \omega_0$.

It can be seen from Fig. 11.13 and eqn (11.61) that the period of the interference pattern formed by the non-monochromatic optical field in the Michelson interferometer is determined by the mean radiation frequency ω_0. As for the visibility of the interference pattern, it decreases as the delay time $\tau = \Delta/c$, determined by the path difference $\Delta = l_1 - l_2$ of the beams in the interferometer, increases. The time τ at which the visibility of the interference pattern becomes zero for the first time can be taken as a characteristic ("critical") time separating the regions in which the interference is well pronounced and in which it is practically absent (Fig. 11.13). According to eqns (11.55) and (11.60), the time τ_c is determined by the condition $\Delta\omega\tau_c/2 = \pi$, whence

$$\tau_c = \frac{2\pi}{\Delta\omega}. \tag{11.62}$$

Note that from the standpoint of the statistical model of the field, the time τ_c characterizes the width of the distribution $r(\tau)$, and, consequently, can be regarded as the *correlation time* of the optical wave. Thus, eqn (11.62) relates the spectral bandwidth $\Delta\omega$ and the correlation time τ_c of a plane non-monochromatic optical wave. This relationship obtained here for the model of a narrow-band optical wave with a rectangular spectrum coincides with the general formula (11.32) which follows from the Wiener–Khinchin theorem.

We can thus conclude that the interference of non-monochromatic light is possible only if the path difference of the beams is small enough. The critical path difference for which the interference pattern is still observable is given by the formula $\Delta_c = c\tau_c = 2\pi c/\Delta\omega$, or

$$\Delta_c = \frac{1}{\delta\nu}, \tag{11.63}$$

where $\delta\nu = \Delta\omega/2\pi c$ is the spectral bandwidth of the radiation, expressed in

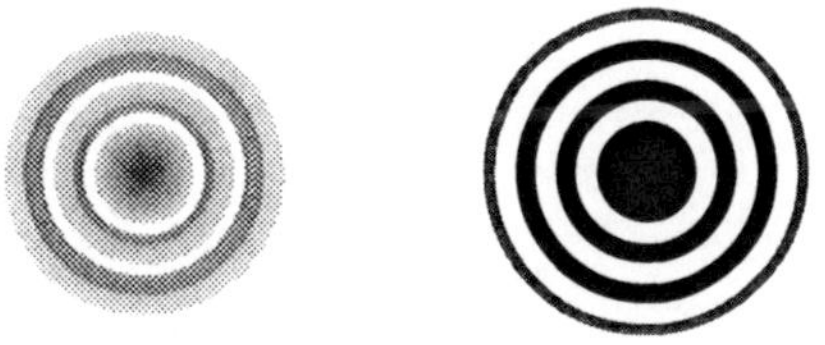

(a) (b)

Fig. 11.14 Newton's rings observed with (a) white and (b) coloured light.

units of cm^{-1}. If, for example, $\delta\nu = 1000\ \text{cm}^{-1}$, the critical path difference of the beams $\Delta_c = 0.001$ cm. Experiment confirms this result. For example, the interference of white light can be observed with *thin films*, such as the film of a soap bubble or an oil film on a water surface. When illuminated by sunlight, such films look brightly coloured. This effect is caused by the interference of light and by non-homogeneity of the film thickness.

Newton's rings can be observed in white light only within a small region near the point at which the lens and the plate touch, i.e. in the region in which the thickness of the air layer and the path difference of the interfering beams are small enough (see Fig. 11.10). However, if we use radiation with a narrower spectrum, e.g. arc-lamp light passed through a red filter, then the visibility and the size of the interference pattern essentially increase (Fig. 11.14). Similar effects can be demonstrated with a thin glass wedge or with soap bubbles.

The interference of non-monochromatic light in the Michelson interferometer is possible only if the optical lengths of its arms are equal with a high degree of accuracy. To equalize the lengths of the interferometer arms, a special *compensation plate* is used (see Fig. 11.15). The compensating plate is not necessary if a laser is used as a source of light, because the interfering radiation has a narrow spectrum and the allowable path difference of the beams becomes quite large. For example, using a helium–neon laser, we have $\delta\nu = 0.04\ \text{cm}^{-1}$ and, consequently, $\Delta_c = 1/\delta\nu = 25$ cm.

Thus, the experiments on the interference of light confirm the basic conclusions of the theory, based on the model of a random optical wave.

11.6.3 Fourier spectroscopy

The foregoing discussion shows that the Michelson interferometer can be used for direct experimental measurement of the correlation function $B(\tau)$ of an optical wave. Using the Wiener–Khinchin theorem, we can calculate the radiation spectrum:

$$G(\omega) = \frac{1}{\pi}\int_{-\infty}^{\infty} B(\tau)\mathrm{e}^{-i\omega\tau}\,d\tau. \tag{11.64}$$

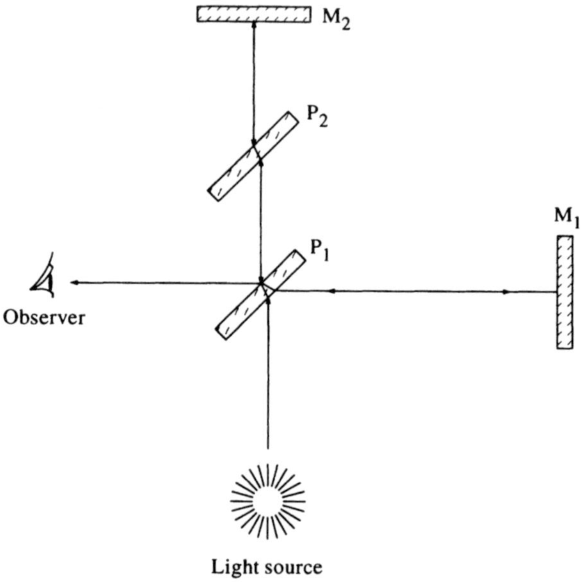

Fig. 11.15 A schematic diagram of the Michelson interferometer with a compensating plate.

This method of measurement of optical spectra is called *Fourier spectroscopy.* It is illustrated in Fig. 11.16. In practice, Fourier spectrometers are used for the investigation of narrow spectral lines, as well as infrared spectral lines, the analysis of which by the usual spectral instruments (prisms and diffraction gratings) presents certain difficulties.

11.7 Multiple-beam interference

The phenomenon of interference of light is also used for diagnostics of optical radiation, including structural analysis of the optical field, spectrum measurements, etc. To improve the sharpness of the interference pattern, one frequently applies the technique of multiple-beam interference, when several optical beams take part in the formation of the interference field. We shall now consider the main features of multiple-beam interference, using the example of the Fabry–Perot interferometer.

Suppose that a plane monochromatic optical wave falls onto a plane parallel glass plate, both sides of which efficiently reflect the light (Fig. 11.17). Penetrat-

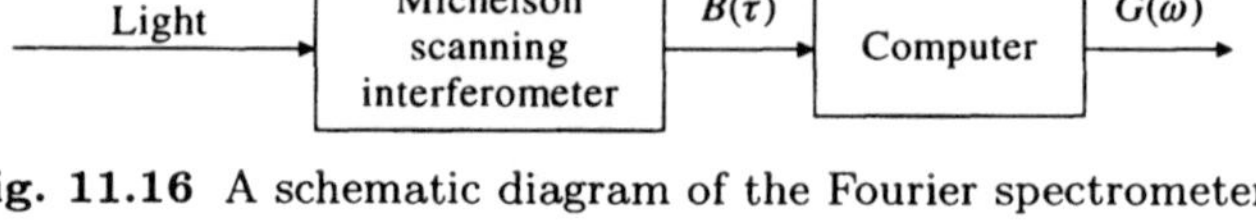

Fig. 11.16 A schematic diagram of the Fourier spectrometer.

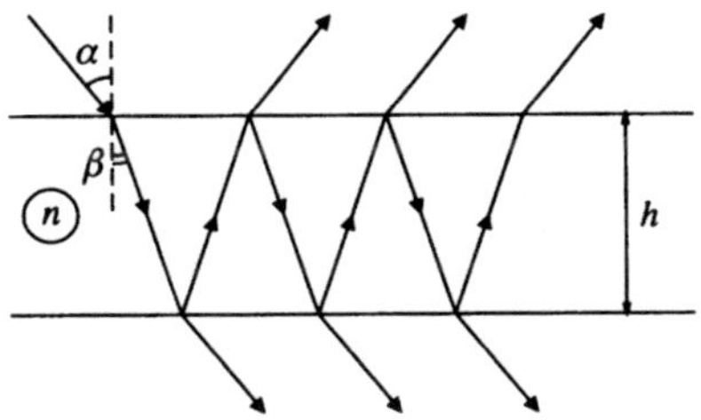

Fig. 11.17 Multiple-beam light interference.

ing partially into the plate, the light is reflected repeatedly by the sides of the plate. At each reflection, a fraction of the radiation escapes the plate, forming a set of plane waves that interfere with one another. Let us calculate the main characteristics of the interference pattern that emerges in this process.

Let R be the reflection coefficient, that is, the ratio of the reflected intensity to the incident one, and T the transmission coefficient, that is, the ratio of the transmitted intensity to the incident one, for each side of the plate. For simplicity, the characteristics of the both sides of the plate are assumed to be the same. Neglecting the absorption of light in the plate, we can write

$$T + R = 1. \tag{11.65}$$

Let us also introduce the coefficients of reflection and transmission of light, ρ and τ, at the sides of the plate in terms of the wave amplitude. Since the intensity of light is proportional to the square of the wave amplitude, we have

$$\rho^2 = R, \qquad \tau^2 = T. \tag{11.66}$$

Denoting the amplitude of the incident wave by A_0, and the amplitude of the wave transmitted by the plate by A, we can write

$$A = A_0\tau^2(1 + \rho^2 e^{i\delta} + \rho^4 e^{2i\delta} + \ldots), \tag{11.67}$$

where δ is the phase shift of the optical wave over the round trip across the plate. Equation (11.67) takes into account that the first beam leaving the plate has crossed the plate boundaries twice, while the second beam has crossed the boundaries twice and reflected from them twice, the third beam has crossed the boundaries twice and reflected from them four times, etc.

Using eqn (11.16), we can present the phase shift δ in the form $\delta = k\Delta$, where $k = 2\pi/\lambda$ is the wave number, $\Delta = 2hn\cos\beta$ is the optical path, n is the refractive index of the glass, h is the plate thickness, and β is the refraction angle of the beam in the plate (see Figs 11.8 and 11.17), related to the angle of incidence α by

$$\sin\alpha = n\sin\beta. \tag{11.68}$$

Thus,

$$\delta = \frac{4\pi}{\alpha} hn \cos\beta. \tag{11.69}$$

Note that the expression in brackets in eqn (11.67) is the sum of a diminishing geometric progression. Using the formula

$$\sum_{n=0}^{\infty} q^n = \frac{1}{1-q}, \tag{11.70}$$

where

$$q = \rho^2 e^{i\delta}, \tag{11.71}$$

we find from eqn (11.67) that

$$A = A_0 \frac{\tau^2}{1 - \rho^2 e^{i\delta}}. \tag{11.72}$$

As the transmission coefficient of the plate $T(\delta) = I/I_0 = |A/A_0|^2$, we obtain

$$T(\delta) = \frac{T^2}{|1 - Re^{i\delta}|^2} \tag{11.73}$$

or, in real variables,

$$T(\delta) = \frac{1}{1 + F\sin^2(\delta/2)}, \tag{11.74}$$

where eqn (11.65) is taken into account and the quantity F, called the *sharpness factor*, is introduced:

$$F = \frac{4R}{(1-R)^2}. \tag{11.75}$$

The meaning of this quantity is clear from Fig. 11.18, which shows the function $T(\delta)$ for two different values of F. It can be seen that the parameter F characterizes the sharpness of the interference pattern.

How large can the value of F be? Equation (11.75) shows that the sharpness factor grows with the reflection coefficient. A mirrored surface can have a reflection coefficient as high as $R = 0.99$. In that case, the sharpness factor $F = 4 \times 10^4$, and the function of the plate transmission $T(\delta)$ represents a set of discrete lines (Fig. 11.19).

11.7.1 A plane parallel plate as a frequency filter

Figure 11.19 shows that a plane parallel glass plate with highly reflecting (mirror) side surfaces transmits radiation only at certain discrete frequencies. In other words, such a plate represents a high-quality frequency filter. Such filters are used for mode selection in lasers (Fig. 11.20) as well as for wavelength calibration in spectroscopic measurements.

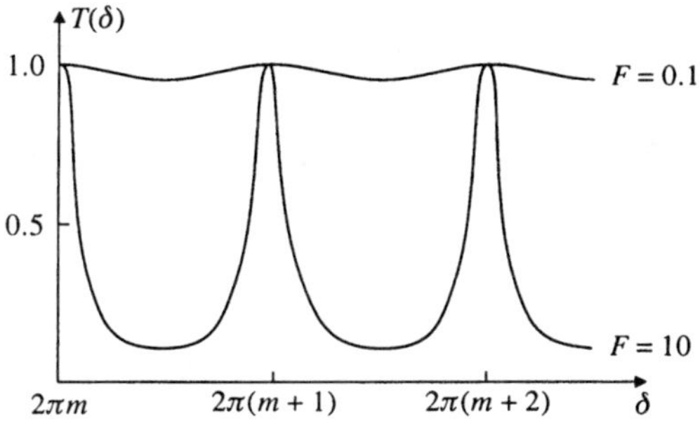

Fig. 11.18 The function $T(\delta)$ calculated from eqn (11.74) for different values of the sharpness factor F.

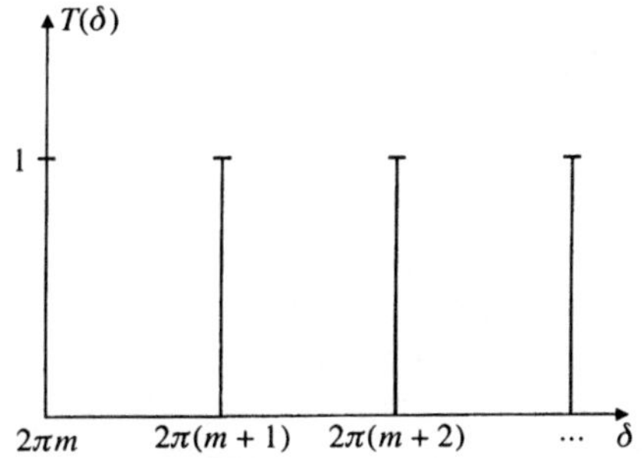

Fig. 11.19 The transmission function of the Fabry–Perot interferometer.

11.8 The Fabry–Perot interferometer

The Fabry–Perot interferometer is intended for investigation of the fine structure of optical spectra. A schematic diagram of the device is shown in Fig. 11.21. Its basic element is a pair of parallel mirror plates, mounted at some distance d from each other. The interferometer plates have a wedge-like form to avoid undesirable interference of the beams reflected from the back sides of the plates.

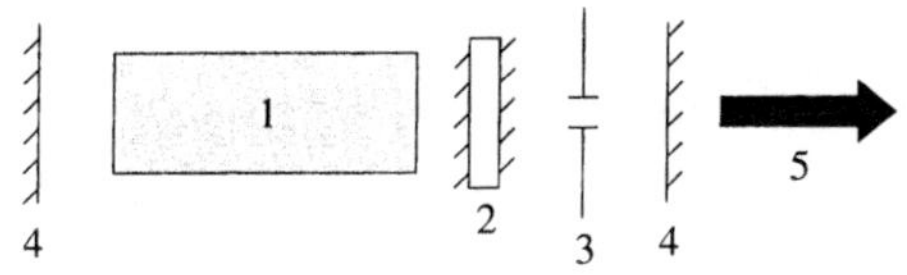

Fig. 11.20 The application of a plane parallel plate for laser mode selection. 1, Active element of the laser; 2, plane parallel plate; 3, diaphragm; 4, resonator mirrors; 5, laser beam.

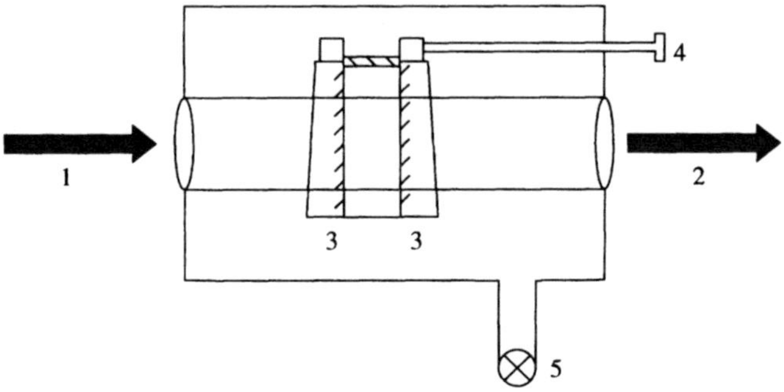

Fig. 11.21 A schematic diagram of the Fabry–Perot interferometer. 1, Input beam; 2, output beam; 3, mirror plates; 4, mirror adjustment screws; 5, gas inlet.

Like the plane parallel plate, the Fabry–Perot interferometer uses multiple-beam interference of light to obtain a sharp interference pattern. However, unlike the plate, the interferometer allows variation of the distance between the plates to adjust the phase shift δ, determined by eqn (11.69). In some cases the inter-plate space is filled with a gas. By varying the gas pressure, one can gradually vary the refractive index n of the gas and, consequently, the phase shift δ (see eqn (11.69)).

When using the Fabry–Perot interferometer, it is conventional to use a laser as a source of light. The interferometer has a rigid construction in which the mirrors are fixed by a special calibrated ring about 5 mm thick. A system of air pumping by a rubber hand pump is provided. The interferometer is illuminated by a diverging beam of monochromatic light produced by passing the helium–neon laser beam through a lens. The focal length of the lens is about 30 cm. As follows from eqns (11.68), (11.69), (11.74), and (11.75), for a fixed radiation wavelength λ, the interferometer is characterized by a discrete set of angles at which it transmits the light. Thus, we observe interference fringes of equal inclination. The pattern on the screen has the form of bright sharp rings separated by wide dark gaps. Thus, the observed interference pattern coincides with that shown in Fig. 11.19. If we pump the air into the space between the plates, the ring radii vary (Fig. 11.22).

Comparing the interference patterns observed using the Michelson interferometer with those observed using the Fabry–Perot interferometer, we can conclude that multiple-beam interference substantially improves the sharpness of the interference pattern. The high sharpness of the interference in the Fabry–Perot interferometer allows one to use it as a spectroscopic device of high resolving power (see Chapter 17).

11.8.1 The finesse of the fringes

Figure 11.23 shows the function $T(\delta)$ for the case in which the width of the interference maxima is comparable to the separation between the maxima, i.e.

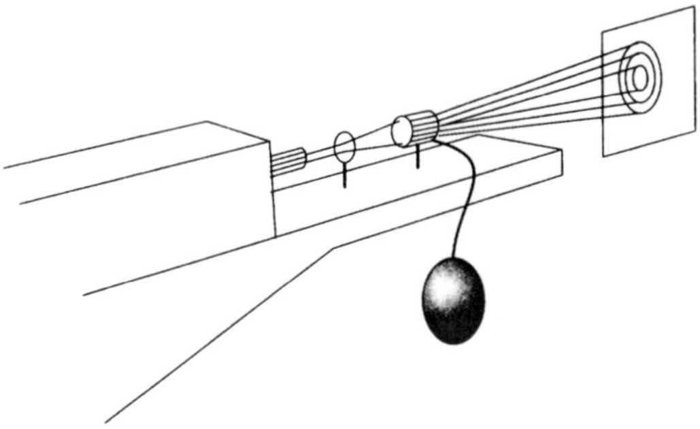

Fig. 11.22 The observation of the interference of laser light with the Fabry–Perot interferometer.

when the sharpness factor F is not too high. The sharpness of the interference pattern can be characterized by the ratio of the separation between adjacent maxima to the width of an individual maximum:

$$\mathcal{F} = 2\pi/\varepsilon. \tag{11.76}$$

The value $\mathcal{F}$ is called *finesse* of the fringes. The finesse can be determined experimentally as the ratio of the distance between adjacent interference rings to the width of an individual ring.

Using eqn (11.74) and assuming that

$$R \approx 1, \qquad F \gg 1, \tag{11.77}$$

it is easy to show that the width of the interference maximum, that is, the full width at half maximum (FWHM) is determined by the formula (see Fig. 11.23)

$$\varepsilon = 4/\sqrt{F} \tag{11.78}$$

or, by virtue of eqns (11.75) and (11.77),

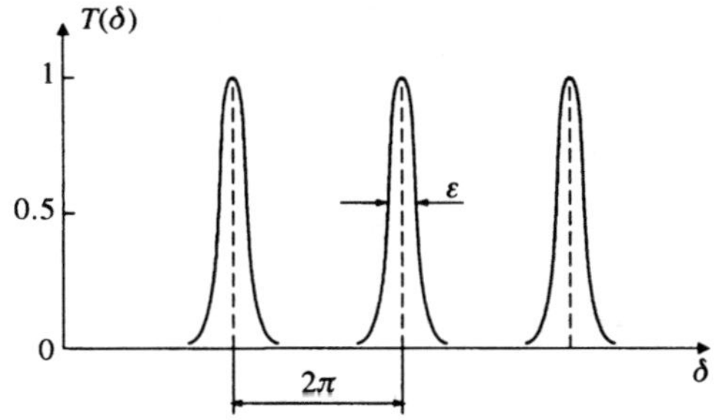

Fig. 11.23 The calculation of the finesse of the fringes.

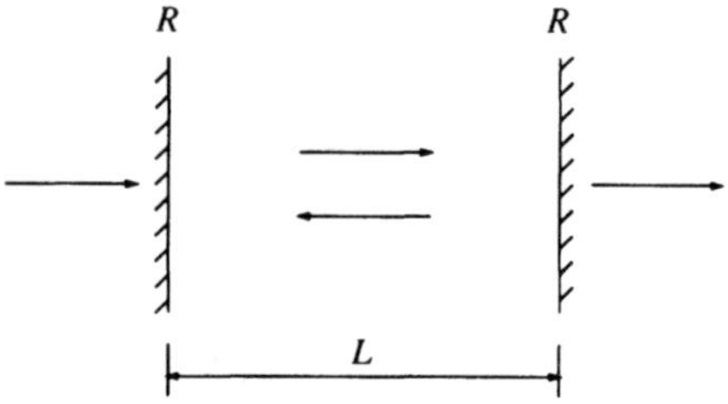

Fig. 11.24 The optical resonator.

$$\varepsilon = 2(1 - R). \tag{11.79}$$

It follows from eqns (11.76) and (11.79) that

$$\mathcal{F} = \pi/(1 - R). \tag{11.80}$$

If, for example, $R = 0.9$, then $\mathcal{F} = 30$. This estimate can be checked experimentally.

11.9 Eigenmodes and the quality factor of a laser resonator

The simplest laser resonator represents a pair of parallel mirrors, placed opposite to each other (Fig. 11.24). It can be considered as a version of the Fabry–Perot interferometer with the parameters

$$n = 1, \qquad \alpha = \beta = 0. \tag{11.81}$$

Let L be the resonator length and let R be the reflection coefficient of the mirrors in terms of the light intensity. Let us assume, for simplicity, that both mirrors are identical. It follows from eqns (11.69) and (11.81) that

$$\delta = 4\pi\nu L, \tag{11.82}$$

where $\nu = 1/\lambda$ is the radiation frequency, expressed in units of cm^{-1}.

According to eqn (11.74), the transmission maxima of the Fabry–Perot interferometer fall at the values of δ determined by

$$\delta_m = 2\pi m, \qquad m = 1, 2, 3, \ldots . \tag{11.83}$$

Equations (11.82) and (11.83) define the discrete set of frequencies

$$\nu_m = m\Delta\nu, \qquad \Delta\nu = 1/2L, \qquad m = 1, 2, \ldots, \tag{11.84}$$

at which the Fabry–Perot interferometer transmission reaches a maximum. The frequencies determined by eqn (11.84) are called the *eigenmode frequencies of the*

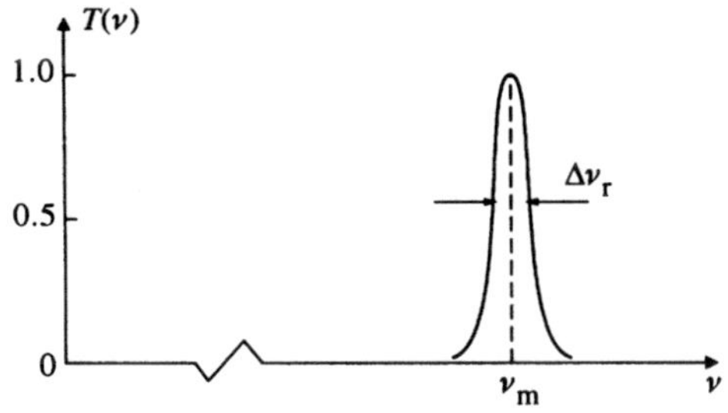

Fig. 11.25 The calculation of the quality factor of an optical resonator.

resonator. It is easy to show that the resonator length accomodates an integer number of half-wavelengths of the eigenmode frequency. Note that eqn (11.84) in fact defines the same frequencies as the phase condition of laser self-excitation does for the laser generation frequencies (see Chapter 10).

11.9.1 Quality factor of laser resonator

Using eqns (11.74) and (11.82), we can plot the transmission coefficient $T(\nu)$ of the Fabry–Perot interferometer as a function of the light frequency. Figure 11.25 shows its typical form near the frequency ν_m of one of the resonator eigenmodes for $F \gg 1$.

Denote the interference maximum width shown in Fig. 11.25 by $\Delta\nu_r$. This quantity can be called the spectral bandwidth of the optical resonator. Let us also introduce the quality factor of the laser resonator:

$$Q = \nu/\Delta\nu_r, \tag{11.85}$$

where $\nu = 1/\lambda$ is the light frequency. It follows from eqns (11.79) and (11.82) that $\Delta\nu_r = \varepsilon/4\pi L$ and

$$\Delta\nu_r = \frac{1-R}{2\pi L}. \tag{11.86}$$

Hence

$$Q = \frac{2\pi L}{(1-R)\lambda}. \tag{11.87}$$

Thus, eqns (11.86) and (11.87) determine the spectral bandwidth and quality factor of the optical resonator. In these equations L is the resonator length, R is the mirror reflection coefficient, and λ is the optical wavelength.

Let us make a numerical estimate. Assuming $L = 50$ cm, $\lambda = 0.5$ μm, and $R = 0.9$, we obtain $\Delta\nu = 1/2L = 10^{-2}$ cm^{-1}, $\Delta\nu_r = 3 \times 10^{-4}$ cm^{-1}, and $Q = 6 \times 10^7$. Thus, the optical resonator represents a high-quality system.

Bibliography

Born, M. and Wolf, E. (1968). *Principles of optics.* Pergamon Press, Oxford.
Demtröder, W. (1982). *Laser spectroscopy.* Springer-Verlag, Berlin.

12

The coherence of light

The temporal coherence of light—coherence time and coherence length. The spatial coherence of light—the coherence radius. A model of a random optical field. Calculation of the interference pattern obtained with the Young interferometer. The measurement of coherence and the coherence of radiation emitted by real sources of light.

The notion of *coherence* of light plays an important part in modern optics. Coherence implies regularity in the structure of the optical field and characterizes its similarity to the structure of an ideal harmonic wave. Coherence is the basic property of laser radiation. It is this property that drastically distinguishes the laser from other sources of light. The coherence of light plays a decisive part in holography, and it is important in the recording and processing of information, optical communications, the formation of optical images, the transmission of optical energy, the concentration of optical energy in time and space, the generation of superstrong optical fields and so on.

Historically, the notion of light coherence emerged in connection with experiments on the interference of light. The ability of light to interfere, i.e. to form a system of dark and bright fringes by superposition of optical beams, was called coherence. Two kinds of coherence, spatial and temporal, should be distinguished. Let us first consider the temporal coherence of light.

12.1 The temporal coherence of light—coherence time and coherence length

Suppose that we observe light interference from a source S with a Michelson interferometer, a schematic diagram of which is shown in Fig. 12.1(a). Figure 12.1(b) shows a fragment of the observed interference pattern and Fig. 12.1(c) demonstrates the distribution of light intensity in the interference pattern. Here I_0 and I are the intensities of light at the entrance and exit of the interferometer, $\tau = \Delta/c$ is the delay time proportional to the optical path difference of the beams $\Delta = l_1 - l_2$ in the interferometer, and c is the velocity of light.

The interference pattern can be characterized quantitatively by the parameter:

$$\gamma = \frac{I_{\max} - I_{\min}}{I_{\max} + I_{\min}}, \tag{12.1}$$

called the *visibility*. Here $I_{\max}$ and $I_{\min}$ are the values of the light intensity in adjacent maxima and minima of the interference pattern.

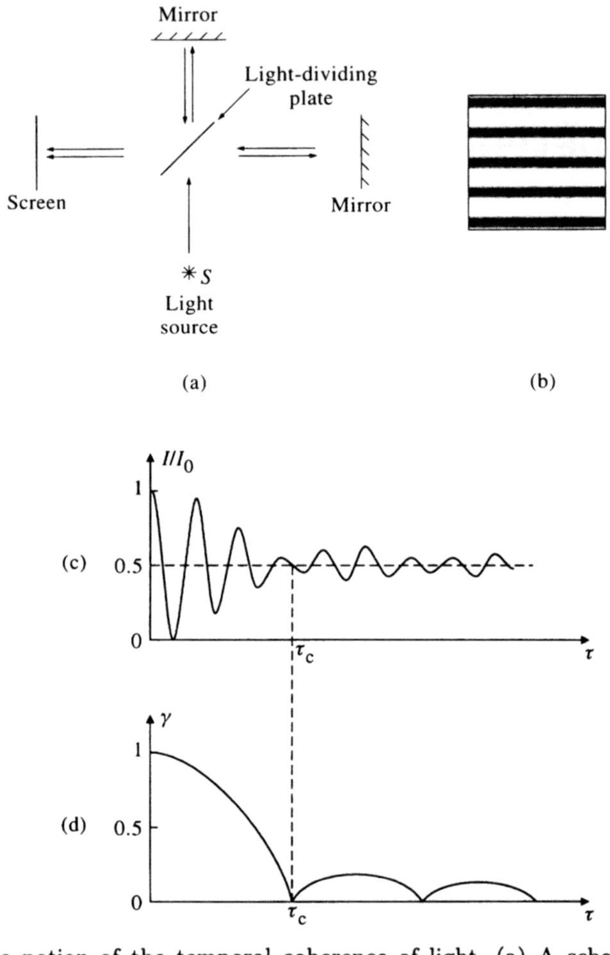

Fig. 12.1 The notion of the temporal coherence of light. (a) A schematic diagram of the Michelson interferometer. (b) A fragment of an interference pattern. (c) The distribution of light intensity in the interference pattern. (d) The visibility of the interference pattern as a function of the delay time τ proportional to the path difference of the beams in the interferometer.

Experiment shows that the visibility of the interference pattern varies with the path difference; a typical dependence $\gamma(\tau)$ has the form shown in Fig. 12.1(d). One can distinguish two characteristic regions depending on the relative delay τ of the beams in the interferometer: (i) the region $\tau < \tau_c$ of well pronounced interference and (ii) the region $\tau > \tau_c$, in which the interference is drastically reduced. The critical value of the delay time τ_c separating these two regions can be found experimentally.

The coherence of light was defined earlier as the ability of light to produce an interference pattern. We can then say that the light is *coherent* for $\tau < \tau_c$, and *non-coherent* for $\tau > \tau_c$. The time τ_c thus characterizes the optical wave itself and called a *coherence time.*

The critical delay time τ_c corresponds to the critical path difference $\Delta_c = c\tau_c$, where c is the velocity of light. This quantity characterizes the optical wave and is called a coherence length. Denoting the coherence length by l_c, we obtain

$$l_c = c\tau_c. \tag{12.2}$$

These are characteristics of the temporal, or longitudinal, coherence of light from the experimental point of view. The theory of temporal coherence treats the optical field as a steady-state *random process*. Writing the field in the form

$$E(t) = \tfrac{1}{2}\mathcal{E}(t)e^{i\omega_0 t} + \text{c. c.} \tag{12.3}$$

and introducing the correlation coefficient of the complex amplitude

$$r(\tau) = \frac{\langle \mathcal{E}(t)\mathcal{E}^*(t+\tau)\rangle}{\langle \mathcal{E}\mathcal{E}^*\rangle}, \tag{12.4}$$

we can calculate the visibility of the interference pattern observed with the Michelson interferometer (see Chapter 11) and show that

$$\gamma = r(\tau). \tag{12.5}$$

Equation (12.5) establishes a relationship between an experimemtally measurable quantity, that is, the visibility of the interference pattern, and a statistical characteristic of light, which is the correlation coefficient of the amplitude of optical oscillations.

The characteristic decay time of the function $r(\tau)$ is called the *correlation time* of the optical oscillations. Let us denote this time by τ_{cor}. Since the coherence time of light τ_c was defined earlier as the characteristic decay time of the function $\gamma(\tau)$, we can obtain from eqn (12.5) the simple relation between the times τ_c and τ_{cor}:

$$\tau_c = \tau_{\text{cor}}. \tag{12.6}$$

Thus, the coherence time τ_c which can be measured experimentally proves to be equal to the correlation time of light, which is a parameter of the theoretical model.

According to the Wiener–Khinchin theorem, the correlation function $B(\tau) = \langle E(t)E(t+\tau)\rangle$ is related to the power spectrum $G(\omega)$ of the optical field by the Fourier transformation

$$B(\tau) = \tfrac{1}{2}\int_{-\infty}^{\infty} G(\omega)e^{-i\omega\tau}\,d\omega, \qquad G(\omega) = \frac{1}{\pi}\int_{-\infty}^{\infty} B(\tau)e^{i\omega\tau}\,d\tau. \tag{12.7}$$

The theorem gives the following relationship between the correlation time of light τ_{cor} and its spectral bandwidth $\Delta\omega$:

$$\tau_{\text{cor}} = 2\pi/\Delta\omega. \tag{12.8}$$

Using eqns (12.6) and (12.8), we obtain a convenient formula for evaluating the coherence time:

$$\tau_{\text{c}} = 2\pi/\Delta\omega. \tag{12.9}$$

This relationship, containing the parameters τ_c and $\Delta\omega$ that can be measured in independent optical experiments, allows a direct experimental verification. The experiment thus confirms the theory based on a statistical model of the optical field.

Using eqns (12.2) and (12.9), we obtain the following expression for the coherence length of light:

$$l_{\text{c}} = 1/\Delta\nu, \tag{12.10}$$

where $\Delta\nu = \Delta\omega/2\pi c$ is the spectral bandwidth, expressed in units of cm^{-1}. Specifically, for white light occupying the visible range from $\lambda_1 = 0.4\ \mu\text{m}$ to $\lambda_2 = 0.8\ \mu\text{m}$, we obtain $\Delta\nu = \nu_1 - \nu_2 = 1/\lambda_1 - 1/\lambda_2 = 10^4\ \text{cm}^{-1}$. Hence, using eqn (12.10), we have that $l_c = 10^{-4}\,\text{cm} = 1\,\mu\text{m}$. Thus, the coherence length of white light turns out to be of the order of an optical wavelength. This estimate explains the results of experiments with *thin films* in which the interference of white light is observed (see Chapter 11).

Let us make a similar estimate for helium–neon laser radiation. In this case the spectral bandwidth of the radiation is determined by the Doppler bandwidth $\Delta\nu = \Delta\nu_{\text{D}} = 0.04\ \text{cm}^{-1}$ (see Chapter 6). Consequently, $l_c = 1/\Delta\nu = 25$ cm. Thus, the coherence of laser radiation greatly exceeds that of natural light. This explains the experiments carried out with the Michelson laser interferometer, in which a distinct interference pattern was observed for a quite substantial optical path difference of about 10 cm.

12.2 The spatial coherence of light—the coherence radius

Suppose that we observe the interference of light with the Young interferometer, a schematic diagram of which is shown in Fig. 12.2(a). Figure 12.2(b) shows a fragment of the observed interference pattern, and Fig. 12.2(c) illustrates the light intensity in the observation plane.

The visibility of the interference pattern can be characterized by the parameter γ, defined by eqn (12.1). Experiment shows that the visibility depends on the distance s between the two pinholes (or slits) in the Young interferometer. A typical form of this dependence is shown in Fig. 12.2(d); the visibility of the interference pattern decreases as the distance s grows.

The ability of light to produce an interference pattern in the Young interferometer is called the *spatial coherence of light*. The character of the dependence $\gamma(s)$ allows us to establish the critical value s_c of the distance s, which separates the regions of strong and weak interference. According to the definition just given, we can say that the light is spatially coherent at $s < s_c$, and non-coherent at $s > s_c$. The critical distance s_c characterizes the optical wave itself. This distance is called a radius of spatial coherence or, simply, a *coherence radius* of

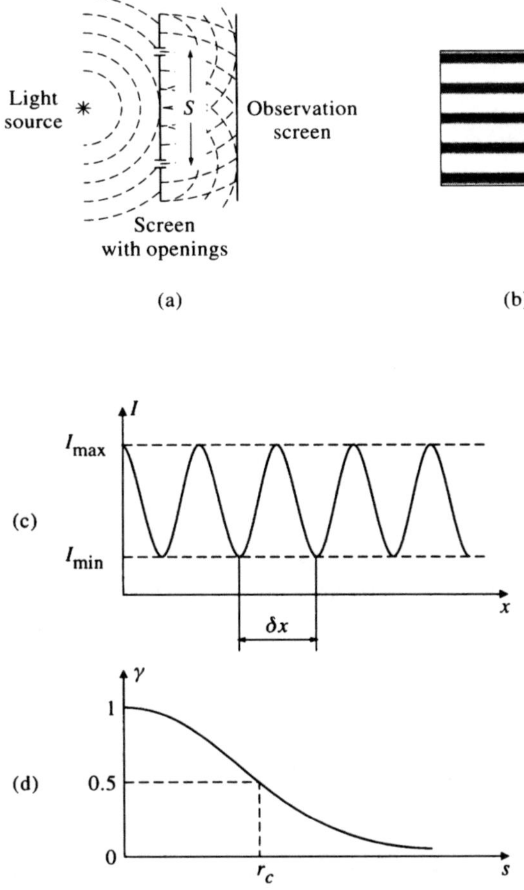

Fig. 12.2 The notion of the spatial coherence of light. (a) A schematic diagram of the Young interferometer. (b) A fragment of an interference pattern. (c) The distribution of light intensity on the observation screen. (d) The visibility of the interference pattern as a function of the distance s between the slits.

light. We shall denote this distance by r_c. Thus, the coherence radius of light, r_c, can be measured experimentally with the help of the Young interferometer.

12.3 A model of a random optical field

The physical meaning of spatial coherence can be clarified by calculating the interference pattern obtained with the Young interferometer. As in the case of the temporal coherence of light, we shall use the notion of a random optical field as a basis for the theory of spatial coherence. This time, however, the optical field is random in terms of spatial coordinates, but not in terms of time.

For simplicity, let us consider a monochromatic optical field. Assuming that the field oscillations at any point of space are harmonic, we can write

$$E(\vec{r}, t) = \tfrac{1}{2}\mathcal{E}(\vec{r})e^{i\omega_0 t} + \text{c. c.} \tag{12.11}$$

The dependence of the complex amplitude $\mathcal{E}(\vec{r})$ on the spatial coordinates implies that the optical oscillations may have different amplitudes and phases at different points in space. If, in particular, the function $\mathcal{E}(\vec{r})$ is *random*, then eqn (12.11) describes a random spatially non-uniform optical field. Such a field can be characterized by the mean intensity

$$\overline{I} = \frac{c}{4\pi}\overline{E^2} \tag{12.12}$$

and by the spatial correlation coefficient of the complex amplitude,

$$b(\vec{s}) = \frac{\langle \mathcal{E}(\vec{r})\mathcal{E}^*(\vec{r}+\vec{s})\rangle}{\langle \mathcal{E}\mathcal{E}^*\rangle}. \tag{12.13}$$

Using the model described by eqns (12.11)–(12.13), we can calculate the interference pattern obtained with the Young interferometer. In so doing, we come to the conclusion that the visibility of the interference pattern γ is equal to the correlation coefficient $b(\vec{s})$:

$$\gamma = b(\vec{s}). \tag{12.14}$$

Thus, we have established the physical meaning of visibility in terms of the structure of the optical field.

12.4 Calculation of the interference pattern obtained with the Young interferometer

Let us derive eqn (12.14). Figure 12.3 shows schematically the interference in the Young interferometer. In this figure L is the distance from the screen with two pinholes to the screen of observation, s is the distance between the pinholes, x is the coordinate of the observation point P, measured from the axis of symmetry, and l_1 and l_2 are the distances from the pinholes to the observation point.

Optical oscillations at the point P result from a superposition of optical waves coming from the pinholes O_1 and O_2:

$$E = E_1 + E_2. \tag{12.15}$$

Usually, the following conditions are satisfied in the experiment:

$$L \gg x \gg s. \tag{12.16}$$

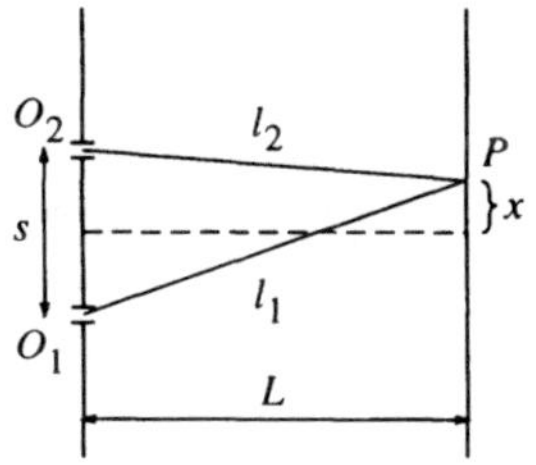

Fig. 12.3 The calculation of the interference pattern in the Young interferometer.

Omitting constant factors that are dependent on the size of the pinholes, we can rewrite the fields E_1 and E_2 in the form

$$E_1 = \tfrac{1}{2}\mathcal{E}_1 e^{i\omega_0(t-l_1/c)} + \text{c. c.}, \qquad E_2 = \tfrac{1}{2}\mathcal{E}_2 e^{i\omega_0(t-l_2/c)} + \text{c. c.}, \tag{12.17}$$

where

$$\mathcal{E}_1 = \mathcal{E}(\vec{r}_1), \qquad \mathcal{E}_2 = \mathcal{E}(\vec{r}_2), \tag{12.18}$$

and $\vec{r}_1$ and $\vec{r}_2$ are the radius vectors of the pinholes O_1 and O_2.

Substituting eqn (12.15) into eqn (12.12), we obtain the following expression for the mean radiation intensity at the point P:

$$\overline{I} = 2I_0(1 + R), \tag{12.19}$$

where

$$R = \frac{\langle E_1 E_2 \rangle}{\langle E^2 \rangle} \tag{12.20}$$

is the spatial correlation coefficient of the optical field, and I_0 is the light intensity corresponding to the case in which only one of the pinholes is open while the other is closed.

Using eqn (12.17) and assuming, for simplicity, that

$$\langle \mathcal{E}_1 \mathcal{E}_2^* \rangle = \langle \mathcal{E}_1^* \mathcal{E}_2 \rangle, \tag{12.21}$$

i.e. that the correlation function of the complex amplitude of the field is real, we obtain

$$\langle E_1 E_2 \rangle = \tfrac{1}{2}\langle \mathcal{E}_1 \mathcal{E}_2^* \rangle \cos(k_0 \Delta), \tag{12.22}$$

where

$$k_0 = \omega_0/c, \qquad \Delta = l_1 - l_2. \tag{12.23}$$

Let

$$\vec{r}_1 = \vec{r}, \qquad \vec{r}_2 = \vec{r} + \vec{s}, \tag{12.24}$$

where $\vec{s}$ is the vector drawn from the pinhole O_1 to the pinhole O_2 (see Fig. 12.3). Then it follows from eqns (12.20), (12.22), and (12.24) that

$$R = b(\vec{s})\cos(k_0\Delta), \tag{12.25}$$

where $b(\vec{s})$ is determined by eqn (12.13).

Using Fig. 12.3 and taking into account the relations (12.16), it is easy to show that the path difference of the beams in the Young interferometer is

$$\Delta = (s/L)x. \tag{12.26}$$

Denoting

$$q = (s/L)k_0 \tag{12.27}$$

and substituting eqn (12.25) into eqn (12.19), we finally obtain

$$\overline{I} = 2I_0\left[1 + b(\vec{s})\cos qx\right]. \tag{12.28}$$

Thus, eqn (12.28) describes the distribution of the mean radiation intensity in the interference pattern observed with the Young interferometer. In this formula I_0 is the intensity of light at the point P when only one of the pinholes, O_1 or O_2, is open, while the other is closed, x is the coordinate in the observation plane, measured from the axis of symmetry of the interferometer (see Fig. 12.3), q is the parameter determined by eqns (12.27) and (12.23), L is the distance from the screen with pinholes to the observation screen, and s is the distance between the pinholes. The function $b(\vec{s})$ in eqn (12.28) is defined by eqn (12.13) and denotes the spatial correlation coefficient of the complex amplitude of the optical field.

Using eqn (12.28), one can easily calculate the visibility of the interference pattern. Indeed, assuming $\cos qx = 1$ and $\cos qx = -1$, we obtain expressions for the light intensity at adjacent maxima and minima of the interference pattern:

$$I_{\max} = 2I_0\left[1 + b(\vec{s})\right], \qquad I_{\min} = 2I_0\left[1 - b(\vec{s})\right]. \tag{12.29}$$

Inserting eqn (12.29) into eqn (12.1), we obtain eqn (12.14). The derivation of eqn (12.14) is thus completed.

Equation (12.14) shows that the visibility of the interference pattern in the Young interferometer is governed by the spatial correlation of the optical field. In addition, the coherence radius coincides with the correlation radius of light.

Using eqns (12.23), (12.27), and (12.28), we can determine the period of the interference pattern:

$$\delta x = \lambda L/s, \tag{12.30}$$

where λ is the radiation wavelength.

A practical demonstration of the Young interferometer involves a helium–neon or argon laser as a source of light. The laser beam falls on a plate in which two narrow slits are cut. The interference pattern is observed on the screen in the form of alternating dark and bright fringes. The basic parameters are $L = 5$ m, $\lambda = 0.5$ μm, and $s = 0.1$ mm. Thus, the period of the interference pattern calculated from eqn (12.30) is $\delta x = 2.5$ cm, which agrees well with the experiment.

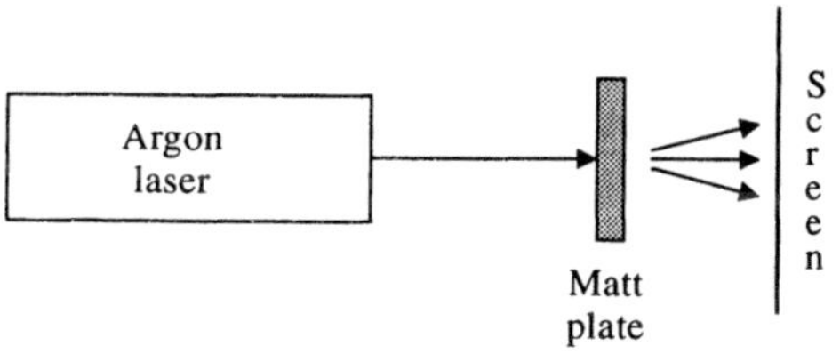

(a) (b)

Fig. 12.4 A demonstration of speckles. (a) A schematic diagram of the experimental set-up. (b) A view of the speckle pattern.

A good example of the monochromatic spatially non-uniform optical field is given by a pattern of *speckles* (spots). Such a pattern can be observed on the screen by passing the argon laser beam through a matt plate (Fig. 12.4). In passing through the matt plate, the laser beam acquires a small-scale transverse non-uniformity. As a result, the spatial coherence decreases, while its angular divergence rises sharply. The pattern observed on the screen has the form of a steady-state speckle field and is called a speckle pattern. This pattern is caused by interference and, consequently, testifies to the spatial coherence of the laser beam. When the matt plate is illuminated by non-coherent light, e.g. by an arc-lamp beam, the speckle pattern is not formed, and the screen looks uniformly illuminated.

Using the notions of the coherence radius r_c and coherence length l_c, we can furnish a visual image of the coherence in terms of the optical beam structure (Fig. 12.5). The parameters r_c and l_c characterize the size of the spatial region within which the light has a structure close to that of the ideal harmonic wave. The optical beam consists of individual coherence regions. If the light is non-coherent, then the coherence regions are very small; their linear dimensions may be of the order of an optical wavelength. On the contrary, if the light is coherent, the coherence regions are large. Specifically, one can imagine an ideal optical beam that is coherent throughout the whole volume (Fig. 12.6).

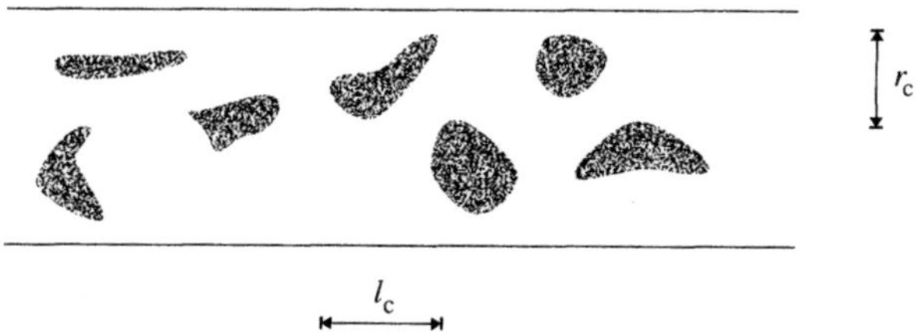

Fig. 12.5 The coherence characteristics and structure of an optical beam.

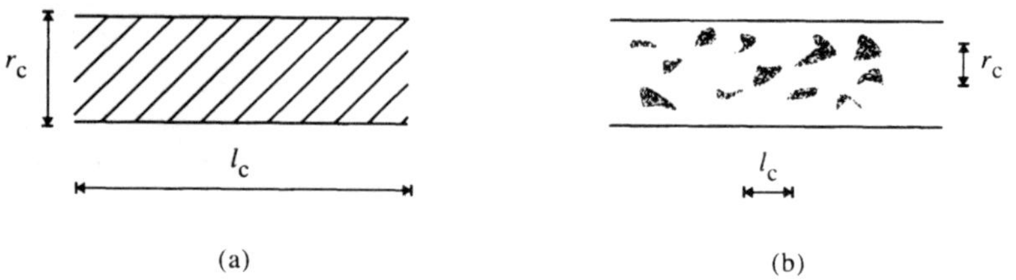

Fig. 12.6 The structure of (a) coherent and (b) non-coherent optical beams.

12.5 The measurement of coherence and the coherence of radiation emitted by real sources of light

The classical interference experiments carried out by Young and Michelson proved to provide direct methods for the measurement of spatial and temporal correlation functions; the distribution of the mean intensity in the interference pattern directly yields the correlation function of the optical field.

As an example, Fig. 12.7 illustrates the spatial correlation functions of laser radiation measured with the the Young interferometer. Different curves in this figure correspond to different lasers. Note that in some cases the value of the coherence radius approaches that of the laser beam radius. Consequently, the structure of the laser light is close to that of an ideal harmonic wave.

12.5.1 A point source of light

The interference of light in the Young interferometer can be observed only if the light incident on the screen with pinholes is sufficiently coherent. In his experiment, Young achieved this by passing sunlight through a pinhole (see Fig. 11.3). The light passing through a small pinhole should obviously possess a structure similar to that of the spherical wave, i.e. it should be spatially coherent.

Let us evaluate the size of the pinhole in screen Q_1 (see Fig. 11.3) needed to obtain a distinct interference pattern in the Young interferometer. To do this, we shall consider the cases in which the screen with pinholes is illuminated (i) by a single point source of light (Fig. 12.8) and (ii) by two point sources of light

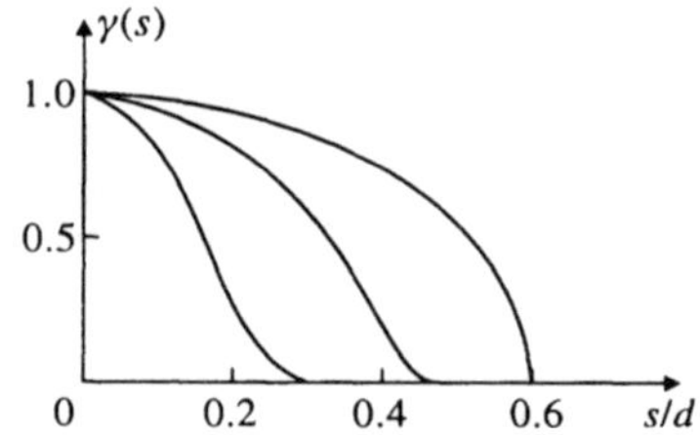

Fig. 12.7 Spatial correlation functions of laser radiation measured with the Young interferometer. s is the distance between the pinholes and d is the laser beam diameter. Different curves relate to different lasers.

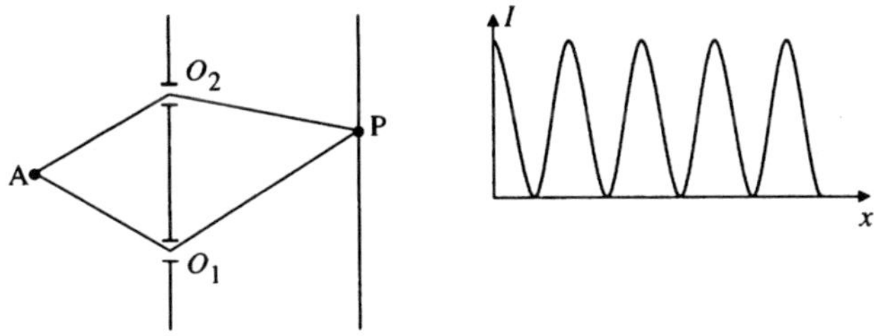

Fig. 12.8 A diagram of point-source radiation interference in the Young interferometer and the distribution of light intensity in the interference pattern.

separated by a distance a (Fig. 12.9).

It can be seen from Fig. 12.8 that in case (i) the visibility of the interference pattern is equal to unity and the light intensity on the observation screen varies sinusoidally. In case (ii) the two sinusoidal intensity distributions shifted relative to each other become superposed on the observation screen (Fig. 12.9). The resulting interference pattern has the same spatial period, but a *lower visibility*. We can thus conclude that the visibility of the interference pattern in the Young interferometer generally depends on the size of the light source.

Let us denote the path difference from source A_1 to the pinholes Q_1 and Q_2 by Δ_1 (Fig. 12.9) and the similar quantity for source A_2 by Δ_2:

$$\Delta_1 = A_1O_1 - A_1O_2, \qquad \Delta_2 = A_2O_1 - A_2O_2. \tag{12.31}$$

Evidently, the finite dimensions of the light source will have no effect on the visibility of the interference pattern, provided that the following condition is satisfied:

$$|\Delta_1 - \Delta_2| \ll \lambda, \tag{12.32}$$

where λ is the optical wavelength. Let a be the size of the light source, let s be the distance between the pinholes in the Young interferometer, and let z be

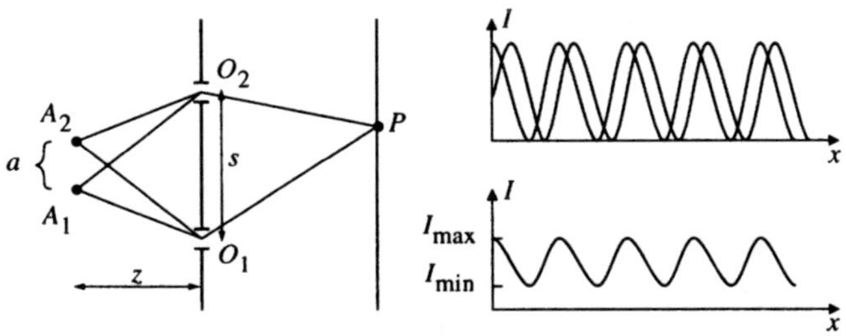

Fig. 12.9 A diagram of the interference of radiation of two point sources in the Young interferometer and the distributions of light intensity in the interference pattern for each source individually (top) and for both sources combined (bottom).

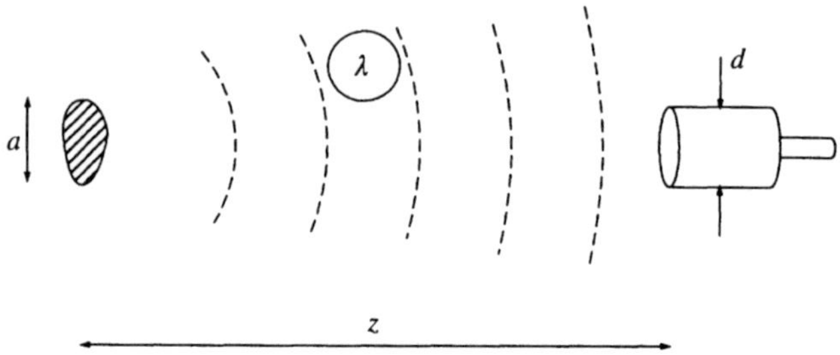

Fig. 12.10 The evaluation of the resolving power of an optical device.

the distance from the source to the screen with pinholes, as shown in Fig. 12.9. Usually, the following conditions are readily met in an experiment:

$$a \ll s \ll z. \tag{12.33}$$

It can easily be shown that, in this case,

$$|\Delta_1 - \Delta_2| \approx as/z. \tag{12.34}$$

Thus, a real source of light can be regarded as a *point source* if its size a satisfies the condition

$$as/z \ll \lambda. \tag{12.35}$$

Here λ is the optical wavelength, z is the distance from the light source to an optical receiver, and s is the receiver aperture.

This simple relation (12.35) has a number of important consequences. Let us consider some of them.

12.5.2 The resolving power of optical devices

Suppose that we observe an object with the help of an optical device, e.g. a telescope (Fig. 12.10). Let a be the size of the object, d the aperture of the device, z the distance from the object to the device, and λ the optical wavelength. It follows from eqn (12.35) that the resolving power of the device is determined by the relation

$$ad/z = \lambda. \tag{12.36}$$

If, however, the object is located far enough away or is so small that condition (12.35) is satisfied, the internal structure of the object cannot be resolved, and its image will be seen as a point. By way of example, Fig. 12.11 shows an image of two shining points near the resolution limit.

Equation (12.36) allows one to estimate the resolving power of an optical device. The resolution limit in terms of the size of object is $a_{\text{min}} = z\lambda/d$; in terms of the angular size it is $\theta_{\text{min}} = a_{\text{min}}/z = \lambda/d$; and in terms of the distance

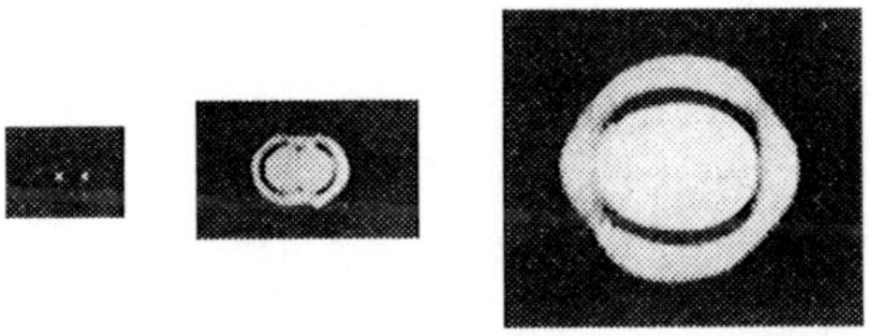

(a) (b) (c)

Fig. 12.11 Variation of the image of two bright points for different observation distances. (a) The distance is sufficiently small; the points are well resolved. (b) A view of the points at the limit of resolution. (c) The distance is too large; the points merge.

from the object to the device it is $z_{\max} = ad/\lambda$. According to these formulae, the resolving power of the device can be improved by increasing the entrance aperture of the device.

Let us make some numerical evaluations.

From what maximum distance can a human eye distinguish two shining points separated by the distance $a = 5$ cm? Assuming the pupil of the eye to have $d = 1$ mm and $\lambda = 0.5$ μm, we obtain $z_{\max} = 100$ m. When observed, using a telescope with an entrance diaphragm of $d = 10$ cm, the same object is resolved from the maximum distance $z_{\max} = 10$ km (eqn (12.36)).

Telescope

One of the largest optical telescopes, mounted in the Northern Caucasus (Zelenchuk), has a main mirror diameter equal to $d = 6$ m. What is the minimum size of an object on the Moon's surface for it still to be discernible through this telescope? The distance from the Earth to the Moon is $z = 384\,000$ km. Assuming $\lambda = 0.5$ μm, we obtain, from eqn (12.36), $a_{\min} = 30$ m. The real resolving power of the telescope is, however, much lower due to the effects of the atmosphere.

Microscope

Estimating the smallest size of the object that can be resolved by a microscope, we obtain from eqn (12.36) $a_{\min} = z\lambda/d$. In practice, the minimum value of the ratio z/d is about unity. Hence we obtain $a_{\min} \approx \lambda \approx 10^{-4}$ cm. Experiment confirms this result. The limiting resolving power of the microscope is indeed of the order of an optical wavelength.

12.5.3 Evaluation of the coherence radius of light

Using Eqs. (12.35) and (12.36), we can estimate the coherence radius of radiation emitted by a non-laser source of light, e.g. a heated body. According to eqn (12.36)

$$d = \lambda z/a. \tag{12.37}$$

This formula provides an estimation of the smallest aperture d of a light receiver for which the optical wave entering the receiver contains information about the size a of the light source. As was mentioned above, the spatial coherence of light is linked to the size of the source. The point source gives perfectly coherent light

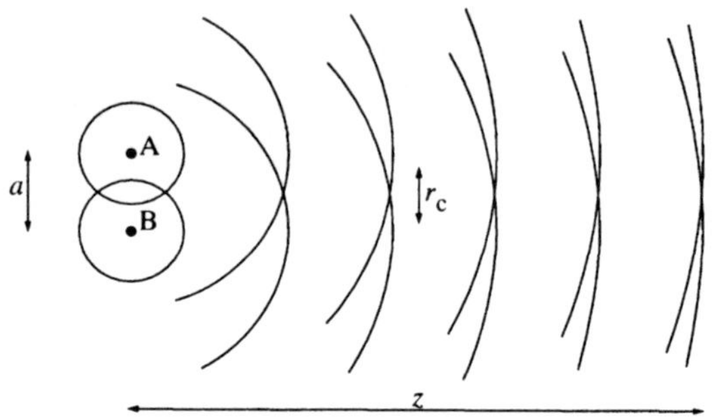

Fig. 12.12 The growth of the coherence radius with the distance traversed by the light.

(see Fig. 12.8), while a source of finite size produces partially coherent light (see Fig. 12.9). Consequently, the size d determined by eqn (12.37) can be taken as a measure of the coherence radius of light. Thus, we obtain

$$r_c = \lambda z / a. \tag{12.38}$$

In this formula, z is the distance from the light source to the observation point, a is the size of the source, and λ is the optical wavelength.

According to eqn (12.38), the coherence radius of light grows proportionally to the distance travelled by the optical wave: $r_c \sim z$. The physical meaning of this result is that the wavefronts of spherical waves emitted by different points of the source approach one another as the distance from the source grows (Fig. 12.12). The structure of the radiation is gradually transformed to that of a spherical wave and the coherence radius increases.

By way of example, let us estimate the coherence radius of sunlight. We shall use eqn (12.38) and the following parameter values: the distance from the Sun to the Earth, $z = 150\,000\,000$ km, the diameter of the Sun, $a = 700\,000$ km, and the optical wavelength, $\lambda = 0.5$ μm. Thus we obtain $r_c = 10^{-2}$ cm. For scattered sunlight, for example, for a cloudy day, we should apparently assume $a/z \approx 1$. Then we obtain $r_c \approx \lambda \approx 10^{-4}$ cm. For comparison, we should remind the reader that laser radiation can be coherent over the whole cross-section of the beam. For example, wide laser beams with a coherence radius of about 1 m are used to produce large holograms.

12.5.4 The Michelson stellar interferometer

Equation (12.38) can be used to estimate the dimensions of a distant source of light. For example, the angular dimension of a star can be found from the formula

$$a/z = \lambda / r_c, \tag{12.39}$$

if the coherence radius of the radiation coming from the star is known. This idea was used by Michelson to estimate the dimension of stars. At the very

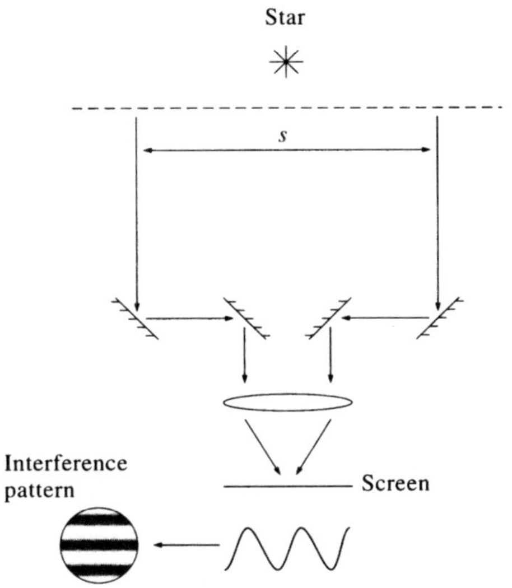

Fig. 12.13 A schematic diagram of the Michelson stellar interferometer.

beginning of the 20th century, he performed a number of measurements with the help of a "stellar interferometer," a schematic of which is shown in Fig. 12.13. To measure the coherence radius r_c, two telescopes separated by a distance s are directed towards the star. Using mirrors, the light beams from the telescopes are superposed, to produce an interference pattern similar to that produced by the Young interferometer. Moving the telescopes further apart and measuring the visibility of the interference pattern as a function of the distance s, one can measure the coherence radius of the star light.

Michelson's measurements showed that the light coming from some giant stars has a coherence radius of the order of 10 m. Consequently, the angular dimensions of the stars are of the order of $\theta = a/z = \lambda/r_c = 5 \times 10^{-8}$ rad. For example, for the star Betelgeuse it was found that $\theta = 0.0047'' = 10^{-8}$ rad. This signifies that the diameter of the star is about 400 000 000 km, which is equal to the diameter of the orbit of Mars around the Sun.

Bibliography

Born, M. and Wolf, E. (1968). *Principles of optics.* Pergamon Press, Oxford.

Steward, E. G. (1983). *Fourier optics: an introduction.* Ellis Horwood, Chichester.

13
The diffraction of light

Diffraction as evidence of the wave nature of light—fundamental experimental facts. Grimaldi's experiment. The Huygens principle. The Huygens–Fresnel principle. Fresnel's diffraction integral. Fresnel zones. Graphical construction of diffraction patterns. Edge diffraction. The diffraction length of an optical beam. Near- and far-field zones of diffraction. The diffraction divergence of a beam in the far-field zone. The focusing of light as a diffraction phenomenon. Kirchhoff's diffraction theory.

13.1 Diffraction as evidence of the wave nature of light—fundamental experimental facts

Many phenomena that we observe in our everyday life indicate that light propagates rectilinearly. A beam of sunlight, the beam of a searchlight, and a laser beam are all associated in the human mind with straight or nearly straight lines. Rectilinear propagation is one of the main and most readily apparent properties of light. Undoubtedly, it is this property that gave rise to the idea that light was a stream of particles. Futhermore, nothing seemed to resemble a wave less than a rectilinear beam of light.

There are, however, a number of fine optical phenomena and experiments that indicate that the law of rectilinear propagation of light does not hold. For example, the beam of sunlight or the laser beam has always a small, but *finite*, divergence. The boundary of an optical beam, or a boundary between light and shadow, is never sharp; it has a small, but finite, width.

In some cases light does not propagate rectilinearly at all. For example, in the Young experiment (see Chapter 11) the light penetrates deeply into the geometric shadow region where the interference pattern is observed. In passing through a pinhole, the optical beam acquires a large angular divergence, which is greater for a smaller hole. A small bright spot (Poisson's spot) is observed at the centre of the geometric shadow produced by a small opaque disc illuminated by coherent light. A laser beam falling on to a diffraction grating produces a widely diverging fan of rays.

All of these facts can be explained by the assumption that light is a *wave*. The rectilinear propagation of light does not contradict its wave nature because the wavelength of light is very small, i.e. less than 10^{-4} cm. Strong wave effects can be expected when the transverse dimensions of an optical beam become comparable to the wavelength. This agrees well with many experiments and, specifically, with the diffraction of a laser beam by a diffraction grating.

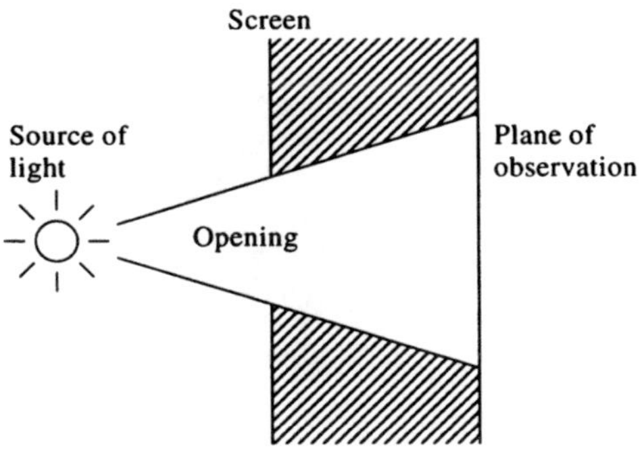

Fig. 13.1 The observation of the diffraction of light.

In optics, the notion of diffraction is associated with violation of the rectilinear propagation of light. Sommerfeld defined diffraction as any deviation of light from rectilinear propagation, not caused by reflection or refraction. In more precise terms, a wave diffracts whenever it bends round an obstacle. Such phenomena are well known for long waves, for example, for sound waves or waves on a water surface; in optics these phenomena imply a penetration of light into the region of geometric shadow. In wave theory, diffraction encompasses all of the phenomena in the wave field that arise in the presence of obstacles hindering the wave propagation. Finally, using the idea of the interference of light, we can say that diffraction is interference in bounded optical beams.

Diffraction is of fundamental significance because it confirms, just as does interference, the wave nature of light. Diffraction is also of great practical importance because it sets an upper limit on the spatial concentration of light, limits the resolving power of optical devices, affects the formation of optical images, etc.

13.2 Grimaldi's experiment

Grimaldi was the first to observe the diffraction of light; his report was published in 1665, soon after his death. A schematic diagram of Grimaldi's experiment is shown in Fig. 13.1. The light from a point source passes through a hole in an opaque screen and falls on to the second screen (observation plane) some distance behind the first one. Grimaldi studied the distribution of illumination over the observation screen and noticed a gradual transition from light to shadow (Fig. 13.2). This result could not be explained in terms of the corpuscular theory of light, according to which the light should propagate rectilinearly, and the image of the hole in the observation plane must have a sharp boundary. A still more surprising result is obtained if we recreate Grimaldi's experiment using a laser as a source of light. This can be demonstrated by observing the diffraction of an argon laser beam by the edge of a razor blade. A basic diagram of this experiment is shown in Fig. 13.3. The distribution of illumination in the observation plane is shown in Fig. 13.4. As in Grimaldi's classic experiment, there is no sharp

Fig. 13.2 The distribution of the illumination in the observation plane near the geometric shadow boundary (non-coherent illumination).

boundary between light and shadow. Moreover, we can see a system of alternating dark and bright fringes, parallel to the boundary of the geometric shadow. Such features of the observed pattern point to the interference nature of diffraction phenomena.

13.3 The Huygens principle

An understanding of the nature of diffraction phenomena is associated with the development of the wave theory of light. The first step forward in this direction was made by Huygens at the end of the 17th century. Using the idea that light is a wave, Huygens was able to explain the mechanism of light propagation. Huygens believed that light propagated from a source like a wave on the water surface. Every point of the wavefront of an optical disturbance generates a secondary spherical wavelet. The wavefront position at a subsequent moment is determined by the envelope of the secondary wavelets.

The Huygens principle is illustrated in Fig. 13.5, which shows the wavefront of the optical disturbance, the secondary wavelets, and the envelope of secondary wavelets.

Using this principle, Huygens could explain phenomena such as the propagation of light from a point source, the propagation of a light beam, and the

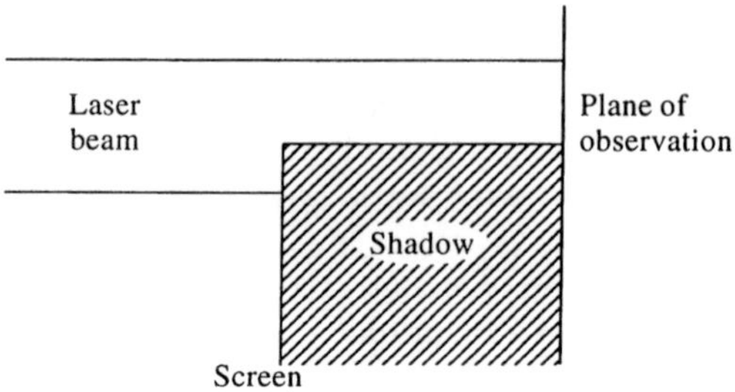

Fig. 13.3 The edge diffraction of a laser beam.

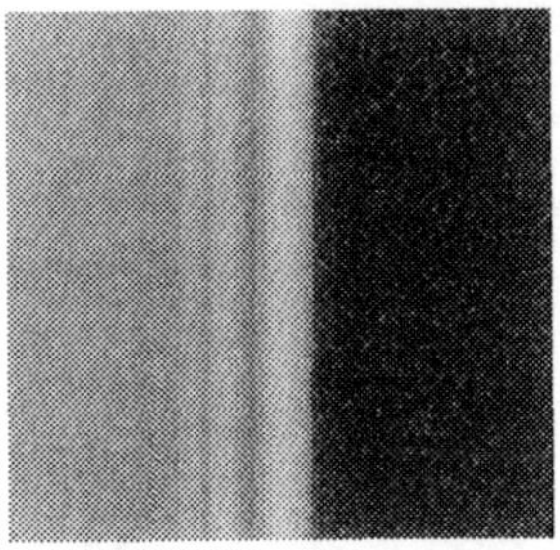

Fig. 13.4 The distribution of the illumination in the observation plane near the geometric shadow boundary (coherent illumination).

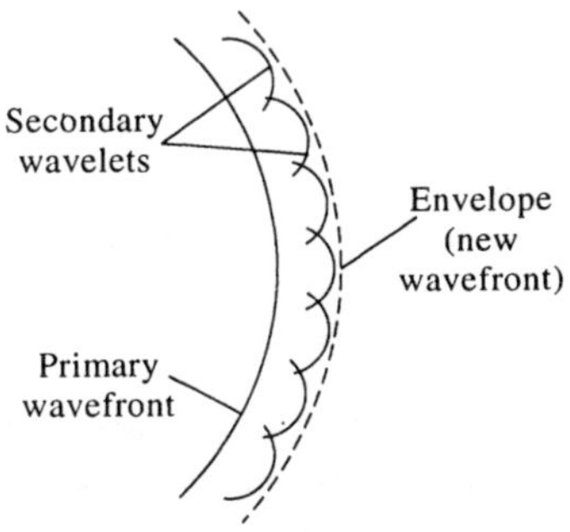

Fig. 13.5 A schematic of the Huygens construction.

reflection and refraction of light. Examples of corresponding constructions are shown in Fig. 13.6.

13.4 The Huygens–Fresnel principle

At the beginning of the 19th century, the ideas of Huygens were promoted by Fresnel. He supplemented the Huygens principle with a postulate that the secondary wavelets can either reinforce or weaken one another: in other words, they can interfere. The optical field is a result of the interference of elementary secondary wavelets emitted by each element of a certain wave surface—this statement constitutes the content of the Huygens–Fresnel principle. Starting from this principle, Fresnel calculated with a high accuracy the distribution of light in diffraction patterns.

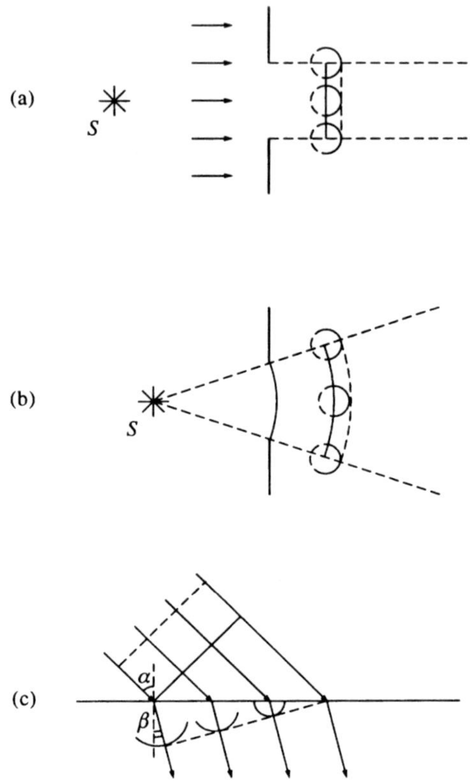

Fig. 13.6 Explanations of the rectilinear propagation of light (a,b) and of the refraction law (c) by Huygens. (a) The transmission of a parallel beam of light through an opening in a screen. The source of light is supposed to be far enough away (e.g. the Sun). (b) The transmission of a diverging beam of light through an opening in a screen. A point source of light is supposed to be located not very far from the screen. (c) An explanation of the refraction of light at the boundary separating two media. The velocity of light in the second medium is supposed to be less than in the first one.

13.5 Fresnel's diffraction integral

The Huygens–Fresnel principle provides the necessary prerequisites for constructing the elementary theory of light diffraction. One of the simplest problems can be formulated as follows. Let S be a point source of light. Suppose that we have to calculate the optical field at a point P under the condition that between the points S and P there is an obstruction for the light propagation, e.g. a screen with holes or an opaque disc (Fig. 13.7).

We shall approach this problem using a mathematical formulation of the Huygens–Fresnel principle. Let us introduce a closed surface Σ surrounding the source of light (Fig. 13.8). Every element $d\sigma$ of the surface is a source of a

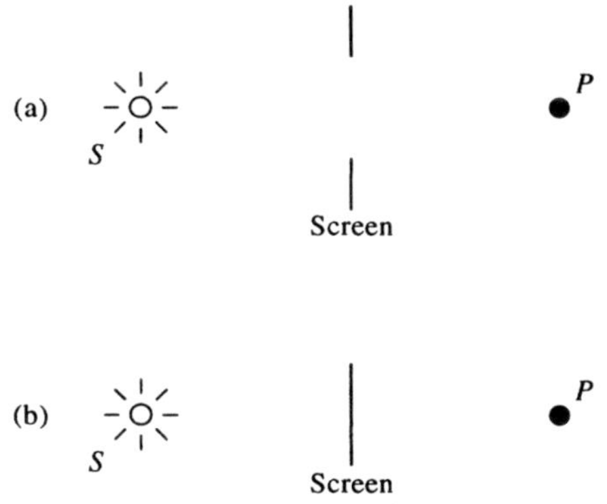

Fig. 13.7 Examples of diffraction problems: (a) diffraction by an opening; (b) diffraction by a disc.

secondary spherical wavelet. Consider a point M on the surface σ. Assuming S to be a point source of light, we shall denote the distance from S to M by ρ_1, and the distance from M to the observation point P by ρ. We shall also introduce the angle φ between the normal $\vec{n}$ to the surface Σ at the point M and the direction from M to P.

Suppose, for simplicity, that the source of light emits a monochromatic optical wave. Then the optical field at any point of space will also be monochromatic and can be presented in the form

$$E = \tfrac{1}{2}\mathcal{E}e^{i\omega t} + \text{c. c.}, \tag{13.1}$$

where ω is the frequency, and $\mathcal{E}$ is the complex amplitude of the oscillations.

According to the Huygens–Fresnel principle, the optical field at the point P results from a superposition of wavelets emitted by the elements of the surface Σ. Consequently, we can write

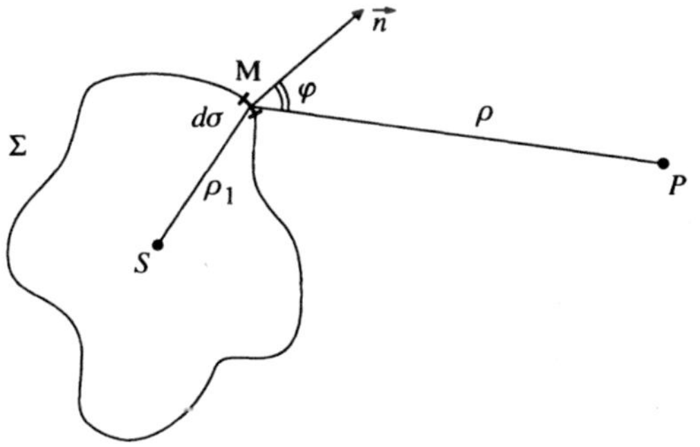

Fig. 13.8 A derivation of the Huygens–Fresnel integral.

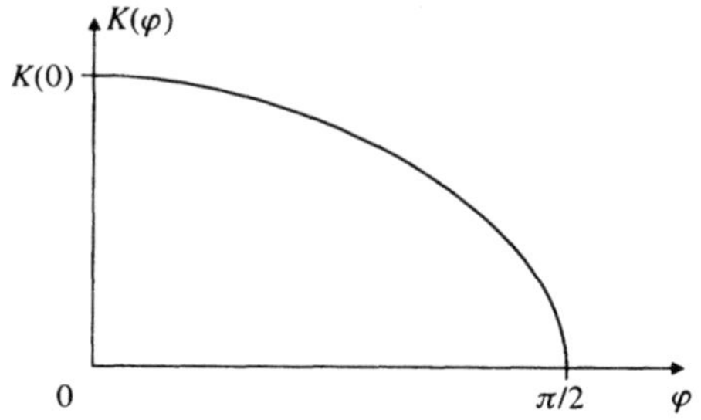

Fig. 13.9 The Fresnel inclination factor.

$$\mathcal{E}(P) = \int_{\Sigma} \mathcal{E}(M) \frac{\exp(-ik\rho)}{\rho} K(\varphi)\, d\sigma. \tag{13.2}$$

Here $\mathcal{E}(P)$ and $\mathcal{E}(M)$ are the complex amplitudes of the field at the points P and M,

$$k = \omega/c = 2\pi/\lambda \tag{13.3}$$

is the wave number of the optical wave, and $K(\varphi)$ is the inclination factor, that allows for the fact that the contribution of the element $d\sigma$ to the resulting field depends on the orientation of the given surface element relative to the direction to the observation point.

The integral (13.2) is called the *Huygens–Fresnel integral.* Note that it is constructed on the basis of qualitative physical considerations. The factor under the integral sign, $(1/\rho)\exp(-ik\rho)$, describes the propagation of the secondary spherical optical wavelet. It is most important that the Huygens–Fresnel integral takes into account the phases of the secondary wavelets coming to the point P from different elements of the surface Σ. In other words, the *interference of secondary waves* is taken into account.

The function $K(\varphi)$ in eqn (13.2) remains indeterminate so far. Fresnel believed that $K(\varphi)$ decreases monotonously from the intial value $K(0)$ to zero as the angle φ grows from zero to $\pi/2$ (Fig. 13.9).

13.6 Fresnel zones

Fresnel proposed an approximate method to calculate diffraction patterns, based on the notion of so-called *half-wave zones* or *Fresnel zones.* The Fresnel zones are introduced as follows. Choose a spherical surface Σ with its centre at the point S. According to the Huygens–Fresnel principle, this spherical surface can be treated as a source of secondary optical wavelets. Construct circular zones on the spherical surface so that distances from the zone boundaries to the observation point differ by a half-wavelength of the optical wave. Denoting the boundaries of the zones by $M_0, M_1, M_2, \ldots,$ we obtain

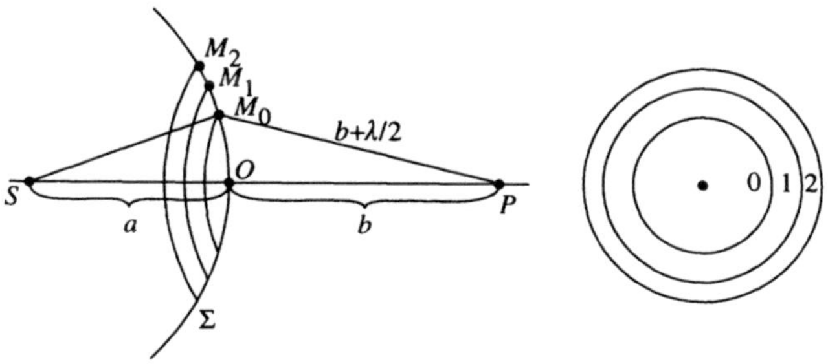

Fig. 13.10 The Fresnel zones.

$$\begin{cases} M_0P = OP + \lambda/2, \\ M_1P = M_0P + \lambda/2, \\ \dots\dots\dots\dots\dots\dots \\ M_nP = M_{n-1}P + \lambda/2, \end{cases} \tag{13.4}$$

where λ is the optical wavelength, P is the observation point, and O is the centre of the zeroth Fresnel zone (Fig. 13.10).

Thus, eqns (13.4) determine the positions of the Fresnel zone boundaries. The main idea of the zone construction is that the phase differences of secondary wavelets arriving at the observation point from a given zone do not exceed π. Such waves interfere constructively, i.e. they reinforce each other. Consequently, every Fresnel zone can be regarded as a source of secondary wavelets having a definite phase. On the contrary, two adjacent Fresnel zones act as sources oscillating in counterphase. Note that the position of the Fresnel zone boudaries depends on the choice of the observation point.

13.6.1 Dimensions of the Fresnel zones

To estimate the relative contributions of the Fresnel zones to the Huygens–Fresnel integral, we shall calculate their radii and areas. Figure 13.11 shows the point source of light S, the observation point P, a fragment of the spherical surface Σ (the source of the secondary waves), and the zeroth Fresnel zone boundary M_0. Let a be the radius of the sphere Σ, b the least distance from the point P to the surface of the sphere, and r_0 the radius of the zeroth Fresnel zone. It can be seen from Fig. 13.11 that

$$r_0^2 = a^2 - (a - x)^2, \tag{13.5}$$

where x is the length of the straight-line section shown in the figure by a bold line. On the other hand,

$$r_0^2 = (b + \lambda/2)^2 - (b + x)^2. \tag{13.6}$$

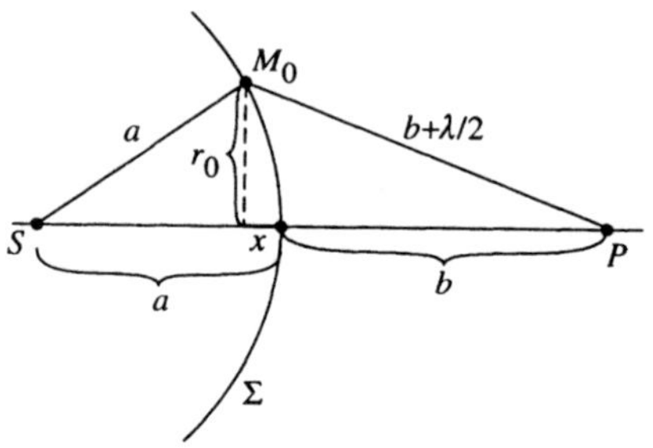

Fig. 13.11 A calculation of the dimensions of the Fresnel zones.

It follows from eqns (13.5) and (13.6) that

$$2ax = b\lambda - 2bx + (\lambda/2)^2. \tag{13.7}$$

As a rule, in optics we are interested in the case in which

$$a, b \gg \lambda, x. \tag{13.8}$$

Hence, neglecting the terms proportional to λ^2 and x^2, we obtain

$$x = \frac{b\lambda}{2(a+b)}$$

and $r_0^2 \approx 2ax$, whence

$$r_0 = \sqrt{\frac{\lambda ab}{a+b}}. \tag{13.9}$$

Thus, eqn (13.9) gives the radius of the zeroth Fresnel zone. Here λ is the optical wavelength, and a and b are the lengths of the line segments shown in Fig. 13.11. Similarly, we find the outer radius of the nth Fresnel zone:

$$r_n = \sqrt{(n+1)\frac{\lambda ab}{a+b}}. \tag{13.10}$$

13.6.2 Diffraction of a plane wave

The physical content of the problem remains just the same, but the formulae become simpler, if we consider the diffraction of a plane optical wave instead of a spherical wave.

Suppose, for example, that a plane monochromatic optical wave is diffracted by a circular opening (Fig. 13.12). We shall try to find how the light intensity varies at a point P on the axis of the beam as the radius of the opening changes.

Let z denote the distance from the observation point to the screen with the opening and r the radius of the opening. For the surface Σ, that is, the source

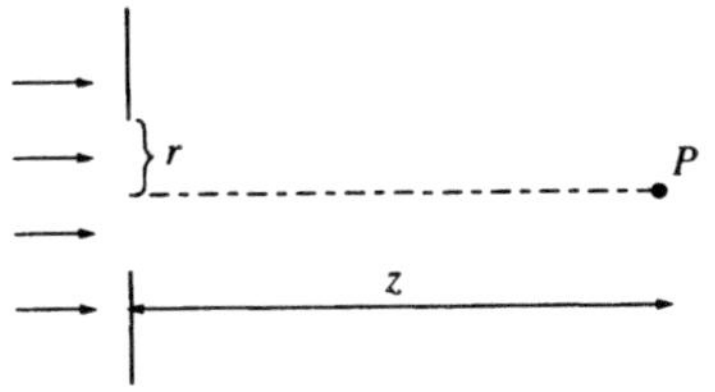

Fig. 13.12 A diagram of the plane wave diffraction by a round aperture.

of secondary wavelets, we shall use a circle of radius r lying in the plane of the screen and coinciding with the opening. Using the definition given earlier, we shall divide the surface Σ into Fresnel zones that have the form of planar rings. Their radii can be calculated by eqn (13.10), assuming $a \to \infty$, $b = z$:

$$r_n = \sqrt{(n+1)\lambda z}. \tag{13.11}$$

It follows from eqn (13.11) that all of the Fresnel zones have equal areas determined by the formula

$$S_n = \pi(r_n^2 - r_{n-1}^2) = \pi\lambda z. \tag{13.12}$$

13.6.3 Fresnel's number

Suppose that the optical wavelength λ, the opening radius r, and the distance z from the screen to the observation point P are known. Then, using eqn (13.11), we can calculate the number N_F of Fresnel zones that get within the opening, or the number of *Fresnel free zones*. This number is called *Fresnel's number*, and plays an important part in diffraction theory. Assuming that

$$r_n = r, \qquad n + 1 = N_F, \tag{13.13}$$

we obtain, from eqns (13.11) and (11.13),

$$N_F = r^2/\lambda z. \tag{13.14}$$

13.7 Graphical construction of diffraction patterns

A graphical method of analysis of diffraction phenomena is based on the use of the so-called *spiral of Fresnel*. The spiral of Fresnel can be constructed as follows.

The calculation of the resulting optical field using the Huygens–Fresnel integral can be reduced to a summation of the waves excited by the elementary secondary sources. From the mathematical point of view, the problem is reduced to the summation of harmonic vibrations having the same frequency, but different amplitudes and phases. An illustrative example of a solution to this problem is the construction of a *vector diagram*.

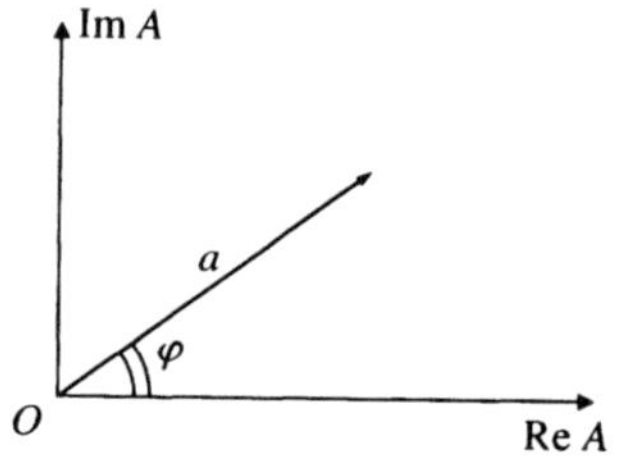

Fig. 13.13 Harmonic vibration on the complex plane.

13.7.1 Vector diagram

It is known that harmonic vibration with an amplitude a and phase φ can be characterized by the complex amplitude $A = a\exp(i\varphi)$, or by the vector on the plane of variables $\mathrm{Re}\,A$, $\mathrm{Im}\,A$, the vector modulus being equal to a, and the angle of its inclination to the axis $\mathrm{Re}\,A$ being equal to φ (Fig. 13.13).

A sum of harmonic vibrations of frequency ω with arbitrary amplitudes and phases is harmonic vibration of the same frequency ω. The real amplitude A and phase Φ of the resulting vibration can be found by summing the vectors representing the individual vibrations. To make the summation, we present the individual harmonic vibration as a vector, the modulus of which is equal to the vibration amplitude and the angle of inclination to the abscissa of which is equal to the phase of the contributing vibration. Having constructed the vector sum, we find the amplitude of the resulting vibration as the modulus of the resulting vector and the phase of the resulting vibration as the angle of inclination of the resulting vector to the abscissa. An example of this summation is shown in Fig. 13.14 for three vibrations.

13.7.2 The spiral of Fresnel

Let us apply the vector diagram method for the calculation of the Huygens–Fresnel diffraction integral. First, we shall evaluate the contribution of only one zone, e.g. the zeroth Fresnel zone.

The corresponding construction is shown in Fig. 13.15(a). First, we subdivide

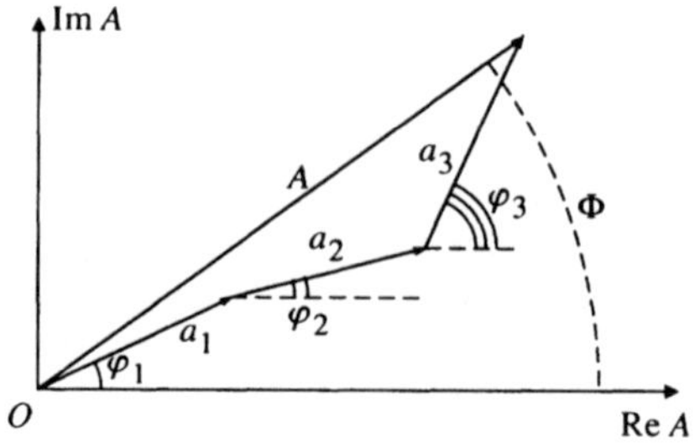

Fig. 13.14 The summation of harmonic vibrations by the vector diagram method.

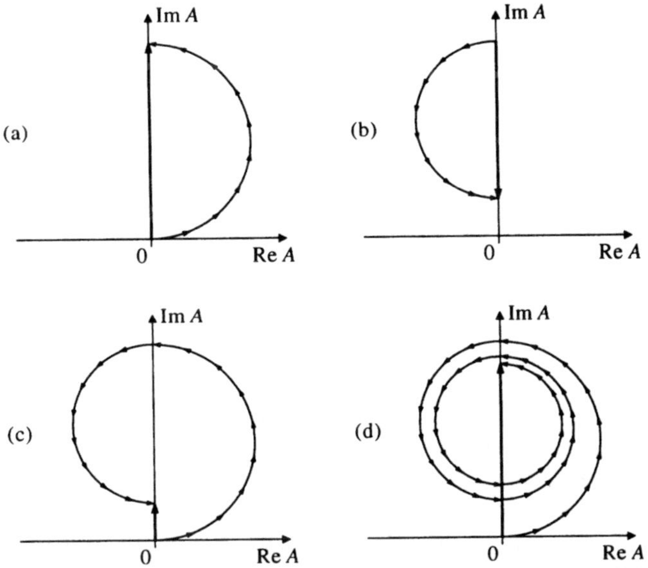

Fig. 13.15 Vector diagram constructions of the complex amplitude A of a diffracted optical field. (a) The zeroth Fresnel zone is open, with the other zones being obstructed. (b) The first Fresnel zone is open. (c) The zeroth and first Fresnel zones are open. (d) The zeroth, first, second, third, and fourth Fresnel zones are open. Short vectors denote the complex amplitudes of the optical fields contributed by individual subzones of the Fresnel zones. Bold lines show the resulting complex amplitude vectors of the optical fields at the observation point.

the Fresnel zone into concentric rings (subzones). Obviously, the division is made so that all subzones have equal areas, and their number is sufficiently large. Then the contributions of the subzones are represented by vectors of approximately equal length and different angles of inclination to the abscissa. The last of these vectors makes an angle with the first one equal to π, in accordance with the definition of the Fresnel zone. Due to the inclination factor, a plot of which is shown in Fig. 13.9, the contribution of the subzone to resulting field (and, consequently, to the modulus of the corresponding vector) decreases slightly as the radius of the subzone and its number increases.

Similarly, we construct the vectors representing the contribution of the first Fresnel zone (Fig. 13.15(b)), and the combined contribution of the zeroth and the first zones (Fig. 13.15(c)). Note that the elementary vectors representing the subzones become shorter as the zone number increases. This accounts for a decrease in the total contribution of the given zone to the resulting diffraction field, caused by the increase of the angle of inclination of the zone, i.e. by the factor $K(\varphi)$.

Continuing the vector diagram construction for larger zone numbers (see,

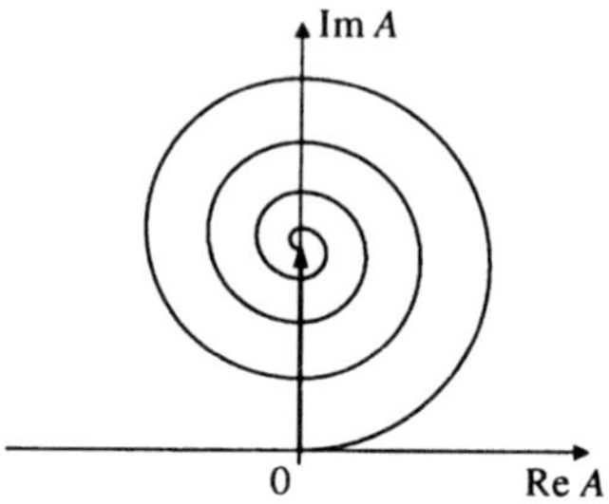

Fig. 13.16 A vector diagram for the limiting case, when all the Fresnel zones are open and the number of subzones of each zone tends to infinity. A is the complex amplitude of the optical field at the point P.

e.g. Fig. 13.15(d)), we obtain a *winding spiral*. It can easily be seen that, as the number of subzones in each zone increases, the polygonal line of the vector diagram is gradually transformed to a smooth curve. In the limiting case in which all of the Fresnel zones are open and the number of subzones in each zone tends to infinity, we obtain the vector diagram shown in Fig. 13.16. This limiting diagram has form of a smooth winding spiral that is called the spiral of Fresnel. Let us consider some specific examples.

Diffraction by a round aperture

Let a plane monochromatic optical wave fall normally on a screen with a round aperture, as shown in Fig. 13.12. Let us choose a point P on the aperture axis and study the dependence of the amplitude A of optical field at the given point on the aperture radius.

Figure 13.17 shows constructions of the spiral of Fresnel carried out for cases in which a different number of Fresnel zones occupy the aperture: (a) one, (b) two, (c) three, and (d) infinity. The last case corresponds to the absence of the screen, i.e. to the free propagation of the optical wave.

Using the data of Fig. 13.17, and eqn (13.11), we can construct the sought-after dependence $|A(r)|$ (see Fig. 13.18). As seen from this figure, Fresnel theory predicts a non-monotonic variation of the field amplitude with an increase in the aperture radius. Figure 13.18 shows that the light intensity reaches a maximum at the observation point, at which the aperture coincides with the zeroth Fresnel zone. Note that in this case the field amplitude is twice and the light intensity is four times higher than in the absence of the screen. This result can be verified experimentally by using a coherent laser beam and an iris diaphragm. If, for example, $\lambda = 0.5\ \mu\text{m}$ and $z = 100$ cm, then from eqn (13.11) we obtain $r_0 = 0.7$ mm, $r_1 = r_0\sqrt{2} = 1$ mm and so on. We can thus expect some noticable diffraction effects if the aperture radius is close to the radius of the zeroth Fresnel zone, i.e. to $r \simeq 1$ mm in our case.

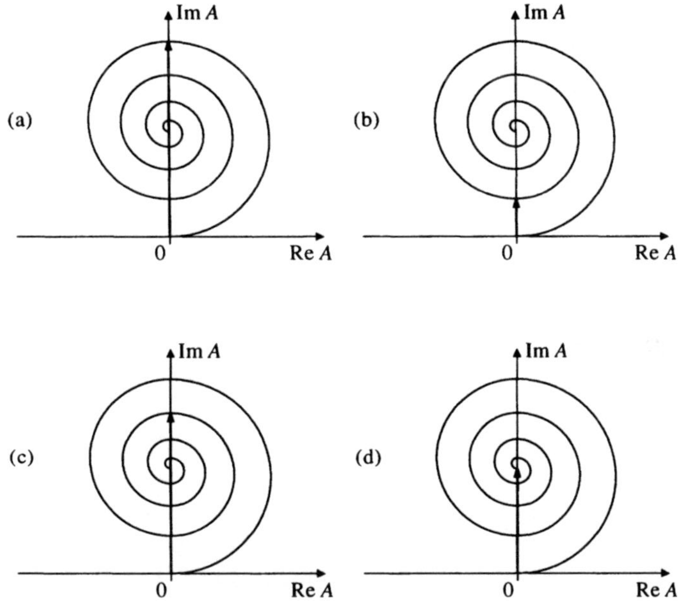

Fig. 13.17 The calculation of diffraction by a round aperture. The field amplitude at the observation point is shown for different radii of the opening: (a) $r = r_0$; (b) $r = r_1$; (c) $r = r_2$; (d) $r = \infty$.

Diffraction by a disc and the Poisson spot

The diffraction by a disc is shown schematically in Fig. 13.19. Let us suppose that a plane monochromatic optical wave falls normally on a round opaque disc of radius r, and the diffraction field is observed at a point P located in the geometric shadow on the axis of the disc.

The complex amplitude vectors of the field constructed using the spiral of

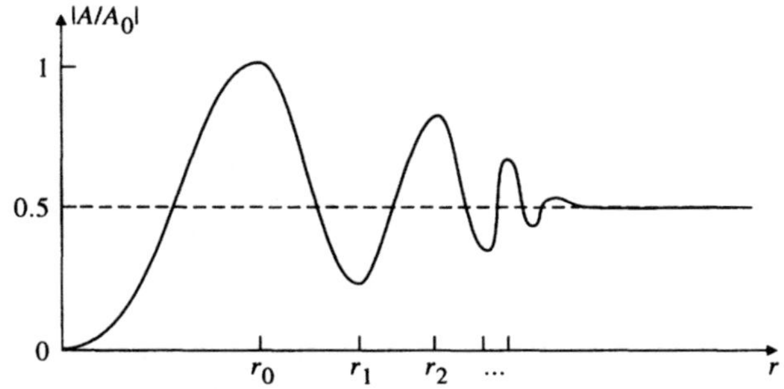

Fig. 13.18 The field amplitude at the observation point as a function of the opening radius; r_0, r_1, r_2, etc. are the radii of the Fresnel zones.

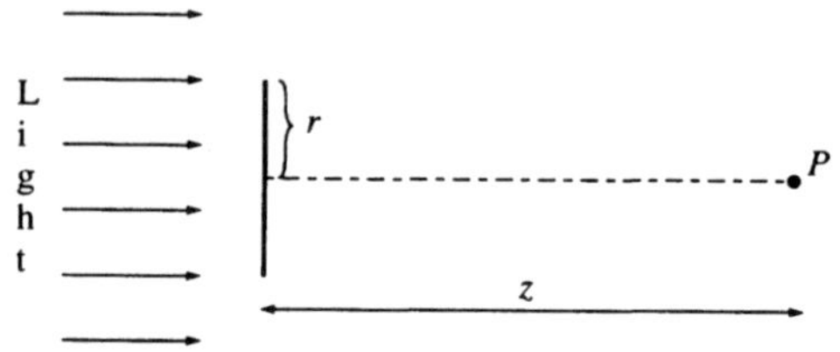

Fig. 13.19 A diagram of diffraction by a disc.

Fresnel are shown in Fig. 13.20. Figure 13.21 shows the dependence of the optical vibration amplitude A at the observation point on the disc radius r. Note that in this case the dependence $|A(r)|$ is monotonously decreasing. This means that, as the disc radius increases, the light intensity at the observation point drops. Nevertheless, Fig. 13.21 shows that even when the disc is so large as to cover a number of the Fresnel zones, the light intensity at the centre of the geometric shadow is non-zero.

Fresnel theory thus predicts a penetration of light into the centre of the geometric shadow of a disc placed in the way of a plane monochromatic optical wave. This result was once considered as an argument against Fresnel theory. However,

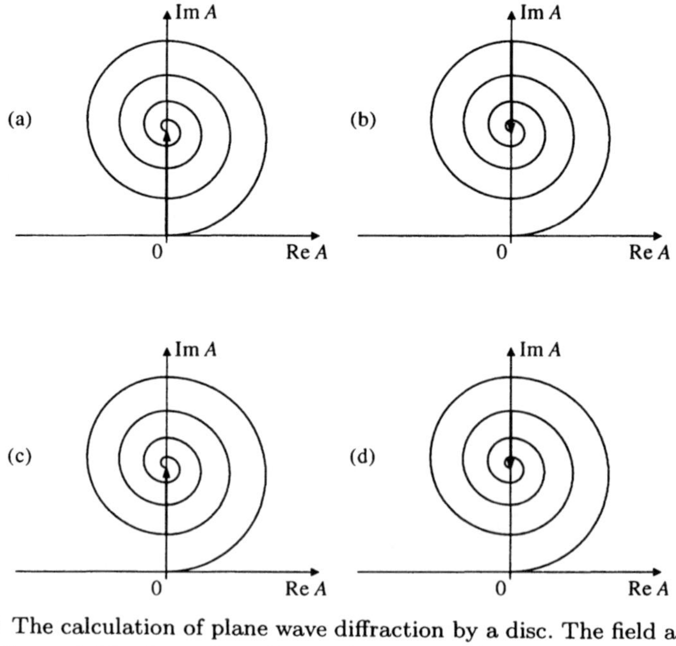

Fig. 13.20 The calculation of plane wave diffraction by a disc. The field amplitude at the observation point is shown for different disc radii: (a) $r = 0$; (b) $r = r_0$; (c) $r = r_1$; (d) $r = r_2$.

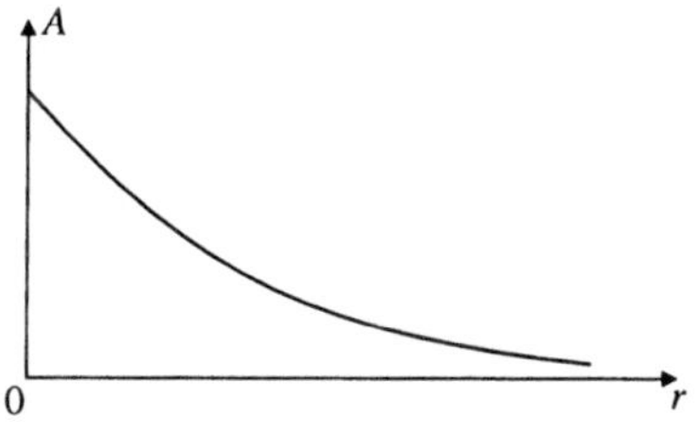

Fig. 13.21 The field amplitude at the observation point as a function of the disc radius.

the experiment performed by Arago confirmed this surprising prediction. It was shown that a bright spot did indeed exist at the centre of the geometric shadow of a small disc illuminated by the light of a point source. This was an impressive result testifying to the wave theory of light. A schematic diagram of the Arago experiment and its result are shown in Fig. 13.22.

A practical demonstration of the Poisson disc employs the continuous radiation of an argon laser. It is directed onto a small steel sphere (about 1 mm in diameter) pressed by a glass plate against the centre of a lens with a focal length of 6 cm. The lens projects a highly magnified diffraction pattern on the screen. The diffraction pattern represents a large round dark spot (the shadow of the sphere) with a bright halo around it and with a small, but clearly visible, bright spot in its centre (Fig. 13.23).

13.8 Edge diffraction

The problem of edge diffraction or, in other words, the question of the transition from light to shadow occurring at the boundary of the geometric shadow, is no doubt one of the key problems of diffraction theory. As mentioned earlier, the notion of the diffraction emerged after Grimaldi's experiment, in which the

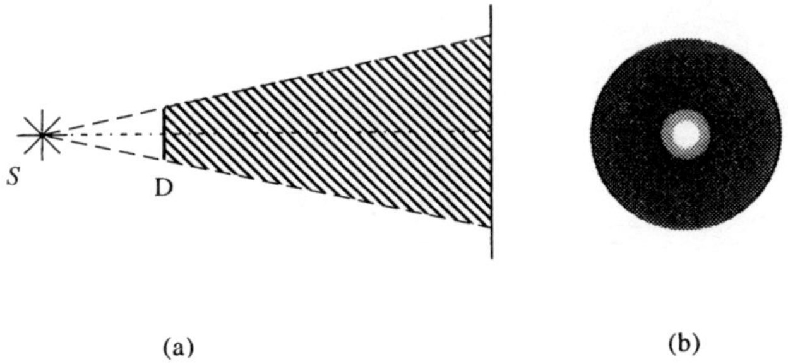

Fig. 13.22 The observation of the Poisson spot in Arago's experiment. (a) A schematic diagram of Arago's experiment. (b) A view of a diffraction pattern: a bright spot (the Poisson spot) is observed in the centre of the geometric shadow.

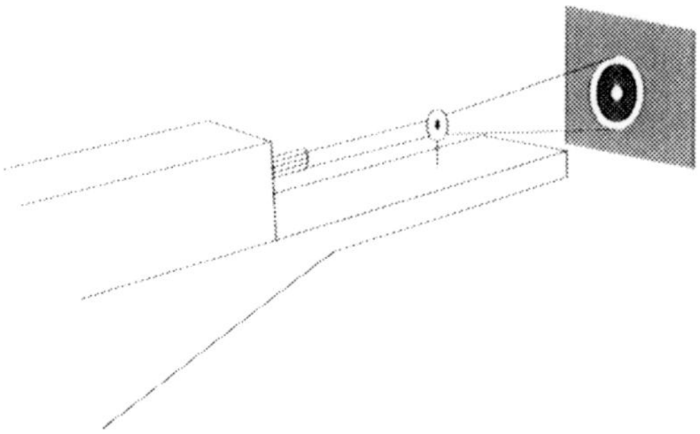

Fig. 13.23 A Poisson disc demonstration.

absence of a sharp boundary between light and shadow was established (see Figs 13.1 and 13.2).

Suppose that a plane monochromatic optical wave falls normally on a semi-infinite opaque plane (screen) with a rectilinear boundary (edge of the screen). Let us find the distribution of light in the observation plane parallel to the screen and located at some distance l behind it (Fig. 13.24).

To construct the diffraction pattern, we shall use the method of the Fresnel zones. For a surface Σ emitting secondary wavelets, we shall choose a half-plane that is an extension of the screen; this surface coincides with the wavefront of the incident plane optical wave. Let us introduce the Fresnel zones on the surface Σ, assuming that the observation point M lies in the observation plane exactly under the edge of the screen (Fig. 13.25). Then the Fresnel zones have the form of plane bands parallel to the edge of the screen. Denoting the zone boundaries by $O_1, O_2, O_3, \ldots$, we can write

$$\begin{cases} OM = l, \\ O_1M = l + \lambda/2, \\ O_2M = O_1M + \lambda/2, \\ \ldots\ldots\ldots\ldots\ldots\ldots, \end{cases} \tag{13.15}$$

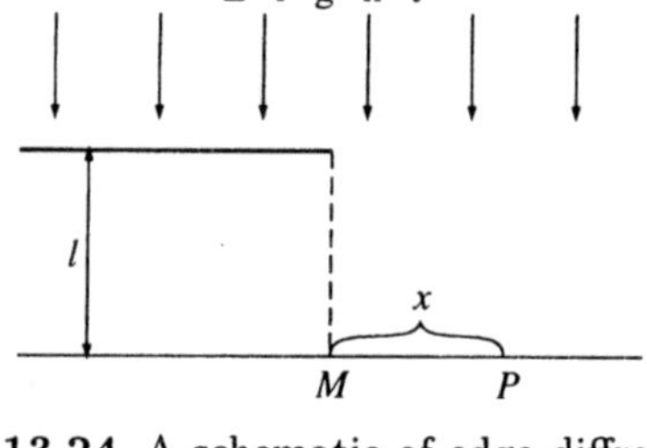

Fig. 13.24 A schematic of edge diffraction.

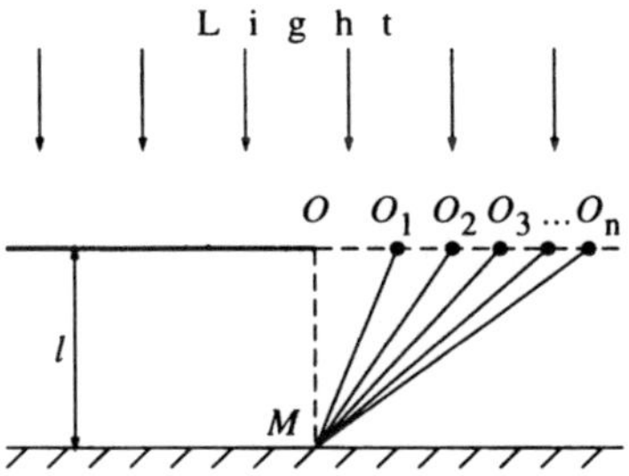

Fig. 13.25 Construction of the Fresnel zones.

where λ is the optical wavelength.

Let d_n denote the distance from the edge of the screen to the origin of the Fresnel zone with number n. It can be seen from Fig. 13.25 that

$$d_n^2 = OO_n^2 = (l + n\lambda/2)^2 - l^2 = nl\lambda + (n\lambda/2)^2.$$

Taking into account that $l \gg \lambda$ and neglecting the last term, we obtain the approximate formula

$$d_n = \sqrt{nl\lambda}. \tag{13.16}$$

The area of the nth Fresnel zone,

$$S_n = (d_{n+1} - d_n)L,$$

where L is the length of the screen edge, can easily be calculated, using eqn (13.16):

$$S_n = (\sqrt{n+1} - \sqrt{n})L\sqrt{\lambda L} = \frac{L\sqrt{\lambda l}}{\sqrt{n+1} + \sqrt{n}} \approx \frac{L}{2}\sqrt{\lambda l/n}.$$

Thus, in the case considered the areas of the Fresnel zones decrease with an increase in the zone number n:

$$S_n \sim \frac{1}{\sqrt{n}}. \tag{13.17}$$

Now let us divide every Fresnel zone into a great (in the limiting case, infinite) number of subzones. To determine the diffraction field at the point P, we need to sum up the secondary wavelets coming to this point from all zones and subzones which are not obstructed by the screen. We will perform the summation by the vector diagram method. To do this, we have to take into account that the zone width and area decrease as the distance from the zone to the screen edge grows. Consequently, the greater the distance from the zone to the screen edge, the less is the modulus of the vector describing the contribution of an individual subzone

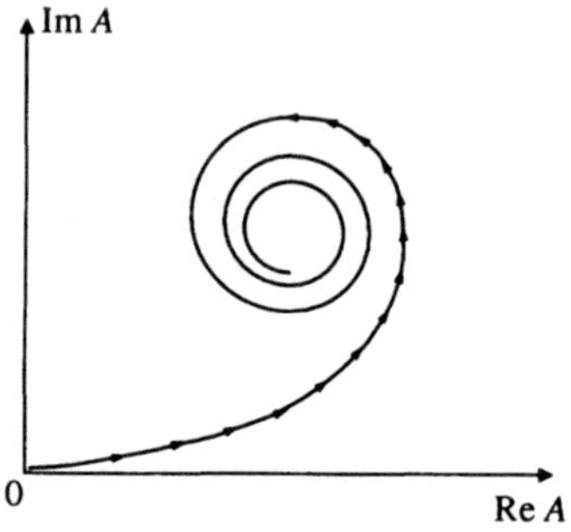

Fig. 13.26 The vector diagram construction for plane wave edge diffraction.

to the resulting field. As a result, the vector diagram takes the form shown in Fig. 13.26.

Now suppose that the observation point M is fixed (Fig. 13.25) while the screen edge starts to move to the left of its original position (point O). It is clear that in this case the Fresnel zones located to the left of point O begin to open. The boundaries of the Fresnel zones are located symmetrically to the left and to the right of point O. Consequently, the corresponding spiral should also be symmetric. The result of the construction is shown in Fig. 13.27.

Using the spiral shown in Fig. 13.27, we can construct the complete pattern of edge diffraction. The construction procedure is illustrated in Figs 13.28(a–c). It is convenient to present the result as the dependence of the light intensity I on the distance x shown in Fig. 13.24. The function $I(x)$ is plotted in Fig. 13.28(d). Note that the result obtained agrees well with the experimental data on the diffraction of the laser beam by the edge of the screen (see Figs 13.3 and 13.4).

The method of Fresnel zones allows one to explain, at least qualitatively, some types of diffraction, such as diffraction by an aperture, by a disc, and by an edge of the screen. We may thus conclude that the diffraction phenomena are indeed caused by the interference of secondary wavelets.

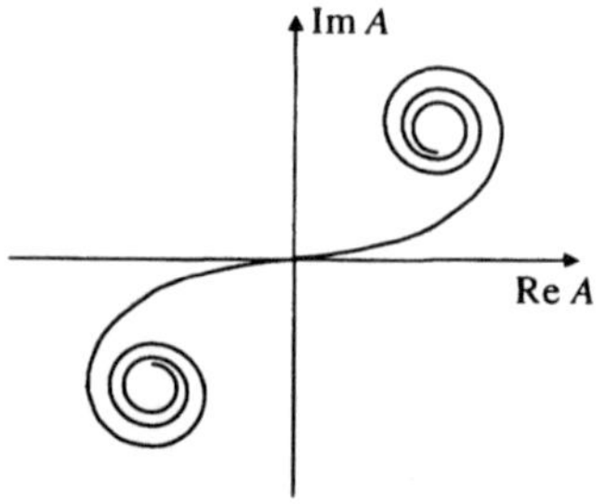

Fig. 13.27 The symmetric spiral of Fresnel.

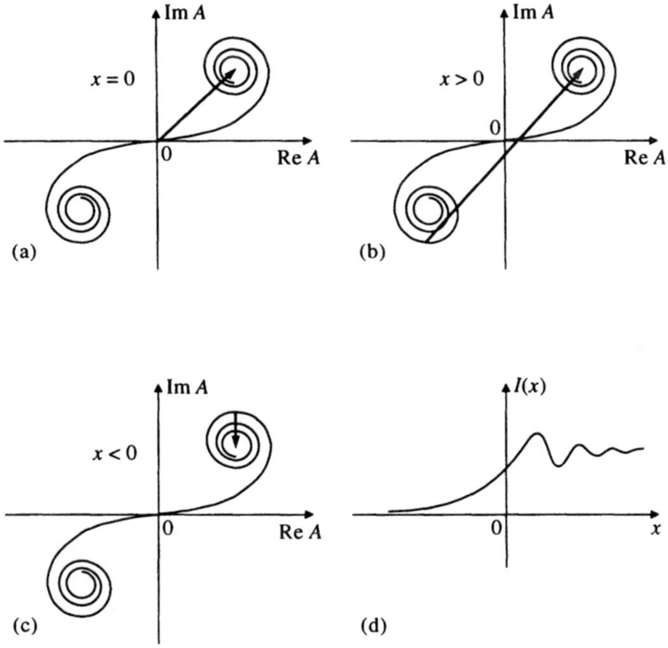

Fig. 13.28 The construction of the function $I(x)$ describing plane wave edge diffraction.

13.8.1 Free propagation of a plane optical wave and normalization of the Huygens–Fresnel integral

The Huygens–Fresnel integral (13.2) can be used for the quantitative calculation of diffraction patterns. As a simple test problem, let us examine the free propagation of a plane optical wave.

We shall introduce the designations shown in Fig. 13.29. Let the z-axis be directed along the wave propagation and let the plane Σ of the wavefront be the source of secondary wavelets. Let P be the observation point, O the origin of coordinates, and M the point lying on the surface Σ at the distance r from the point O. We shall also introduce the distance z from the point O to the point P and the distance ρ from the point M to the point P. As can be seen from Fig. 13.29, $\rho^2 = r^2 + z^2$. Then, for $z \gg r$, we obtain the approximate expression

$$\rho = z + r^2/2z. \tag{13.18}$$

Inserting eqn (13.18) into eqn (13.2), we obtain

$$\mathcal{E}(P) = \mathcal{E}_0 \frac{e^{-ikz}}{z} \int_\Sigma e^{-ikr^2/2z} K(\varphi)\, d\sigma, \tag{13.19}$$

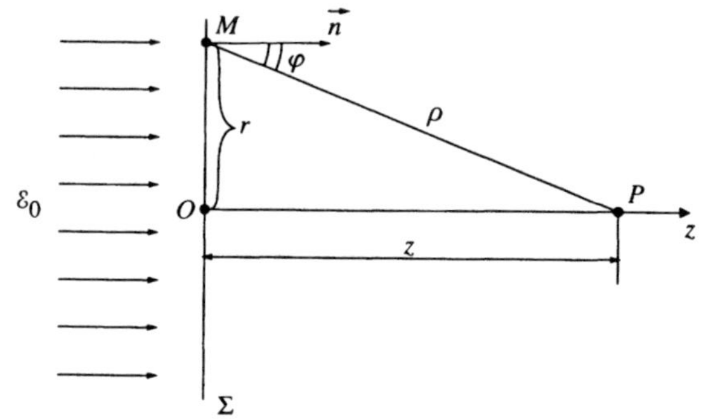

Fig. 13.29 The calculation of the free propagation of a plane wave.

where $\mathcal{E}_0$ is the optical wave amplitude in the plane $z = 0$. We have neglected the difference between the values of ρ and z in the integrand.

Let us introduce polar coordinates in the plane Σ, with the point O being the origin of the coordinates. Then $d\sigma = 2\pi r\, dr$ and the inclination factor $K(\varphi)$ can be treated as a function of the argument r. Equation (13.19) then takes the form

$$\mathcal{E}(P) = \mathcal{E}_0 e^{-ikz} \frac{2\pi}{z} \int_0^\infty e^{-ikr^2/2z} K(r) r\, dr. \tag{13.20}$$

We shall set

$$K(r) = K(0) f(r), \tag{13.21}$$

where $K(0)$ is the constant and $f(r)$ is a function decreasing from 1 to 0 as the argument r varies from 0 to ∞. Let us choose this function so that the integral (13.20) takes the simplest form. Setting, for example,

$$f(r) = \exp(-\alpha r^2), \tag{13.22}$$

we obtain

$$\mathcal{E}(P) = \mathcal{E}_0 e^{-ikz} K(0) \frac{\pi}{z(\alpha + ik/2z)}. \tag{13.23}$$

Passing to the limit $\alpha \to 0$, we obtain

$$\mathcal{E}(P) = \mathcal{E}_0 e^{-ikz} K(0) \frac{2\pi}{ik}. \tag{13.24}$$

On the other hand, the following condition must be met when a plane wave propagates in free space (see Fig. 13.29)

$$\mathcal{E}(P) = \mathcal{E}_0 e^{-ikz}. \tag{13.25}$$

Comparing eqns (13.24) and (13.25), we find that $K(0) = ik/2\pi$ or

$$K(0) = i/\lambda. \tag{13.26}$$

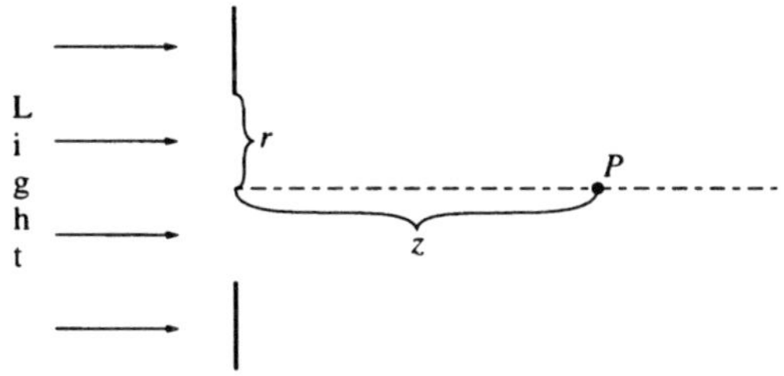

Fig. 13.30 A schematic diagram of light diffraction by a round aperture.

Thus, we have shown that the Huygens–Fresnel diffraction integral correctly describes the free propagation of a plane optical wave, if the condition (13.26) is satisfied.

Assuming that $K(\varphi) = K(0) = i/\lambda$, we can write the integral (13.2) in the form

$$\mathcal{E}(P) = \frac{i}{\lambda} \int_{\Sigma} \mathcal{E}(M) \frac{e^{-ik\rho}}{\rho} \, d\sigma. \tag{13.27}$$

Later, we shall use the Huygens–Fresnel integral written in the form (13.27) for quantitative calculations of diffraction patterns (see Chapters 14 and 15).

13.9 The diffraction length of an optical beam—near- and far-field zones of diffraction

Let us turn once again to the problem of diffraction of a plane optical wave by a round aperture and try to clarify how the light intensity I varies on the aperture axis as a function of the distance z from the screen (Fig. 13.30).

If the distance z is fixed, the radii of the Fresnel zones are given by eqn (13.11):

$$r_n = \sqrt{(n+1)\lambda z}, \tag{13.28}$$

where n is the zone number, and λ is the optical wavelength. Now suppose that the aperture radius r is fixed and the observation point moves away from the screen along the z-axis. The periphery Fresnel zones will then start leaving the aperture, until only the zeroth zone remains within it. Using the spiral of Fresnel (Fig. 13.31), we find that in this position, the light intensity at the observation point reaches its maximum. The further increase in the distance z lowers the light intensity I (Fig. 13.32).

The distance z, for which the aperture coincides with the zeroth Fresnel zone, is called the *diffraction length* and usually denoted by z_d. It can be seen from Fig. 13.32 that the diffraction length determines a boundary between two different zones. The zone for which

$$z \ll z_\mathrm{d} \tag{13.29}$$

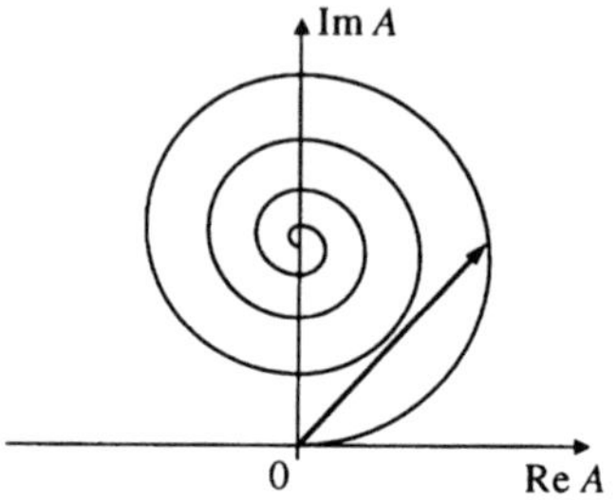

Fig. 13.31 The spiral of Fresnel.

is called the *near-field zone of diffraction.* In this zone the optical beam retains its structure, defined by the shape of the aperture, and the light intensity at the beam axis is approximately equal to the intensity of the original optical wave. If the observation point is in the near-field zone, then many Fresnel zones are located within the aperture, and the beam profile remains invariable due to the interference of elementary secondary wavelets coming from different Fresnel zones.

The zone for which

$$z \gg z_d \tag{13.30}$$

is called the *far-field zone of diffraction.* In this zone the light intensity at the beam axis is much lower than the intensity of the original wave (see Fig. 13.32) and, consequently, the beam expands. If the observation point is in the far-field zone, then only a central part of the zeroth Fresnel zone is located within the aperture. The interference of the elementary secondary wavelets is less pronounced. It fails to sustain the original beam profile any longer, so that the beam diverges.

Using eqn (13.28) and the definition of the diffraction length z_d given earlier, we can show that

$$z_d = r^2/\lambda, \tag{13.31}$$

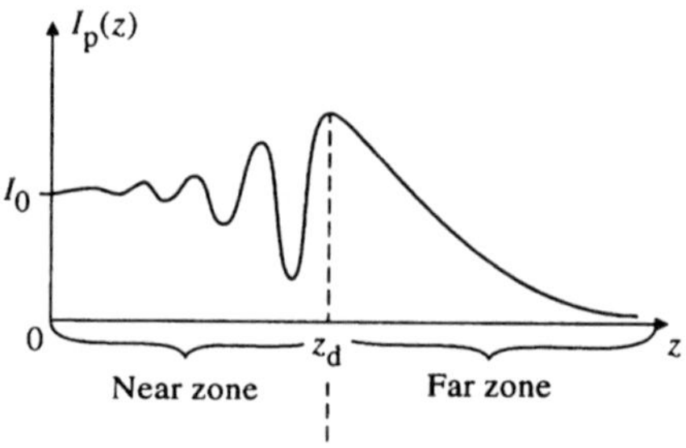

Fig. 13.32 The light intensity on the axis of an opening, plotted as a function of the distance from the screen. z_d is the diffraction length of the optical beam.

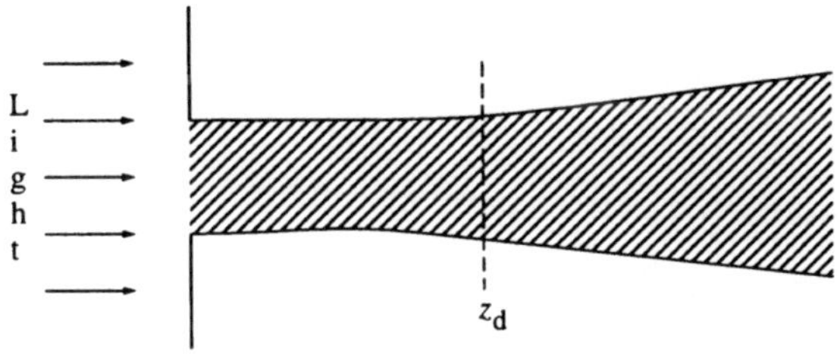

Fig. 13.33 The diffraction of an optical beam.

where r is the original radius of the beam (aperture radius), and λ is the optical wavelength. The diffraction length is connected with Fresnel's number

$$N_F = r^2/\lambda z, \tag{13.32}$$

by the relation

$$N_F = z_d/z. \tag{13.33}$$

It follows from eqns (13.29), (13.30), and (13.33) that in the near-field zone

$$N_F \gg 1, \tag{13.34}$$

while in the far-field zone

$$N_F \ll 1. \tag{13.35}$$

13.10 The diffraction divergence of an optical beam in the far-field zone

A typical variation of the transverse size of an optical beam due to diffraction is shown in Fig. 13.33. Note that the beam expands in the far-field diffraction zone. Let us estimate the divergence angle θ_d (Fig. 13.34).

Starting from the idea of the interference of secondary wavelets, it is natural to assume that the position of the optical beam boundary is determined by the condition

$$\Delta = \lambda/2, \tag{13.36}$$

where λ is the optical wavelength, and Δ is the path difference of the beams coming to the given point from the opposite edges of the aperture (Fig. 13.35).

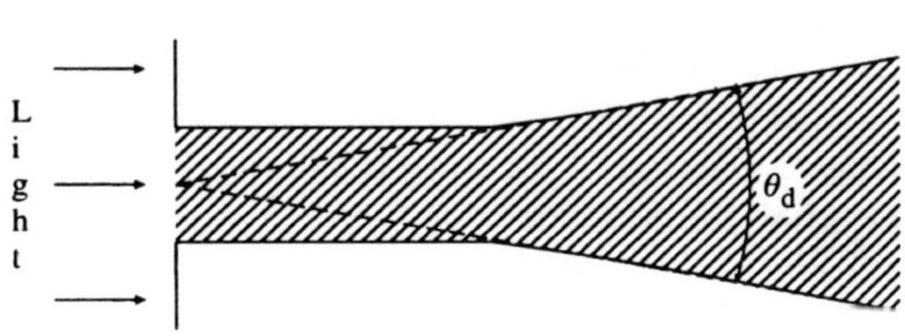

Fig. 13.34 A definition of the diffraction divergence angle of an optical beam.

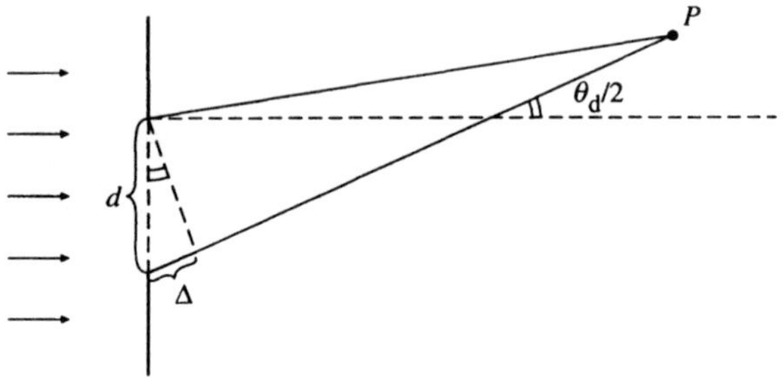

Fig. 13.35 The calculation of the diffraction divergence of an optical beam.

It can be seen from Fig. 13.35 that $\Delta = d\sin(\theta_d/2)$, where d is the aperture diameter and θ_d is the beam divergence angle. As a rule, the beam divergence is rather small ($\theta_d \ll 1$) and, consequently, we can write the approximate relation $\Delta = d\theta_d/2$, from which it follows that

$$\theta_d = \lambda/d, \tag{13.37}$$

where λ is the optical wavelength and d is the original diameter of the beam (the aperture diameter). Thus, eqn (13.37) determines the diffraction divergence of the optical beam in the far-field zone.

The beam diameter in the far-field zone can be expressed as

$$d(z) = \frac{\lambda z}{d}, \tag{13.38}$$

where z is the coordinate measured along the beam axis.

Equation (13.37) shows that the diffraction divergence is higher for more narrow beams. We can verify this result experimentally using a laser beam and an iris diaphragm, or a slit of tunable width, as an aperture. The experiment confirms the dependence expressed by eqn (13.37).

Let us estimate the diffraction length z_d and the angular divergence θ_d for the helium–neon laser beam. Setting $d = 2$ mm and $\lambda = 0.6$ μm, we obtain $z_d = 1.5$ m; $\theta_d = 3 \times 10^{-4}$ rad. The correctness of the estimate obtained can be proved by observing the free diffraction of the laser beam.

13.11 The focusing of light as a diffraction phenomenon

Equation (13.37) allows us to evaluate the minimum transverse dimension of the beam that can be obtained by focusing the light by a lens. This parameter is important for physical optics because it sets a limit for the concentration of light in space. A schematic diagram of focusing is shown in Fig. 13.36.

Let us assume that the focusing diagram is symmetric relative to the focal plane of the lens. The validity of this assumption is confirmed by experiment, by the rules of ray optics, and by the theory based on the solution of the wave equation (see Chapter 14).

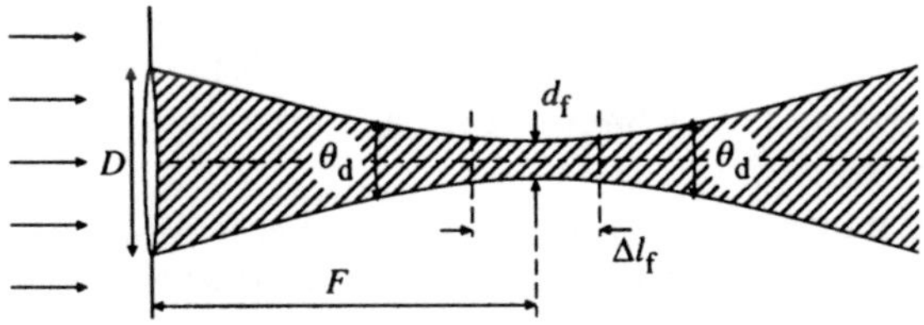

Fig. 13.36 The focusing of an optical beam.

It is evident that the beam divergence to the right of the focal plane is caused by diffraction. Using eqn (13.37), we can write the divergence angle in the form

$$\theta_d = \lambda / d_f, \tag{13.39}$$

where d_f is the beam diameter in the focal plane. On the other hand, the same angle can be expressed via the focal length F of the lens and the incident beam diameter D:

$$\theta_d = D/F. \tag{13.40}$$

We find from eqns (13.39) and (13.40) that

$$d_f = \lambda F/D. \tag{13.41}$$

Thus, the focal spot diameter is determined by eqn (13.41). In this equation λ is the optical wavelength, F is the focal length of the lens, and D is the diameter of the beam incident on the lens.

What is the minimum possible value of d_f? Equation (13.41) shows that when the original beam diameter is equal to the lens diameter, the value d_f is smaller for higher ratios of the lens diameter to its focal length. The parameter D/F is called the relative aperture of the lens. For real lenses, this parameter does not exceed unity. In the most favourable case, when $D/F = 1$, eqn (13.41) yields

$$d_f^{\min} = \lambda, \tag{13.42}$$

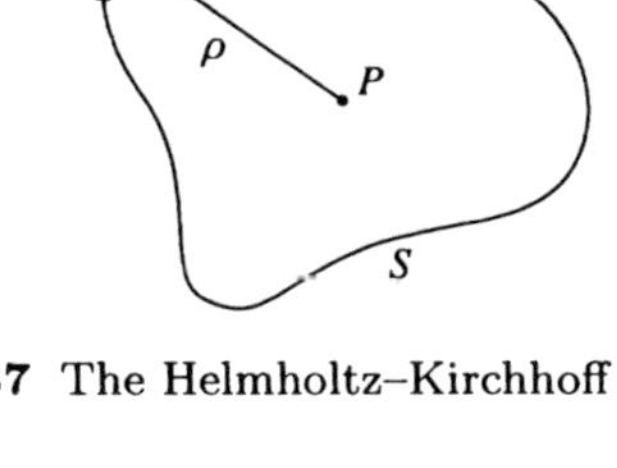

Fig. 13.37 The Helmholtz–Kirchhoff theorem.

i.e. the focal spot diameter is of the order of an optical wavelength.

Note that eqn (13.41) is valid only for spatially coherent beams. In the case of a non-coherent beam the quantity D in eqn (13.14) should be replaced by the coherence radius r_c (see Chapter 12). As this takes place, the value of d_f generally increases.

Let us now estimate the length Δl_f of the beam waist at the focus. Figures 13.33 and 13.36 show that the value of Δl_f is twice the diffraction length of the beam with the original diameter d_f:

$$\Delta l_f = 2z_d, \qquad z_d = d_f^2/4\lambda.$$

Thus,

$$\Delta l_f = d_f^2/2\lambda \tag{13.43}$$

or, in view of eqn (13.41),

$$\Delta l_f = \lambda F^2/2D^2. \tag{13.44}$$

Thus, the diameter and length of the beam waist are governed by eqns (13.41) and (13.43). These formulae should be regarded as approximations based on physical reasoning. More accurate expressions can be obtained on the basis of the mathematical theory of diffraction, i.e. by solving the wave equation under appropriate boundary conditions (see Chapter 14).

Let us make one numerical evaluation. Suppose that a coherent laser beam with a wavelength of $\lambda = 0.5$ μm and a diameter of $d = 2.5$ mm is focused by a lens with a focal length of $F = 16$ cm. Using eqns (13.41) and (13.43), we obtain $d_f = \lambda F/d = 3 \times 10^{-3}$ cm, and $\Delta l_f = d_f^2/2\lambda = 0.1$ cm.

13.12 Kirchhoff's diffraction theory

The basic problem in diffraction theory is to find the structure of the optical field in the presence of obstacles to wave propagation. After the discovery of the electrodynamics equations and of the electromagnetic nature of light, the mathematical problem of diffraction was formulated as a problem of solving the wave equation under definite boundary conditions:

$$\begin{cases} \Delta E - \dfrac{1}{c^2}\dfrac{\partial^2 E}{\partial t^2} = 0 \\ + \text{boundary conditions.} \end{cases} \tag{13.45}$$

A general method for solving this problem was proposed by Kirchhoff in 1882. He succeeded in showing that the Huygens–Fresnel diffraction integral can be regarded as an approximate solution to the diffraction problem (13.45). Fresnel diffraction theory was thus put on a sound mathematical basis. Let us briefly consider the basic ideas of Kirchhoff's theory.

13.12.1 The Helmholtz equation

Let us consider a monochromatic optical field

$$E(\vec{r},t) = \tfrac{1}{2}\mathcal{E}(\vec{r})e^{i\omega t} + \text{c. c.}, \tag{13.46}$$

where ω is the frequency and $\mathcal{E}(\vec{r})$ is the complex amplitude of the field. Substituting eqn (13.46) into the wave equation (13.45), we obtain the *Helmholtz equation* for $\mathcal{E}(\vec{r})$:

$$\Delta\mathcal{E} + k^2\mathcal{E} = 0, \tag{13.47}$$

where $k = \omega/c$ is the wave number, c is the velocity of light, and Δ is the Laplace operator. The simplest solutions to this equation are the plane wave

$$\mathcal{E} = \mathcal{E}_0 e^{-i\vec{k}\vec{r}} \tag{13.48}$$

and the spherical wave

$$\mathcal{E} = A\frac{e^{-ikr}}{r} \tag{13.49}$$

(see Chapter 1). Since the Helmholtz equation is linear, any set of plane waves propagating in all possible directions and any set of spherical waves emerging at different points of space also obey this equation. Consequently, the diffraction problem (13.45) can be solved by the method of spectral expansion, according to which any optical field can be represented as a set of plane or spherical waves.

In Kirchhoff's theory the spherical optical wave is adopted as a standard wave, so that any real optical field can be represented as a set of spherical waves.

13.12.2 The Helmholtz–Kirchhoff integral theorem

According to this theorem the optical field amplitude at an arbitrary point P can be calculated if the field amplitude $\mathcal{E}$ and its normal derivative $\partial\mathcal{E}/\partial n$ are known at all points of an arbitrary surface S surrounding the point P (Fig. 13.37):

$$\mathcal{E}(P) = \frac{1}{4\pi}\oint_S \left(G\frac{\partial\mathcal{E}}{\partial n} - \mathcal{E}\frac{\partial G}{\partial n}\right) dS, \tag{13.50}$$

where

$$G = \frac{e^{-ik\rho}}{\rho} \tag{13.51}$$

is the point-source function or the Green function for the Helmholtz equation.

13.12.3 Kirchhoff's boundary conditions

To solve the diffraction problem in the case in which a screen or a system of screens obstruct the propagation of the optical wave, Kirchhoff proposed the following approximate boundary conditions. The optical field within the openings

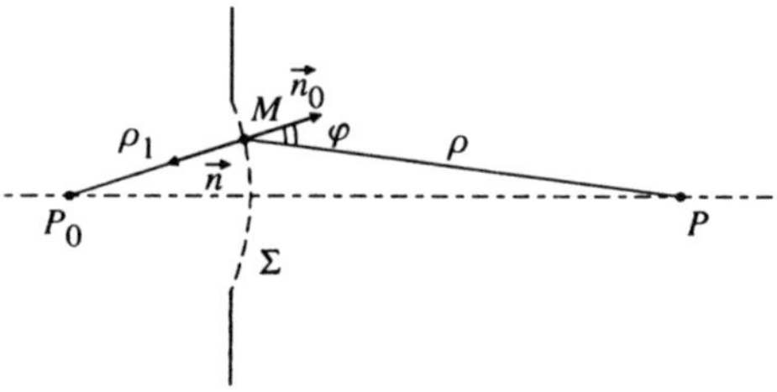

Fig. 13.38 The calculation of spherical wave diffraction by an opening.

is assumed to be the same as if the screens were absent, and the field on the shadow side of the screens to be equal to zero:

$$\begin{cases} \text{at the shadow side of the screen } \mathcal{E} = 0, \ \partial\mathcal{E}/\partial n = 0, \\ \text{within the openings } \mathcal{E} \text{ and } \partial\mathcal{E}/\partial n \text{ are the same as} \\ \qquad \text{if the screens were absent.} \end{cases} \tag{13.52}$$

Despite the obviously approximate character of these conditions, they prove to be a good model in optics, mainly due to the small value of the optical wavelength.

13.12.4 The Helmholz–Kirchhoff diffraction integral

Kirchhoff theory gives the following result for the problem of diffraction of a spherical wave by an opening (Fig. 13.38):

$$\mathcal{E}(P) = \int_\Sigma \mathcal{E}(M)\frac{e^{-ik\rho}}{\rho}K(\varphi)\,d\sigma, \tag{13.53}$$

where

$$K(\varphi) = \frac{i}{\lambda}\left(\frac{1+\cos\varphi}{2}\right). \tag{13.54}$$

Note that the integral (13.53) coincides exactly with the Huygens–Fresnel diffraction integral (13.2). However, the form of the function $K(\varphi)$ proposed by Fresnel (see Fig. 13.9) is not exact. Kirchhoff theory shows that the inclination factor $K(\varphi)$ is correctly determined by eqn (13.54); the graph of this function is shown in Fig. 13.39.

The main contribution to the diffraction integral is due to the Fresnel central zones, i.e. the zones located near the optical beam axis, for which $\varphi \ll 1$. Setting $K(\varphi) = K(0) = i/\lambda$, we obtain

$$\mathcal{E}(P) = \frac{i}{\lambda}\int_\Sigma \mathcal{E}(M)\frac{e^{-ik\rho}}{\rho}\,d\sigma. \tag{13.55}$$

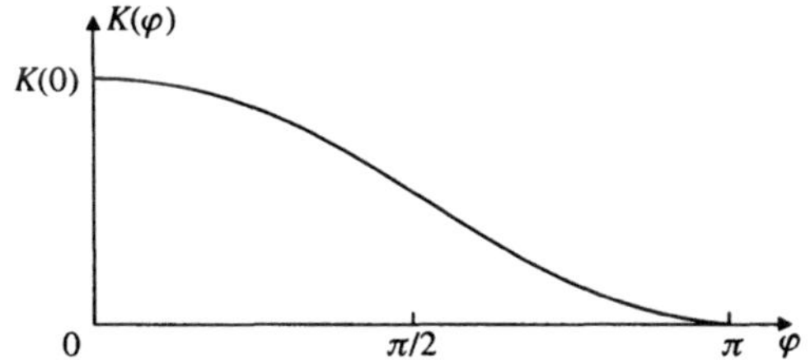

Fig. 13.39 The inclination factor in Kirchhoff theory.

The diffraction integral written in such a form coincides with the integral (13.27). Kirchhoff theory thus justifies and corrects Fresnel theory.

Let us briefly clarify the derivation of eqns (13.53) and (13.54). Suppose that a point monochromatic source of light is located at a point P_0. We shall calculate the optical field at a point P, assuming that there is an obstruction of the optical wave propagation, e.g. a screen with an opening, located between the points P_0 and P (Fig. 13.38).

According to the Helmholz–Kirchhoff theorem, the diffraction field at the point P is determined by the integral (13.50) taken over an arbitrary surface S surrounding this point. Suppose that the surface S consists of three parts: (i) the surface Σ subtending the opening; (ii) the surface S_1 of the shaded (non-illuminated) side of the screen; and (iii) the spherical surface S_2 of sufficiently large radius centred at the point P. The surface S is illustrated in Fig. 13.40.

Physical considerations suggest that the main contribution to the optical field at point P is made by the integral over the surface Σ, because the light reaches the point P through the opening in the screen. Consequently,

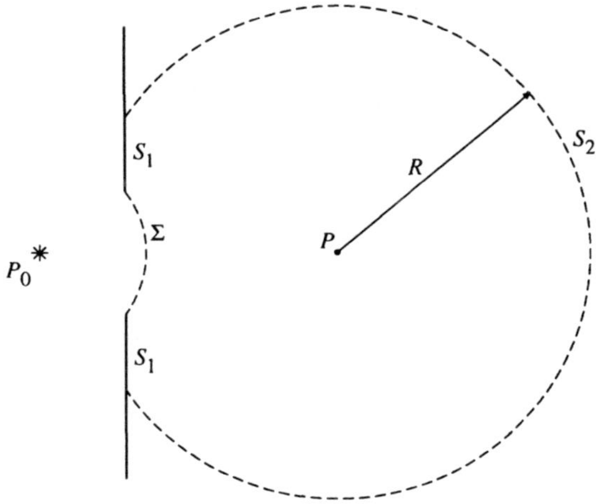

Fig. 13.40 The calculation of the diffraction integral in Kirchhoff theory.

$$\mathcal{E}(P) = \frac{1}{4\pi} \int\limits_{\Sigma} \left(G \frac{\partial \mathcal{E}}{\partial n} - \mathcal{E} \frac{\partial G}{\partial n} \right) d\sigma, \tag{13.56}$$

where

$$G = \frac{\mathrm{e}^{-ik\rho}}{\rho} \tag{13.57}$$

is the Green function, k is the wave number of the optical wave, ρ is the distance between the points M and P, and $\vec{n}$ is the unit vector of the normal to the surface Σ at the point M. In this case it is convenient to choose the surface Σ as part of a spherical surface centred at the point P_0 at which the light source is located. A mathematical justification of the transition from the integral over the closed surface S to the integral over the surface Σ is given, for example, in Born and Wolf (1968).

Thus, the diffraction optical field at the point P is described by eqns (13.56) and (13.57). Now let us transform the expression for $\mathcal{E}(P)$. We shall begin with the calculation of the derivative $\partial G/\partial n$. As can be seen from Fig. 13.38, $\vec{n} = -\vec{n}_0$. Consequently, $\partial G/\partial n = -\partial G/\partial n_0$. Since the Green function determined by eqn (13.30) depends on n_0 implicitly, we shall make use of the rule for indirect differentiation: $\partial G/\partial n_0 = (\partial G/\partial \rho)(\partial \rho/\partial n_0)$. By virtue of eqn (13.57),

$$\frac{\partial G}{\partial \rho} = -\left(ik + \frac{1}{\rho} \right) G \approx -ikG.$$

Usually, this approximation, referred to as *optical* one, ensures a sufficient accuracy because in optics, as a rule, $\rho \gg \lambda$, and therefore, $k = 2\pi/\lambda \gg 1/\rho$.

The calculation of the directional derivative $\partial \rho/\partial n_0$ is illustrated in Fig. 13.41. As can be seen from this figure,

$$\frac{\partial \rho}{\partial n_0} = \lim_{\Delta n_0 \to 0} \frac{\Delta \rho}{\Delta n_0} = -\cos \varphi.$$

Thus,

$$\frac{\partial G}{\partial n} = -ikG \cos \varphi. \tag{13.58}$$

Now let us calculate the derivative $\partial \mathcal{E}/\partial n$. As this derivative is taken at the point M of the surface $\sum$, we have

$$\mathcal{E} = \mathcal{E}(M) = A_0 \frac{\mathrm{e}^{-ik\rho_1}}{\rho_1},$$

where A_0 is a constant and ρ_1 is the distance between the points P_0 and M. The calculations are carried out similarly to the earlier calculations of the derivative

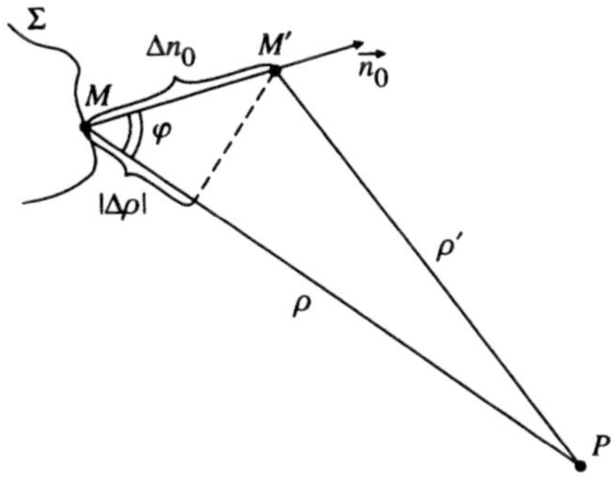

Fig. 13.41 The calculation of the directional derivative.

$\partial G/\partial n$. We obtain $\partial\mathcal{E}/\partial n = (\partial\mathcal{E}/\partial\rho_1)(\partial\rho_1/\partial n)$, $\partial\rho_1/\partial n = -1$, and $\partial\mathcal{E}/\partial\rho_1 = -(ik + 1/\rho_1)\mathcal{E} \approx -ik\mathcal{E}$. Hence,

$$\frac{\partial\mathcal{E}}{\partial n} = ik\mathcal{E}. \tag{13.59}$$

Inserting eqns (13.57)–(13.59) into eqn (13.56), we obtain eqns (13.53) and (13.54), which proves the statement.

Thus, the approximate solution of the wave equation for the optical field, given by Kirchhoff, confirms the Fresnel diffraction theory. Fresnel guessed the form of the diffraction integral correctly. At present, Fresnel theory retains its value primarily as a system of descriptive models revealing the physics of the diffraction of light. The mathematical formulation of the diffraction problem (13.45), based on the Maxwell theory, allows one to use the advanced apparatus of mathematical physics to solve specific diffraction problems. In particular, the method of spectral expansion and the method of the parabolic equation can be applied (see Chapters 14 and 17), as well as the powerful and universal methods of numerical simulation.

Bibliography

Born, M. and Wolf, E. (1968). *Principles of optics.* Pergamon Press, Oxford.

Goodman, J. W. (1968). *Introduction to Fourier optics.* McGraw-Hill, San Francisco.

14

The diffraction of paraxial beams

Fresnel's approximation in diffraction theory. Fresnel's integrals and Cornu's spiral. Fresnel diffraction by unidimensional structures. Fresnel diffraction by two-dimensional structures. The parabolic equation and quasi-optical approximation. Solution of the parabolic equation. The propagation and diffraction of a Gaussian beam. The focusing of a Gaussian beam. The application criterion for quasi-optical approximation.

14.1 Fresnel's approximation in diffraction theory

Two basic approximations, Fresnel's and Fraunhofer's, are widely used for the calculation of diffraction patterns. Fresnel's approximation describes the diffraction of paraxial, that is, weakly diverging optical beams, and the Fraunhofer approximation describes the diffraction in the far-field zone. Here we shall consider Fresnel's approximation.

Let us consider the diffraction of a plane monochromatic optical wave by an aperture. The general solution to this problem is given by the Huygens–Fresnel integral (see Chapter 13)

$$\mathcal{E}(P) = \frac{i}{\lambda} \int\limits_{\Sigma} \mathcal{E}(M) \frac{\mathrm{e}^{-ik\rho}}{\rho} d\sigma, \tag{14.1}$$

where $\mathcal{E}$ is the complex amplitude of the field, Σ is the surface subtending the aperture, P is the observation point, M is a point on the surface Σ, ρ is the distance between the points M and P, λ is the optical wavelength, and $k = 2\pi/\lambda$ is the wave number.

Let us introduce the coordinates x and y within the plane of the screen with the aperture, and the coordinates x_0 and y_0 in the observation plane. The planes are parallel and are separated by the distance z (Fig. 14.1). In this case the diffraction integral can be written in the form

$$\mathcal{E}(x_0, y_0, z) = \frac{i}{\lambda} \int\limits_{-\infty}^{\infty} \int\limits_{-\infty}^{\infty} \mathcal{E}_0(x, y) \frac{\mathrm{e}^{-ik\rho}}{\rho} dx\, dy, \tag{14.2}$$

where

$$\rho = \sqrt{z^2 + (x - x_0)^2 + (y - y_0)^2}. \tag{14.3}$$

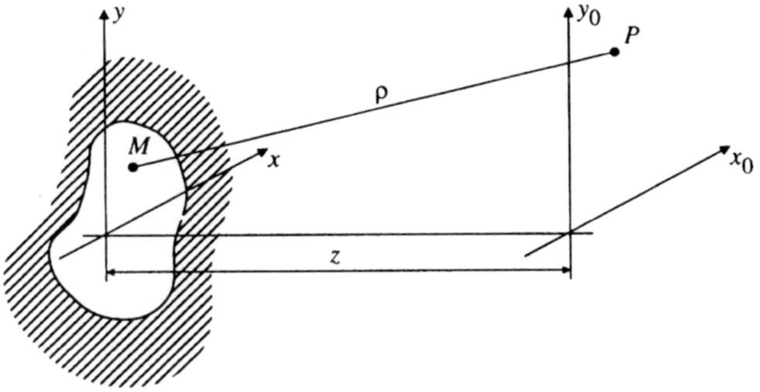

Fig. 14.1 The formulation of a diffraction problem.

Usually in optics (especially in laser optics) we deal with narrow and weakly diverging beams of light, for which the following inequalities are well satisfied:

$$z \gg x, y, x_0, y_0. \tag{14.4}$$

These inequalities allow us to write the following approximate expression for ρ:

$$\rho = z + \frac{(x - x_0)^2 + (y - y_0)^2}{2z}. \tag{14.5}$$

Substituting eqn (14.5) into eqn (14.2) and neglecting the difference between ρ and z in the denominator of the integrand, we obtain

$$\mathcal{E}(x_0, y_0, z) = \frac{i}{\lambda z} \times$$
$$\times e^{-ikz} \int_{-\infty}^{\infty} \int_{-\infty}^{\infty} \mathcal{E}_0(x, y) \exp\left\{ -\frac{ik}{2z} \left[(x - x_0)^2 + (y - y_0)^2 \right] \right\} dx\, dy \tag{14.6}$$

where $\mathcal{E}(x_0, y_0, z)$ is the distribution of the field complex amplitude in the observation plane, $\mathcal{E}_0(x, y)$ is the distribution of field amplitude in the plane of the screen with the aperture, z is the distance between the plane of the screen and the observation plane, λ is the optical wavelength, and $k = 2\pi/\lambda$ is the wave number. Equation (14.6) provides a solution to the diffraction problem in the Fresnel approximation. Let us now discuss the physical content and the application criterion of the Fresnel approximation.

From the physical point of view, the basic formula (14.5) of the Fresnel approximation implies the substitution of the parabolic surfaces for the spherical wavefronts of the Huygens secondary wavelets. In general, such an approximation imposes definite limitations on the allowable dimensions of the aperture and on the position of the observation point. Specifically, the Fresnel approximation excludes from consideration the region of space immediately adjacent

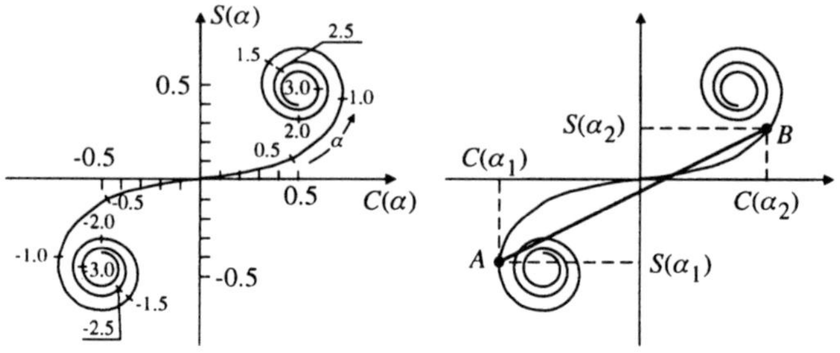

(a) (b)

Fig. 14.2 Cornu's spiral (a) and a construction using the spiral (b).

to the screen with the aperture and the region removed from the beam axis by large distances. However, it is obvious that in the former case the optical field is nearly identical to the field of the incident wave, while in the latter case the light hardly reaches the indicated regions. Accordingly, we will not consider all of these regions although, in principle, the corresponding calculations could be made using the general equations (14.2) and (14.3).

14.2 Fresnel's integrals and Cornu's spiral

In practice, the calculations based on eqn (14.6) can be reduced to the calculation of the integrals

$$C(\alpha) = \int_0^\alpha \cos(\pi t^2/2)\, dt, \qquad S(\alpha) = \int_0^\alpha \sin(\pi t^2/2)\, dt, \tag{14.7}$$

called *Fresnel's integrals*. The values of the special functions $C(\alpha)$ and $S(\alpha)$ are tabulated and can be found, for example, in Janke, Emde, and Losch (1960). There is also an exquisite graphical technique for constructing diffraction patterns, based on the use of the so-called *Cornu's spiral*. This curve can be constructed as follows. In the x–y plane, we mark the points the Cartesian coordinates of which are equal to the values of the Fresnel integrals:

$$x = C(\alpha), \qquad y = S(\alpha). \tag{14.8}$$

With the parameter α gradually varying, the points form a smooth curve called Cornu's spiral (Fig. 14.2(a)). An example of constructing Cornu's spiral is shown in Fig. 14.2(b). The length of the section AB in this figure is determined by the formula

$$l^2(\alpha_1, \alpha_2) = [C(\alpha_1) - C(\alpha_2)]^2 + [S(\alpha_1) - S(\alpha_2)]^2 . \tag{14.9}$$

14.3 Fresnel diffraction by unidimensional structures

If the original field distribution $\mathcal{E}_0(x, y)$ depends only on one spatial variable, e.g.

$$\mathcal{E}_0(x, y) = \mathcal{E}_0(x), \tag{14.10}$$

then we say that diffraction by a unidimensional structure takes place. Substituting eqn (14.10) into eqn (14.6) and using the formulae

$$\int_{-\infty}^{\infty} \exp\left(-\frac{ik}{2z}y^2\right) dy = \sqrt{\frac{2\pi z}{ik}} = (1 - i)\sqrt{\frac{\pi z}{k}}, \tag{14.11}$$

we obtain

$$\mathcal{E}(x_0, z) = \frac{(1 + i)}{\sqrt{2\lambda z}} e^{-ikz} \int_{-\infty}^{\infty} \mathcal{E}_0(x) \exp\left[-\frac{ik}{2z}(x - x_0)^2\right] dx. \tag{14.12}$$

The diffraction integral is thus reduced to the single integral.

14.3.1 Edge diffraction

Let us direct the x-axis perpendicular to the edge of the screen. We can then write

$$\mathcal{E}_0(x) = \mathcal{E}_0 \begin{cases} 1, & x \geq 0, \\ 0, & x < 0. \end{cases} \tag{14.13}$$

Substituting eqn (14.13) into eqn (14.12), we obtain

$$\mathcal{E}(x_0, z) = \mathcal{E}_0 \frac{(1 + i)}{\sqrt{2\lambda z}} e^{-ikz} \mathcal{J}(x_0), \tag{14.14}$$

where

$$\mathcal{J}(x_0) = \int_0^{\infty} \exp\left[-\frac{ik}{2z}(x - x_0)^2\right] dx. \tag{14.15}$$

Let us express the integral (14.15) through the Fresnel integrals (14.8). To do this, we shall introduce the variable

$$\xi = (x - x_0)\sqrt{k/\pi z}. \tag{14.16}$$

Then

$$\mathcal{J}(x_0) = \sqrt{\pi z/k} \int_{\xi_1}^{\infty} \exp(-i\pi\xi^2/2)\, d\xi, \tag{14.17}$$

where

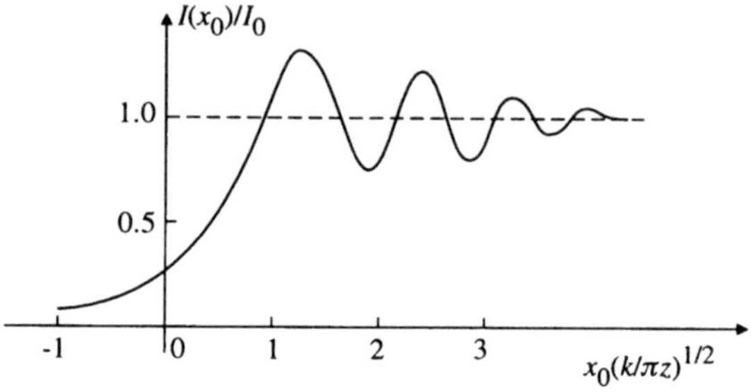

Fig. 14.3 The diffraction of a plane optical wave by a straight edge: the distribution of the light intensity in the diffraction pattern.

$$\xi_1 = -x_0\sqrt{k/\pi z}. \tag{14.18}$$

Using the designation (14.9), we can describe the light intensity at the observation point as

$$I(x_0, z) = \tfrac{1}{2} I_0 l^2(\xi_1, \infty), \tag{14.19}$$

where I_0 is the intensity of the incident plane wave.

The result of a construction made with Cornu's spiral (see Fig. 14.2) is shown in Fig. 14.3. Note that the calculation result agrees well with the experimental data (see Figs 13.3 and 13.4).

14.3.2 Diffraction by a slit

Let us denote the width of the slit by d and direct the x-axis perpendicular to the slit, as shown in Fig. 14.4. Then

$$\mathcal{E}_0(x) = \mathcal{E}_0 \begin{cases} 1, & |x| \le d/2, \\ 0, & |x| > d/2. \end{cases} \tag{14.20}$$

Inserting eqn (14.20) into eqn (14.12), we obtain

$$\mathcal{E}(x_0, z) = \mathcal{E}_0 \frac{(1+i)}{\sqrt{2\lambda z}} \mathrm{e}^{-ikz} \mathcal{J}_1(x_0), \tag{14.21}$$

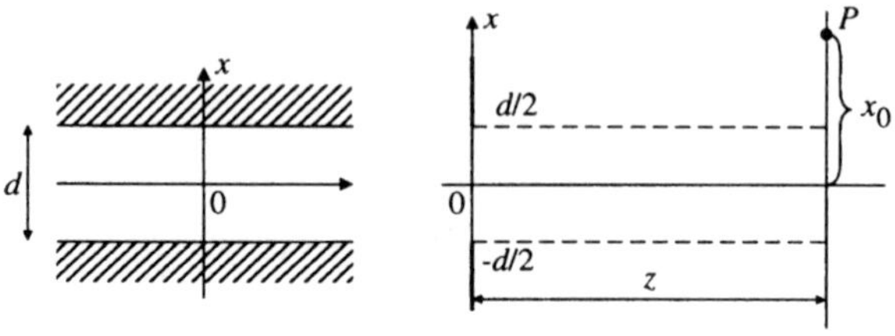

Fig. 14.4 A calculation of plane wave diffraction by a slit.

where

$$\mathcal{J}_1(x_0) = \int_{-d/2}^{d/2} \exp\left[-\frac{ik}{2z}(x-x_0)^2\right] dx. \tag{14.22}$$

Introduce the variable

$$\xi = (x - x_0)\sqrt{k/\pi z}. \tag{14.23}$$

Then

$$\mathcal{J}_1(x_0) = \sqrt{\pi z/k} \int_{\xi_1}^{\xi_2} \exp(-i\pi\xi^2/2)\, d\xi, \tag{14.24}$$

where

$$\xi_1 = -\left(\frac{d}{2} + x_0\right)\sqrt{k/\pi z}, \qquad \xi_2 = \left(\frac{d}{2} - x_0\right)\sqrt{k/\pi z}. \tag{14.25}$$

Using the designation (14.9), we can write the light intensity distribution in the diffraction pattern as

$$I(x_0, z) = \tfrac{1}{2} I_0 l^2(\xi_1, \xi_2), \tag{14.26}$$

where I_0 is the intensity of the incident wave.

To make the construction of the graphs more convenient, we shall rewrite eqns (14.25) and (14.26) in the form

$$I(x_0)/I_0 = \tfrac{1}{2} l^2(\xi_2, \xi_2) \tag{14.27}$$

and

$$\xi_1 = -\alpha(1+p), \qquad \xi_2 = \alpha(1-p). \tag{14.28}$$

Here the dimensionless parameters

$$\alpha = \sqrt{\frac{kd^2}{4\pi z}} = \sqrt{2N_{\mathrm{F}}}, \qquad p = 2x_0/d \tag{14.29}$$

and Fresnel's number (see Chapter 13)

$$N_{\mathrm{F}} = (d/2)^2/\lambda z \tag{14.30}$$

are introduced.

The diffraction pattern drawn by using Cornu's spiral is shown in Fig. 14.5. It can be seen from this figure that the diffraction properties change as Fresnel's number N_{F} varies. Thus, if $N_{\mathrm{F}} \gg 1$ (see Fig. 14.5(a)), the diffraction hardly manifests itself, and the ray optics approximation provides a sufficient accuracy. The radiation intensity profile is then nearly rectangular, the width of the beam equals the width of the slit, and the light intensity on the beam axis coincides

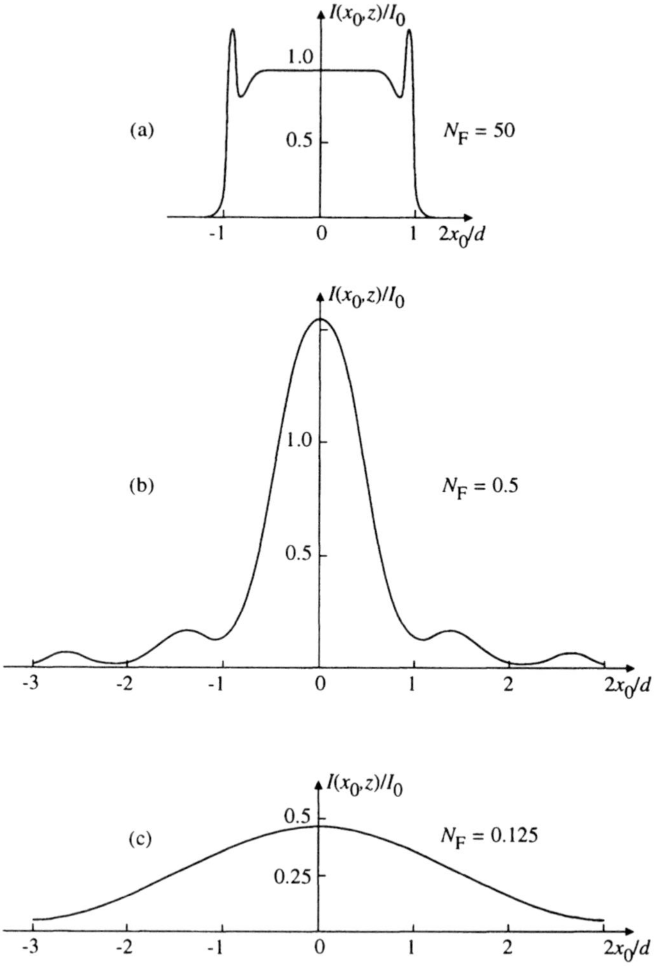

Fig. 14.5 The diffraction of a plane wave by a slit: the distribution of the light intensity in the diffraction pattern.

with the incident wave intensity. The effect of diffraction in this case is appreciable only near the geometric shadow boundary where intensity oscillations are observed, and the light penetrates slightly into the region of geometric shadow.

As the point of observation moves away from the screen with the slit, Fresnel's number N_F decreases. In the region in which $N_F \simeq 1$ (see Fig. 14.5(b)) the light intensity on the beam axis shows appreciable oscillations (along the beam axis), and lateral intensity maxima appear, but the optical beam width is still nearly equal to the width of the slit. However, in the region in which $N_F \ll 1$ (the far-field zone of diffraction) the optical beam broadens, and its transverse profile differs drastically from the original (rectangular) profile (Fig. 14.5(c)).

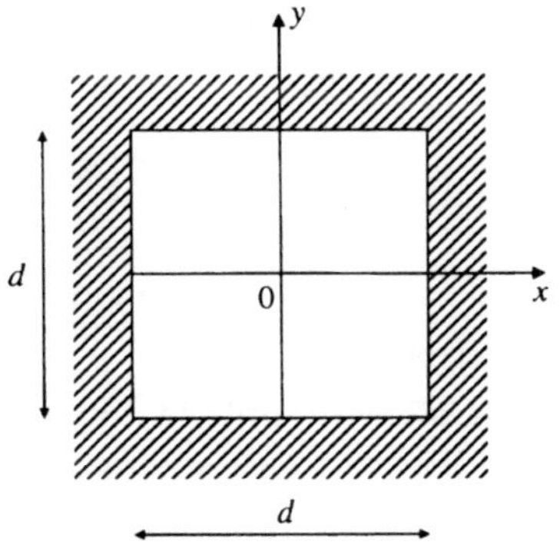

Fig. 14.6 A calculation of plane wave diffraction by a square aperture.

14.4 Fresnel diffraction by two-dimensional structures

To describe diffraction by a two-dimensional structure, we shall make use of the general formula (14.6). This time the calculations are generally more complicated than in the case of diffraction by unidimensional structures. Let us consider some specific examples.

14.4.1 Diffraction by a rectangular aperture

Suppose that the aperture has the form of a square with side d (Fig. 14.6). Then

$$\mathcal{E}_0(x, y) = \mathcal{E}_0 \begin{cases} 1, & \text{if } |x|, |y| \le d/2, \\ 0, & \text{outside this region.} \end{cases} \tag{14.31}$$

Substituting eqn (14.31) into eqn (14.6), we obtain

$$\mathcal{E}(x_0, y_0, z) = \mathcal{E}_0 \frac{i}{\lambda z} \mathrm{e}^{-ikz} \mathcal{J}(x_0) \mathcal{J}(y_0), \tag{14.32}$$

where

$$\begin{aligned} \mathcal{J}(x_0) &= \int_{-d/2}^{d/2} \exp\left[-\frac{ik}{2z}(x - x_0)^2\right] dx, \\ \mathcal{J}(y_0) &= \int_{-d/2}^{d/2} \exp\left[-\frac{ik}{2z}(y - y_0)^2\right] dy. \end{aligned} \tag{14.33}$$

Let us introduce the variables

$$\xi = (x - x_0)\sqrt{k/\pi z}, \qquad \eta = (y - y_0)\sqrt{k/\pi z}. \tag{14.34}$$

Then

$$\mathcal{J}(x_0) = \sqrt{\frac{\pi z}{k}} \int_{\xi_1}^{\xi_2} \exp(-i\pi\xi^2/2)\, d\xi,$$

$$\mathcal{J}(y_0) = \sqrt{\frac{\pi z}{k}} \int_{\eta_1}^{\eta_2} \exp(-i\pi\eta^2/2)\, d\eta,$$

where

$$\begin{cases} \xi_1 = -\sqrt{k/\pi z}\left(\dfrac{d}{2} + x_0\right), & \eta_1 = -\sqrt{k/\pi z}\left(\dfrac{d}{2} + y_0\right), \\ \xi_2 = \sqrt{k/\pi z}\left(\dfrac{d}{2} - x_0\right), & \eta_2 = \sqrt{k/\pi z}\left(\dfrac{d}{2} - y_0\right). \end{cases}$$

Using the designation (14.9), we can represent the distribution of the light intensity in the diffraction pattern as

$$I(x_0, y_0, z) = \tfrac{1}{4} I_0 l^2(\xi_1, \xi_2) l^2(\eta_1, \eta_2).$$

In particular, for the light intensity on the beam axis, i.e. at the points with the coordinates $x_0 = y_0 = 0$, we obtain

$$I(0, 0, z) = \tfrac{1}{4} I_0 \left[l^2(-A, A)\right]^2,$$

where the dimensionless parameter

$$A = (d/2)\sqrt{k/\pi z} = \sqrt{z_\mathrm{d}/z} = \sqrt{N_\mathrm{F}}$$

and the diffraction length (see Chapter 13),

$$z_\mathrm{d} = kd^2/4\pi,$$

are introduced.

The distribution of light intensity along the beam axis, constructed with the use of Cornu's spiral, is shown in Fig. 14.7. The analysis of the intensity distribution within the plane perpendicular to the beam axis shows that at small distances from the screen with the aperture the diffraction effects are appreciable only near the geometric shadow boundary. However, in the far-field zone of diffraction the optical beam loses its original profile; its transverse dimension increases in proportion to the distance z from the screen, while the light intensity on the beam axis decreases monotonously. Experimental observations confirm these conclusions.

14.4.2 Diffraction by a round aperture

Let us denote the aperture radius by R and introduce the polar coordinates r, φ in the x–y plane (Fig. 14.8), and r_0, φ_0 in the x_0–y_0 plane. As before, to calculate the diffraction field, we shall use eqn (14.6). Making the transition to the polar coordinates via the formulae

$$\begin{cases} x = r\cos\varphi, & y = r\sin\varphi, \\ x_0 = r_0\cos\varphi_0, & y_0 = r_0\sin\varphi_0 \end{cases}$$

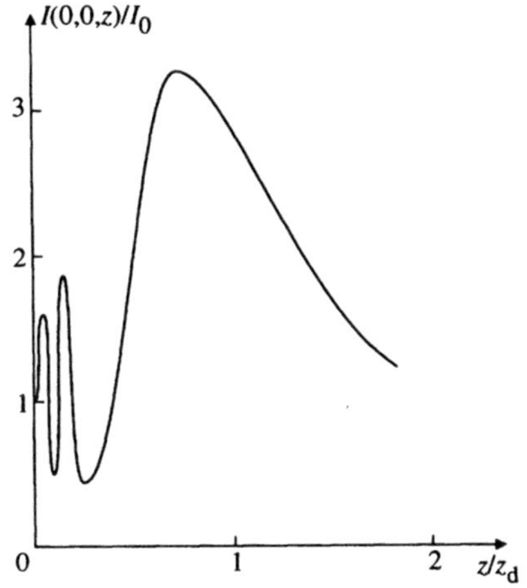

Fig. 14.7 The diffraction of a plane optical wave by a square aperture: the distribution of the light intensity along the beam axis.

and writing the surface element area in the form $d\sigma = r\,dr\,d\varphi$, we obtain

$$\mathcal{E}(r_0, \varphi_0, z) = \frac{i}{\lambda z} e^{-ikz} \int_0^\infty r\,dr \int_0^{2\pi} d\varphi \mathcal{E}_0(r, \varphi) e^{-\frac{ik}{2z}\left[r_0^2 + r^2 - 2rr_0 \cos(\varphi - \varphi_0)\right]},$$

where $\mathcal{E}_0(r, \varphi)$ is the field amplitude distribution in the plane of the screen with the aperture. By virtue of the axial symmetry of this distribution, $\mathcal{E}_0(r, \varphi) = \mathcal{E}_0(r)$. Hence

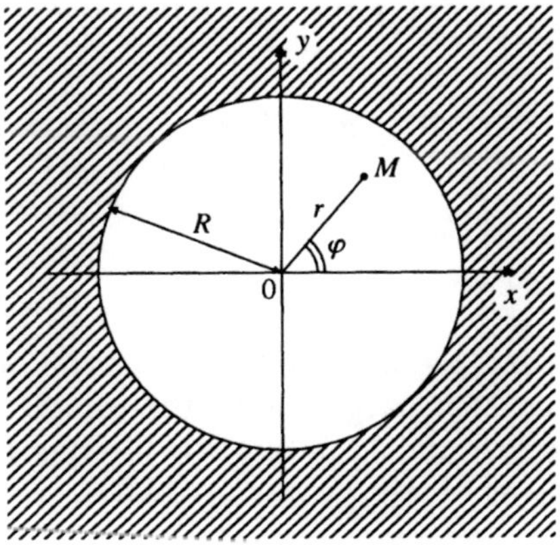

Fig. 14.8 A calculation of plane wave diffraction by a round aperture.

$$\mathcal{E}(r_0, \varphi_0, z) = \frac{i}{\lambda z} \exp\left[-ik\left(z + \frac{r_0^2}{2z}\right)\right] \int_0^\infty \mathcal{E}_0(r)\mathcal{J}(r, r_0)\mathrm{e}^{-\frac{ikr^2}{2z}} r\, dr,$$

where

$$\mathcal{J}(r, r_0) = \int_0^{2\pi} \exp\left[\frac{ikrr_0}{z}\cos(\varphi - \varphi_0)\right] d\varphi.$$

The last integral can be expressed via the Bessel function of zeroth order, $J_0(\alpha)$. By definition,

$$J_0(\alpha) = \frac{1}{2\pi}\int_{-\pi}^{\pi} \mathrm{e}^{i\alpha \sin t} dt.$$

Introducing a new variable, $\psi = \pi/2 - \varphi_0 + \varphi$, we obtain

$$\mathcal{J}(r, r_0) = \int_0^{2\pi} \exp\left[\frac{ikrr_0}{z}\sin\psi\right] d\psi = 2\pi J_0\left(\frac{krr_0}{z}\right)$$

and, consequently,

$$\mathcal{E}(r_0, \varphi_0, z) = \mathcal{E}(r_0, z)$$
$$= \frac{2\pi i}{\lambda z} \exp\left[-ik\left(z + \frac{r_0^2}{2z}\right)\right] \int_0^\infty \mathcal{E}_0(r) J_0\left(\frac{krr_0}{z}\right) \mathrm{e}^{-\frac{ikr^2}{2z}} r\, dr. \quad (14.35)$$

This formula allows one to calculate the diffracted optical field for any original axial-symmetrical distribution $\mathcal{E}_0(r)$. In particular, assuming

$$\mathcal{E}_0(r) = \mathcal{E}_0 \begin{cases} 1, & \text{if } r \le R \\ 0, & \text{if } r > R \end{cases}$$

(diffraction by a round aperture), we obtain

$$\mathcal{E}(r_0, z) = \frac{2\pi i \mathcal{E}_0}{\lambda z} \exp\left[-ik\left(z + \frac{r_0^2}{2z}\right)\right] \int_0^R J_0\left(\frac{krr_0}{z}\right) \mathrm{e}^{-\frac{ikr^2}{2z}} r\, dr. \quad (14.36)$$

Equation (14.36) describes the amplitude distribution of the diffracted optical field $\mathcal{E}(r_0, z)$ throughout the space. Due to the complexity of eqn (14.36), we shall consider only the field on the beam axis. Setting $r_0 = 0$ and taking into account that $J_0(0) = 1$, we obtain, from eqn (14.36),

$$\mathcal{E}(o, z) = 2i\mathcal{E}_0 \exp\left[-ik\left(z + \frac{R^2}{4z}\right)\right] \sin(kR^2/4z).$$

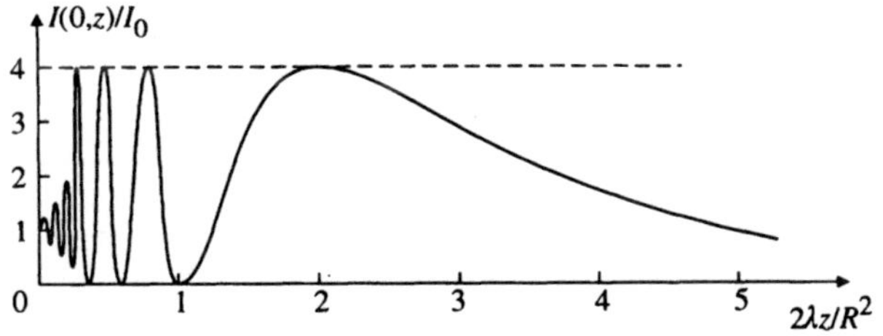

Fig. 14.9 The diffraction of a plane optical wave by a round aperture: the distribution of the light intensity along the beam axis. z is the distance from the screen with the aperture, R is the aperture radius, I_0 is the intensity of the incident wave, and λ is the optical wavelength.

Correspondingly, the radiation intensity on the beam axis is determined by the formula

$$I(0, z) = 4I_0 \sin^2(kR^2/4z), \tag{14.37}$$

where

$$I_0 = \frac{c}{8\pi}|\mathcal{E}_0|^2$$

is the intensity of the incident plane wave.

Equation (14.37) shows that the light intensity on the beam axis oscillates (as a function of the distance from the screen) within the range from zero to a value exceeding the intensity of the incident plane wave by a factor of four. The graph of function $I(0, z)$ is shown in Fig. 14.9.

14.4.3 Diffraction of a Gaussian beam

We shall analyse the diffraction of a Gaussian optical beam using the general formula (14.35). Setting

$$\mathcal{E}_0(r) = \mathcal{E}_0 \exp(-r^2/2\rho_0^2), \tag{14.38}$$

we obtain

$$\mathcal{E}(r_0, z) = \frac{\mathcal{E}_0 \exp(-ikz)}{(1 + z/ik\rho_0^2)} \exp\left(-\frac{r_0^2/2\rho_0^2}{1 + z/ik\rho_0^2}\right). \tag{14.39}$$

Equation (14.39) describes the diffraction field throughout the space. In particular, on the beam axis

$$\mathcal{E}(0, z) = \frac{\mathcal{E}_0 \exp(-ikz)}{(1 + z/ik\rho_0^2)}. \tag{14.40}$$

The radiation intensity on the beam axis is thus

$$I(0, z) = \frac{I_0}{1 + (z/z_d)^2}, \tag{14.41}$$

where

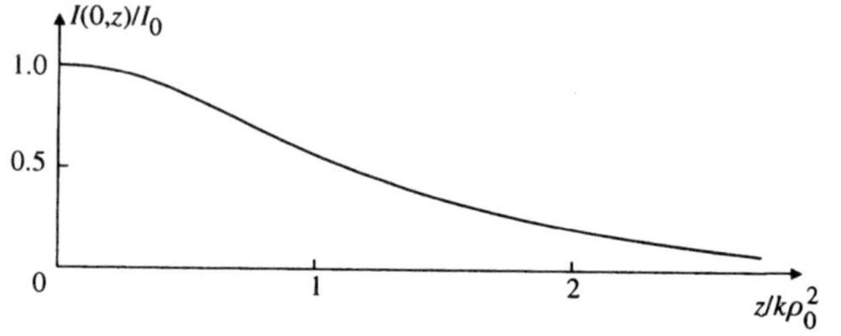

Fig. 14.10 The diffraction of a Gaussian beam: the distribution of the light intensity along the beam axis. z is the coordinate measured along the beam axis, ρ_0 is the original radius of the beam, I_0 is the light intensity on the beam axis in the plane $z = 0$, $k = 2\pi/\lambda$ is the wave number, and λ is the optical wavelength.

$$z_{\mathrm{d}} = k\rho_0^2 \tag{14.42}$$

is the diffraction length of the beam, and I_0 is the intensity determined by the formula $I_0 = c|\mathcal{E}_0|^2/8\pi$.

The graph of the function $I(0, z)$ is shown in Fig. 14.10. Note that in contrast to the diffraction by a round aperture, in this case the light intensity does not oscillate along the beam axis, but decreases monotonously with the distance travelled by the wave.

14.5 The parabolic equation and quasi-optical approximation

Let us describe the propagation of light by the scalar wave equation (see Chapter 1):

$$\Delta E - \frac{1}{c^2}\frac{\partial^2 E}{\partial t^2} = 0. \tag{14.43}$$

We shall consider a bounded optical beam possessing axial symmetry. Then, to describe the propagation process, two cylindrical coordinates are sufficient; that is, the distance z along the beam axis and the distance r from the beam axis. Let us write the field of the beam in the form

$$E(r, z) = a(r, z)\cos[\omega t - kz + \varphi(r, z)], \tag{14.44}$$

where $a(r, z)$ and $\varphi(r, z)$ are the real amplitude and phase of the wave in the beam. Equation (14.44) expresses the fact that the optical wave is harmonic, and the beam axis is directed along the z-axis. A field of the form (14.44) with a variable amplitude and phase is called a spatially modulated or *quasi-plane* wave. Note that such a field is not an exact solution to the wave equation (14.43); here we shall construct only approximate solutions.

It is convenient to use a complex-variable description

$$E = \tfrac{1}{2}A(r, z)\mathrm{e}^{i(\omega t - kz)} + \text{c. c.}, \tag{14.45}$$

where $A(r, z)$ is the complex amplitude of the field. The complex amplitude contains information about both the real amplitude and the real phase of the wave, because of the relations

$$A = ae^{i\varphi}, \qquad a = |A|, \qquad \varphi = \arg A. \tag{14.46}$$

According to eqn (14.45)

$$\frac{\partial^2 E}{\partial t^2} = -\omega^2 E = -\omega^2 \tfrac{1}{2} A e^{i(\omega t - kz)} + \text{c.c.} \tag{14.47}$$

and

$$\frac{\partial^2 E}{\partial z^2} = \tfrac{1}{2}\left(\frac{\partial^2 A}{\partial z^2} - 2ik\frac{\partial A}{\partial z} - k^2 A\right) e^{i(\omega t - kz)} + \text{c.c.} \tag{14.48}$$

Let us assume that

$$\left|\frac{\partial^2 A}{\partial z^2}\right| \ll k \left|\frac{\partial A}{\partial z}\right|. \tag{14.49}$$

We can then neglect the term $\partial^2 A/\partial z^2$ in eqn (14.48). Substituting eqns (14.47) and (14.48) into eqn (14.43) and taking into account the formula $\omega^2/c^2 = k^2$ and condition (14.49), we obtain the following equation for the complex amplitude of the wave beam:

$$\frac{\partial A}{\partial z} + \frac{i}{2k}\Delta_\perp A = 0, \tag{14.50}$$

where

$$\Delta_\perp = \frac{\partial^2}{\partial x^2} + \frac{\partial^2}{\partial y^2} \tag{14.51}$$

is the Laplace operator with respect to the transverse coordinates in the beam cross-section, $k = 2\pi/\lambda$ is the wave number, and λ is the optical wavelength. Equation (14.50) is called a parabolic equation.

The approximation under which the parabolic equation (14.50) is valid is called a *quasi-optical approximation.* According to the inequality (14.49), this approximation is justified if the derivative of the optical wave amplitude $\partial A/\partial z$ varies weakly over the distance of the order of an optical wavelength. A more convenient application criterion will be given later. Here we only note that a solution to the parabolic equation (14.50) should in any case obey the condition (14.49).

14.6 Solution of the parabolic equation

Let us represent the complex amplitude of the wave beam in the form of a spatial Fourier expansion over the transverse coordinates x, y of the beam:

$$A(\vec{r}, z) = \int_{-\infty}^{\infty} A_{\vec{\kappa}}(z) e^{i\vec{\kappa}\vec{r}} d^2\kappa,$$
$$A_{\vec{\kappa}}(z) = \left(\frac{1}{2\pi}\right)^2 \int_{-\infty}^{\infty} A(\vec{r}, z) e^{-i\vec{\kappa}\vec{r}} d^2 r \tag{14.52}$$

Here $\vec{r}$ is the radius vector lying in the plane of the beam cross-section, and $\vec{\kappa}$ is the transverse component of the wave vector (with respect to the beam axis). In more detailed notation,

$$A(x, y, z) = \iint_{-\infty}^{\infty} A_{\kappa_x, \kappa_y}(z) e^{i(\kappa_x x + \kappa_y y)} d\kappa_x \, d\kappa_y,$$
$$A_{\kappa_x, \kappa_y}(z) = \left(\frac{1}{2\pi}\right)^2 \iint_{-\infty}^{\infty} A(x, y, z) e^{-i(\kappa_x x + \kappa_y y)} dx \, dy, \tag{14.53}$$

Inserting eqn (14.52) into eqn (14.50), we obtain the following equation for the Fourier amplitude $A_{\vec{\kappa}}$ of the beam:

$$\frac{\partial A_{\vec{\kappa}}}{\partial z} - \frac{i\kappa^2}{2k} A_{\vec{\kappa}} = 0, \tag{14.54}$$

where $\kappa^2 = \kappa_x^2 + \kappa_y^2$. The solution to eqn (14.54) takes the form

$$A_{\vec{\kappa}}(z) = A_{\vec{\kappa}}(z = 0) \exp(i\kappa^2 z / 2k). \tag{14.55}$$

Equation (14.55) describes the transformation of the spatial Fourier amplitudes of the beam in the course of its propagation in vacuum. In this formula, $A_{\vec{\kappa}}(z = 0)$ is the spectral amplitude of the wave beam in the original cross-section $z = 0$. This amplitude can be expressed via the initial distribution of the beam complex amplitude $A(\vec{r}, z = 0) = A_0(\vec{r})$ as

$$A_{\vec{\kappa}}(z = 0) = \left(\frac{1}{2\pi}\right)^2 \int_{-\infty}^{\infty} A_0(\vec{r}) e^{-i\vec{\kappa}\vec{r}} d^2 r. \tag{14.56}$$

Substituting eqn (14.56) into eqn (14.55) and then eqn (14.55) into eqn (14.52), we obtain the formula relating the complex amplitude $A_0(\vec{r})$ of the beam in the original cross-section $z = 0$ with the amplitude $A(\vec{r}, z)$ at the arbitrary distance z:

$$A(\vec{r}, z) = \int_{-\infty}^{\infty} A_0(\vec{r}\,') H(\vec{r} - \vec{r}\,', z) \, d^2 r'. \tag{14.57}$$

Here the function $H(\vec{r}, z)$ is called the *Green function* of the free space and is determined as

$$H(\vec{r}, z) = \frac{ik}{2\pi z} \exp\left(-i\frac{kr^2}{2z}\right), \tag{14.58}$$

where $r^2 = x^2 + y^2$. Equations (14.57) and (14.58) provide the general solution to the parabolic equation (14.50).

14.7 The propagation and diffraction of a Gaussian beam

Let us now specify the form of the complex amplitude of the beam in the original cross-section $z = 0$. We shall assume that

$$A_0(\vec{r}) = A_0 \exp(-r^2/2\rho_0^2), \qquad A_0 = \text{constant}. \tag{14.59}$$

According to eqn (14.59), the original beam has a plane wavefront and a Gaussian cross-sectional intensity distribution:

$$I_0(\vec{r}) = I_0 \exp(-r^2/\rho_0^2). \tag{14.60}$$

Such a beam is called *Gaussian*. The Gaussian model is most convenient for calculations.

The beam width can be characterized by a radius (i.e. a half-width) for the intensity level equal to e^{-1} of the maximum intensity. Thus, according to eqn (14.60), the original beam radius

$$\rho_0 = HW\mathrm{e}^{-1}M. \tag{14.61}$$

The right-hand side contains the abbreviation for the expression "half-width e^{-1} maximum".

Now we shall determine how the beam changes as it propagates in the free space. Substituting eqns (14.58) and (14.59) into eqn (14.57) and performing the integration, we obtain

$$A(\vec{r}, z) = \frac{A_0}{(1 + z/ik\rho_0^2)} \exp\left[-\frac{r^2}{2\rho_0^2(1 + z/ik\rho_0^2)}\right]. \tag{14.62}$$

Equation (14.62) allows us to calculate the beam characteristics (e.g. the shape of a wavefront, the intensity profile, and the radius) at an arbitrary point z. The quantity

$$k\rho_0^2 = z_\mathrm{d} \tag{14.63}$$

is the *diffraction length* of the Gaussian beam. The dimensionless quantity

$$D = z/z_\mathrm{d} \tag{14.64}$$

is called the *wave parameter*, and its reciprocal

$$N_{\mathrm{f}} = z_{\mathrm{d}}/z \tag{14.65}$$

is *Fresnel's number* (see Chapter 13). According to eqn (14.62), the real amplitude a and phase φ of the beam are

$$a = |A_0|\frac{\rho_0}{\rho(z)}\exp\left[-\frac{r^2}{2\rho^2(z)}\right], \qquad \varphi = \alpha - k\frac{r^2}{2R(z)}, \tag{14.66}$$

where the following designations are introduced:

$$\begin{aligned} \rho(z) &= \rho_0\sqrt{1+(z/z_{\mathrm{d}})^2}, & z_{\mathrm{d}} &= k\rho_0^2, \\ R(z) &= z(1+z_{\mathrm{d}}^2/z^2), & \alpha &= \arctan(z/z_{\mathrm{d}}). \end{aligned} \tag{14.67}$$

The total electric field of the beam is

$$E(r,z) = \tfrac{1}{2}A_0\frac{\rho_0}{\rho(z)}\exp\left[-\frac{r^2}{2\rho^2(z)}\right]\exp\left\{i\left[\omega t - kz - k\frac{r^2}{2R(z)} + \alpha\right]\right\} + \mathrm{c.\,c.} \tag{14.68}$$

and the radiation intensity is

$$I(r,z) = I_0\frac{\rho_0^2}{\rho^2(z)}\exp\left[-\frac{r^2}{\rho^2(z)}\right]. \tag{14.69}$$

Equations (14.68) and (14.69) express the main result of this section. They describe the electric field and intensity of the Gaussian optical beam at the arbitrary point with coordinates r, z. It follows from eqns (14.68) and (14.69) that the beam retains the Gaussian shape of its intensity profile, i.e. it remains Gaussian at any distance z. The beam radius $\rho(z)$ increases monotonously with the distance z (i.e. the beam broadens—see Fig. 14.11), whereas the intensity decreases so that the total power of the beam remains constant:

$$P = \int_0^\infty I(r,z)2\pi r\,dr = \pi\rho_0^2 I_0 = P_0. \tag{14.70}$$

Let us show that the parameter $R(z)$ in eqns (14.66)–(14.68) signifies the curvature radius of the beam wavefront near the axis. Consider the spherical wave

$$E = \frac{\mathrm{constant}}{R}\exp[i(\omega t - kR)] + \mathrm{c.\,c.}, \qquad R = \sqrt{z^2+r^2}. \tag{14.71}$$

In the region in which

$$z^2 \gg r^2 \tag{14.72}$$

the following approximate formulae are valid: $R \approx z$ and

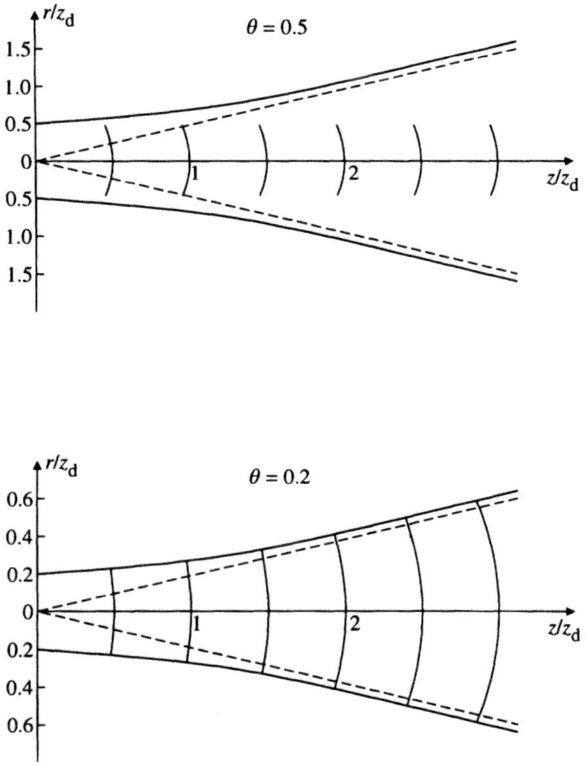

Fig. 14.11 Diffraction spreading and wavefront transformation for a Gaussian beam propagating in free space. $\theta = (k\rho_0)^{-1}$ is the diffraction divergence angle of the beam in the far-field zone, $z_d = k\rho_0^2$ is the diffraction length of the beam, ρ_0 is the original radius of the beam, $k = 2\pi/\lambda$ is the wave number, and λ is the optical wavelength.

$$E = \frac{\text{constant}}{R} \exp[i(\omega t - kz - k\frac{r^2}{2R})] + \text{c. c.} \tag{14.73}$$

Comparing eqns (14.68) and (14.73), we see that the parameter $R(z)$ in eqn (14.68) signifies the curvature radius of the wavefront. The same conclusion can be drawn after the formal calculation of the curvature radius from eqn (14.68). The wavefront curvature as a function of the distance z, calculated using eqn (14.67), is shown in Fig. 14.12.

14.8 The focusing of a Gaussian beam

The effects produced by a thin spherical lens on an optical beam can be mathematically described by the complex transfer coefficient $\Lambda(r)$, which depends on the transverse coordinate r of the beam. We can write

$$A_1(r) = \Lambda(r) A_0(r), \tag{14.74}$$

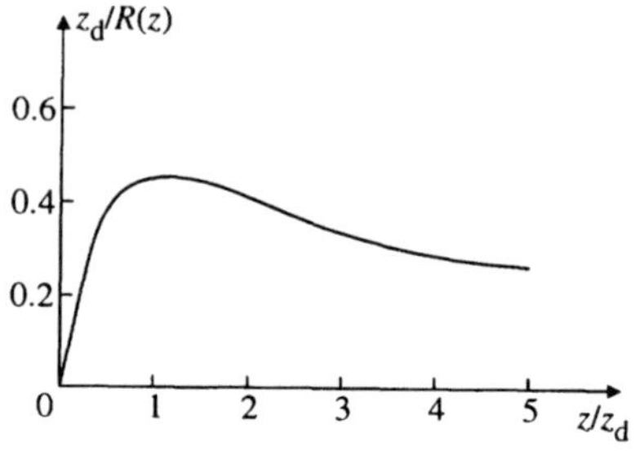

Fig. 14.12 The variation of the wavefront curvature of a Gaussian beam propagating in free space.

where $A_0(r)$ and $A_1(r)$ are the cross-sectional distributions of the complex amplitudes of the optical wave at the face and rear surfaces of the lens. Since the lens does not change the intensity distribution and only slightly bends the beam wavefront, we obtain

$$\Lambda(r) = \exp\left(ikr^2/2f\right), \tag{14.75}$$

where k is the wave number of the optical wave and f is the focal length of the lens. The factor featuring in eqn (14.75) is written by analogy with the factor $\exp(-ikr^2/2R)$ describing the wavefront curvature in eqn (14.68). It is reasonable to assume that the radius of the wavefront curvature introduced by the lens is equal to its focal length. The "+" sign in the exponent in eqn (14.75) corresponds to the concave shape of the focused beam wavefront, i.e. it describes the effect of the focusing (convex) lens.

Let us consider the focusing of a Gaussian optical beam with a plane wavefront and a complex amplitude determined by eqn (14.59). Suppose that the beam falls from the left onto a lens located in the plane $z = 0$. In accordance with eqns (14.74) and (14.75), the complex amplitude of the beam at the exit from the lens is then

$$A_1(r) = A_0 \exp\left[-\frac{r^2}{2\rho_0^2} + ik\frac{r^2}{2f}\right] \tag{14.76}$$

or

$$A_1(r) = A_0 \exp\left(-\frac{r^2}{2\rho_1^2}\right), \tag{14.77}$$

where

$$\frac{1}{\rho_1^2} = \frac{1}{\rho_0^2} - \frac{ik}{f}. \tag{14.78}$$

Thus, the effect of the lens is reduced to a substitution of the complex quantity ρ_1 defined by eqn (14.78) for the real radius ρ_0 of the beam. The optical field of the focused beam can now be calculated by eqns (14.62), and (14.66)–(14.68), the substitution

$$\rho_0 \rightarrow \rho_1 \tag{14.79}$$

having been made. For example, making this substitution in eqn (14.62), we obtain the following expression for the complex amplitude of the focused beam:

$$A(r,z) = \frac{A_0}{(1+z/ik\rho_1^2)} \exp\left[-\frac{r^2}{2\rho_1^2(1+z/ik\rho_1^2)}\right]. \tag{14.80}$$

Furthermore, substituting eqn (14.78) into eqn (14.80), we find that

$$A(r,z) = A_0\frac{\rho_0}{\rho(z)} \exp\left[-\frac{r^2}{2\rho^2(z)}\right] \exp\left\{i\left[\alpha - k\frac{r^2}{2R(z)}\right]\right\}, \tag{14.81}$$

where

$$\rho^2(z) = \rho_0^2\left[\left(1-\frac{z}{f}\right)^2 + \left(\frac{z}{z_\mathrm{d}}\right)^2\right], \qquad z_\mathrm{d} = k\rho_0^2 \tag{14.82}$$

and

$$\frac{1}{R(z)} = \frac{1}{z_\mathrm{d}}\left[\frac{\dfrac{z}{z_\mathrm{d}} - \dfrac{z_\mathrm{d}}{f}\left(1-\dfrac{z}{f}\right)}{\left(1-\dfrac{z}{f}\right)^2 + \left(\dfrac{z}{z_\mathrm{d}}\right)^2}\right], \qquad \alpha = \arctan\left(\frac{z/z_\mathrm{d}}{1-z/f}\right). \tag{14.83}$$

The electric field and radiation intensity can be written in the form (14.68) and (14.69), where the parameters $\rho(z)$, $R(z)$, and α are now expressed by eqns (14.82) and (14.83). Thus, we have completely described the focusing of a Gaussian beam. In practice, it is convenient to write the formula for the radius of a focused Gaussian beam in the form

$$\rho^2(z) = \rho_0^2(1-z/f)^2 + (z/k\rho_0)^2, \qquad k = 2\pi/\lambda. \tag{14.84}$$

Generalizing this formula to the case of a spatially non-coherent Gaussian beam, it becomes

$$\rho^2(z) = \rho_0^2(1-z/f)^2 + (z/k_0)^2(1/\rho_0^2 + 1/\rho_\mathrm{c}^2). \tag{14.85}$$

where ρ_0 is the radius and ρ_c is the transverse coherence radius of the incident beam.

14.8.1 The dimensions of the focal region of the lens

As seen from eqn (14.82), the focused beam radius reaches its minimum (the beam waist) at the point $z = z_\mathrm{w}$, at which

$$z_\mathrm{w} = \frac{f}{1+(f/z_\mathrm{d})^2}. \tag{14.86}$$

Thus, the beam waist is located slightly to the left of the focal point. According to eqn (14.83), the wavefront curvature of the beam changes to zero at the same point. The radius of the beam waist

$$\rho_{\mathrm{w}} = \rho_0 \frac{f/z_{\mathrm{d}}}{\sqrt{1 + (f/z_{\mathrm{d}})^2}}, \tag{14.87}$$

while at the focal point

$$\rho_{\mathrm{f}} = \rho_0 \frac{f}{z_{\mathrm{d}}}. \tag{14.88}$$

Usually in optics the following condition is satisfied:

$$(f/z_{\mathrm{d}})^2 \ll 1. \tag{14.89}$$

Thus, to a good degree of approximation, the waist is located at the focal point. The focal spot radius is given by eqn (14.88) or, in view of eqn (14.82), by the formula

$$\rho_{\mathrm{f}} = \frac{f}{k\rho_0}. \tag{14.90}$$

Correspondingly, the focal spot area is

$$S_{\mathrm{f}} = \pi \rho_{\mathrm{f}}^2. \tag{14.91}$$

As the observation plane moves away from the focal point along the beam axis, the beam cross-section $S(z) = \pi\rho^2(z)$ grows, to become twice the focal spot area at the points

$$z_{1,2} = f \pm \frac{b}{2} \tag{14.92}$$

i.e.

$$\rho^2(z_1) = \rho^2(z_2) = 2\rho_{\mathrm{f}}^2. \tag{14.93}$$

The quantity b in eqn (14.92) is determined by the formula

$$b = 2\frac{f^2}{k\rho_0^2} \tag{14.94}$$

and called the *confocal parameter* of the beam. The physical meaning of this quantity is the distance between the planes at which the cross-sectional area of the beam is twice that of the beam waist. The general picture of the focusing of a Gaussian beam is shown in Fig. 14.13. In conclusion, let us consider a numerical example.

Example

Let $\lambda = 0.5\ \mu$m, $f = 10$ cm, and $\rho_0 = 0.1$ cm. Then $k = 2\pi/\lambda = 1.3 \times 10^5$ cm^{-1} and, using eqns (14.82), (14.90), and (14.94), we obtain $z_{\mathrm{d}} = 13$ m, $\rho_{\mathrm{f}} = 0.8 \times 10^{-3}$ cm, and $b = 0.16$ cm.

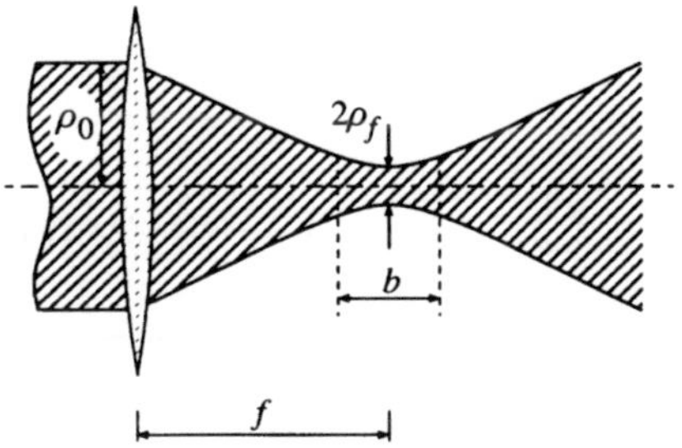

Fig. 14.13 The focusing of a Gaussian optical beam.

14.9 The application criterion for quasi-optical approximation

Using the the solution to the parabolic equation obtained earlier, we can show that the condition (14.49) is satisfied for not too narrow and for sufficiently coherent optical beams the radius ρ and transverse coherence radius ρ_c of which exceed the optical wavelength significantly:

$$\rho,\ \rho_c \gg \lambda. \tag{14.95}$$

For sources of thermal radiation possessing a small coherence radius, condition (14.95) is not always met. As for coherent laser radiation, this condition is usually satisfied. The quasi-optical approximation is thus applied primarily for solving problems in laser optics.

Bibliography

Born, M. and Wolf, E. (1968). *Principles of optics.* Pergamon Press, Oxford.

Goodman, J. W. (1968). *Introduction to Fourier optics.* McGraw-Hill, San Francisco.

Abramowitz, M. and Stegun, I. A. (eds). (1964). *Handbook of mathematical functions. Applied Mathematics Series 55.* National Bureau of Standards, Washington, D. C.

Janke, E., Emde, F., and Losch, F. (1960). *Tafeln Hoherer Functionen.* B. G. Teubner Verlagsgesellschaft, Stuttgart.

Yariv, A. and Yeh, P. (1984). *Optical waves in crystals.* John Wiley, New York.

15

Diffraction in the far-field zone

The formation of a stable diffraction pattern in the far-field zone. Fraunhofer diffraction as a spatial Fourier transformation. Fraunhofer diffraction by unidimensional structures—slit diffraction. Fraunhofer diffraction by two-dimensional structures. Diffraction by a rectangular aperture. Diffraction by a round aperture. Diffraction by a Gaussian beam.

15.1 The formation of a stable diffraction pattern in the far-field zone

Experiments on the diffraction of optical beams show that the angular intensity distribution of light in the far-field diffraction zone does not depend on the coordinate z measured along the beam axis. The diffraction pattern acquires a stable structure, the form of which depends only on the distribution of the field in the original cross-section. The diffraction in the far-field zone is called *Fraunhofer diffraction*. Let us consider the distinguishing features of the far-field zone diffraction in the context of the theory formulated earlier (see Chapters 13 and 14).

Suppose that a plane monochromatic optical wave falls normally onto a screen with an aperture located in the plane $z = 0$ (Fig. 15.1). Calculate the intensity distribution in the plane x_0–y_0 parallel to the screen with an aperture and located at a sufficiently long distance z from it.

Using eqns (14.2) and (14.3), we shall write the diffraction optical field in the form

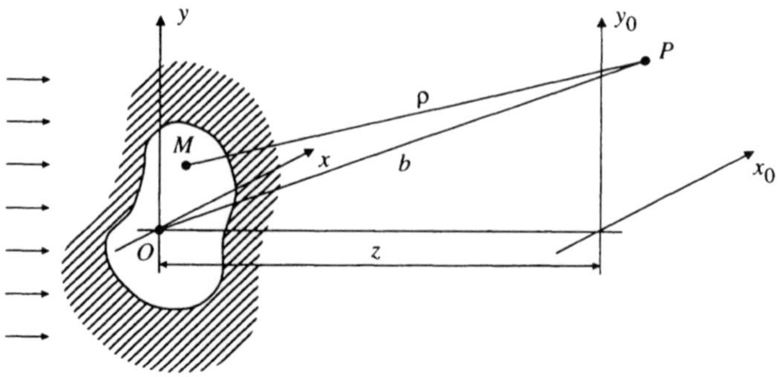

Fig. 15.1 Diffraction in the far-field zone.

$$\mathcal{E}(x_0, y_0, z) = \frac{i}{\lambda} \int_{-\infty}^{\infty} \int_{-\infty}^{\infty} \mathcal{E}_0(x, y) \frac{e^{-ik\rho}}{\rho} dx\, dy, \tag{15.1}$$

where $\mathcal{E}_0(x, y)$ is the field distribution in the original cross-section $z = 0$ specified by the aperture shape, λ is the wave number, and $k = 2\pi/\lambda$ is the optical wavelength. The parameter ρ is defined by the formula

$$\rho = \sqrt{z^2 + (x - x_0)^2 + (y - y_0)^2}, \tag{15.2}$$

where x, y are the coordinates of a point M in the plane of the screen with aperture, and x_0, y_0, z are the coordinates of the observation point.

Let O be the point in the plane of the screen with the aperture, taken as the origin of coordinates, and let b be the distance from the point O to the observation point P. As seen from Fig. 15.1,

$$b = \sqrt{z^2 + x_0^2 + y_0^2}. \tag{15.3}$$

Consequently, in the paraxial approximation,

$$z \gg x, y, x_0, y_0, \tag{15.4}$$

we can write

$$\rho = b + \frac{x^2 + y^2}{2b} - \frac{xx_0 + yy_0}{b}. \tag{15.5}$$

Equation (15.5) differs from eqn (14.5) only in that the value z is replaced by b as the zero approximation of the quantity ρ. Such a refinement is necessary because the dimensions of the diffraction pattern in the far-field zone are generally quite large, and the difference between the values b and z therefore becomes substantial.

Substituting eqn (15.5) into eqn (15.1), we obtain

$$\mathcal{E}(x_0, y_0, z) = \frac{i}{\lambda b} e^{-ikb} \int_{-\infty}^{\infty} \int_{-\infty}^{\infty} \mathcal{E}_0(x, y) e^{-\frac{ik}{2b}(x^2+y^2)} e^{\frac{ik}{b}(xx_0+yy_0)} dx\, dy. \tag{15.6}$$

Here, as usual, we have neglected the difference between ρ and b in the denominator of the integrand. In particular, for diffraction by unidimensional structures,

$$\mathcal{E}(x_0, z) = \frac{(i+1)}{\sqrt{2\lambda b}} e^{-ikb} \int_{-\infty}^{\infty} \mathcal{E}_0(x) e^{-\frac{ik}{2b}x^2} e^{\frac{ik}{b}xx_0}\, dx \tag{15.7}$$

(cf. eqn (14.12)) or

$$\mathcal{E}(\theta, z) = \frac{(i+1)}{\sqrt{2\lambda b}} e^{-ikb} \int_{-\infty}^{\infty} \mathcal{E}_0(x) e^{-\frac{ik}{2b}x^2} e^{ikx \sin\theta}\, dx, \tag{15.8}$$

where the angle θ, defined by the formula

$$\sin\theta = x_0/b, \tag{15.9}$$

is introduced. The angle θ denotes the angular coordinate of the observation point.

Equations (15.6)–(15.8) correspond to the Fresnel approximation. It follows from eqn (15.8) that the angular distribution of the field in the diffraction pattern generally varies as the distance z traversed by the beam grows. However, for large values of z, this variation becomes weaker and, finally, at

$$kd^2/2b \ll 1, \tag{15.10}$$

where d is the transverse dimension of the original beam, the stable angular distribution of the field

$$\mathcal{E}(\theta) = \frac{(i+1)}{\sqrt{2\lambda b}} e^{-ikb} \int_{-\infty}^{\infty} \mathcal{E}_0(x) e^{ikx\sin\theta} dx. \tag{15.11}$$

is reached. Using eqns (15.3) and (15.4), we can rewrite the inequality (15.10) as

$$z \gg z_{\rm d}, \tag{15.12}$$

where the parameter

$$z_{\rm d} = kd^2/2 \tag{15.13}$$

is called the *diffraction length of the beam.* The region of space specified by the condition (15.12) is called the *far-field diffraction zone*, or the *Fraunhofer zone.* Thus we have shown that a stable angular distribution of the field formed in the far-field zone remains invariable during the further propagation of the beam. The expression (15.11) for the diffracted optical field is known as the Fraunhofer approximation. Note that the given approximation is valid in the far-field diffraction zone.

15.2 Fraunhofer diffraction as a spatial Fourier transformation

It is self-evident that the stable diffraction pattern emerging in the far-field zone is of particular interest for optics primarily because such a stable pattern is easier to observe experimentally. From the mathematical point of view, the expression (15.11) is the *spatial Fourier integral.* By analogy with the Fourier integral with respect to time, the quantity $k\sin\theta$ is called the spatial frequency. The physical meaning of this notion becomes clear from Fig. 15.2, in which one can see that

$$k_x = k\sin\theta \tag{15.14}$$

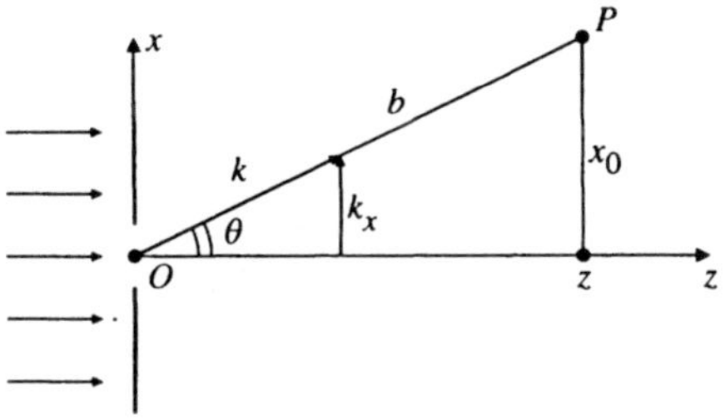

Fig. 15.2 The notion of spatial frequency.

is the transverse component of the wave vector directed from the point O (the aperture) to the observation point P. Equation (15.14) shows that the spatial frequency k_x and the angular coordinate θ of the observation point correspond reciprocally to each other. This allows us to write the complex amplitude of the field at the observation point as

$$\mathcal{E}(P) = \frac{(1+i)}{\sqrt{2\lambda b}} e^{-ikb} \mathcal{E}_0(k_x), \tag{15.15}$$

where

$$\mathcal{E}_0(k_x) = \int_{-\infty}^{\infty} \mathcal{E}_0(x) e^{ik_x x} dx \tag{15.16}$$

is the *spatial spectral amplitude* corresponding to the field distribution $\mathcal{E}_0(x)$. Thus, the diffraction field in the far-field zone is proportional to the spatial Fourier amplitude of the original beam. Using eqn (15.15), it is easy to calculate the light intensity distribution in the far-field zone:

$$I(P) = \frac{c}{8\pi} |\mathcal{E}(P)|^2. \tag{15.17}$$

Substituting eqn (15.15) into eqn (15.17), we obtain

$$I(P) = \frac{c}{8\pi} \frac{1}{\lambda b} S_0(k_x), \tag{15.18}$$

where

$$S_0(k_x) = |\mathcal{E}_0(k_x)|^2 \tag{15.19}$$

is the *spatial spectral density*, or the angular spectrum of radiation.

Thus, the angular distribution of the radiation intensity in the far-field zone reproduces the shape of the angular spectrum of the radiation. This conclusion discloses the physical meaning of Fraunhofer diffraction as a spatial resolution of a finite optical beam into plane waves. As an example, Fig. 15.3 shows the transformation of the angular spectrum accompanying the diffraction of a plane

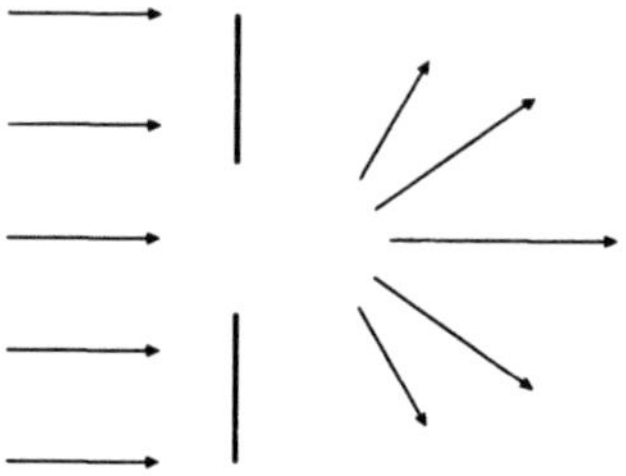

Fig. 15.3 The transformation of the angular spectrum of a plane wave diffracted by an aperture.

wave by the aperture. According to general properties of spectral expansion, the transverse component of the wave vector arises due to restrictions that the aperture imposes on the transverse dimensions of the beam. The representation of an optical beam as a set of plane waves propagating in different directions is quite similar to the presentation of an optical pulse as a set of the harmonic vibrations of different frequencies.

15.3 Fraunhofer diffraction by unidimensional structures—slit diffraction

Let us now turn to specific examples of Fraunhofer diffraction. We shall begin by calculating the diffraction of the plane wave by a unidimensional structure; that is, by a slit having the width d (Fig. 15.4). Setting

$$\mathcal{E}_0(x) = \mathcal{E}_0 \begin{cases} 1, & |x| \le d/2, \\ 0, & |x| > d/2, \end{cases} \tag{15.20}$$

we obtain, from eqn (15.16),

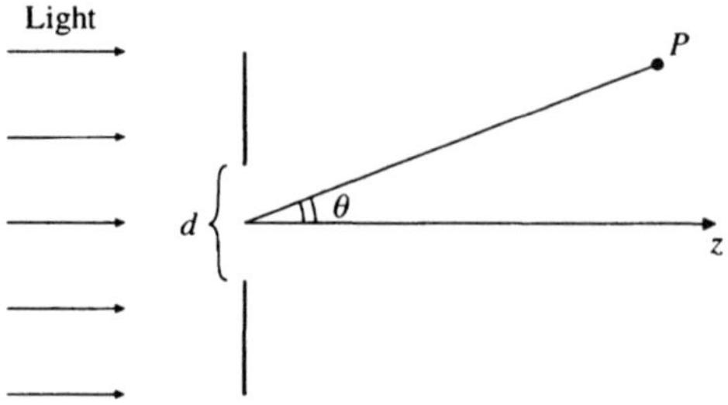

Fig. 15.4 The calculation of slit diffraction.

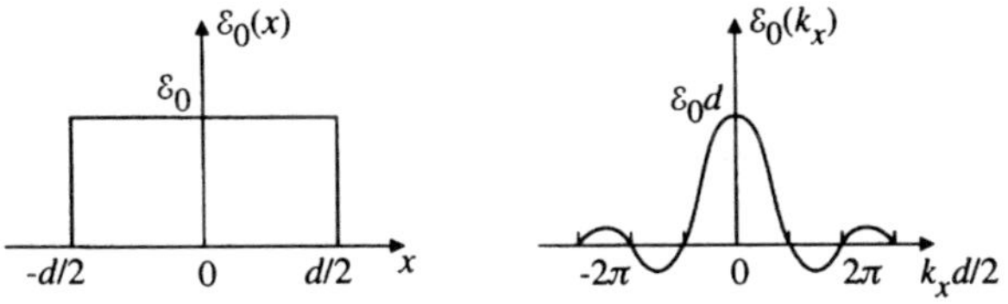

Fig. 15.5 The original field amplitude distribution $\mathcal{E}_0(x)$ and the spatial spectral amplitude $\mathcal{E}_0(k_x)$ of a plane wave diffracted by a slit of width d.

$$\mathcal{E}_0(k_x) = \mathcal{E}_0 \int_{-d/2}^{d/2} e^{ik_x x} dx = \mathcal{E}_0 d \operatorname{sinc}(k_x d/2), \tag{15.21}$$

where the standard designation

$$\operatorname{sinc}(x) \equiv \frac{\sin(x)}{x}. \tag{15.22}$$

is used. The functions $\mathcal{E}_0(x)$ and $\mathcal{E}_0(k_x)$ are plotted in Fig. 15.5. Using eqns (15.15) and (15.17), we find the field amplitude,

$$\mathcal{E}(P) = \mathcal{E}_0 \frac{(1+i)d}{\sqrt{2\lambda b}} e^{-ikb} \operatorname{sinc}(k_x d/2), \tag{15.23}$$

and the light intensity at the point P,

$$I(P) = I_{\text{max}} \operatorname{sinc}^2(k_x d/2). \tag{15.24}$$

Here we have used the following notation:

$$I_{\text{max}} = I_0 d^2/\lambda b, \qquad I_0 = \frac{c}{8\pi}|\mathcal{E}_0|^2, \tag{15.25}$$

where I_0 is the incident wave intensity. Substituting eqn (15.14) into eqn (15.24), we find the angular distribution of the radiation intensity in the far-field zone:

$$I(\theta) = I_{\text{max}} \operatorname{sinc}^2(\tfrac{1}{2} kd \sin\theta) \tag{15.26}$$

or

$$I(\theta) = I_{\text{max}} \operatorname{sinc}^2\left(\frac{\pi d \sin\theta}{\lambda}\right). \tag{15.27}$$

This distribution is plotted in Fig. 15.6. Note that the diffraction beam divergence in the far-field zone is of the order of

$$\Delta\theta = \lambda/d, \tag{15.28}$$

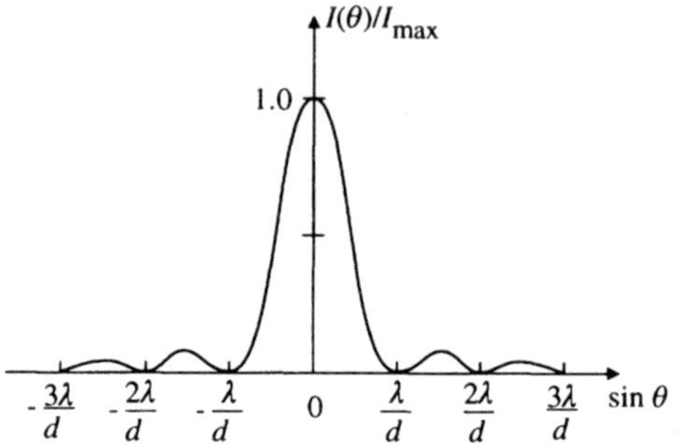

Fig. 15.6 The diffraction of a plane wave by a slit: the angular distribution of the light intensity in the far-field zone. d is the width of the slit and λ is the optical wavelength.

in accordance with the general result obtained earlier (see Chapter 13).

A theoretically calculated angular distribution of radiation intensity, shown in Fig. 15.6, can be verified experimentally, using an argon laser beam diffracted by a slit (Fig. 15.7). The diffraction pattern observed on the screen in a darkroom has the form of a central bright band and, to the side, alternating dark and light bands of diminishing brightness. In a practical demonstration we can gradually vary the width of the slit by a special micrometer screw. By doing so, we can observe a variation of the diffraction pattern, which takes place according to the prediction of the theory: the narrower the slit is, the broader are the bright bands on the screen and, consequently, the larger is the angular divergence of the radiation in the far-field zone. Thus, the Fraunhofer diffraction theory presented above can be confirmed experimentally.

15.4 Fraunhofer diffraction by two-dimensional structures

The theory formulated earlier can be generalized to the diffraction of an optical wave by a two-dimensional structure (Fig. 15.1). The diffraction optical field in the Fresnel approximation (a paraxial beam) is described by eqn (15.6). In the far-field zone defined by the condition

$$z \gg z_{\mathrm{d}} = kd^2/2, \tag{15.29}$$

where z_{d} is the diffraction length, k is the wave number, and d is the maximum linear dimension of the aperture in the screen, eqn (15.6) is reduced to a simpler form:

$$\mathcal{E}(P) = \frac{i}{\lambda b}\mathrm{e}^{-ikb} \iint_{\infty}^{\infty} \mathcal{E}_0(x, y)\mathrm{e}^{\frac{ik}{b}(xx_0+yy_0)}dx\,dy. \tag{15.30}$$

Let us define the angular coordinates θ and ψ of the observation point by the relations (see Fig. 15.8)

$$\sin\theta = x_0/b, \qquad \sin\psi = y_0/b. \tag{15.31}$$

(a)

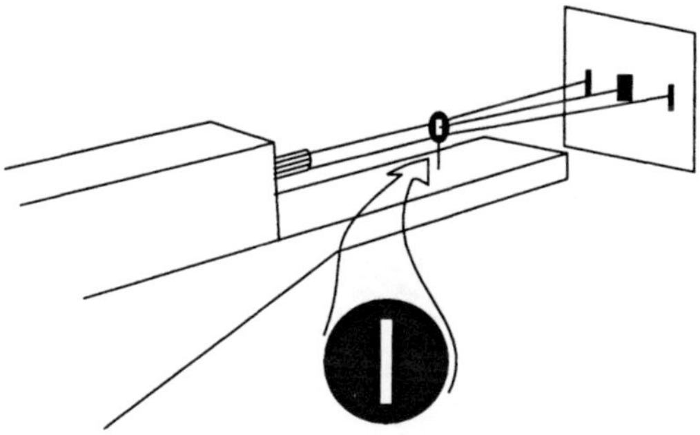

(b) (c)

Fig. 15.7 An experimental study of slit diffraction. The experimental set-up (a) and the diffraction patterns for narrow (b) and wide (c) slits.

We shall also introduce the spectral frequencies

$$k_x = k \sin\theta, \qquad k_y = k \sin\psi. \tag{15.32}$$

We can then rewrite eqn (15.30) as

$$\mathcal{E}(P) = \frac{i}{\lambda b} \mathrm{e}^{-ikb} \mathcal{E}_0(k_x, k_y), \tag{15.33}$$

where

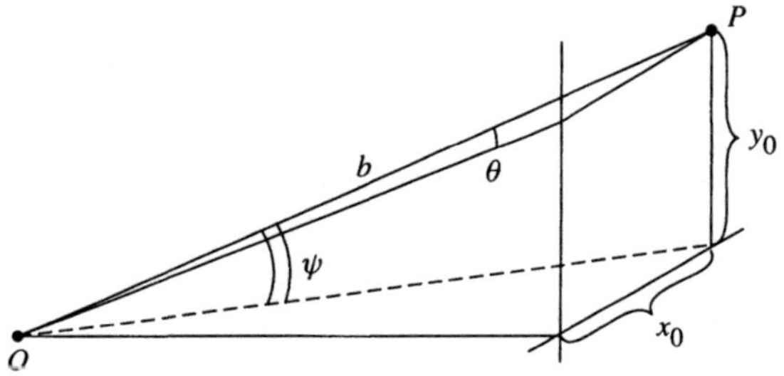

Fig. 15.8 The angular coordinates of the observation point.

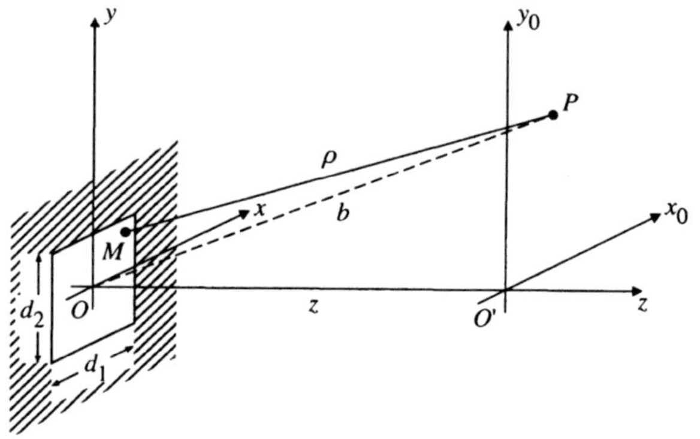

Fig. 15.9 The calculation of diffraction by a rectangular aperture.

$$\mathcal{E}_0(k_x, k_y) = \int_{-\infty}^{\infty}\int_{-\infty}^{\infty} \mathcal{E}_0(x, y) e^{i(k_x x + k_y y)} dx\, dy \tag{15.34}$$

is the spatial spectral amplitude corresponding to the original two-dimensional field distribution $\mathcal{E}_0(x, y)$. Hence, in the case of diffraction by a two-dimensional structure, the field distribution in the far-field zone has the form of a two-dimensional Fourier transformation of the original field distribution $\mathcal{E}_0(x, y)$.

In accordance with eqns (15.17) and (15.33), the light intensity at the observation point is

$$I(P) = \frac{c}{8\pi}\frac{1}{(\lambda b)^2} S_0(k_x, k_y), \tag{15.35}$$

where

$$S_0(k_x, k_y) = |\mathcal{E}_0(k_x, k_y)|^2 \tag{15.36}$$

is the spatial spectral density or the angular radiation spectrum. Thus, as in the case of a unidimensional structure, the spatial distribution of the radiation intensity in the far-field zone has the form of the angular spectrum of the radiation.

15.5 Diffraction by a rectangular aperture

Suppose that a plane monochromatic optical wave is diffracted by a rectangular aperture with sides d_1 and d_2 (Fig. 15.9). Writing the original distribution of the field amplitude in the form

$$\mathcal{E}_0(x, y) = \mathcal{E}_0 \begin{cases} 1, & |x| \le d_1/2, |y| \le d_2/2, \\ 0, & \text{outside this region,} \end{cases} \tag{15.37}$$

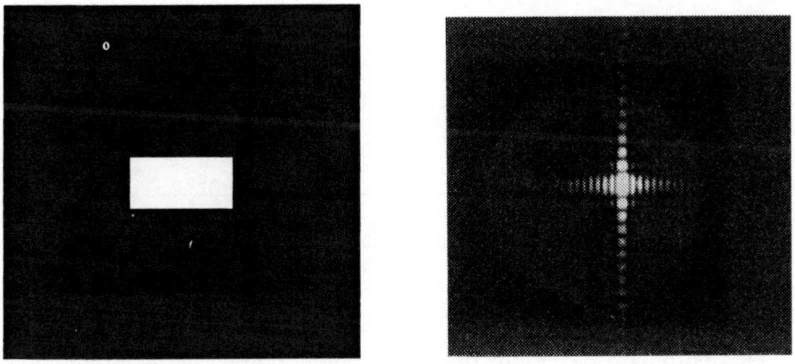

(a) (b)

Fig. 15.10 Fraunhofer diffraction by a rectangular aperture: (a) the aperture shape; (b) the diffraction pattern.

and substituting eqn (15.37) into eqn (15.34), we obtain

$$\mathcal{E}_0(k_x, k_y) = \mathcal{E}_0 \int\limits_{-d_1/2}^{d_1/2} dx \int\limits_{-d_2/2}^{d_2/2} dy\, \mathrm{e}^{i(k_x k + k_y y)}$$
$$= \mathcal{E}_0 d_1 d_2 \operatorname{sinc}(k_x d_1/2) \operatorname{sinc}(k_y d_2/2). \quad (15.38)$$

Furthermore, inserting eqn (15.38) into eqns (15.35) and (15.36) and taking into account eqn (15.32), we obtain the angular distribution of the radiation intensity in the far-field zone:

$$I(\theta, \psi) = I_{\max} \operatorname{sinc}^2\left(\frac{\pi d_1 \sin\theta}{\lambda}\right) \operatorname{sinc}^2\left(\frac{\pi d_2 \sin\psi}{\lambda}\right), \quad (15.39)$$

where

$$I_{\max} = I_0 (d_1 d_2/\lambda b)^2, \quad (15.40)$$

and I_0 is the intensity of the incident wave.

Diffraction by a rectangular aperture can be experimentally observed with an argon laser as a source of light and two crossed gratings with slits, the width of which can be varied with the help of a micrometer screw. The aperture shape and a typical diffraction pattern are shown in Fig. 15.10. Note that the experimental data completely confirm the results of the theoretical calculation.

15.6 Diffraction by a round aperture

Let us introduce polar coordinates r, φ in the aperture plane x–y and r_0, φ_0 in the observation plane x_0–y_0 (Fig. 15.11). We can then write

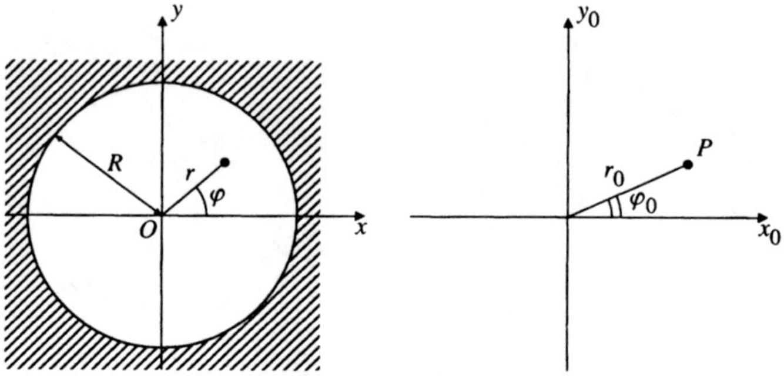

Fig. 15.11 Coordinate system for the calculation of diffraction by a round aperture.

$$\begin{aligned} x &= r\cos\varphi, \quad x_0 = r_0\cos\varphi_0, \\ y &= r\sin\varphi, \quad y_0 = r_0\sin\varphi_0. \end{aligned} \tag{15.41}$$

Furthermore, we shall introduce the angle θ between the beam z-axis and the line joining the aperture centre and the observation point (Fig. 15.12). It can be seen from this figure that

$$\sin\theta = r_0/b. \tag{15.42}$$

Denote

$$k\sin\theta = k_\perp. \tag{15.43}$$

Then the quantities k_x and k_y in eqn (15.34) can be written as

$$k_x = kx_0/b = k_\perp\cos\varphi_0, \qquad k_y = ky_0/b = k_\perp\sin\varphi_0. \tag{15.44}$$

It follows from eqns (15.41) and (15.44) that

$$k_x x + k_y y = k_\perp r\cos(\varphi - \varphi_0). \tag{15.45}$$

In terms of the new variables, the spatial spectral amplitude of the field is

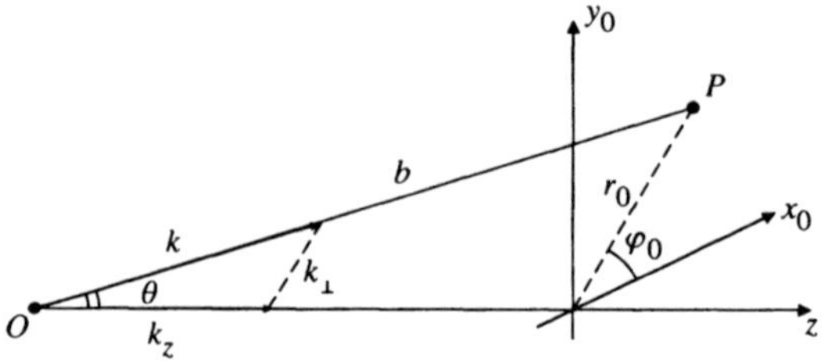

Fig. 15.12 The calculation of diffraction by a round aperture.

$$\mathcal{E}(k_\perp, \varphi_0) = \int_0^\infty r\,dr \int_0^{2\pi} \mathcal{E}_0(r,\varphi) e^{ik_\perp r}\cos(\varphi - \varphi_0) d\varphi. \tag{15.46}$$

Consider further the case of the axially symmetrical original distribution of the field. Setting

$$\mathcal{E}_0(r,\varphi) = \mathcal{E}_0(r) \tag{15.47}$$

and inserting eqn (15.47) into eqn (15.46), we obtain

$$\mathcal{E}(k_\perp, \varphi_0) = \int_0^\infty \mathcal{E}_0(r) r\,dr \int_0^{2\pi} e^{ik_\perp r\cos(\varphi-\varphi_0)} d\varphi. \tag{15.48}$$

Using the tabulated integral

$$\int_0^{2\pi} e^{i\alpha\cos(\varphi-\varphi_0)} d\varphi = 2\pi J_0(\alpha), \tag{15.49}$$

where $J_0(\alpha)$ is the Bessel function of zeroth order, we obtain

$$\mathcal{E}(k_\perp) = 2\pi \int_0^\infty \mathcal{E}_0(r) J_0(k_\perp r) r\,dr. \tag{15.50}$$

The transformation (15.50) is known as the *Fourier–Bessel transformation*, or the *Hankel transformation* of zeroth order. We have thus shown that the two-dimensional Fourier transformation reduces to the Hankel transformation if the original field distribution is axially symmetrical.

The function $\mathcal{E}_0(r)$ in eqn (15.50) describes the original axially symmetrical field distribution. In particular, for a round aperture of radius R, we have

$$\mathcal{E}_0(r) = \mathcal{E}_0 \begin{cases} 1, & r \le R, \\ 0, & r > R. \end{cases} \tag{15.51}$$

Then

$$\mathcal{E}(k_\perp) = \mathcal{E}_0 2\pi \int_0^R J_0(k_\perp r) r\,dr = \mathcal{E}_0 2\pi R^2 \frac{J_1(k_\perp R)}{k_\perp R}. \tag{15.52}$$

Here we have used the tabulated integral

$$\int_0^x x J_0(x)\,dx = x J_1(x). \tag{15.53}$$

where $J_1(x)$ is the Bessel function of the first order.

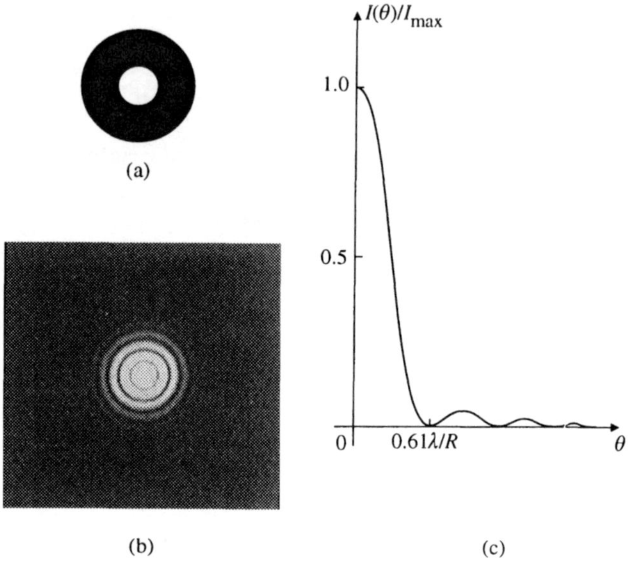

Fig. 15.13 Fraunhofer diffraction by a round aperture: (a) the aperture shape; (b) the diffraction pattern ('Airy pattern'); (c) the calculated angular distribution of light intensity in the diffraction pattern.

Let us calculate the angular distribution of the light intensity in the diffraction pattern. Using eqns (15.35), (15.36), and (15.52), and assuming that $\theta \ll 1$, and $k_\perp = k\sin\theta = k\theta$, we obtain

$$I(\theta) = I_{\max}\left[\frac{J_1(2\pi\theta R/\lambda)}{\pi\theta R/\lambda}\right]^2, \tag{15.54}$$

where

$$I_{\max} = I_0\left(\frac{\pi R^2}{\lambda b}\right)^2, \tag{15.55}$$

and I_0 is the intensity of the incident wave. The distribution (15.54) illustrated in Fig. 15.13 is called the *Airy pattern*. The total width (at the base level) of the central maximum of the diffraction pattern is

$$\Delta\theta = 1.22\lambda/R, \tag{15.56}$$

where R is the aperture radius and λ is the optical wavelength.

Fraunhofer diffraction by rectangular and round apertures can be demonstrated by the experiment shown schematically in Fig. 15.7, in which a continuous-wave argon laser is used as a source of light. The rectangular aperture produces a diffraction pattern in the form of a cross-like grid of spots, the brightness of which diminishes towards the periphery of the pattern (Fig. 15.10). The round aperture produces a pattern consisting of a central round bright spot surrounded by concentric dark and light rings of diminishing brightness (Fig. 15.13). In both cases, the observable patterns agree well with the results of the theoretical calculation.

15.7 Diffraction of a Gaussian beam

As a rule, laser beams have a Gaussian intensity distribution over the cross-section. Let us calculate the far-field zone diffraction pattern for the Gaussian beam. Setting

$$\mathcal{E}_0(r) = \mathcal{E}_0 \exp(-r^2/2\rho_0^2), \tag{15.57}$$

we obtain, from eqn (15.50),

$$\mathcal{E}(k_\perp) = \mathcal{E}_0 2\pi\rho_0^2 \exp(-k_\perp^2\rho_0^2/2). \tag{15.58}$$

Assuming further that $\theta \ll 1$ and $k_\perp = k\theta$, we shall find the angular distribution of the radiation intensity in the far-field zone:

$$I(\theta) = I_{\max} \exp\left[-(2\pi\theta\rho_0/\lambda)^2\right], \tag{15.59}$$

where

$$I_{\max} = I_0 \left(\frac{2\pi\rho_0^2}{\lambda b}\right)^2, \qquad I_0 = \frac{c}{8\pi}|\mathcal{E}_0|^2. \tag{15.60}$$

I_0 is the intensity on the beam axis for $z = 0$, and ρ_0 is the original radius of the beam. Thus, the intensity profile of the Gaussian beam keeps its form in the diffraction process. The radius of the propagating beam grows gradually while the field intensity on its axis diminishes. This conclusion is confirmed by experimental observations of the free diffraction of the laser beam.

Bibliography

Handbook of mathematical functions. Ed. by M. Abramowitz and I. A. Stegun. National Bureau of standards. Applied mathematic series, 55. Issued June 1964.

Born, M. and Wolf, E. (1968). *Principles of optics.* Pergamon Press, Oxford.

Goodman, J. W. (1968). *Introduction to Fourier optics.* McGraw-Hill, San Francisco.

Janke, E., Emde, F., and Losch. F. (1960). *Tafeln hoherer Functionen.* B. G. Teubner Verlagsgesellschaft, Stuttgart.

16

Diffraction by periodic structures

Diffraction gratings. The physics of light diffraction by a grating. The equation of a diffraction grating. A mathematical description of diffraction of a plane wave by a grating. Diffraction by two-dimensional periodic structures. Diffraction by three-dimensional periodic structures. X-ray diffraction in perfect crystals. X-ray diffraction analysis.

The diffraction of light by periodic structures is distinguished by the sharpness of the observable diffraction pattern. This phenomenon is applied for optical wavelength measurements and for analysis of the spectral composition of radiation. In experiments on laser beam diffraction by a diffraction grating, we can clearly see the effects of Fraunhofer diffraction; that is, the formation in the far-field zone of a stable angular light distribution reproducing the shape of the angular spectrum of the beam. The diffraction of X-rays in crystals proves the wave nature of these rays and provides a powerful tool for the investigation of crystal structure.

16.1 Diffraction gratings

A diffraction grating represents a spatial periodic structure, the interval of which is comparable to the optical wavelength. As the spatial periodic structure is characterized by a discrete angular spectrum, one may expect that a plane wave transmitted through the grating becomes transformed into a set of plane waves propagating at different angles to the incident wave direction. Experiments on laser beam diffraction by a diffraction grating confirm this conclusion: a typical diffraction pattern has the characteristic form of a discrete set of rays, "a fan" (Fig. 16.1).

There are transmitting and reflecting, as well as amplitude and phase, diffraction gratings. The transmitting gratings transmit light, while the reflecting ones reflect it. The amplitude gratings spatially modulate the amplitude, while the phase gratings modulate the phase of the incident optical wave. The simplest amplitude transmitting grating consists of a system of slits in an opaque screen (Fig. 16.2(a)). The reflecting amplitude grating is produced by ruling lines on a plane or concave mirror (Fig. 16.2(c)). The phase grating can be a glass plate with a periodically varying thickness (see the transmitting grating in Fig. 16.2(b)) or a mirror with a similar profile (see the reflecting grating in Fig. 16.2(d)).

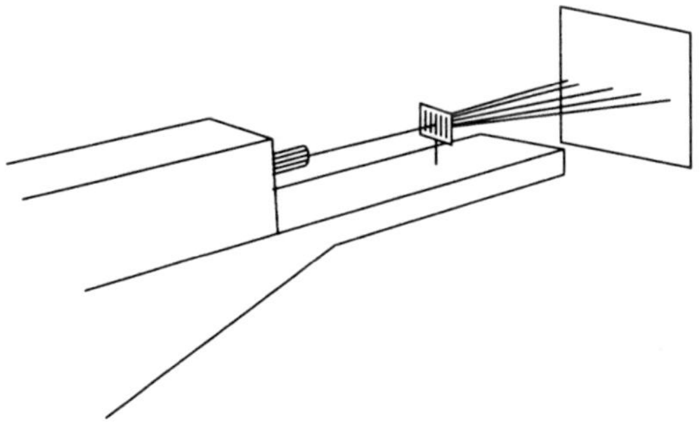

Fig. 16.1 The observation of diffraction of a laser beam by a diffraction grating.

16.2 The physics of light diffraction by a grating

The elementary theory of light diffraction by a grating can be presented on the basis of the Huygens–Fresnel ideas about the secondary wavelets interference. Let us consider the diffraction of a plane monochromatic wave by the periodic system of slits forming the amplitude transmitting grating (Fig. 16.3). The diffraction process consists of the following. The incident optical wave produces in the slits coherent sources of secondary wavelets. The resulting optical field is formed by superposition and interference of these waves. Figure 16.3 shows the wavefronts of Huygens' secondary wavelets. Continuing the construction, one can notice that the wavefront intersection points, i.e. the points at which the fields add in phase, align in straight lines, forming a discrete set of directions along which the secondary wavelets reinforce one another. These are the directions in which the principal maxima of the diffraction pattern are formed. Along any other direction, the secondary wavelets interfere destructively, i.e. the phase relations between the waves are such that the waves suppress each other. As a result, the narrow principal maxima of the diffraction pattern turn out to be separated by broad dark gaps. A "fan" of rays thus arises, that can be observed experimentally with a laser beam and a diffraction grating (Fig. 16.1).

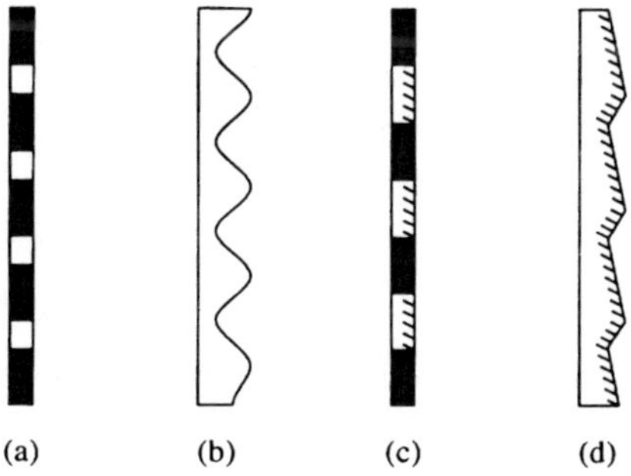

Fig. 16.2 Types of diffraction gratings.

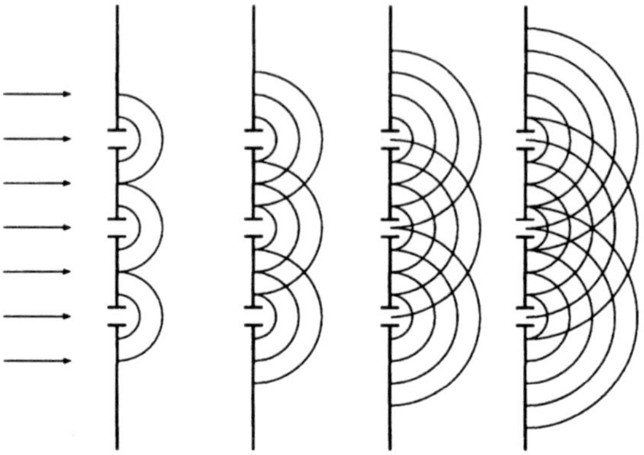

Fig. 16.3 The mechanism of the diffraction of light by a grating.

16.3 The equation of a diffraction grating

Using Fresnel's ideas about the interference of secondary wavelets, it is easy to find the directions to the principal maxima of the diffraction pattern. These are apparently the directions for which the path difference of the beams from adjacent slits is a multiple of the wavelength. Denote the grating interval by d. It can be seen from Fig. 16.4 that the path difference between the rays coming in the direction θ from two adjacent slits of the grating is equal to $d\sin\theta$. Consequently, the directions of principal maxima are determined by the equation

$$d\sin\theta = m\lambda, \qquad m = 0, \pm 1, \pm 2, \dots , \tag{16.1}$$

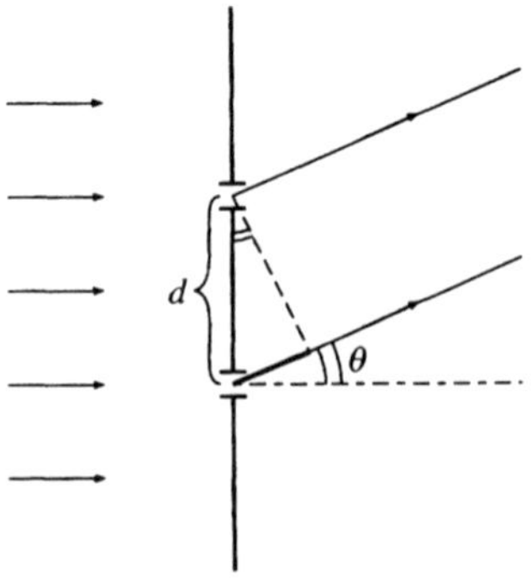

Fig. 16.4 The calculation of the directions towards the principal maxima of the diffraction pattern. The bold line indicates the path difference between the rays diffracted in the direction θ from the two adjacent slits of the grating.

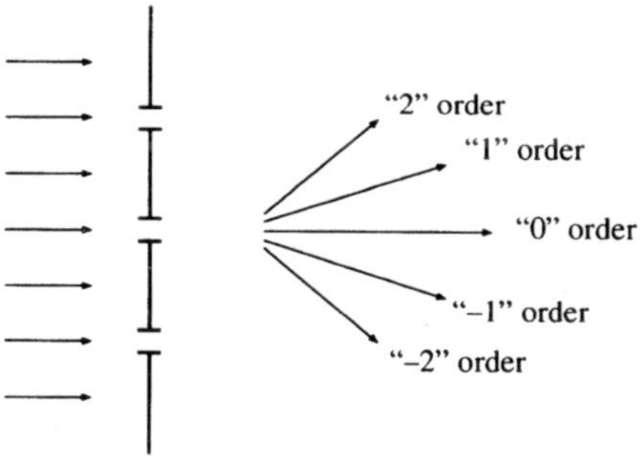

Fig. 16.5 The diffraction of a plane wave by a transmitting grating: the directions towards the principal maxima of the diffraction pattern.

where λ is the optical wavelength. Equation (16.1) is called the equation of the diffraction grating.

The diffraction of a plane wave by a transmitting grating is schematically shown in Fig. 16.5. The rays propagating in different directions correspond to different values of m in eqn (16.1) or to different orders of diffraction. For example, the light ray passing directly through the grating is called the ray of the zeroth order, the two nearest deflected rays are called the rays of the first and minus first orders, and so on. As seen from the eqn (16.1), the order of diffraction characterizes the positions of the maxima of illumination according to the number of wavelengths fitting the path difference between the rays diffracted in the given direction θ from the two adjacent slits of the grating.

The equation of the diffraction grating (16.1) allows us to draw some important conclusions. First, it follows from this equation that the grating can produce appreciable diffraction, i.e. significant angles θ of deflection, only if the grating interval is comparable to an optical wavelength, i.e.

$$d \simeq \lambda \simeq 10^{-4}\ \text{cm}. \tag{16.2}$$

Consequently, the optical diffraction grating should have $10^2 - 10^3$ lines (slits) per millimetre. It also follows from eqn (16.1) that a grating with a small interval should deflect beams more strongly than a grating with a large interval. On the other hand, a grating with too small an interval (less than an optical wavelength) will produce no diffraction at all because, in accordance with eqn (16.1), only a directly propagating wave can pass through such a grating ($m = 0$, $\theta = 0$).

Second, it follows from eqn (16.1) that the positions of the principal maxima of the diffraction pattern depend on the wavelength of the incident radiation. Therefore, if we illuminate a grating using a beam of non-monochromatic radiation, the grating will deflect the different spectral components of the radiation through different angles. The diffraction grating can thus be used as a spectral device which, like a prism, performs the spatial spectral decomposition of non-monochromatic radiation. Both of these conclusions are confirmed by an ex-

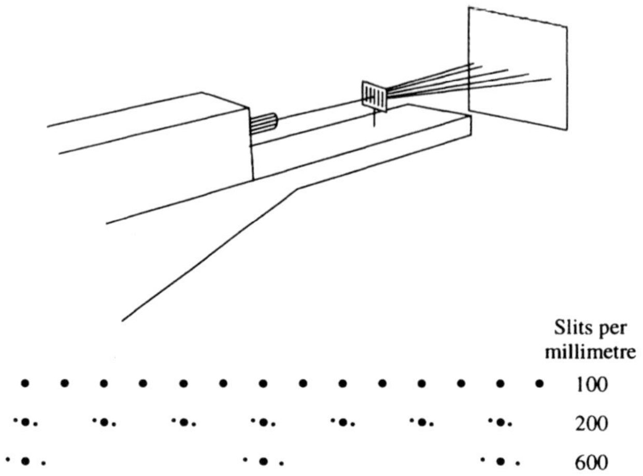

Fig. 16.6 The diffraction of an argon laser beam by a transmitting grating: the diffraction pattern observed with gratings having different numbers of slits per mm.

periment in which an argon laser beam passes through diffraction gratings with different numbers of lines per millimetre (Fig. 16.6).

Putting the different gratings into the laser beam in succession, we can observe an increase in the angular deflection of the diffracted beams as the grating interval decreases. For example, if we use a 50 lines per mm grating, then we observe on the screen (or the ceiling of the lecture room) a great number of diffraction maxima separated by small gaps. If we blow smoke near the grating, a "fan" of optical beams, propagating from the grating in different directions, becomes clearly visible. A 200 lines per mm grating generates a smaller number of visible maxima. Finally, a 600 lines per mm grating produces only two or three visible maxima, separated by broad gaps. In the last case we can see the spectral structure of the argon laser radiation (green and yellow–green spots on the screen). This demonstration shows that a diffraction grating with a sufficiently small interval can operate as a spectral device, or a spectrum analyser, with a high resolving power. Diffraction gratings are used in this capacity in modern spectrometers.

The diffraction grating was invented by Fraunhofer in 1821. He made his first gratings by winding a fine wire round two screws placed in parallel. Those gratings had from 40 to 340 lines per inch. Using these, Fraunhofer managed to measure the wavelength of the sodium D line, which is equal to 588.6 nm. Later, Fraunhofer started to rule the lines on a thin gold film deposited on a glass plate. Finally, he mastered ruling the lines directly onto the glass surface, using a diamond as a ruling point. The best grating manufactured by Fraunhofer was half an inch wide and had an interval of about 3 μm, or 8000 lines per inch. Fraunhofer pointed to the possibility of making reflecting gratings as well, although all of his gratings operated as transmitting ones.

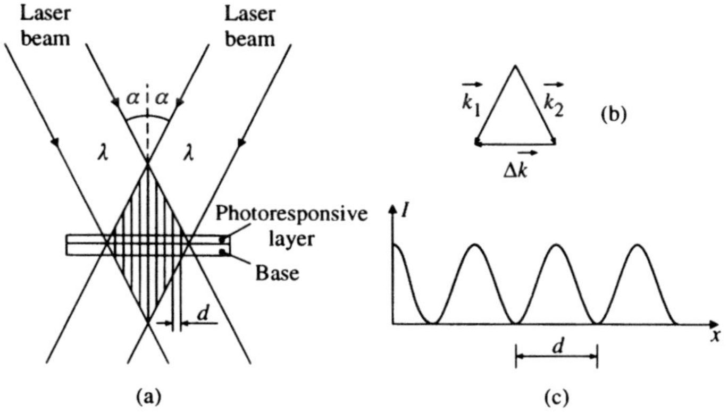

Fig. 16.7 A schematic diagram showing the production of a holographic diffraction grating.

The transition from primitive Fraunhofer gratings to modern diffraction gratings posed a formidable technical challenge, to be taken up by many scientists. An important step was taken by Rowland, who constructed special ruling machines to produce the finest gratings of a large size. Rowland was the first to make concave reflecting gratings that acted simultaneously as a grating and as a convergence lens. High-quality Rowland gratings had up to 20 000 lines per inch and were up to 10 cm wide. Further improvements on Rowland's ruling machines were made by Michelson, Wood and others.

Most of the modern spectrometers utilize reflecting gratings with a triangular groove profile, so-called echelettes (Fig. 16.2(d)), which concentrate up to 80% of the incident light into a few orders, or even into a single, non-zero order, of the spectrum. Different gratings are manufactured for different spectral regions from the infrared ($\lambda \approx 1\ \mu$m) to the ultraviolet ($\lambda \approx 100$ nm), with dimensions of up to 400×400 mm^2, and have between four and 3600 lines per mm. At present, replica gratings are widely used. These are made using a special plastic cast taken from a ruled surface and covered with a metal reflecting coating. The replicas have almost the same quality as the originals.

In the 1970s a new technology for making high-quality diffraction gratings was developed. It is based on a photographic registration of the interference pattern produced by coherent laser beams. Gratings of this type are called *holographic*. They are produced for the visible and ultraviolet regions, and have 600–6000 lines per mm and dimensions of up to 600×400 mm^2. The technique for the production of a holographic diffraction grating is shown schematically in Fig. 16.7.

It is easy to calculate the interval of a holographic grating. Let us write the electric field of two non-collinear plane waves (laser beams) in the form

$$E = A\cos(\omega t - \vec{k}_1\vec{r}) + A\cos(\omega t - \vec{k}_2\vec{r}).$$

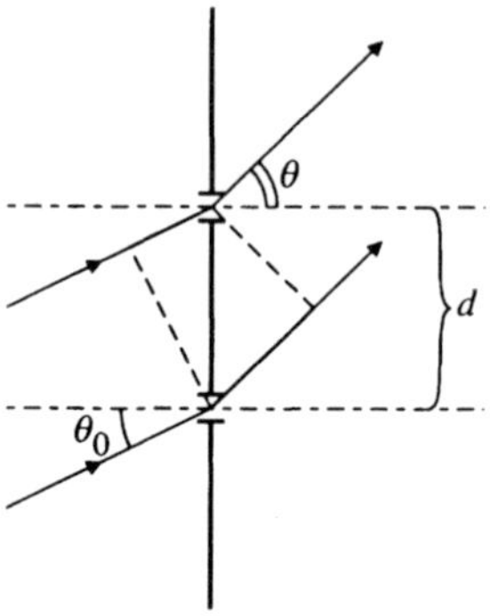

Fig. 16.8 The analysis of the diffraction pattern obtained for oblique incidence of light on the grating.

Then, the resulting field intensity

$$I = 2I_0[1 + \cos(\Delta\vec{k}\vec{r})],$$

where $\Delta\vec{k} = \vec{k}_1 - \vec{k}_2$. From this,

$$d = \frac{\lambda}{2\sin\alpha},$$

where α is the half-angle between the laser beams and λ is the optical wavelength. As can be seen from this formula, the interval of the holographic grating is of the order of an optical wavelength and can be varied by changing the angle α. It is clear that a highly coherent laser radiation is required to produce holographic gratings of high quality.

16.3.1 Oblique incidence of rays on a grating

We have considered earlier the normal incidence of a plane wave on a diffraction grating. Let us now generalize our consideration to the case of oblique incidence. Let θ_0 denote the angle of incidence and θ the angular coordinate of the observation point (Fig. 16.8). Setting the path difference between rays diffracted in the direction θ from two adjacent slits of the grating equal to the whole number of wavelengths, we obtain the following equation:

$$d(\sin\theta - \sin\theta_0) = m\lambda, \qquad m = 0, \pm 1, \pm 2, \ldots, \tag{16.3}$$

where d is the grating interval.

Equation (16.3) determines the directions of the principal maxima of the diffraction pattern in the case of oblique incidence. It follows from this equation that when the light falls on the grating at a glancing angle, i.e. the angle of incidence is close to a right angle,

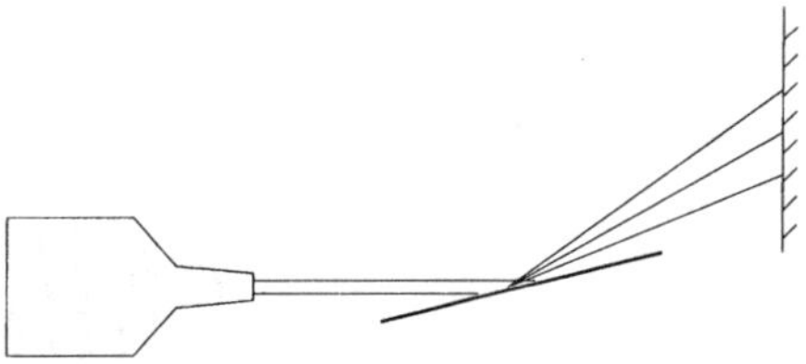

Fig. 16.9 The observation of diffraction by a phonograph disc.

$$\theta_0 \approx \pi/2, \tag{16.4}$$

then even a rough grating the interval of which is much more than the optical wavelength can produce noticeable diffraction. Indeed, setting $d \gg \lambda$, we obtain from eqn (16.3) $\theta \approx \theta_0$ and, consequently,

$$\sin\theta - \sin\theta_0 = 2\sin\left(\frac{\theta-\theta_0}{2}\right)\cos\left(\frac{\theta+\theta_0}{2}\right) \approx (\theta-\theta_0)\cos\theta_0.$$

The equation of the grating (16.3) takes the form

$$\theta - \theta_0 = m\lambda/d_e,$$

where $d_e = d\cos\theta_0$ is the effective interval of the grating. By virtue of eqn (16.4), we have $d_e \ll d$, i.e. the effective interval can be much smaller than d. This provides the conditions for diffraction.

The above conclusion is confirmed by the experiment illustrated in Fig. 16.9, in which light diffraction by a phonograph disc is observed. In this experiment, the concentrically located grooves of the disc play the role of a reflecting diffraction grating. The interval of this grating substantially exceeds an optical wavelength. Nevertheless, setting the disc almost in parallel with the incident optical beam (produced by an arc lamp), we can observe the diffraction. The diffraction manifests itself by rainbow colouring of the beam reflected by the disc on the demonstration screen.

16.3.2 The reflecting grating

The reflecting grating is a mirror with grooves ruled on its surface. Such gratings are easier to manufacture than transmitting gratings, and therefore are most commonly used. Reflecting gratings, just as transmitting ones, can be either amplitude or phase gratings. Examples of reflecting gratings are illustrated in Figs 16.2(c) and (d).

The mathematical description of diffraction by a reflecting grating is similar to that given earlier for the transmitting grating. In particular, the equation of the reflecting grating has the form of eqn (16.3), where θ_0 is the angle of incidence and θ is the angle of reflection of light (Fig. 16.10).

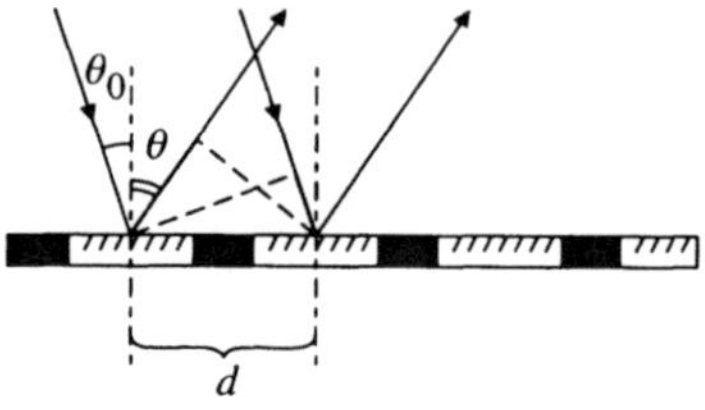

Fig. 16.10 The derivation of the equation of the reflecting diffraction grating.

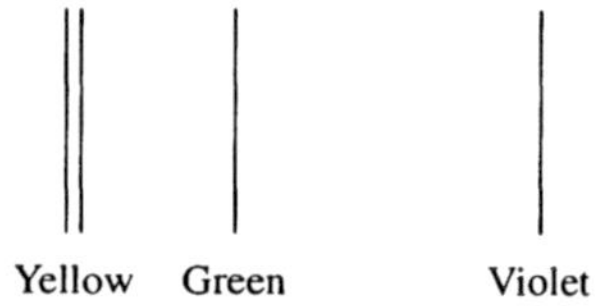

Fig. 16.11 The mercury lamp spectrum obtained with a reflecting diffraction grating.

Reflecting gratings have the following useful feature. By choosing an appropriate groove profile, one can redistribute the energy among different orders of diffraction. Specifically, it becomes possible to direct most of the diffracted light into the first or second order of diffraction. Such gratings, called *echelettes*, are widely used in optical spectroscopy. As an example, Fig. 16.11 shows the spectrum of a mercury-discharge lamp, obtained with the use of a reflecting phase grating. The grating used in this demonstration has an interval of 0.002 mm. The groove profile is chosen so that most of the diffracted light is directed into the second order of diffraction. When this grating is illuminated by the light of a mercury-discharge lamp, several discrete coloured spectral lines of mercury, including the doublet of yellow lines, are clearly seen on the screen.

Figure 16.12 shows schematically how the diffraction grating can be utilized as a dispersive element in the laser resonator. In a dye laser, one can gradually vary the laser radiation wavelength by rotating the grating with respect to the resonator axis.

Thus, the most important features of light diffraction by a diffraction grating can be understood on the basis of Fresnel's classical ideas about the interference

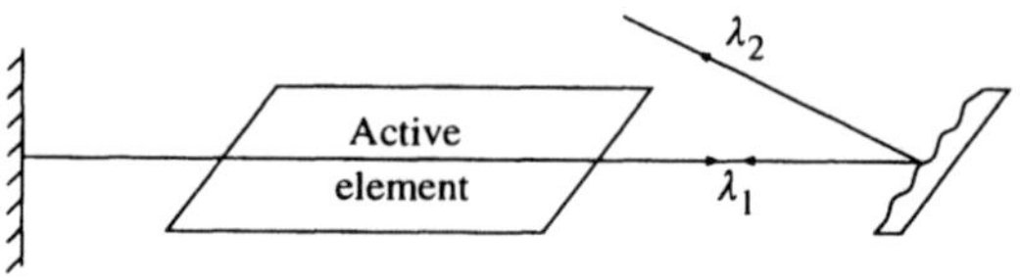

Fig. 16.12 The application of the diffraction grating as an element to control the dye laser wavelength.

of secondary wavelets. Now let us turn our attention to a more comprehensive mathematical description of light diffraction by a diffraction grating.

16.4 A mathematical description of diffraction of a plane wave by a grating

Let us restrict ourselves to the consideration of diffraction in the far-field zone. A general approach to the calculation of Fraunhofer diffraction is described in Chapter 15. Specifying an initial distribution of the field amplitude $\mathcal{E}_0(x)$, we find the spatial spectral amplitude $\mathcal{E}_0(k_x)$, then the angular spectrum of radiation $S(k_x)$, and finally the angular intensity distribution $I(\theta)$. The working formulae have the following form:

$$\mathcal{E}_0(k_x) = \int_{-\infty}^{\infty} \mathcal{E}_0(x)e^{ik_x x}dx,$$

$$S(k_x) = |\mathcal{E}_0(k_x)|^2, \quad k_x = k\sin\theta, \tag{16.5}$$

$$I(P) = \frac{c}{8\pi}\frac{1}{\lambda b}S(k_x).$$

For the sake of definiteness, we shall consider the transmitting diffraction grating. The grating is characterized by the *complex transmission function*, defined as the ratio of the complex amplitude $\mathcal{E}_0(x)$ of the transmitted wave to the amplitude $\mathcal{E}_0$ of the incident wave:

$$T(x) = \mathcal{E}_0(x)/\mathcal{E}_0. \tag{16.6}$$

First, we consider an elementary spatial periodic structure, i.e. a *sinusoidal grating*.

16.4.1 The sinusoidal grating

This is a grating the transmission function of which has the form

$$T(x) = \cos(\kappa x), \tag{16.7}$$

where $\kappa = 2\pi/d$, d is the grating interval. Substituting eqns (16.6) and (16.7) into eqn (16.5), we obtain

$$\mathcal{E}_0(k_x) = \pi\mathcal{E}_0[\delta(k_x + \kappa) + \delta(k_x - \kappa)]. \tag{16.8}$$

Thus, the spectral amplitude of the diffracted wave is a sum of two delta-functions, i.e. a pair of infinitely narrow spectral lines located at the spatial frequencies $\pm\kappa$. The angles of deflection,

$$\theta = \pm\lambda/d, \tag{16.9}$$

correspond to these spatial frequencies. Here d is the grating interval and λ is the optical wavelength. The pattern of plane wave diffraction due to a sinusoidal grating is shown in Fig. 16.13.

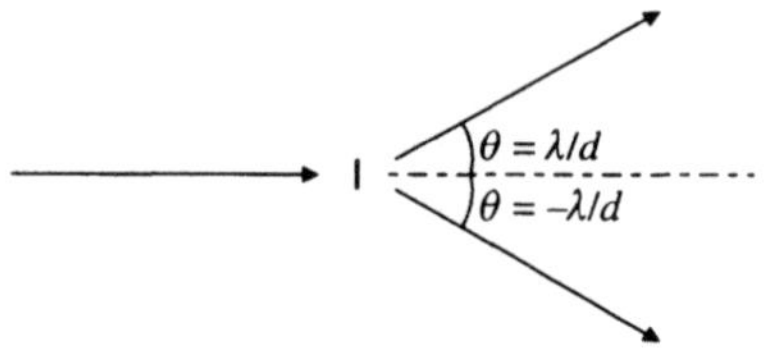

Fig. 16.13 The diffraction of a plane wave by a sinusoidal grating.

16.4.2 A finite sinusoidal grating

Let us clarify how the diffraction pattern is affected by a finite dimension of the grating. This is important, because real diffraction gratings and incident optical beams have finite transverse dimensions. Denoting the overall width of the grating by D, we shall write the original distribution of the field amplitude in the form

$$\mathcal{E}_0(x) = \mathcal{E}_0 \begin{cases} \cos(\kappa x), & |x| \le D/2, \\ 0, & |x| > D/2. \end{cases} \tag{16.10}$$

Substituting eqn (16.10) into eqn (16.5) and assuming that $D \gg d$, $\theta \ll 1$, we obtain the following approximate expression for $I(\theta)$:

$$I(\theta) \approx I_m \left\{ \mathrm{sinc}^2 \left[\left(\theta + \frac{\lambda}{d} \right) \frac{\pi D}{\lambda} \right] + \mathrm{sinc}^2 \left[\left(\theta - \frac{\lambda}{d} \right) \frac{\pi D}{\lambda} \right] \right\}, \tag{16.11}$$

where

$$I_m = I_0 \frac{D^2}{4\lambda b}, \qquad I_0 = \frac{c}{8\pi} |\mathcal{E}_0|^2 \tag{16.12}$$

and $\mathrm{sinc}(x) \equiv \sin x / x$.

The graph of the function $I(\theta)$ is plotted in Fig. 16.14. The diffraction pattern is seen to have two principal maxima corresponding to the angles specified by eqn (16.9). However, these maxima now have a finite angular width, determined by the formula

$$\Delta\theta = \lambda / D. \tag{16.13}$$

Here D is the total width of the grating (the aperture) and λ is the optical wavelength.

16.4.3 The rectangular amplitude grating

This grating consists of a succession of slits in an opaque screen. The transmission function has the form of a rectangle for each interval (Fig. 16.15), so that such a grating is referred to as rectangular. The transmission function of the rectangular grating can be written in the form

$$T(x) = \sum_{n=0}^{N} t(x - x_n), \tag{16.14}$$

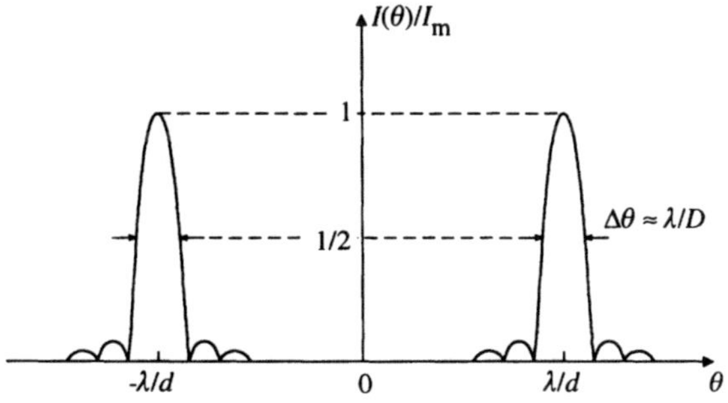

Fig. 16.14 The angular distribution of the radiation intensity in the far-field zone for diffraction of a plane wave by a finite sinusoidal grating.

where N is the number of slits,

$$x_n = nd, \qquad n = 0, 1, \ldots, N \tag{16.15}$$

are the coordinates of the centres of the slits, d is the interval of the grating, and

$$t(x) = \begin{cases} 1, & |x| \leq l/2, \\ 0, & |x| > l/2 \end{cases} \tag{16.16}$$

is the transmission function of a single slit, the width of which is denoted by l. Substituting eqns (16.14)–(16.16) into eqn (16.5), we obtain

$$\mathcal{E}_0(k_x) = \mathcal{E}_1(k_x) f_N(k_x), \tag{16.17}$$

where

$$\mathcal{E}_1(k_x) = \mathcal{E}_0 l \operatorname{sinc}(k_x l/2) \tag{16.18}$$

is the spatial spectral amplitude of the wave diffracted by the individual slit and

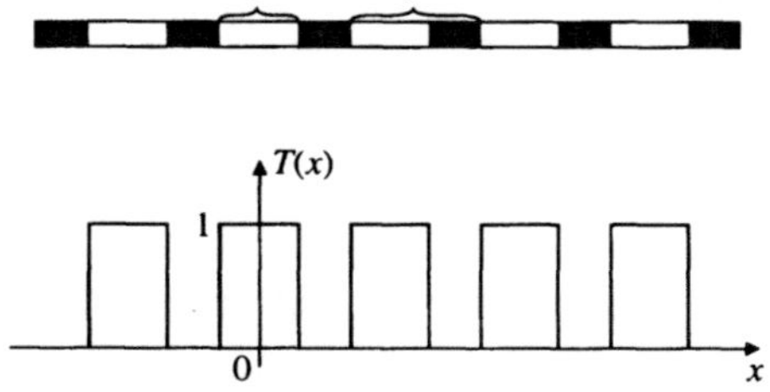

Fig. 16.15 The rectangular amplitude grating and its transmission function; l is the width of the slit and d is the interval of the grating.

$$f_N(k_x) = \sum_{n=0}^{N} e^{ik_x nd} \tag{16.19}$$

is the multiple-beam interference factor describing the cooperative action of all slits. Calculating this factor using the formula for the sum of a finite geometrical progression,

$$\sum_{n=0}^{N} e^{in\varepsilon} = \frac{1-\exp(iN\varepsilon)}{1-\exp(i\varepsilon)}, \tag{16.20}$$

we finally obtain

$$I(\theta) = I_m F_1(\theta) F_N(\theta), \tag{16.21}$$

where

$$I_m = I_0 \frac{l^2}{\lambda b}, \quad I_0 = \frac{c}{8\pi}|\mathcal{E}_0|^2,$$
$$F_1(\theta) = \mathrm{sinc}^2\left(\frac{1}{2}kl\sin\theta\right), \tag{16.22}$$
$$F_N(\theta) = \left[\frac{\sin\left(\frac{1}{2}Nkd\sin\theta\right)}{\sin\left(\frac{1}{2}kd\sin\theta\right)}\right]^2.$$

The functions $F_1(\theta)$, $F_N(\theta)$, and $I(\theta)$ are plotted in Fig. 16.16. In addition, we will indicate the main characteristics of the diffraction pattern.

Principal maxima directions:

$$\sin\theta = \lambda m/d, \qquad m = 0, \pm 1, \pm 2, \ldots \tag{16.23}$$

Angular width of principal maxima:

$$\Delta\theta = \lambda/Nd. \tag{16.24}$$

Light intensity at a maximum:

$$I_{\max} = I_m N^2. \tag{16.25}$$

Number of principal maxima:

$$M = d/l. \tag{16.26}$$

The last formula shows that a diffraction pattern can have either one or a great number of principal maxima, depending on the ratio of the parameters d and l.

16.5 Diffraction by two-dimensional periodic structures

A two-dimensional periodic structure ("two-dimensional grating") can be obtained by the superposition of two crossed diffraction gratings. In particular, if the slits of one grating are perpendicular to the slits of the other, a rectangular

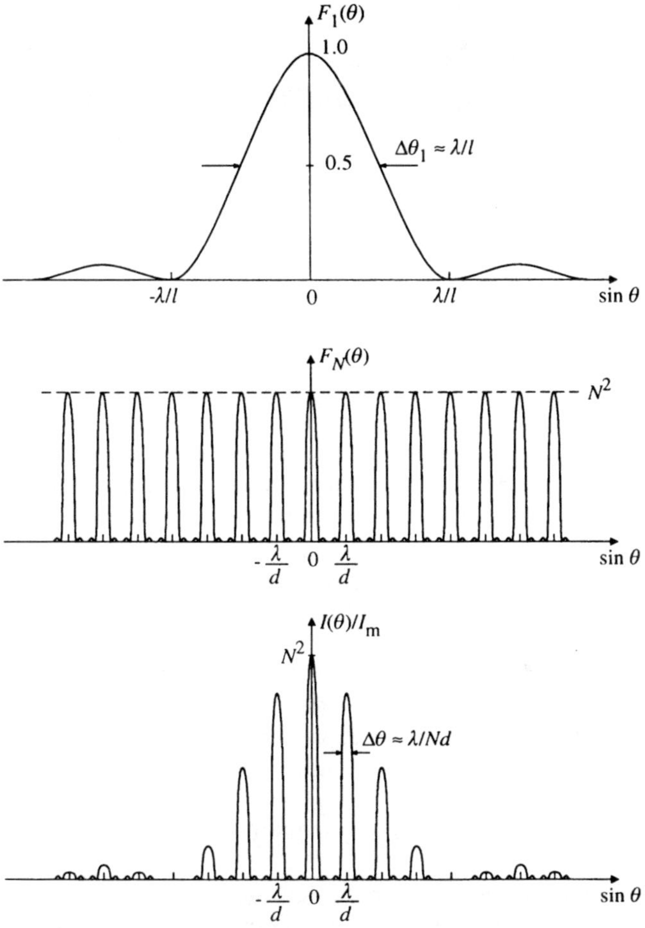

Fig. 16.16 The pattern of a plane wave diffracted by a rectangular amplitude grating.

structure is formed, as shown in Fig. 16.17. A view of the diffraction pattern appearing when a laser beam is diffracted by crossed gratings is shown in Fig. 16.18. Such a pattern can be demostrated using an argon laser and a pair of diffraction gratings mounted in the rotating rims.

A qualitative explanation of the observed diffraction pattern is given in Fig. 16.17. The first diffraction grating, the slits of which are directed vertically, decomposes the laser beam into a "fan" of rays lying in the horizontal plane. Then the second grating with horizontal slits decomposes each of the incident rays into a "fan" of rays lying in the vertical plane. As a result, the two-dimensional diffraction pattern shown in Fig. 16.18 is formed.

It is easy to find the directions towards the principal maxima of the diffraction pattern. Let us introduce the Cartesian x_0, y_0 and angular θ, ψ coordinates of

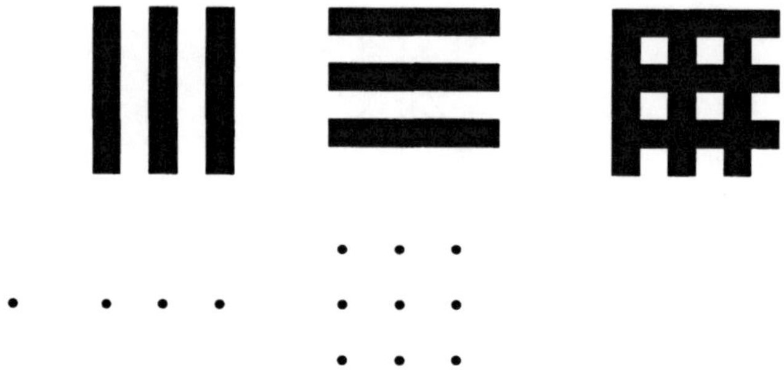

Fig. 16.17 The construction of a two-dimensional periodic structure by the superposition of crossed gratings; diffraction patterns produced by the first grating and by both gratings together when illuminated by a laser beam.

the observation point P, as shown in Fig. 16.19. Then, in accordance with what was established earlier for unidimensional gratings, the directions towards the principal maxima of the diffraction pattern can be determined from the formulae

$$\sin\theta = m_1\lambda/d_1, \qquad \sin\psi = m_2\lambda/d_2, \tag{16.27}$$

where d_1 is the grating interval along the x-axis (the interval of the vertical slits), d_2 is the grating interval along the y-axis (the interval of the horizontal slits), λ is the optical wavelength, and m_1 and m_2 are whole numbers. The angles θ and ψ shown in Fig. 16.19 are determined as follows:

$$\sin\theta = x_0/b, \qquad \sin\psi = y_0/b. \tag{16.28}$$

Now, for the sake of convenience, we shall introduce the other angular coordinates of the point P; that is, the angles α, β, γ between the straight line OP and the coordinate axes x, y, z. These angles are shown in Fig. 16.20. We can write the following expressions for the angles α, β, and γ:

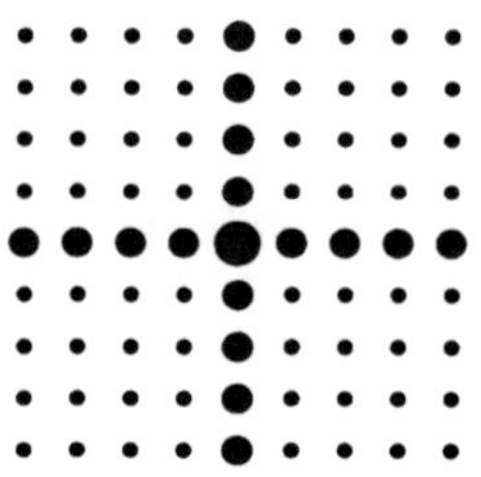

Fig. 16.18 The diffraction pattern produced by crossed transmitting gratings when illuminated by an argon laser beam.

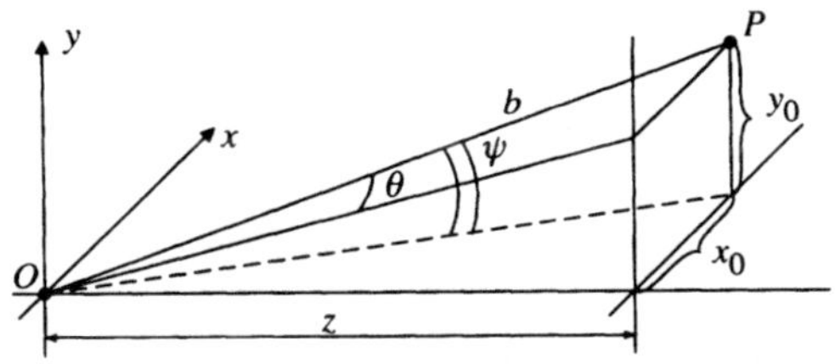

Fig. 16.19 The diffraction of light by crossed gratings. The crossed gratings are located in the x–y plane (at the point O); P is the observation point.

$$\begin{gathered}\cos\alpha = x_0/b = \sin\theta, \quad \cos\beta = y_0/b = \sin\psi, \quad \cos\gamma = z/b, \\ \cos^2\alpha + \cos^2\beta + \cos^2\gamma = 1.\end{gathered} \tag{16.29}$$

The last formula expresses the Pythagorean theorem and shows that only two angles from the three, e.g. α, β, and γ, are independent. The directions towards the principal maxima of the diffraction pattern can be expressed by the formulae

$$d_1 \cos\alpha = m_1\lambda, \qquad d_2 \cos\beta = m_2\lambda. \tag{16.30}$$

16.5.1 The mathematical description of diffraction of a plane wave by crossed gratings

Using the general formulae describing Fraunhofer diffraction by arbitrary two-dimensional structures (see Chapter 15), one can easily calculate the diffraction pattern appearing in the far-field zone as a result of diffraction of a plane wave by crossed rectangular gratings. The calculation includes the following steps. Specifying the original distribution of the field amplitude $\mathcal{E}_0(x, y)$, we calculate the spatial spectral amplitude of the field $\mathcal{E}_0(k_x, k_y)$, the spectral density $S(k_x, k_y)$, and, finally, the angular distribution of the radiation intensity $I(\theta, \psi)$. The working formulae have the form

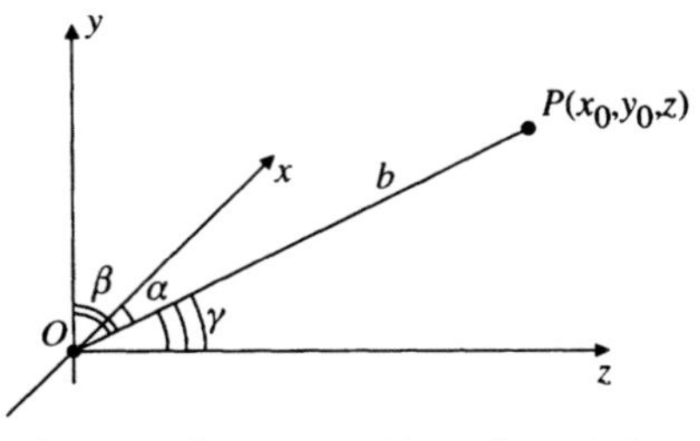

Fig. 16.20 The angular coordinates α, β, and γ of the observation point P.

$$\begin{aligned}
\mathcal{E}_0(k_x, k_y) &= \int_{-\infty}^{\infty}\int_{-\infty}^{\infty} \mathcal{E}_0(x, y) e^{i(k_x x + k_y y)} dx\, dy, \\
S(k_x, k_y) &= |\mathcal{E}_0(k_x, k_y)|^2, \\
I(P) &= \frac{c}{8\pi}\frac{1}{(\lambda b)^2} S(k_x, k_y), \\
k_x = k\sin\theta = k\cos\alpha, &\qquad k_y = k\sin\psi = k\cos\beta.
\end{aligned} \tag{16.31}$$

The original field distribution can be presented as

$$\mathcal{E}_0(x, y) = \mathcal{E}_0 T(x, y), \tag{16.32}$$

where $\mathcal{E}_0$ is the amplitude of the incident plane wave and $T(x, y)$ is the transmission function of the two-dimensional grating. In particular, for the case of gratings crossing at right angles, we can write

$$T(x, y) = T_1(x) T_2(y), \tag{16.33}$$

where $T_1(x)$ and $T_2(y)$ are the transmission functions of the unidimensional gratings. In this case the double integral in eqn (16.31) reduces to the product of two single integrals, and the further calculation in fact duplicates the calculation made earlier for the unidimensional grating. Using designations the meaning of which is clear from Fig. 16.21, we finally obtain

$$\begin{aligned}
I(\theta, \psi) &= I_0(\theta, \psi) F(\theta, \psi), \\
I_0(\theta, \psi) &= I_m F_1(\theta, \psi), \\
I_m = I_0 \left(\frac{l_1 l_2}{\lambda b}\right)^2, &\qquad I_0 = \frac{c}{8\pi}|\mathcal{E}_0|^2, \\
F(\theta, \psi) &= \left[\frac{\sin\left(\frac{1}{2}N_1 k d_1 \sin\theta\right)}{\sin\left(\frac{1}{2}k d_1 \sin\theta\right)} \frac{\sin\left(\frac{1}{2}N_2 k d_2 \sin\psi\right)}{\sin\left(\frac{1}{2}k d_2 \sin\psi\right)}\right]^2, \\
N_1 d_1 = D_1, &\qquad N_2 d_2 = D_2, \\
F_1(\theta, \psi) &= \mathrm{sinc}^2\left(\tfrac{1}{2}k l_1 \sin\theta\right) \mathrm{sinc}^2\left(\tfrac{1}{2}k l_2 \sin\psi\right).
\end{aligned} \tag{16.34}$$

These formulae can be verified experimentally, and they adequately describe the experimental data.

16.6 Diffraction by three-dimensional periodic structures—X-ray diffraction in perfect crystals

Crystals are natural three-dimensional periodic structures with a spatial interval of $d \simeq 1$ Å. Since the wavelength of visible light (10^{-4} cm) essentially exceeds the crystal lattice period (10^{-8} cm), visible light does not diffract in crystals. A

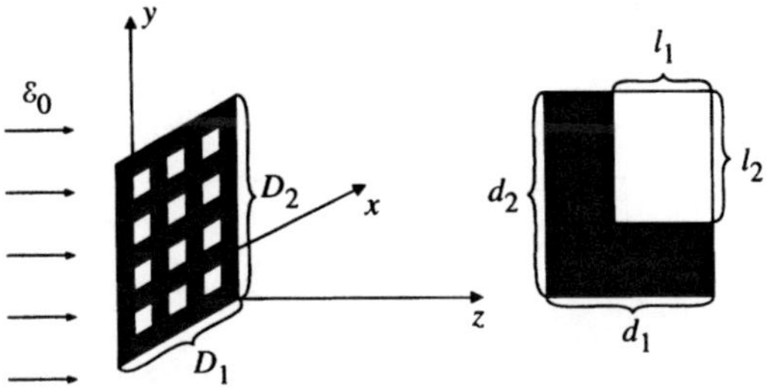

Fig. 16.21 The geometric parameters of a two-dimensional diffraction grating.

different situation arises in the case of X-rays, the wavelength of which is comparable with the crystal lattice period. According to the equation of a diffraction grating, such radiation will undergo strong diffraction in crystals.

Let us examine the basic properties of the diffraction of an electromagnetic wave by a three-dimensional periodic structure. Suppose, for simplicity, that the elementary crystalline cell has the form of a rectangular parallelepiped with the sides d_1, d_2, and d_3 (Fig. 16.22).

Let this crystal be illuminated by a plane monochromatic electromagnetic wave, as shown in Fig. 16.22. We shall find the directions towards the principal maxima of the diffraction pattern. The lattice can obviously be regarded as a succession of two-dimensional plane gratings spaced by the distance d_3. Let us use the coordinate system shown in Fig. 16.20. The directions towards the principal maxima arising due to diffraction of the wave by an individual two-dimensional grating are then determined by eqn (16.30). However, it is not sufficient to satisfy these conditions for the diffraction maxima to arise in the directions specified by the angles α and β. The formation of maxima depends essentially on the interference of waves diffracted by different atomic planes.

Interference reinforcement of the waves diffracted by two adjacent atomic planes takes place if the following condition is met (Fig. 16.23):

$$d_3 - d_3 \cos\gamma = m_3\lambda, \qquad (16.35)$$

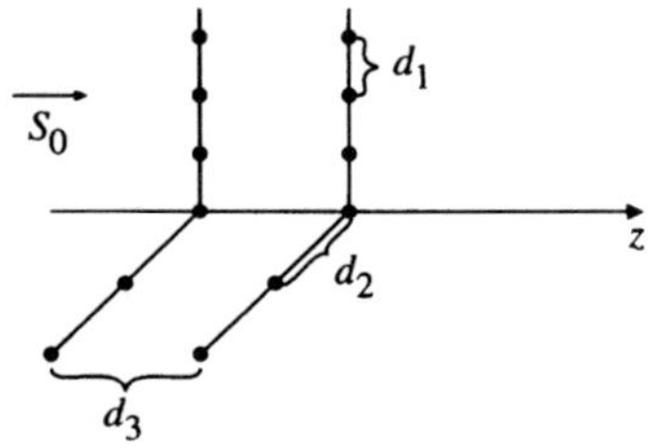

Fig. 16.22 The structure of a crystal lattice. d_1, d_2, and d_3 are the lattice periods, and $\vec{s}_0$ is the unit vector along the incident wave direction.

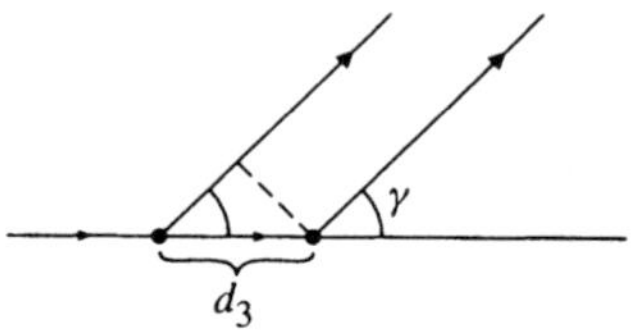

Fig. 16.23 The condition for interference reinforcement of the diffracted waves in a crystal.

where m_3 is an integer. Equation (16.35) expresses the requirement that the path difference between the rays coming in the direction γ from two adjacent atomic planes is equal to a whole number of wavelengths. If this requirement is satisfied, then the waves diffracted by different atomic planes of the crystal will interfere constructively and reinforce each other. On the other hand, in the directions γ not satisfying the condition (16.35) the waves interfere destructively. The complete set of equations for the angles α, β, and γ specifying the directions towards the principal maxima of the diffraction pattern produced by plane waves diffracted in the crystal thus has the form

$$\begin{aligned} d_1 \cos\alpha &= m_1\lambda, \\ d_2 \cos\beta &= m_2\lambda, \\ d_3(1 - \cos\gamma) &= m_3\lambda, \\ \cos^2\alpha + \cos^2\beta + \cos^2\gamma &= 1. \end{aligned} \tag{16.36}$$

Equations (16.36) are called *Laue's equations.* Here they are written for the case in which the incident wave propagates along the z-axis perpendicular to the atomic planes of the crystal. It is easy to generalize these equations to the case of an arbitrary direction of the incident wave (specified by the angles α_0, β_0, and γ_0) with respect to the crystal axes x, y, and z. In this case Laue's equations take the form

$$\begin{aligned} d_1(\cos\alpha - \cos\alpha_0) &= m_1\lambda, \\ d_2(\cos\beta - \cos\beta_0) &= m_2\lambda, \\ d_3(\cos\gamma - \cos\gamma_0)) &= m_3\lambda, \\ \cos^2\alpha + \cos^2\beta + \cos^2\gamma &= 1. \end{aligned} \tag{16.37}$$

The set of equations (16.37) is consistent only for some definite values of the radiation wavelength λ or for definite values of the angles α_0, β_0, and γ_0 specifying the crystal orientation with respect to the incident wave. In practice, the diffraction pattern can therefore be obtained either using X-rays with a broad frequency spectrum or by rotating the crystal relative to the monochromatic X-ray beam with the use of a special goniometric device. A schematic diagram

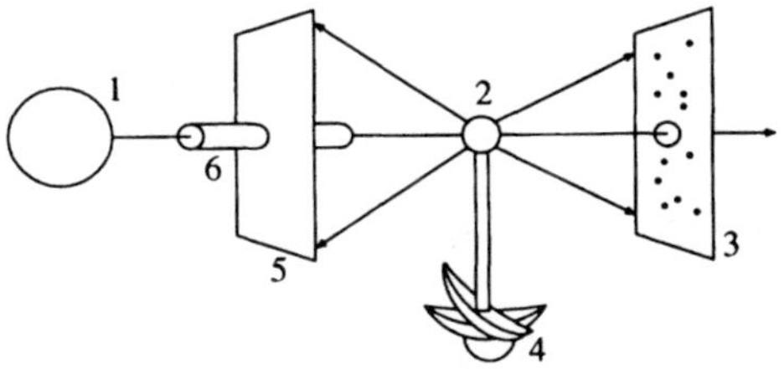

Fig. 16.24 A schematic diagram of X-ray diffraction analysis. 1, X-ray source; 2, a crystal under investigation; 3, the photoplate position for direct photography; 4, the goniometric device; 5, the photoplate position for back photography; 6, the collimator.

of the set-up for obtaining Laue diffraction patterns is shown in Fig. 16.24. A typical example of a diffraction pattern is presented in Fig. 16.25.

X-ray diffraction by crystals was first observed by Laue, Friedrich, and Knipping in 1912. Their experiments proved the wave nature of X-rays. At present, X-ray diffraction is widely used for the investigation of crystal structure.

16.7 X-ray diffraction analysis

X-ray diffraction by crystals makes it possible to formulate the inverse diffraction problem, i.e. the determination of crystal structure from the diffraction pattern observed. This problem is solved by the formulae describing Fraunhofer diffraction by a three-dimensional structure:

$$I(\vec{k}) \sim |\mathcal{E}_0(\vec{k})|^2, \tag{16.38}$$

$$\mathcal{E}_0(\vec{k}) = \int\limits_{-\infty}^{\infty} \mathcal{E}_0(\vec{r}) e^{i\vec{k}\vec{r}} d^3r. \tag{16.39}$$

Here $I(\vec{k})$ is the X-ray intensity distribution in the diffraction pattern, $\vec{k}$ is the vector directed from the object under investigation to the observation point, and

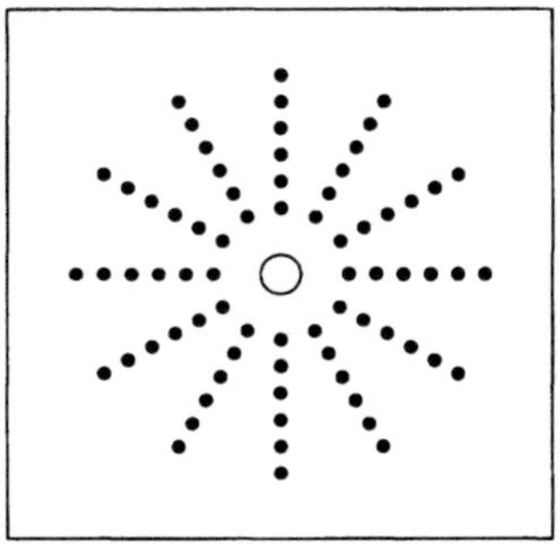

Fig. 16.25 An example of an X-ray diffraction pattern.

$\mathcal{E}_0(\vec{r})$ is the original spatial field distribution. The last quantity is determined from the structure of the object diffracting the X-rays; that is, from the spatial distribution of the electron density $\rho(\vec{r})$ in the crystal: $\mathcal{E}_0(\vec{r}) \sim \rho(\vec{r})$. Equations (16.38) and (16.39) can be considered as generalizations of eqns (16.5) and (16.31) to the case of the three-dimensional original structure of the field. Writing the formula inverse to eqn (16.39), we obtain

$$\mathcal{E}_0(\vec{r}) = \left(\frac{1}{2\pi}\right)^3 \int_{-\infty}^{\infty} \mathcal{E}_0(\vec{k}) e^{-i\vec{k}\vec{r}} d^3k. \tag{16.40}$$

Thus eqn (16.39) describes the Fourier analysis of the electron density of the crystal, while eqn (16.40) describes the Fourier synthesis of the crystal structure. The calculation by eqn (16.40) can be carried out numerically with the use of a computer. At present, X-ray diffraction analysis is a subtle sophisticated technique capable of determining the structure of many minerals and large complicated molecules, including biological molecules (e.g. the haemoglobin molecule).

Bibliography

Blakemore, J. S. (1985). *Solid state physics.* Cambridge University Press, Cambridge.

Born, M. and Wolf, E. (1968). *Principles of optics.* Pergamon Press, Oxford.

Goodman, J. W. (1968). *Introduction to Fourier optics.* McGraw-Hill, San Francisco.

17

The analysis, transformation, and synthesis of optical fields

The spectral description of the spatial structure of the field. The lens as an element performing Fourier's spatial transformation. Optical image formation and Abbe's theory. Porter's experiments. The dark-field technique. The phase-contrast technique. The resolving power of a microscope and telescope. Holography: the recording and reconstruction of optical fields. The diffraction grating and the Fabry–Perot interferometer as spectroscopic instruments. Fourier spectroscopy. Light-beat spectroscopy.

The methods of linear systems theory, which are widely applied in radiophysics for the investigation of temporal signals, also prove to be very useful in optics, in which light fields are studied. Spectral expansion based on Fourier transformation allows one to represent an arbitrary optical field with a complicated spatial–temporal structure as a superposition of plane monochromatic waves. Due to the linearity of the wave equation, each of these waves propagates in vacuum or in a linear optical medium independently of the others, so that the analysis of the transformation of complicated fields is reduced to the simpler problem of elementary wave transformation. The resultant field can then be found by summing up the plane monochromatic waves passing through the transforming system.

The procedures mentioned, that is, the analysis, transformation, and synthesis of optical fields, are not only mathematical operations, but are often, in fact, real physical processes. For example, the free diffraction of an optical beam forms a stable spatial distribution of radiation intensity in the far-field zone, reproducing the shape of the angular spectrum of the beam. The field distribution in the focal plane of a lens has the same shape, but much smaller dimensions. In both cases the spatial spectral decomposition of the field is actually realized. Various operations with optical fields performed in the focal plane of the lens, using different screens, masks, and phase plates, can be regarded as spatial analogues of the frequency filtering of electric oscillations in radiophysics.

The ideas and methods associated with the direct control of the optical field structure are combined in a special branch of optics called Fourier optics. This term stresses the fundamental role of spectral decomposition in solving problems connected with the analysis, transformation, and synthesis of optical fields.

17.1 The spectral description of the spatial structure of the field

Let us illustrate the spectral approach by solving the following diffraction problem. The electric field of an optical wave propagating in free space obeys the wave equation (see Chapter 1)

$$\Delta E - \frac{1}{c^2}\frac{\partial^2 E}{\partial t^2} = 0, \tag{17.1}$$

where

$$\Delta = \frac{\partial^2}{\partial x^2} + \frac{\partial^2}{\partial x^2} + \frac{\partial^2}{\partial z^2} \tag{17.2}$$

is the Laplace operator. Let us consider the diffraction of a monochromatic wave. Assuming

$$E(\vec{r}, t) = \tfrac{1}{2}\mathcal{E}(\vec{r})\mathrm{e}^{i\omega t} + \text{c. c.} \tag{17.3}$$

we obtain the Helmholtz equation for the complex amplitude $\mathcal{E}(\vec{r})$:

$$\Delta\mathcal{E} + k^2\mathcal{E} = 0, \tag{17.4}$$

where $k = \omega/c$. The diffraction problem consists in finding a solution to eqn (17.4) satisfying the boundary condition

$$\mathcal{E}(x, y, z = 0) = \mathcal{E}_0(x, y). \tag{17.5}$$

Such a condition arises, for example, when light passes through a "black-mask" screen, a planar phase transparency, etc.

Earlier, we solved a diffraction problem using the Green theorem and the Helmholtz–Kirchhoff integral theorem. We could mathematically validate the Huygens–Fresnel principle, write the diffraction integral in the Fresnel and Fraunhofer approximations and solve a number of diffraction problems (see Chapters 13–15). Let us now show that the diffraction problem can be solved in a different way; that is, using a *spectral approach*. This technique turns out to be simpler.

Let us expand the two-dimensional optical field $\mathcal{E}_0(x, y)$, given in the original section $z = 0$, into the Fourier integral

$$\mathcal{E}_0(x, y) = \left(\frac{1}{2\pi}\right)^2 \iint\limits_{-\infty}^{\infty} \mathcal{E}_0(k_x, k_y)\mathrm{e}^{-i(k_x x + k_y y)}\,dk_x\,dk_y. \tag{17.6}$$

Here the spatial spectral amplitude $\mathcal{E}_0(k_x, k_y)$ is determined by the inverse Fourier transformation

$$\mathcal{E}_0(k_x, k_y) = \iint\limits_{-\infty}^{\infty} \mathcal{E}_0(x, y)\mathrm{e}^{i(k_x x + k_y y)}\,dx\,dy. \tag{17.7}$$

We shall seek the optical field $\mathcal{E}(x, y, z)$ at an arbitrary point in space in a form analogous to (17.6), but with the spectral amplitude depending on z:

$$\mathcal{E}(x,y,z) = \left(\frac{1}{2\pi}\right)^2 \iint\limits_{-\infty}^{\infty} \mathcal{E}_0(k_x, k_y, z) e^{-i(k_x x + k_y y)} dk_x \, dk_y. \tag{17.8}$$

Substituting eqn (17.8) into eqn (17.4), we find the following equation for the spectral amplitude $\mathcal{E}(k_x, k_y, z)$:

$$\frac{\partial^2 \mathcal{E}}{\partial z^2} + (k^2 - k_x^2 - k_y^2)\mathcal{E} = 0. \tag{17.9}$$

The solution to this equation is

$$\mathcal{E}(k_x, k_y, z) = \mathcal{E}_0(k_x, k_y) e^{-i\sqrt{k^2 - k_x^2 - k_y^2} z}. \tag{17.10}$$

Inserting eqn (17.10) into eqn (17.8), we obtain

$$\mathcal{E}(x,y,z) = \left(\frac{1}{2\pi}\right)^2 \iint\limits_{-\infty}^{\infty} \mathcal{E}_0(k_x, k_y) e^{-i(k_x x + k_y y + \sqrt{k^2 - k_x^2 - k_y^2} z)} dk_x \, dk_y. \tag{17.11}$$

or, taking into account eqn (17.7),

$$\mathcal{E}(x,y,z) = \iint\limits_{-\infty}^{\infty} \mathcal{E}_0(x', y') h(x - x', y - y', z) dx' \, dy', \tag{17.12}$$

where the function

$$h(x,y,z) = \left(\frac{1}{2\pi}\right)^2 \iint\limits_{-\infty}^{\infty} e^{-i(k_x x + k_y y + \sqrt{k^2 - k_x^2 - k_y^2} z)} dk_x \, dk_y \tag{17.13}$$

is introduced. This function is called the *Green function* of the given linear system. Let us also introduce the function

$$\chi(k_x, k_y, z) = e^{-i\sqrt{k^2 - k_x^2 - k_y^2} z}, \tag{17.14}$$

called the frequency transmission coefficient of the system. It follows from eqns (17.10) and (17.14) that

$$\mathcal{E}(k_x, k_y, z) = \mathcal{E}_0(k_x, k_y) \, \chi(k_x, k_y, z). \tag{17.15}$$

Additionally, by virtue of eqns (17.13) and (17.14), we obtain

$$h(x, y, z) = \left(\frac{1}{2\pi}\right)^2 \iint_{-\infty}^{\infty} \chi(k_x, k_y, z) \mathrm{e}^{-i(k_x x + k_y y)} dk_x \, dk_y \tag{17.16}$$

and, consequently,

$$\chi(k_x, k_y, z) = \iint_{-\infty}^{\infty} h(x, y, z) \mathrm{e}^{i(k_x x + k_y y)} dx \, dy. \tag{17.17}$$

The frequency transmission coefficient and the Green function are thus connected by the Fourier transformation. This is a general property of all linear systems.

Substituting eqns (17.12) and (17.13) into eqns (17.4) and (17.5), it is easy to verify that the *exact* solution to the diffraction problem has been obtained. The simplicity of the solution indicates that the mathematical approach taken, i.e. spectral expansion, is appropriate to the physical content of the diffraction problem.

17.1.1 Propagation invariants of the optical beam

Let us introduce the intensity I, power P, and angular spectrum $S(k_x, k_y)$ of the optical beam, defining them by the formulae

$$I(x, y, z) = \frac{c}{8\pi} |\mathcal{E}(x, y, z)|^2, \tag{17.18}$$

$$P = \iint_{-\infty}^{\infty} I(x, y, z) \, dx \, dy, \tag{17.19}$$

$$S(k_x, k_y) = |\mathcal{E}(k_x, k_y)|^2. \tag{17.20}$$

According to the Parseval theorem,

$$\iint_{-\infty}^{\infty} |\mathcal{E}(x, y, z)|^2 dx \, dy = \left(\frac{1}{2\pi}\right)^2 \iint_{-\infty}^{\infty} S(k_x, k_y) \, dk_x \, dk_y. \tag{17.21}$$

It follows from eqns (17.14), (17.15), and (17.18)–(17.21) that the angular spectrum and the total power of the optical beam do not depend on z, i.e. they are the propagation invariants:

$$S(k_x, k_y, z) = S_0(k_x, k_y) = \text{inv}, \tag{17.22}$$

$$P(z) = P_0 = \text{inv}. \tag{17.23}$$

It is easy to verify that the specific results obtained earlier (see Chapters 14 and 15) satisfy the general relations (17.22) and (17.23). According to eqn (17.22), the Fraunhofer diffraction pattern formed in the far-field zone is stable because it reproduces the shape of the beam angular spectrum, and the angular spectrum is the propagation invariant.

17.1.2 The Fresnel approximation

If the optical beam is weakly diverging, then the following inequalities hold:

$$k_x^2, k_y^2 \ll k^2, \tag{17.24}$$

i.e. the angular spectrum is quite narrow. In this case we can write the approximate identity

$$\sqrt{k^2 - k_x^2 - k_y^2} = k - (k_x^2 + k_y^2)/2k. \tag{17.25}$$

The frequency transmission function takes the form

$$\chi(k_x, k_y, z) = \exp\left[-ikz + i\left(\frac{k_x^2 + k_y^2}{2k}\right) z\right]. \tag{17.26}$$

By substituting eqn (17.26) into eqn (17.16) and integrating, we obtain the following expression for the Green function:

$$h(x, y, z) = \frac{i}{\lambda z} \exp\left[-ik\left(z + \frac{x^2 + y^2}{2z}\right)\right]. \tag{17.27}$$

Substitution of eqn (17.27) into eqn (17.12) leads to the formula

$$\mathcal{E}(x, y, z) = \frac{i}{\lambda z} e^{-ikz} \iint_{-\infty}^{\infty} \mathcal{E}_0(x', y') \exp\left\{-\frac{ik}{2z}\left[(x - x')^2 + (y - y')^2\right]\right\} dx'\, dy'. \tag{17.28}$$

The obtained expression coincides with the diffraction integral in the Fresnel approximation (see eqn (15.6)). Thus, in the context of the spectral approach, all of the results of Fresnel diffraction theory can be reproduced. Additionally, the spectral concept allows one to gain a more penetrating insight into the physical meaning of a notion such as the diffraction divergence of the optical beam.

17.1.3 The diffraction divergence of radiation

For definiteness, let us consider the diffraction of a plane wave by a slit (Fig. 15.2). According to the spectral concept, the limitation of the transverse dimension of the optical beam leads to broadening of its angular spectrum (Fig. 15.3). The relation between the transverse dimension Δx of the beam and the width of the angular spectrum Δk_x is expressed by the formula

$$\Delta x \Delta k_x = 2\pi, \tag{17.29}$$

which is quite similar to the formula $\Delta t \Delta \omega = 2\pi$, relating the duration Δt and frequency bandwidth $\Delta \omega$ of an optical pulse.

The spatial frequency k_x can be expressed through the angle θ, shown in Fig. 15.2 as $k_x = k \sin\theta$, where $k = 2\pi/\lambda$ is the wave number and λ is the optical

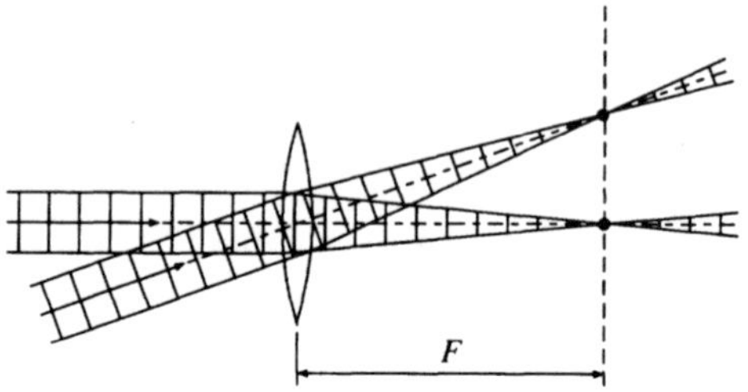

Fig. 17.1 The formation by a lens of the spatial spectrum of optical field.

wavelength. For small angles θ we can therefore write $k_x = k\theta$ and $\Delta k_x = k\Delta\theta$, where $\Delta\theta$ is the width of the angular spectrum, or the angular divergence of the beam. Let the slit width be $\Delta x = d$. Then, from eqn (17.29), it follows that

$$\Delta\theta = \lambda/d, \tag{17.30}$$

in accordance with eqn (15.10) derived earlier. Thus, the diffraction beam divergence can be evaluated in the spectral theory from the fundamental formula (17.29) relating the transverse size of the beam and the width of its angular spectrum.

17.2 The lens as an element performing Fourier's spatial transformation

It is known from ray optics that the lens converges a parallel beam of light into a point located in the focal plane (Fig. 17.1). An arbitrary optical field can be represented as a set of plane waves (parallel beams) incident on the lens at different angles. A definite point in the focal plane is then assigned by the lens to each of the plane waves. This signifies that the light distribution in the focal plane takes the form of a *spatial spectrum* of the incident optical field. In this sense the lens acts as an element performing the spatial spectral decomposition of light, or *Fourier spatial transformation.* It is easy to describe the indicated property of the lens in terms of wave optics. To do this, we shall introduce the transmission function of the lens and use Fresnel diffraction theory.

17.2.1 The transmission function of a thin spherical lens

Let us consider a thin spherical lens, i.e. a lens for which a ray entering the front surface at a point with coordinates x, y exits the lens at a point with the same coordinates on the other side. Obviously, this condition is not met in real lenses. However, it is satisfied with a higher degree of accuracy as the lens becomes thinner. This explains the term a "thin" lens. We shall define the complex transmission function $t(x, y)$ of the lens as the ratio of complex amplitudes of the transmitted and incident waves:

$$t(x, y) = \frac{\mathcal{E}_1(x, y)}{\mathcal{E}_0(x, y)} \tag{17.31}$$

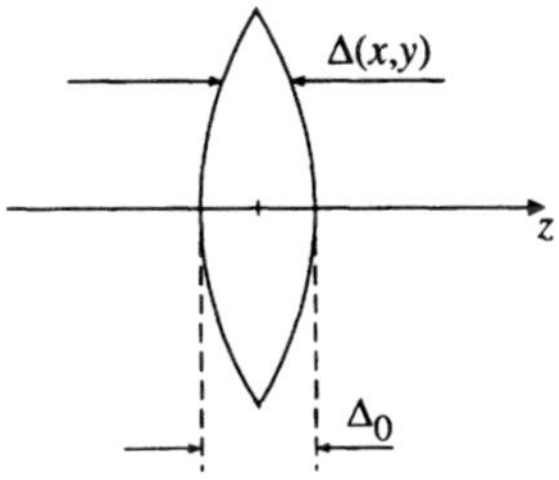

Fig. 17.2 The calculation of the transmission function of a lens.

To calculate this characteristic of the lens, we need to describe the propagation of light in glass. The propagation of light in different optical media will be considered in detail later (see Part IV). Here we shall use only one simple result that is necessary for our calculation. This result is that an optical wave propagates in a transparent linear isotropic optical medium with the velocity

$$v = c/n, \tag{17.32}$$

where c is the velocity of light in vacuum and n is a real positive value exceeding unity, called the *index of refraction.*

Taking into account eqn (17.32), we can describe the optical wave propagating in the glass as

$$E(\vec{r}, t) = \tfrac{1}{2}\mathcal{E}(\vec{r})\mathrm{e}^{i\omega t} + \text{c. c.}, \tag{17.33}$$

where

$$\mathcal{E}(\vec{r}) = A\exp(-iknz) \tag{17.34}$$

is the wave complex amplitude, and $k = \omega/c$.

Let $\Delta(x, y)$ be the lens thickness at a point with coordinates x, y and let Δ_0 be its maximum thickness (Fig. 17.2). It is easy to show that the lens transmission function is then

$$t(x, y) = \exp[-i\varphi(x, y)], \tag{17.35}$$

where

$$\varphi(x, y) = k\Delta_0 + (n-1)k\Delta(x, y). \tag{17.36}$$

Now let us calculate the function $\Delta(x, y)$, taking as an example the plane-convex lens shown in Fig. 17.3. As can be seen from this figure,

$$\Delta(x, y) = \Delta_0 - R + \sqrt{R^2 - x^2 - y^2}, \tag{17.37}$$

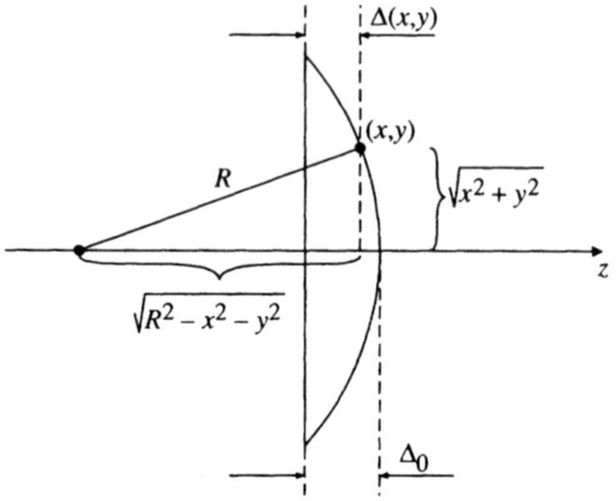

Fig. 17.3 The calculation of the thickness function $\Delta(x, y)$ of a lens.

where R is the curvature radius of the spherical surface of the lens. Assuming that

$$x^2, y^2 \ll R^2 \tag{17.38}$$

(the paraxial approximation), we rewrite eqn (17.37) in the form

$$\Delta(x, y) = \Delta_0 - (x^2 + y^2)/2R. \tag{17.39}$$

The lens transmission function then takes the form

$$t(x, y) = \exp\left[i(n - 1)k(x^2 + y^2)/2R\right]. \tag{17.40}$$

For the sake of simplicity, we have omitted the constant phase factor here. The focal length f of the lens is determined by the formula

$$\frac{1}{f} = \frac{n - 1}{R}. \tag{17.41}$$

It follows from eqns (17.40) and (17.41) that

$$t(x, y) = \exp\left[ik(x^2 + y^2)/2f\right]. \tag{17.42}$$

It should be noted that the transmission function written in the form (17.42) describes not only the plane-convex lens, but any thin lens.

17.2.2 The light distribution in the focal plane of a lens

Let us now calculate the light intensity distribution in the focal plane of a lens. Suppose that a lens with focal length f is illuminated by a monochromatic optical wave with complex amplitude $\mathcal{E}_0 = \mathcal{E}_0(x, y)$. The output wave complex amplitude is then

$$\mathcal{E}_1(x,y) = \mathcal{E}_0(x,y)\exp\left[ik(x^2+y^2)/2f\right]. \tag{17.43}$$

The further propagation of the wave is governed by the laws of diffraction. The diffraction field in the Fresnel approximation (see Chapter 14) is

$$\mathcal{E}(x,y,z) = \frac{i}{\lambda z}\mathrm{e}^{-ikz}\times \\ \times \iint\limits_{-\infty}^{\infty} \mathcal{E}_1(x',y')\exp\left\{-\frac{ik}{2z}\left[(x-x')^2+(y-y')^2\right]\right\}dx'\,dy', \tag{17.44}$$

Inserting eqn (17.43) into eqn (17.44), we obtain

$$\mathcal{E}(x,y,z) = \frac{i}{\lambda z}\exp\left[-ik\left(z+\frac{x^2+y^2}{2z}\right)\right]\iint\limits_{-\infty}^{\infty}\mathcal{E}_0(x',y')\times \\ \times\exp\left[-ik\left(\frac{x'^2+y'^2}{2}\right)\left(\frac{1}{z}-\frac{1}{f}\right)+\frac{ik}{z}(xx'+yy')\right]dx'\,dy'. \tag{17.45}$$

This formula describes the field amplitude distribution for any distance z from the lens. In particular, setting $z = f$, we find the field distribution in the focal plane of the lens:

$$\mathcal{E}(x,y,f) = \frac{i}{\lambda f}\exp\left[-ik\left(f+\frac{x^2+y^2}{2f}\right)\right]\times \\ \times\int\limits_{-\infty}^{\infty}\int\limits_{-\infty}^{\infty}\mathcal{E}_0(x',y')\exp\left[\frac{ik}{f}(xx'+yy')\right]dx'\,dy', \tag{17.46}$$

or

$$\mathcal{E}(x,y,f) = \frac{i}{\lambda f}\exp\left[-ik\left(f+\frac{x^2+y^2}{2f}\right)\right]\mathcal{E}_0(k_x,k_y)\Big|_{\substack{k_x=kx/f\\ k_y=ky/f}}, \tag{17.47}$$

where

$$\mathcal{E}_0(k_x,k_y) = \iint\limits_{-\infty}^{\infty}\mathcal{E}_0(x,y)\mathrm{e}^{i(k_x x+k_y y)}dx\,dy. \tag{17.48}$$

Equation (17.47) shows that field distribution in the lens focal plane is determined by the Fourier transform of the field incident on the lens. The light intensity distribution in the focal plane has the shape of the angular spectrum of the incident wave, i.e.

(a) (b)

Fig. 17.4 An example of an optical image (a) and its spatial spectrum (b).

$$I(x, y, f) = \frac{c}{8\pi}\frac{1}{\lambda^2 f^2} S_0(k_x, k_y)\bigg|_{\substack{k_x = kx/f \\ k_y = ky/f}} . \qquad (17.49)$$

17.2.3 Spectral analysis of optical images

The fact that the light intensity distribution in the lens focal plane reproduces the spatial spectrum of the incident wave can be used to obtain Fourier spectra of optical images. It is sufficient to place a transparency with an optical image in front of the lens and illuminate the transparency using a coherent beam of light. According to the results obtained earlier, the light intensity distribution in the lens focal plane would have the form of the spatial spectrum of the image. This distribution can be recorded on a photographic plate. The Fourier transform of the optical image can thus be obtained. The procedure is illustrated in Fig. 17.4, showing a sample of an optical image and its Fourier transform obtained by the optical method.

17.3 Optical image formation and Abbe's theory

According to the theory proposed by Abbe, the process of optical image formation by a lens can be divided into two stages: the *Fourier analysis* of the wave field generated by the object, and the *Fourier synthesis* of the image. The second focal plane of the lens plays an important role in this process, because the field distribution proportional to the Fourier transform of the original field is formed in this plane.

The schematic diagram of Abbe optical image formation is shown in Fig. 17.5. A diffraction grating is chosen as an object, the image of which is built up by the lens. As the light passing through the grating has a discrete angular spectrum, it is possible to trace the individual spatial spectral components of the field. As can

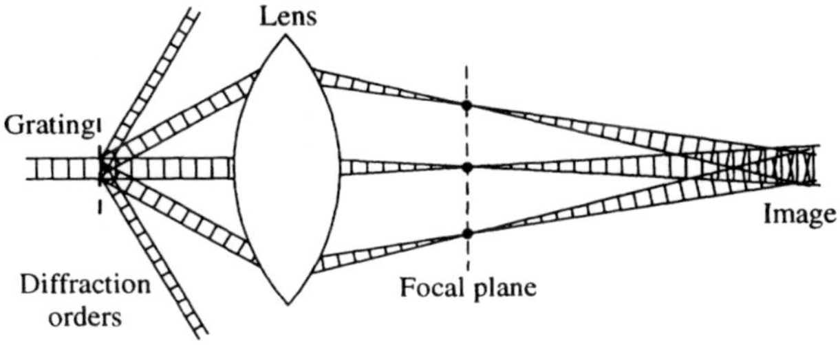

Fig. 17.5 Optical image formation according to Abbe.

be seen from the figure, at the first stage the lens performs the Fourier analysis of the wave field produced by the object. This operation is performed in the space between the lens and its second focal plane, and is mathematically described by Fourier transformation. The light intensity distribution in the second focal plane of the lens is the Fourier transform of the field produced by the object. In the second stage, Fourier synthesis of the image is performed during the free diffraction process. This operation occurs in the space between the second focal plane of the lens and the image plane, and is also mathematically described by Fourier transformation. As a result, an optical field, the structure of which reproduces the structure of the object, is formed in the image plane.

Of course, the Abbe theory does not contradict the principles of ray optics, which have been thoroughly tested experimentally. However, it provides insight into the physics of optical image formation, helps to estimate the limiting resolving power of optical instruments and, additionally, introduces the fruitful idea of image processing via spatial filtering of radiation. For example, by placing a diaphragm, screen or phase plate in the lens focal plane, we can transform the radiation angular spectrum so that the image will change in a controlled manner. Spatial filtering of the light field, that is, the operation similar to filtering of the temporal signals applied in radiophysics, can thus be realized in optics.

Let us consider some examples illustrating the principles of Fourier optics.

17.4 Porter's experiments

A schematic diagram of Porter's experiments is shown in Fig. 17.6. An object (a thin wire grid) is illuminated by coherent light. The light passing through the grid falls on the lens, which produces an image of the grid. The image is observed on the screen.

Figure 17.7(a) shows a photograph of the original spectrum of the grid and Fig. 17.7(b) the grid's corresponding image. Due to the periodic structure of the grid, its Fourier transform consists of discrete spots. The individual spot dimensions are governed by the overall size of the grid, while the spot separation is determined by the grid period (see Chapter 16). The vertical columns of the spots correspond to the vertical non-homogeneity of the grid, i.e. to the system

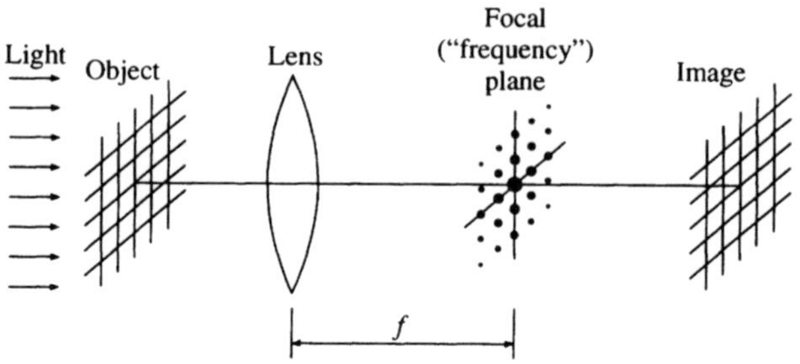

Fig. 17.6 The schematics of the Abbe–Porter experiment.

of its horizontal wires. Accordingly, the horizontal lines of the spots correspond to the system of vertical wires.

The prospects for image spatial filtering can be demonstrated by placing a narrow slit aperture in the second focal plane of the lens. The slit width is adjusted so that only one line of the spectral components is allowed to pass through it. Figure 17.8(a) shows the spectrum of radiation allowed to pass by the horizontal slit. The corresponding image shown in Fig. 17.8(b) contains a system of vertical lines. The horizontal lines are absent in the image because they are suppressed by the spatial filtering.

If we turn the slit through 90° so that only the vertical column of spots is allowed to pass (Fig. 17.9(a)), we will observe the image that contains only the horizontal lines (Fig. 17.9(b)). Now the vertical lines are suppressed by the spatial filtering.

We can also observe other interesting effects. For example, we can place an iris diaphragm in the second focal plane of the lens so that only the central

(a) (b)

Fig. 17.7 The Fourier spectrum (a) and optical image (b) of a wire grid.

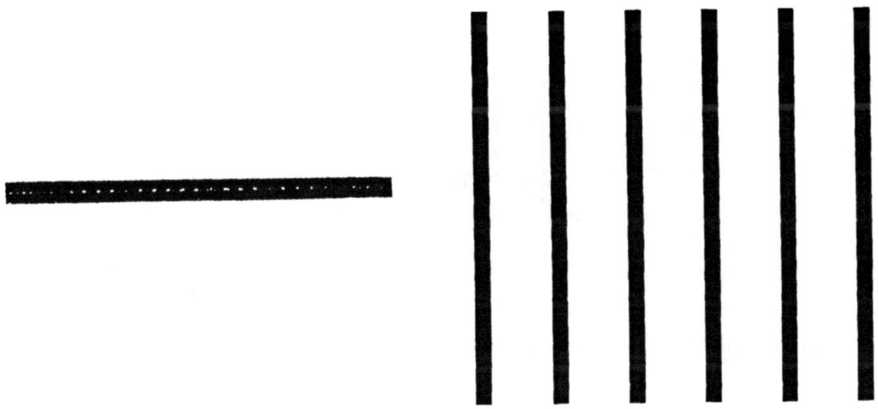

(a) (b)

Fig. 17.8 The grid spectrum filtered by an horizontal slit (a) and a corresponding image (b).

(zero-order) component of the Fourier spectrum is free to pass through it. No image would be seen on the screen in this case. However, by gradually opening the diaphragm, we can trace the Fourier synthesis of the image of the grid step by step. Finally, replacing the diaphragm by a small opaque disc to cover only the central spot of the Fourier spectrum, we can observe the image of the grid with inverse contrast.

(a) (b)

Fig. 17.9 The grid spectrum filtered by a vertical slit (a) and a corresponding image (b).

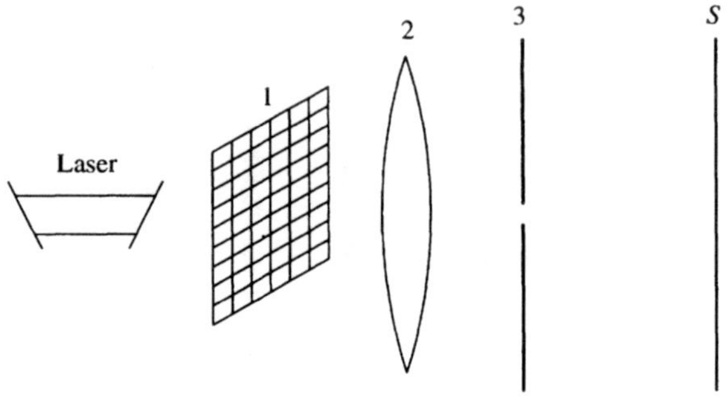

Fig. 17.10 The laser set-up for demonstration of Porter's experiments.

A laser version of Porter's experiment can be demonstrated using the experimental set-up shown schematically in Fig. 17.10. The helium–neon laser beam falls normally on a grid (1) made of wire 0.1 mm thick. The grid mesh is a square with a 0.4 mm side. The grid is used as an original. The convergent lens (2) with a focal length of $F = 5$ cm is placed close to the grid to produce a magnified image of the grid on the screen S located at a distance of 5 m from the lens. During the experiment the lens (2) can be moved along the beam axis to obtain a sharp image of the grid. A slit aperture (3), the width of which can be controlled by a micrometer screw, is placed in the focal plane of the lens. Mounted in a round rotating holder, the slit can perform spatial filtering along different directions.

At the beginning of the demonstration the slit is fixed vertically so that by decreasing its width one can observe a blurring of the image of the grid in the horizontal direction. If the slit width is sufficiently small, then we can see the image shown in Fig. 17.9(b). Then, turning the slit holder around the beam axis, we can observe how the image changes. Finally, when the slit occupies the horizontal position, we observe on the screen the image shown in Fig. 17.8(b).

17.5 The dark-field technique

The dark-field technique is a method of microscope illumination to increase the visibility of small objects, as well as objects that weakly absorb light, e.g. biological tissues, cells, etc. Figure 17.11 illustrates the basic idea of the method.

A parallel beam of light passes through the investigated object O and then through the lens L_1. A small opaque disc D is placed into the focal plane of the lens to prevent direct light from entering another lens L_2, which projects the image of the object onto the screen. In the absence of the disc D, an almost uniformly illuminated light field is seen on the screen. On insertion of the disc, the illumination of the screen decreases sharply. A bright image is thus formed against the dark background.

The experiment is explained as follows. The non-homogeneous optical density (refractive index), or non-homogeneous thickness, of a transparent object causes

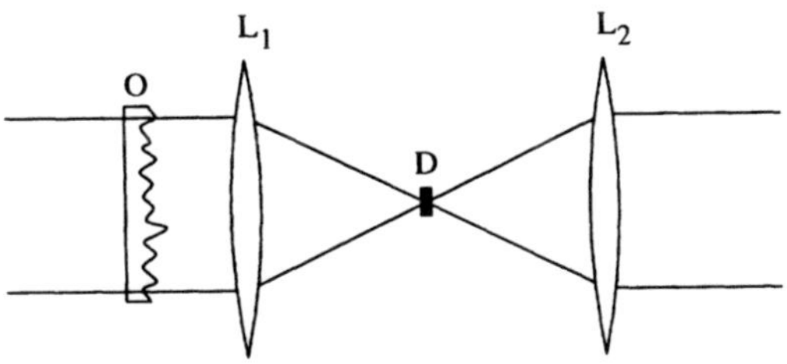

Fig. 17.11 The dark-field technique. O, Object; D, opaque disc; L_1, L_2, lenses.

refraction of light and gives rise to refracted rays. These rays, carrying information about the structure of the object (a useful signal), cannot be stopped by the disc D, the size of which is chosen to be sufficiently small. At the same time, the direct rays which do not carry any information about the object are stopped by the disc D.

The dark-field technique can be demonstrated using an arc lamp as a source of light and a glass stick or tube as an object. Although the stick is transparent, its image is clearly seen on the screen because the light refracted by the stick bypasses the disc. A more impressive experiment can be carried out. We place a cuvette filled with water in front of the lens L_1 and dip a piece of sugar suspended by a thin wire into the water. Then we see on the screen a brightly shining stream of the solution running off the dissolving sugar. In the course of the experiment, we can move the piece of sugar inside the cuvette and thus violate the regularity of the solution flow. We can also stir the heavier solution near the cuvette bottom with a glass stick.

Under laboratory conditions, this method is much more sensitive, so that one can observe, for example, the flow of warm air rising from a human hand. Similarly, it is possible to demonstrate, on a screen, a shining stream of hot air rising from a lit match or a gas burner.

17.6 The phase-contrast technique

This method, proposed by Zernike in 1935, is also used in microscopy to obtain images of transparent and colourless objects. An object of this type, e.g. a living cell, is characterized by non-uniform optical thickness. This causes the phase disturbance of the optical wave passing through the object. The wave undergoes different phase shifts at different points of the object, i.e. the wave acquires a phase relief. In the phase-contrast method the phase changes introduced by the object are transformed into optical intensity changes, i.e. the *phase relief* is transformed into *amplitude relief*. This is achieved by applying a special phase plate, that is, a thin plate of transparent material, which is placed at the second focal plane of the microscope objective.

A simplified diagram of the method is shown in Fig. 17.12. It is similar to the diagram for the dark-field method, but this time a thin glass disc (a phase plate) substitutes for the opaque disc at the lens focal plane. The plate thickness

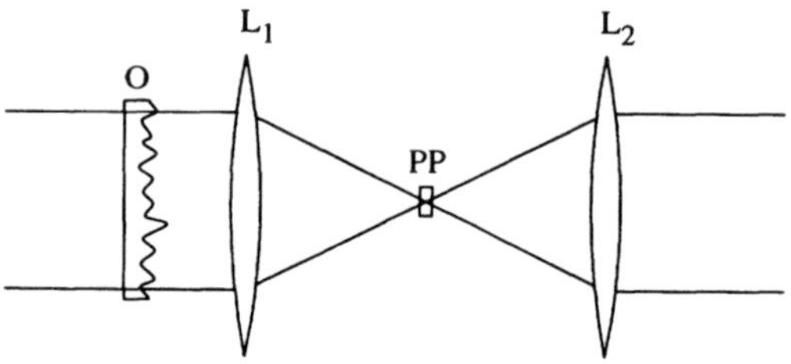

Fig. 17.12 The phase-contrast technique. O, Object; PP, phase plate; L_1, L_2, lenses.

is chosen so that it introduces a relative phase shift of the optical wave equal to $\pi/2$.

As in the dark-field technique, the plate is illuminated by light which has not been refracted by the object. This light, which carries no information about the object structure and corresponds to zero-order diffraction, is converged by the lens at the focal point at which it meets the phase plate and passes through it. At the same time the rays refracted, or diffracted, by the object (a useful signal) bypass the phase plate, the size of which is chosen to be sufficiently small. Then the background wave, the phase of which has been shifted by $\pi/2$, and the signal wave interfere to produce the high-contrast image of the object.

The effect can be explained as follows. Suppose that the object is characterized by the complex transmission function

$$T(x, y) = \exp[i\varphi(x, y)], \tag{17.50}$$

where $\varphi(x, y)$ is a real function with

$$|\varphi(x, y)| \ll 1. \tag{17.51}$$

We can then write the approximation

$$T(x, y) = 1 + i\varphi(x, y). \tag{17.52}$$

Let $\mathcal{E}_0$ denote the complex amplitude of the wave illuminating the object. The amplitude of the wave passing through the object is then

$$\mathcal{E}(x, y) = \mathcal{E}_0 T(x, y) = \mathcal{E}_0 + i\mathcal{E}_0\varphi(x, y). \tag{17.53}$$

In the absence of the phase plate, the field of the amplitude (17.53) is directly projected on to the screen or observed through the eyepiece of a microscope. Then the observed intensity distribution

$$I(x, y) = I_0[1 + \varphi^2(x, y)] \tag{17.54}$$

is nearly uniform and the structure of the object is indiscernible. A different situation takes place when the phase plate shifting the phase of the background wave by $\pi/2$ is placed in the focal plane. In this case the term $\mathcal{E}_0$ on the right-hand side of eqn (17.53) should be replaced by $i\mathcal{E}_0$. As for the second term, $i\mathcal{E}_0\varphi(x, y)$, describing the wave refracted by the object, this does not change because the refracted waves bypass the phase plate, the size of which is sufficiently small. The resultant wave amplitude takes the form

$$\mathcal{E}(x, y) = i\mathcal{E}_0[1 + \varphi(x, y)], \tag{17.55}$$

and the corresponding intensity distribution is

$$I(x, y) \approx I_0[1 + 2\varphi(x, y)]. \tag{17.56}$$

Comparing eqns (17.54) and (17.56) and allowing for the condition (17.51), we see that a phase plate placed at the focal plane should considerably increase the contrast of the image. This conclusion is confirmed by experiment (see, for example, Steward 1985).

17.7 The resolving power of a microscope and telescope

It follows from the Abbe theory that even the most perfect lens builds up an image containing error. This is connected with the fact that the optical field emerging from the object generally has an arbitrary spatial spectrum, while the lens is capable of catching only a finite fraction of this spectral bandwidth due to the finite size of the lens aperture. Not all of the rays emitted by the object are gathered by the lens. The rays corresponding to high spatial frequencies bypass the lens and, consequently, do not take part in image formation. For this reason, the image formed by the lens is not an identical copy of the object; the image does not contain the smallest details of the object structure.

This consideration becomes most clear if we take a diffraction grating as an object, the image of which is built up by a lens. Suppose that only the zeroth and the first diffraction orders formed by the grating reach the lens, i.e. the situation illustrated in Fig. 17.5 takes place. What will the image of the grating be in this case? According to the equation of the diffraction grating,

$$d \sin\theta = m\lambda, \qquad m = 0, \pm 1, \pm 2, \dots, \tag{17.57}$$

the information about the grating interval is contained in the diffraction angle of the first maximum. This information will thus be present in the image. However, small details of the grating structure will be distorted in the image because the information about them is contained in the second and higher diffraction orders which the lens cannot gather. If, for example, the original grating is rectangular (see Chapter 16), the image will show the correct number of lines per unit length, but the sharp edges of slits will be smeared so that an image reminiscent of that of a sinusoidal grating will emerge.

This example allows us to estimate the limiting resolving power of the lens and the microscope. The minimum interval d of the grating which can be resolved

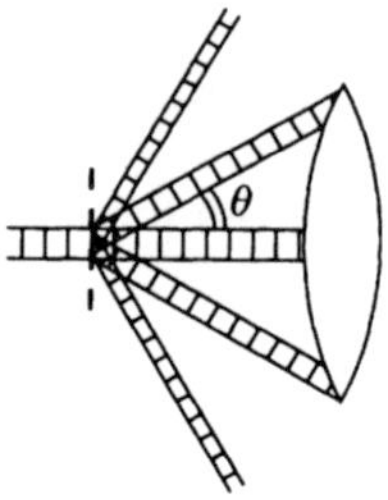

Fig. 17.13 Estimating the resolving power of the lens.

by the lens corresponds to the case in which the first-order diffraction maximum is directed exactly to the edge of the lens (see Fig. 17.13). It can be seen from the figure that the larger the lens diameter and the smaller the distance separating the lens and the grating are, the finer is the grating that can be resolved. However, if the diffraction angle θ of the first maximum is close to $\pi/2$, then this maximum inevitably gets beyond the lens and, consequently, the grating cannot be resolved. We can then write

$$\theta_{\text{max}} = \pi/2 \tag{17.58}$$

and, in view of eqn (17.57),

$$d_{\text{min}} = \lambda. \tag{17.59}$$

Thus, the resolving power of the lens is limited primarily by the optical wavelength. This is the fundamental limit of resolving power in optics.

17.7.1 The electron microscope

A radical way to increase the resolving power lies outside optics. This is *electron microscopy*. It is known that under certain conditions the electron and other elementary particles demonstrate wave properties. An impressive example is provided by the electron diffraction discovered by Davisson and Germer in 1927. The electron wavelength, called also the De Broglie wavelength, depends on the electron energy and is several orders of magnitude smaller the optical wavelength for an energy of about a few electron-volts. The idea of building a microscope in which electron waves would be used instead of visible light came naturally. Since electrons possess a very small wavelength, such a microscope would have an extremely high resolving power. The realization of this idea has led to the construction of the electron microscope.

Let us estimate the resolving power of the electron microscope. We shall assume that, just as for the optical microscope, the resolving power of the electron microscope is governed by eqn (17.59), but with the De Broglie wavelength of the electron replacing λ. Consider, for simplicity, a non-relativistic electron, i.e. an electron the velocity of which is much less than the velocity of light c. Then the momentum p and kinetic energy W of the electron are

$$p = mv, \qquad W = mv^2/2,$$

and the De Broglie wavelength is

$$\lambda = \frac{h}{mv} = \frac{h}{\sqrt{2mW}},$$

where h is Planck's constant, m is the mass of the electron, and v is its velocity. Let us make a numerical estimate. Assuming $W = 100$ eV, 1 eV$= 1.6\times10^{-12}$ erg, $h = 6 \times 10^{-27}$ erg s, and $m = 9.1 \times 10^{-28}$ g, we obtain

$$v = \sqrt{2W/m} = 0.59 \times 10^9 \text{ cm/s} \quad \text{and} \quad \lambda = 1.2 \times 10^{-8} \text{ cm} = 1.2\,\text{Å}. \qquad (17.60)$$

Thus, according to our estimate, the resolving power of the electron microscope is three to four orders of magnitude higher than that of the optical microscope.

In practice, modern electron microscopes ensure a resolving power of about 1–10 Å with electron energies of 10–100 keV. In solid state microscopy it is possible to observe groups of atoms and even individual atoms!

17.7.2 The telescope

This device is intended for the observation of distant objects such as stars, planets, asteroids, etc. One of the main characteristics of the telescope is its *resolving power*, defined as the minimum angular separation between two point objects that can still be resolved with the telescope. As with the microscope, the principal restriction on the resolving power is caused by diffraction.

Let us estimate the resolving power of the telescope. Suppose that the telescope objective is illuminated by the light emitted by two remote sources (stars). We shall find the minimum angle between the directions to the stars at which the stars can still be distinguished. Suppose that the light coming from the first star falls normally onto the objective lens, while the light of the second star falls at an angle ψ. Owing to the great distance to the stars, the incident light from each of them can be treated as a parallel beam (Fig. 17.14). According to ray optics, a parallel beam of light incident on the lens is focused at a point located in the focal plane. Because of diffraction, the actual focal spot for either beam will have a finite size. For sufficiently small angles ψ, the images of the stars will therefore overlap and no longer be resolved as individual components. As is shown in Chapter 13, the focal spot has the size

$$d = \lambda F/D, \qquad (17.61)$$

where F and D are the focal length and aperture of the lens, and λ is the optical wavelength. As seen from Fig. 17.14, the condition for resolution of the stars is

$$x_{\min} = d, \qquad (17.62)$$

where

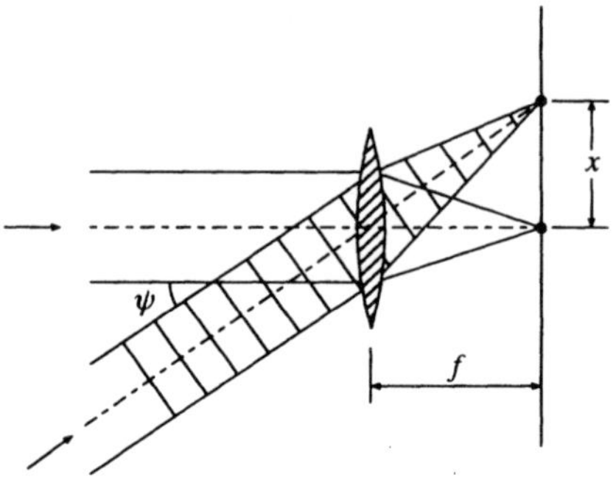

Fig. 17.14 Estimating the resolving power of the telescope.

$$x_{\min} = F\psi_{\min}. \tag{17.63}$$

Hence

$$\psi_{\min} = \lambda/D, \tag{17.64}$$

which coincides with the estimate obtained in Chapter 12.

Thus, the resolving power of the telescope is determined by eqn (17.64), where D is the telescope aperture and λ is the optical wavelength. The largest telescope in existence (on the Zelenchuk river in the Northern Caucasus) has a main mirror diameter of $D = 6$ m. The theoretical limit of resolution for this telescope is consequently $\psi_{\min} = 10^{-7}$ rad.

17.8 Holography: the recording and reconstruction of optical fields

In usual photography, only part of information about the optical field, namely, the spatial distribution of the light intensity, is recorded on the photographic plate. The information about the spatial distribution of the field phase is completely lost in this process. A question arises: Is it possible to arrange the recording process so that the complete information about the optical field, i.e. information about its amplitude and phase, can be preserved? It turns out that such a possibility does exist. The relevant method of optical field recording is called *holography.*

The basic idea of holography is rather simple. In contrast to photographing the optical field emerging from the object, holography registers the pattern of *interference* of this field with a coherent reference wave. The pattern of interference of the object wave and the reference wave, recorded on a photographic plate, is called a *hologram.* Since the interference pattern depends not only on the amplitudes but also on the phases of the interfering waves, all of the information about the object wave, including its amplitude and phase, turns out to be recorded on the hologram. To reconstruct the object wave, one only needs to illuminate the hologram using the reference beam.

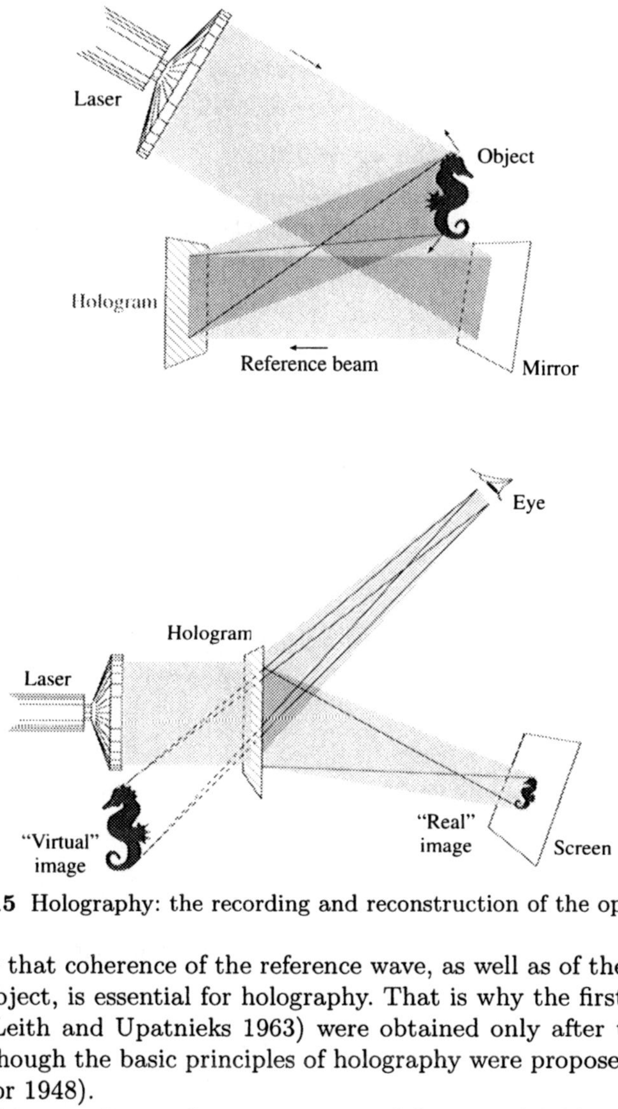

Fig. 17.15 Holography: the recording and reconstruction of the optical field.

It is clear that coherence of the reference wave, as well as of the wave illuminating the object, is essential for holography. That is why the first high-quality holograms (Leith and Upatnieks 1963) were obtained only after the advent of the laser, although the basic principles of holography were proposed many years earlier (Gabor 1948).

Holographic recording and reconstruction of the optical field is schematically shown in Fig. 17.15. A coherent laser beam is subdivided into two parts. One of the beams (the "signal beam") illuminates the object, while the other (the "reference beam") falls directly onto the photographic plate. The light reflected from the object forms the "object beam", which is also directed towards the plate. Thus the reference beam and the object beam are superposed on the photographic plate to form an interference pattern. This pattern is recorded by the plate which, when developed, is called a hologram.

To reconstruct the optical field emerging from the object, the hologram is illuminated by the reference beam. Being diffracted by the hologram, this beam forms a number of beams, one of which provides an exact copy of the original object beam. Thus the hologram reconstructs the optical field.

Now we shall illustrate these processes by simple formulae.

17.8.1 The recording of the optical field

Let $\mathcal{E}_o$ denote the complex amplitude of the object wave and $\mathcal{E}_r$ the complex amplitude of the reference wave. Separating the real amplitudes and phases of these waves and assuming the reference wave to be plane, we can write

$$\mathcal{E}_o = |\mathcal{E}_o(x,y)|e^{i\varphi_o(x,y)}, \qquad \mathcal{E}_r = |\mathcal{E}_r|e^{-i\vec{k}_O\vec{r}}. \tag{17.65}$$

The complex amplitude of the resultant field in the plane of the hologram is

$$\mathcal{E} = \mathcal{E}_o + \mathcal{E}_r, \tag{17.66}$$

while the intensity distribution is

$$I(x,y) = \frac{c}{8\pi}|\mathcal{E}|^2 = \frac{c}{8\pi}\left(|\mathcal{E}_o|^2 + |\mathcal{E}_r|^2 + \mathcal{E}_o\mathcal{E}_r^* + \mathcal{E}_o^*\mathcal{E}_r\right). \tag{17.67}$$

This distribution is fixed on the photographic plate. It is important that this expression features terms containing information about the phase of the object wave.

17.8.2 Field reconstruction

The field is reconstructed by illuminating the hologram using the reference wave. On its back side the hologram produces the field with complex amplitude

$$\mathcal{E}_1 = \mathcal{E}_r T(x,y), \tag{17.68}$$

where $T(x,y)$ is the light transmission function of the hologram. We shall write this function in the form

$$T(x,y) = 1 - D(x,y), \tag{17.69}$$

where

$$D(x,y) = \alpha I(x,y) \tag{17.70}$$

is the function proportional to the radiation intensity distribution recorded on the hologram. The coefficient α depends on the sensitivity of the photographic plate. Substituting eqns (17.67), (17.69) and (17.70) into eqn (17.68), we obtain

$$\mathcal{E}_1 = (1 - \beta|\mathcal{E}_o|^2 - \beta|\mathcal{E}_r|^2)\mathcal{E}_r - \beta|\mathcal{E}_r|^2\mathcal{E}_o - \beta\mathcal{E}_r^2\mathcal{E}_o^*, \tag{17.71}$$

where $\beta = \alpha c/8\pi$.

Equation (17.71) proves the above statement. According to this equation, one of the waves arising due to the diffraction of the reference wave by the hologram, that is, the wave the amplitude of which is proportional to $\mathcal{E}_o$, exactly reproduces the wave field that was coming from the object during the hologram recording. This wave carries a *virtual image* of the object. Observation of the virtual image, as shown in Fig. 17.15, creates an illusion of actually seeing the real object that seems to be present behind the hologram. In particular, by changing the eye position with respect to the hologram, we can view the different sides of the "object" just as if we were observing not a holographic image, but the real three-dimensional object.

Another wave the amplitude of which is proportional to $\mathcal{E}_o^*$ creates the *real image* of the object which can be observed on the screen. Additionally, as can be seen from eqn (17.71), the reference wave, being diffracted by the hologram, generates other waves; in particular, the reference wave running in the forward direction. These waves hamper the observation. In practice, however, the recording and reconstruction of the optical field can be contrived so that such hindrances do not play an essential role.

Figure 17.16 shows a hologram. Note that it does not look like the object image at all. Under conventional (non-coherent) illumination, the hologram looks like a monochromatic, slightly turbid, transparency. Nevertheless, the hologram contains a fine holographic image of an object. This image is recreated on illumination of the hologram with a coherent laser beam.

17.9 The diffraction grating and the Fabry–Perot interferometer as spectroscopic instruments

Instruments employing the effects of light interference and diffraction, i.e. interferometers and diffraction spectrometers, are widely used in modern optical spectroscopy. The most important characteristics of spectroscopic instruments are the resolving power A, defined as the ratio of the optical wavelength λ to the minimum wavelengh difference $\Delta\lambda$ which can still be resolved by the instrument

$$A = \lambda/\delta\lambda, \tag{17.72}$$

and the dispersion range, defined as the wavelength interval accessible to measurement:

$$G = \Delta\lambda. \tag{17.73}$$

Let us calculate these characteristics for the diffraction grating and the Fabry–Perot interferometer.

17.9.1 Diffraction grating

Figure 17.17 shows a pattern of the diffraction of a plane monochromatic optical wave by a grating. The light transmitted or reflected by the grating is a discrete set of rays, the directions of which are determined by the equation of the diffraction grating:

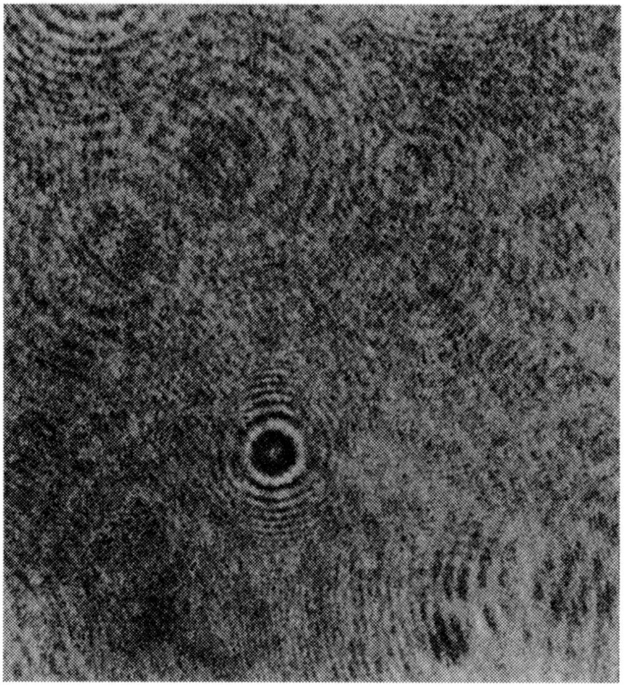

Fig. 17.16 A view of a hologram in non-coherent light.

$$d \sin \theta = m\lambda. \tag{17.74}$$

Here d is the grating period, λ is the optical wavelength, and $m = 0, \pm 1, \pm 2, \ldots$ is an integer called the diffraction order. The path difference between the rays diffracted by two adjacent slits of the grating in the direction θ is expressed by the formula $\Delta = d \sin \theta$. Consequently, the diffraction order $m = \Delta/\lambda$ denotes the ratio of the path difference between the rays to the optical wavelength.

Now suppose that the grating is illuminated by a *non-monochromatic* optical wave. Then, as can be seen from eqn (17.74), the radiation spectral components with wavelengths λ_1 and λ_2 will be diffracted by the grating at different angles according to the equations

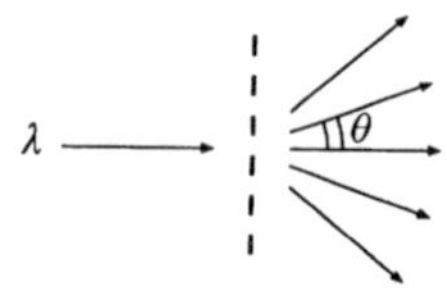

Fig. 17.17 The diffraction of a plane monochromatic wave by a diffraction grating.

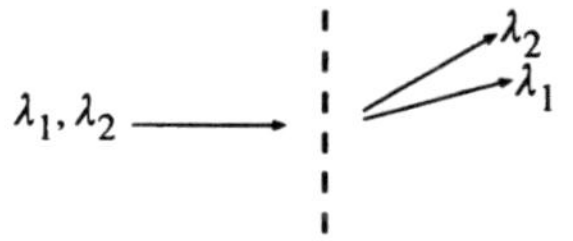

Fig. 17.18 The diffraction of a plane non-monochromatic wave by a diffraction grating: spectral decomposition.

$$\sin\theta_1 = m\lambda_1/d, \qquad \sin\theta_2 = m\lambda_2/d, \tag{17.75}$$

i.e. the grating will perform spatial decomposition of the spectrum (Fig. 17.18). It is this effect that provides the basis for a diffraction spectrometer.

The diffraction of an argon laser beam can be demonstrated by the use of a diffraction grating. The laser radiation contains several spectral components, in particular the components with wavelengths $\lambda_1 = 488$ nm and $\lambda_2 = 514.5$ nm, which differ slightly in colour. The grating spatially separates these components, as can be clearly seen in the experiment.

Another experiment demonstrates the diffraction of mercury lamp light by a reflecting diffraction grating. The radiation of the mercury-discharge lamp contains a number of spectral components with colours ranging from yellow to violet, so that the light incident on the grating looks like white light. The experiment shows that spectral decomposition is absent in the zeroth order of diffraction. The ray reflected by the grating in this direction looks like white light. Spectral decomposition is clearly seen in the first order of diffraction. In the second order of diffraction the spectrum is seen just as well, but with a higher dispersion. The last effect is particularly noticeable due to the splitting of the yellow spectral line, which is not resolved in the first order spectrum. The experiment demonstrates the wide dispersion range of the diffraction grating. The groove profile of the reflecting grating used in this experiment is chosen so that the most energy is directed towards the second order spectrum. This is very convenient for spectral observations.

Figure 17.19 shows the schematic diagram of a spectrometer with a diffraction grating. With the help of a pair of concave mirrors and a diffraction grating, the spectrometer forms images of the entrance slit, which are shifted relative to each other in the lateral direction for different wavelengths λ of the incident radiation.

The physical factor limiting the resolving power of the diffraction grating is evidently the finite width of the diffraction maximum ε (Fig. 17.20). The form of the diffraction maximum can be called an apparatus function of the grating.

It was shown in Chapter 16 that

$$\varepsilon = \lambda/Nd, \tag{17.76}$$

where N is the total number of slits (grooves) in the grating. It follows from eqns (17.72), (17.75), and (17.76) that

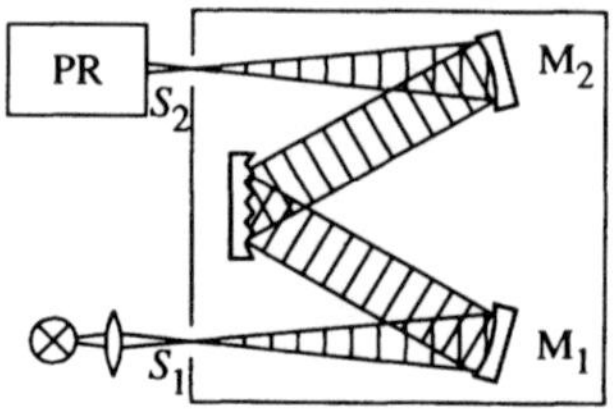

Fig. 17.19 The diffraction grating spectrometer.

$$A = mN. \tag{17.77}$$

Thus, the resolving power of the grating is equal to the product of the order number and the number of grooves.

The dispersion range of the grating is determined by the condition of superposition (overlapping) for two adjacent diffraction maxima:

$$m(\lambda + \Delta\lambda) = (m + 1)\lambda. \tag{17.78}$$

Hence

$$G = \lambda/m. \tag{17.79}$$

Let, for example, $\lambda = 5000$ Å, $m = 2$, and $N = 10^4$. Then $A = 2 \times 10^4$, $\delta\lambda = \lambda/A = 0.25$ Å, and $G = 2500$ Å (Fig. 17.21).

17.9.2 The Fabry–Perot interferometer

The plane Fabry–Perot plane interferometer consists of two wedge-like glass or quartz plates (Fig. 17.22). The inner surfaces of the plates are coated with high-reflectance film and are parallel, while the outer surfaces are made slightly non-parallel to avoid the undesirable interference of rays reflected from the back surfaces of the plates.

The equation of the Fabry–Perot interferometer can be written in a form similar to the equation of the diffraction grating:

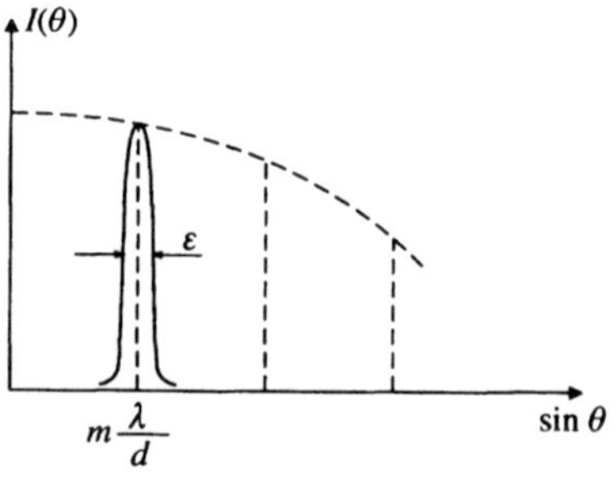

Fig. 17.20 The diffraction of a plane monochromatic wave by a diffraction grating: the angular distribution of the light intensity (a fragment).

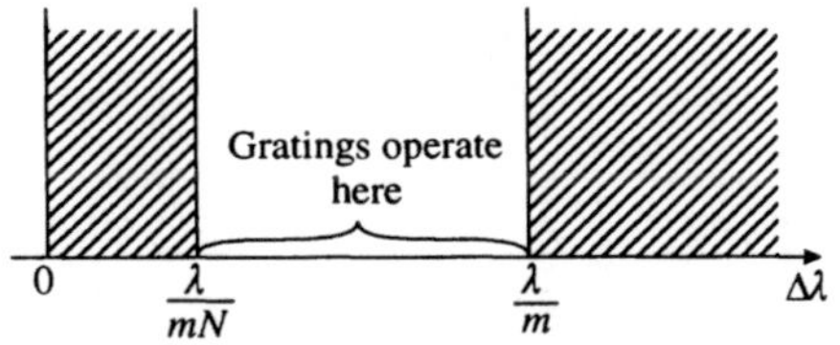

Fig. 17.21 The range of application of the diffraction grating as a spectral instrument.

$$2nl\cos\theta = m\lambda. \tag{17.80}$$

Here l is the separation between the interferometer plates, n is the refractive index of the medium filling the space between the plates, and θ is the angle between the ray and the normal to the plate (Fig. 17.22(b)). The quantity $\Delta = 2nl\cos\theta$ is the path difference between two adjacent rays in the interferometer, and $m = \Delta/\lambda$ is the order of interference.

Equation (17.80) determines the directions of the principal maxima of the interference pattern. If the interferometer is illuminated by a diverging beam of monochromatic light, then a pattern of sharp interference fringes is formed at its exit (Fig. 17.23). In this case, the different fringes correspond to different values of the number m in eqn (17.80).

Now suppose that the interferometer is illuminated by non-monochromatic light; for example, by a pair of optical waves with wavelengths λ_1 and λ_2. Then each of the waves will produce its own system of fringes. For a fixed value of the number m, the directions to interference maxima for different spectral compo-

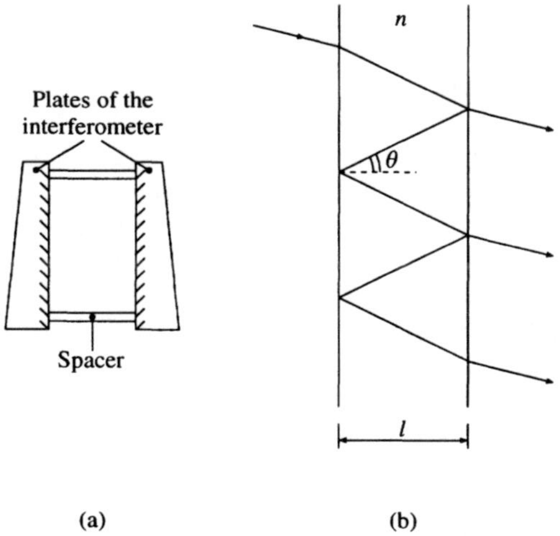

Fig. 17.22 The Fabry–Perot interferometer: (a) schematics; (b) the path of the light rays.

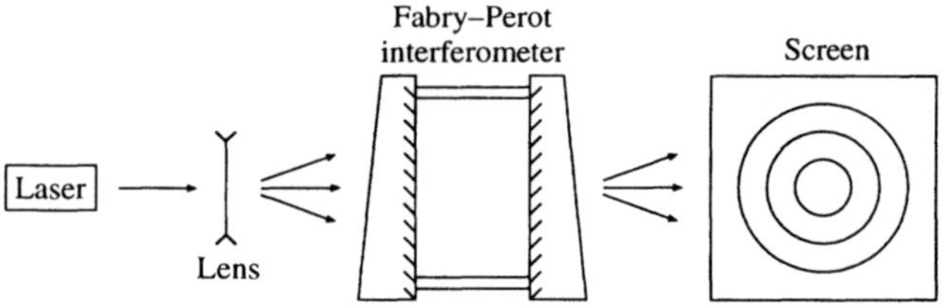

Fig. 17.23 The observation of interference fringes with the Fabry–Perot interferometer.

nents will be different, i.e. the interferometer will perform the spatial decomposition of the spectrum.

It follows from eqn (17.80) that the dispersion range of the Fabry–Perot interferometer is determined by eqn (17.79), as in the case of the diffraction grating. The resolving power of the interferometer is determined by the width of an individual interference fringe. Writing the condition for the resolution of two close spectral lines in the form

$$\delta_1 - \delta_2 = \varepsilon \tag{17.81}$$

(see Fig. 13.7), where $\delta = k\Delta = 2\pi\Delta/\lambda$, we obtain $\delta\lambda = \lambda^2\varepsilon/2\pi\Delta$ or, because of eqn (13.12), $\delta\lambda = \lambda^2/\mathcal{F}\Delta$, where $\mathcal{F}$ is the finesse of the interferometer. Consequently, the resolving power of the instrument is

$$A = \mathcal{F}\Delta/\lambda. \tag{17.82}$$

Taking into account that $\Delta/\lambda = m$, and that the finesse $\mathcal{F}$ is approximately equal to the effective number N of beams in the interferometer, we can reduce eqn (17.82) to the form (17.77), which is convenient for making estimates.

As an example, let us consider a Fabry–Perot interferometer with the following parameters: $l = 2.5$ cm and $R = 0.9$. Setting $\lambda = 5000$ Å and making use of eqn (13.16), we obtain $\Delta = 2l = 5$ cm, $m = \Delta/\lambda = 10^5$, $N = \mathcal{F} = 30$, $A = 3 \times 10^6$, $\delta\lambda = \lambda/A = 0.0017$ Å, and $G = \lambda/m = 0.05$ Å. These estimates show that the Fabry–Perot interferometer has a very high resolving power, but a very narrow dispersion range. In practice, the Fabry–Perot interferometer is often used in combination with a diffraction spectrometer to carry out spectral measurements at a high resolution throughout the broad spectral range.

17.9.3 The Fabry–Perot scanning interferometer

Let us return to eqn (17.80) determining the direction θ to the interference maximum of order m. We have already considered the case in which the parameters n and l of the Fabry–Perot interferometer are fixed. In that case, the different angles θ correspond to different wavelengths λ, i.e. the interferometer performs spatial decomposition of the spectrum. The spectral density $S(\omega)$ of the incident radiation can be found from the angular distribution $I(\theta)$ of the light intensity

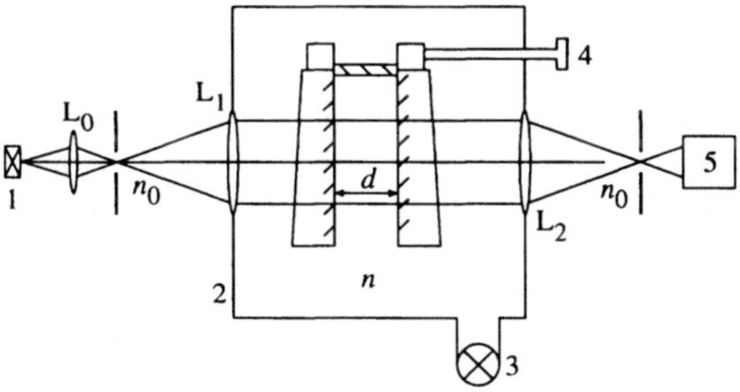

Fig. 17.24 The Fabry–Perot scanning interferometer. 1, Source of light; 2, airtight chamber; 3, gas admission valve; 4, mirror adjustment screw; 5, photoreceiver.

in the interference pattern.

Equation (17.80) implies another possibility for measuring $S(\omega)$. The radiation spectrum can be measured at the fixed direction of observation $\theta =$ constant, provided that the parameters n or l are varied during the measurement process. This can be done by moving the interferometer plates with the aid of a special motor, or by varying the gas (air) pressure in space between the plates. A Fabry–Perot interferometer with variable parameters n or l is called a *scanning interferometer*. Because of the necessity of keeping the plates parallel to a high accuracy, interferometers with a fixed plate separation, secured by a ring spacer (see Fig. 17.22), are usually used in practice, with scanning effected by varying the gas pressure in the inter-plate space. A schematic diagram of such an interferometer is shown in Fig. 17.24.

The pressure-scanned Fabry–Perot interferometer can be demonstrated by directing a helium–neon laser beam towards the interferometer. A pattern of sharp and bright interference fringes is observed on the screen. If the air pressure in the space between the plates is kept constant, then the pattern on the screen is stationary. As the air pressure is varied (by rubber hand pumping), the variation of the refractive index brings about a shift of the fringes on the screen.

17.10 Fourier spectroscopy

Using the Michelson scanning interferometer, one can measure the correlation function of optical radiation (see Chapter 11). Performing Fourier transformation of the correlation function, one can obtain the radiation spectrum. This is the main idea behind Fourier spectroscopy, which is applied successfully in infrared high-resolution spectroscopy.

The effective spectral resolution of the Fourier spectrometer can be estimated as $\delta\omega = 2\pi/\tau$ or $\delta\nu = 1/\Delta$, where τ is the maximum delay time and Δ is the maximum path difference between the rays in the interferometer. The resolving power is

$$A = \Delta/\lambda. \tag{17.83}$$

If, for example $\lambda = 0.5\ \mu\text{m}$ and $\Delta = 100$ cm, then we obtain $A = 2 \times 10^6$.

17.11 Light-beat spectroscopy

Even the best optical spectrometers, using the Fabry–Perot interferometer, cannot resolve two spectral lines if their frequencies differ by less than 10 MHz. There are, however, some problems for which a spectral resolution of 10 Hz to 10 MHz is required. These are, specifically, the problems of optical diagnostics in biology and medicine, including the measurement of diffusion coefficients, velocities of directed transport and migration motion, intramolecular and intracellular mobility parameters, etc. From these measurable characteristics, one can estimate the dimensions, mass, and other characteristics of biological microscopic objects. Light-beat spectroscopy is successfully used for solving similar problems.

The main idea of this method is as follows. Suppose that there are two monochromatic optical waves with close frequencies ω_1 and ω_2:

$$E_1 = A_1 \cos(\omega_1 t - k_1 z), \qquad E_2 = A_2 \cos(\omega_2 t - k_2 z).$$

The frequency difference between these waves, $\Delta\omega = \omega_1 - \omega_2$, can be measured by observing the intensity beatings of the resultant optical field $E = E_1 + E_2$. The intensity of the resultant optical wave is

$$I = \frac{c}{4\pi}\tilde{E^2} = I_1 + I_2 + 2\sqrt{I_1 I_2}\cos(\Delta\omega t - \Delta k z), \tag{17.84}$$

where $I_i = cA_i^2/8\pi$ are the intensities of individual waves ($i = 1, 2$) and $\Delta k = k_1 - k_2$. It can be seen from eqn (17.84) that the resultant field intensity is modulated by a frequency $\Delta\omega$ equal to the frequency difference between the optical waves. This modulation, provided that its frequency is not too high, can be measured by radio engineering methods.

The basic diagram of light-beat spectroscopy is shown in Fig. 17.25. Its main elements are a photodetector and a spectrum analyser. The photodetector, illuminated by optical waves, generates electric current proportional to the intensity of the incident light: $i(t) \sim I(t)$. Then the spectrum analyser produces the spectrum of the photocurrent oscillations.

The effective resolving power of light-beat spectroscopy, $A = \omega/\Delta\omega$, can be extremely high, being limited, in principle, only by the time of observation of the signal $I(t)$. Additionally, the smaller the value of $\Delta\omega$ is, the simpler is the measurement of the beat signal due to less strict requirements for the frequency response of the photodetector.

The first experiment on light-beat spectroscopy was reported by Forrester, Gudmundsen, and Johnson in 1955. They detected photomultiplier current beats caused by the splitting of the mercury spectral line (546.1 nm) into two components due to the Zeeman effect. The frequency difference between the components was estimated to be about 10^{10} Hz. The photodetector current contained an alternating component the frequency of which was also about 10^{10} Hz.

The invention of the laser gave an impetus to the further development of optical beat spectroscopy. At the beginning of the 1960s, some experiments were carried out that demonstrated that the laser light scattered by a moving object

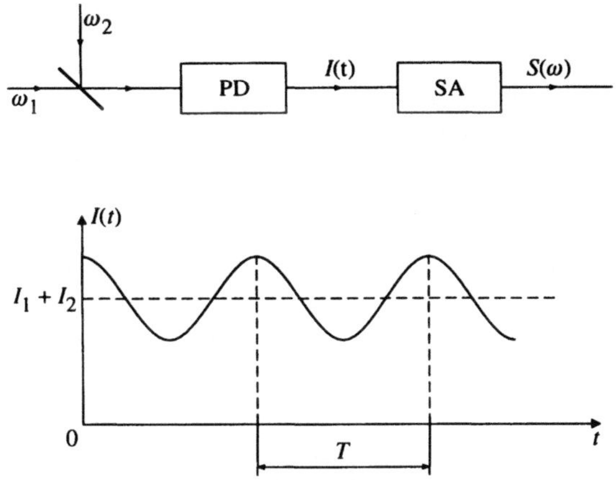

Fig. 17.25 Light-beat spectroscopy. PD, Photodetector; SA, spectrum analyser. The intensity beating period T is connected with the frequency difference between the optical waves $\Delta\omega$ by the formula $T = 2\pi/\Delta\omega$.

can be mixed with the original (non-shifted in frequency) light to generate a beat signal the frequency of which is equal to the Doppler frequency shift. For example, the light scattered by a moving liquid can be mixed using a photomultiplier with the reference optical signal. The frequency difference is the Doppler shift characterizing the flow velocity (laser Doppler anemometry).

Velocity measurements with a magnitude of about 0.1 $\mathrm{mm\,s^{-1}}$ has been reported. These data were obtained with the use of the laser Doppler microscope (Fig. 17.26). It is easy to show that the Doppler frequency shift corresponding to the indicated velocity is as small as 200 Hz (assuming an optical wavelength to be equal to 500 nm). Consequently, the effective resolving power reaches 3×10^{12}.

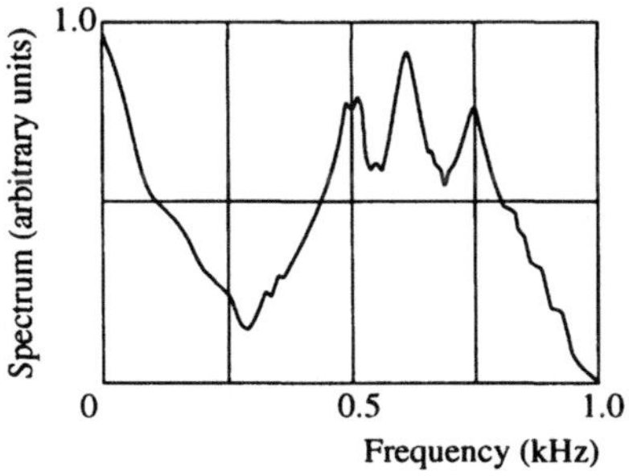

Fig. 17.26 An example of a spectrum obtained by light-beat spectroscopy. The central peak corresponds to zero velocity, and the two symmetric peaks correspond to flows in opposite directions (from Priezzhev, A. V., Proskurin, S. G., and Romanovsky, Yu. M. (1991). *Proceedings SPIE*, **1402**, 107–113).

Bibliography

Born, M. and Wolf, E. (1968). *Principles of optics.* Pergamon Press, Oxford.

Photon correlation and light beating spectroscopy. Ed. by H. Z. Cummins and E. R. Pike. Plenum, N. Y., (1974).

Demtröder, W. (1982). *Laser spectroscopy.* Springer-Verlag, Berlin.

Forrester, A. T., Gudmundsen, R. A., and Johnson, P. O. (1955). *Physical Review,* **99**, 1691–1700.

Gabor, D. (1948). *Nature,* **161**, 777–778.

Goodman, J. W. (1968). *Introduction to Fourier optics.* McGraw-Hill, San Francisco.

Leith, E. N. and Upatnieks, J. (1963). *Journal of the Optical Society of America,* **53**, 522.

Reynolds, G. O., De Velis, J. B., Parrent, G. B., and Thompson, B. J. (1989). *Physical optics notebook: tutorials in Fourier optics.* SPIE–AIP, New York.

Steward, E. G. (1985). *Fourier optics: an introduction.* John Wiley, New York.

Wood, R. W. (1923). *Physical optics.* Macmillan, New York.

Part IV

Light and matter

18

The physics of the interaction of light with matter

A model of a continuum and Maxwell's equations. The material equations. Classification of media. A plane monochromatic optical wave in a linear homogeneous isotropic medium. Complex dielectric permittivity, linear optical susceptibility, and the complex index of refraction. The classical oscillator model of a medium.

18.1 A model of a continuum and Maxwell's equations

The physical picture of the interaction of light with matter is as follows. An incident optical wave excites electron vibrations in atoms. Transferring its energy to the electrons, the light is absorbed, while the moving charges become sources of secondary light wavelets. The resultant optical field in the medium is formed due to interference between the incident radiation and the secondary optical waves.

Under normal conditions, the number of atoms is large and they are located so close to one another that the discrete nature of the medium does not manifest itself. The substance behaves as a *continuum* and characteristics such as polarization (dipole moment per unit volume), electric displacement, and magnetic induction can be introduced.

The theory of interaction of light with matter rests on the basis of the electromagnetic field equations (Maxwell's equations) and equations describing the properties of the medium (material equations).

The Maxwell equations have the form

$$\begin{aligned} \operatorname{rot}\vec{E} &= -\frac{1}{c}\frac{\partial \vec{B}}{\partial t}, \qquad && \operatorname{div}\vec{D} = 4\pi\rho, \\ \operatorname{rot}\vec{H} &= \frac{1}{c}\frac{\partial D}{\partial t} + \frac{4\pi}{c}\vec{j}, \qquad && \operatorname{div}\vec{B} = 0. \end{aligned} \tag{18.1}$$

Here $\vec{E}$ and $\vec{D}$ are the vectors of the electric field and the electric displacement, $\vec{H}$ and $\vec{B}$ are the vectors of the magnetic field and the magnetic induction, ρ is the electric charge density, $\vec{j}$ is the electric current density, and c is the electrodynamic constant, equal to the velocity of light in vacuum (see Chapter 1). Equations (18.1) are written in the so-called Gaussian system of units, in which electrical quantities ($\vec{E}$, $\vec{D}$, $\vec{j}$, and ρ) are expressed in electrostatic units, and magnetic quantities ($\vec{H}$ and $\vec{B}$) in electromagnetic units.

18.2 The material equations

Equations (18.1) are supplemented by the material equations:

$$\begin{aligned} \vec{D} &= \vec{E} + 4\pi\vec{P}, \qquad \vec{P} = \vec{P}(\vec{E}), \\ \vec{B} &= \vec{H} + 4\pi\vec{M}, \qquad \vec{M} = \vec{M}(\vec{H}), \\ \vec{j} &= \vec{j}(\vec{E}). \end{aligned} \tag{18.2}$$

These equations describe the response of the medium to the electromagnetic field, which is the appearance of polarization (electric dipole moment per unit volume) $\vec{P}$, magnetization (magnetic dipole moment per unit volume) $\vec{M}$, and the conduction current $\vec{j}$.

How do the polarization, magnetization, and induction current behave in a medium excited by an optical wave? This is a central question in the physics of the interaction between light and matter. It should be noted that many significant results in this field have been obtained during the past 30 years, i.e. since the advent of the laser. The physics of the interaction of light with matter remains one of the most dynamic branches of physical optics.

For radio-frequency fields of moderate strength, the material equations are usually written in the form

$$\vec{D} = \varepsilon\vec{E}, \qquad \vec{B} = \mu\vec{H}, \qquad \vec{j} = \sigma\vec{E}, \tag{18.3}$$

i.e. in the form adopted for static fields. Here ε is the dielectric permittivity, μ is the magnetic permeability, and σ is the specific conductivity of the medium. Equations (18.3) imply that the polarization and magnetization of the medium can be represented as

$$\vec{P} = \kappa_e\vec{E}, \qquad \vec{M} = \kappa_m\vec{H}, \tag{18.4}$$

According to eqns (18.2)–(18.4), the dielectric and magnetic susceptibilities of the medium, ε and μ, are related to κ_e and κ_m by the formulae

$$\varepsilon = 1 + 4\pi\kappa_e, \qquad \mu = 1 + 4\pi\kappa_m. \tag{18.5}$$

It follows from eqn (18.4) that the electric and magnetic responses of the medium are proportional to the electric and magnetic field strengths. From the physical point of view, eqns (18.4) imply that the medium response to the optical field is: (i) instantaneous, that is, the polarization and magnetization depend at any moment only on the instantaneous values of the fields; (ii) local, that is, the values of $\vec{E}$ and $\vec{H}$ at a given point in space are determined only by the values of the local fields; (iii) linear, that is, the values of $\vec{P}$ and $\vec{M}$ depend linearly on the field strengths; and (iv) isotropic, that is, the susceptibilities are described by scalars.

What about the interaction of the electromgnetic field and matter in optics? Can one use the material equations in the form (18.3) and (18.4) for an optical field? The detailed answer to this question will be given in Chapters 19–23. Here we only note that there are situations in which the assumptions formulated above are satisfied in optics too. Nevertheless, eqns (18.4) should not be regarded as universal. There are many cases in which the response of the medium to the optical field can not be adequately described by eqns (18.4). Such a situation takes place, for example, for resonant optical processes. If the optical field frequency approaches the frequency of a quantum transition in an atom, a molecule, or a condensed medium, then the inertia of the medium response becomes important, and the material equations are replaced by the formula

$$\vec{P}(t) = \int_0^\infty \kappa_e(\tau)\vec{E}(t-\tau)\, d\tau. \tag{18.6}$$

According to eqn (18.6), the polarization of the medium depends at any moment not only on the instantaneous value of the field, but also on all of its previous values, i.e. the polarization becomes history-dependent.

In the intense laser field, the medium response ceases to be linear with respect to the optical field strength, so that linear eqn (18.4) should be replaced by a power series in the electric field strength:

$$\vec{P} = \kappa_e\vec{E} + \chi^{(2)}\vec{E}\vec{E} + \chi^{(3)}\vec{E}\vec{E}\vec{E} + \dots . \tag{18.7}$$

Strictly speaking, expansion (18.7) should contain not only the electric, but also the magnetic field, so that in the general case $\vec{P} = \vec{P}(\vec{E}, \vec{H})$.

The coefficients $\chi^{(2)}$ and $\chi^{(3)}$ in eqn (18.7) are called the *non-linear optical susceptibilities of the medium.* These are the non-linear susceptibilities that often determine the behaviour of a powerful optical beam in the medium. When powerful laser radiation is resonant with the intrinsic vibrations of the atoms or molecules of the medium, then one needs to take into account both the non-linearity and the inertia of the medium. In this case the material equation becomes more complicated.

Finally, it should be noted that the assumption about the locality of the medium response is also an approximation. The response of a real physical system to an optical disturbance is generally not local; generalizing eqn (18.4) for the polarization one needs to write

$$\vec{P}(t,\vec{r}) = \iint_0^\infty \kappa(\tau,\vec{r},\vec{r}')\vec{E}(t-\tau,\vec{r}')\, d\tau\, d^3\vec{r}'. \tag{18.8}$$

This equation shows that the medium response at the moment of time t and at the point $\vec{r}$ is determined by the value of the field not only at $(t, \vec{r})$, but also by the values of the field at the preceding moments of time and at other (at least, nearby) points in space. There are a number of optical phenomena that are caused by non-locality of the response. The natural optical activity, i.e. the rotation of the polarization plane of a linearly polarized wave in the medium, is one such phenomenon.

18.2.1 The wave equation for light in a medium

From a variety of media, we shall single out a few which are the most important for optics. These are non-conducting ($\sigma = 0$), non-magnetic ($\mu = 1$), and neutral media, in which

$$\vec{j} = 0, \qquad \rho = 0, \qquad \vec{B} = \vec{H}. \tag{18.9}$$

For such media the material equation can be written as

$$\vec{D} = \vec{D}(\vec{E}), \tag{18.10}$$

and Maxwell's equations take the form

$$\begin{aligned} \operatorname{rot} \vec{E} &= -\frac{1}{c}\frac{\partial \vec{H}}{\partial t}, \qquad \operatorname{div} \vec{D} = 0, \\ \operatorname{rot} \vec{H} &= \frac{1}{c}\frac{\partial D}{\partial t}, \qquad \operatorname{div} \vec{H} = 0. \end{aligned} \tag{18.11}$$

Equations (18.9) define a wide class of media. For example, media such as air, water, crystals, glasses, plastic materials, etc., all belong to this class.

In view of the relation

$$\vec{D} = \vec{E} + 4\pi\vec{P}, \tag{18.12}$$

the material equation, equivalent to eqn (18.10), can be written in the form of the dependence of the polarization on the field strength:

$$\vec{P} = \vec{P}(\vec{E}). \tag{18.13}$$

Hereinafter we will use the term "material equation" to mean the equation of the form (18.13).

Using eqns (18.11) and (18.12), one can derive the following equation:

$$\operatorname{rot}\operatorname{rot} \vec{E} + \frac{1}{c^2}\frac{\partial^2 \vec{E}}{\partial t^2} = -\frac{4\pi}{c^2}\frac{\partial^2 \vec{P}}{\partial t^2}, \tag{18.14}$$

called the *wave equation* for light in the medium. According to this equation, the optical polarization of the medium is a source of the optical field. The polarization, in turn, is induced by the field of the optical wave falling on the medium. Since the polarization is an electric dipole moment per unit volume of the medium, the material equation is derived by solving the equation describing the motion of the charged particles of the medium. This is Newton's equation in the classical model and the Schrödinger equation in quantum theory. Equations (18.13) and (18.14) form a closed set of equations completely describing, when supplemented by appropriate initial and boundary conditions, the light propagation in a neutral, dielectric, and non-magnetic medium.

18.3 Classification of media

A formal classification of media can be given on the basis of the material equation (18.13). Media in which the dependence $\vec{P}(\vec{E})$ is local and instantaneous, i.e. the value of polarization of the medium at any point in space and at any moment of time is determined by the value of the field at the same point and at the same moment, are called *non-dispersive media*. Non-locality of the response of the medium to the optical field leads to *spatial dispersion* of the medium, while the inertia of the response, i.e. the time delay of $\vec{P}$ with respect to $\vec{E}$, leads to *temporal* or *frequency dispersion*.

Media for which the dependence of the polarization on the field is expressed by the linear operator

$$\vec{P} = \hat{L}(\vec{E}). \tag{18.15}$$

are called *linear media*. The operator $\hat{L}$ can be, for example, a linear retarded functional or a tensor operator. Correspondingly, the media for which the dependence of $\vec{P}$ on $\vec{E}$ is non-linear, i.e.

$$\vec{P} = \widehat{NL}(\vec{E}). \tag{18.16}$$

are called *non-linear media*.

Media in which the polarization vector is parallel to the vector of the field, i.e.

$$\vec{P} \parallel \vec{E}. \tag{18.17}$$

are called *isotropic media*. Finally, media the vector $\vec{P}$ of which is not parallel to the vector $\vec{E}$, i.e.

$$\vec{P} \nparallel \vec{E}. \tag{18.18}$$

are called *anisotropic media*. Note that the properties of dispersion, linearity, and isotropy are not interrelated. For example, a linear medium can be either dispersive or non-dispersive, and isotropic or anisotropic. The same is true for non-linear media.

A number of important properties of the medium can be established directly from the definitions given above, without specifying the model of the medium. Let us consider these properties.

18.4 A plane monochromatic optical wave in a linear homogeneous isotropic medium

Consider first a linear optical medium. Due to the linearity of Maxwell's equations and the material equation, the *optical superposition principle*, according to which different optical waves interact with the medium independently of one another, is applicable to linear media. The superposition principle remarkably simplifies the description of the interaction of light with the medium. Indeed, using the spectral expansion technique, any optical field can be represented as a set of plane monochromatic optical waves. In a linear medium, each of these waves propagates independently. It is sufficient, therefore, to consider the propagation of an elementary optical wave, that is, a plane monochromatic wave, in order to calculate the propagation of an arbitrary field, e.g. of an optical beam or a pulse, using the spectral expansion technique and the superposition principle.

Let a plane monochromatic optical wave

$$\vec{E} = \tfrac{1}{2}\vec{\mathcal{E}}\exp\left[i(\omega t - \vec{k}\vec{r})\right] + \text{c. c.} \tag{18.19}$$

propagate in a linear medium. Here ω is the frequency, $\vec{k}$ is the wave vector, and $\vec{\mathcal{E}}$ is the complex wave amplitude. Due to the linearity of eqns (18.11), (18.12), and (18.15), the magnetic field $\vec{H}$, the electric displacement $\vec{D}$, and the optical polarization $\vec{P}$ will also have the structure of plane monochromatic waves with frequency ω and wave vector $\vec{k}$:

$$\begin{aligned} \vec{H} &= \tfrac{1}{2}\vec{\mathcal{H}}\exp\left[i(\omega t - \vec{k}\vec{r})\right] + \text{c. c.}, \\ \vec{D} &= \tfrac{1}{2}\vec{\mathcal{D}}\exp\left[i(\omega t - \vec{k}\vec{r})\right] + \text{c. c.}, \\ \vec{P} &= \tfrac{1}{2}\vec{\mathcal{P}}\exp\left[i(\omega t - \vec{k}\vec{r})\right] + \text{c. c.}, \end{aligned} \tag{18.20}$$

where $\vec{\mathcal{H}}$, $\vec{\mathcal{D}}$, and $\vec{\mathcal{P}}$ are the complex wave amplitudes.

Substituting eqns (18.19) and (18.20) into Maxwell's equations (18.11), we obtain a set of algebraic equations for the complex amplitudes:

$$\begin{aligned} [-i\vec{k}, \vec{\mathcal{E}}] &= -(i\omega/c)\vec{\mathcal{H}}, \quad (-i\vec{k}, \vec{\mathcal{D}}) = 0, \\ [-i\vec{k}, \vec{\mathcal{H}}] &= (i\omega/c)\vec{\mathcal{D}}, \quad (-i\vec{k}, \vec{\mathcal{H}}) = 0 \end{aligned} \tag{18.21}$$

or

$$\begin{aligned} [\vec{k}, \vec{\mathcal{E}}] &= (\omega/c)\vec{\mathcal{H}}, \qquad (\vec{k}, \vec{\mathcal{D}}) = 0, \\ [\vec{k}, \vec{\mathcal{H}}] &= -(\omega/c)\vec{\mathcal{D}}, \qquad (\vec{k}, \vec{\mathcal{H}}) = 0. \end{aligned} \tag{18.22}$$

To eliminate $\vec{\mathcal{H}}$ from these equations, we shall multiply the first equation vectorially by $\vec{k}$. We thus obtain

$$\left[\vec{k}, [\vec{k}, \vec{\mathcal{E}}]\right] = \frac{\omega}{c}[\vec{k}, \vec{\mathcal{H}}] = -\frac{\omega^2}{c^2}\vec{\mathcal{D}}$$

or

$$\left[\vec{k}, [\vec{k}, \vec{\mathcal{E}}]\right] + \frac{\omega^2}{c^2}\vec{\mathcal{D}} = 0. \tag{18.23}$$

Expanding the double vector product, we obtain

$$(\vec{k}, \vec{\mathcal{E}})\vec{k} - k^2\vec{\mathcal{E}} + \frac{\omega^2}{c^2}\vec{\mathcal{D}} = 0. \tag{18.24}$$

This equation is valid for both isotropic and anisotropic linear media. If the optical wave propagates in vacuum, then $\vec{\mathcal{D}} = \vec{\mathcal{E}}$ and it follows from eqns (18.22) and (18.24) that $k = \omega/c$.

In an isotropic medium, the vector $\vec{P}$ is parallel to the vector $\vec{E}$. Then, as follows from eqn (18.12), the vector $\vec{D}$ is also parallel to the vector $\vec{E}$, and the Maxwell equation $\operatorname{div}\vec{D} = 0$ leads to the equation $\operatorname{div}\vec{E} = 0$, whence

$$\operatorname{rot}\operatorname{rot}\vec{E} = \operatorname{grad}\operatorname{div}\vec{E} - \Delta\vec{E} = -\Delta\vec{E}. \tag{18.25}$$

Taking into account eqn (18.25), we transform the general wave equation (18.14) to the form

$$\Delta\vec{E} - \frac{1}{c^2}\frac{\partial^2\vec{E}}{\partial t^2} = \frac{4\pi}{c^2}\frac{\partial^2\vec{P}}{\partial t^2}. \tag{18.26}$$

This equation is valid for both linear and non-linear isotropic media and for any optical field, but not only for plane monochromatic waves.

18.5 Complex dielectric permittivity, linear optical susceptibility, and the complex index of refraction

Let us return to eqn (18.24) and apply it to the case of propagation of a plane monochromatic optical wave in a linear isotropic medium. In such a medium, the vectors $\vec{E}$, $\vec{P}$, and $\vec{D}$ must be parallel. Then the complex amplitude vectors $\vec{\mathcal{E}}$, $\vec{\mathcal{P}}$, and $\vec{\mathcal{D}}$ which do not depend on time must be also parallel. This suggests that

$$\vec{\mathcal{D}} = \varepsilon(\omega)\vec{\mathcal{E}} \tag{18.27}$$

and

$$\vec{\mathcal{P}} = \kappa(\omega)\vec{\mathcal{E}}, \tag{18.28}$$

where $\varepsilon(\omega)$ and $\kappa(\omega)$ are scalars which may depend on the frequency of light. The quantity $\varepsilon(\omega)$, which is the coefficient of proportionality between the complex amplitudes of the electric displacement and the electric field strength, is called the *complex dielectric permittivity* of the medium. The quantity $\kappa(\omega)$, defined as the coefficient of proportionality between the complex amplitudes of polarization and the field, is called the *linear optical susceptibility* of the medium.

Inserting eqn (18.27) into eqn (18.22) $(\vec{k}, \vec{\mathcal{D}}) = 0$, we obtain $(\vec{k}, \vec{\mathcal{E}}) = 0$. Equation (18.24) then takes the form

$$-k^2\vec{\mathcal{E}} + (\omega^2/c^2)\varepsilon(\omega)\vec{\mathcal{E}} = 0, \tag{18.29}$$

whence

$$k^2 = (\omega^2/c^2)\varepsilon(\omega) \tag{18.30}$$

or

$$k = (\omega/c)n(\omega), \qquad (18.31)$$

where

$$n(\omega) = \sqrt{\varepsilon(\omega)}. \qquad (18.32)$$

Here, the quantity $n(\omega)$ is called the *complex index of refraction.*

Equation (18.31), relating the frequency ω and the wave number k of the optical wave, is called the *dispersion equation.* This equation is valid for linear isotropic media. If the optical wave propagates in vacuum, then $\varepsilon = n = 1$ and from eqn (18.31) we obtain $k = \omega/c$.

Equation (18.31) allows one to write the wave vector for an optical wave propagating in the medium as follows:

$$\vec{k} = \frac{\omega}{c} n\vec{m}, \qquad (18.32_1)$$

where $\vec{m} = \vec{k}/k$ is the unit vector directed along the $\vec{k}$ vector.

18.5.1 The intensity of an optical wave propagating in a medium

The intensity of an optical wave propagating in vacuum or in a medium is determined by the mean value of the Poynting vector (see Chapter 3), $I = \langle S \rangle$, where

$$\vec{S} = \frac{c}{4\pi}[\vec{E}, \vec{H}],$$

and the angular brackets denote averaging over an optical vibration period. Using eqns (18.19) and (18.20), we can express $\langle \vec{S} \rangle$ via the complex amplitudes of the electric and magnetic fields:

$$\langle \vec{S} \rangle = \frac{c}{16\pi}[\vec{\mathcal{E}}^*, \vec{\mathcal{H}}] + \text{c. c.} \qquad (18.32\text{b})$$

It follows from eqns (18.22) and (18.32a) that

$$\vec{\mathcal{H}} = n[\vec{m}, \vec{\mathcal{E}}]. \qquad (18.32\text{c})$$

Substituting eqn (18.32c) into eqn (18.32b), we obtain

$$\langle \vec{S} \rangle = \frac{cn}{8\pi}|\mathcal{E}|^2 \vec{m},$$

the intensity being given by

$$I = \frac{cn}{8\pi}|\mathcal{E}|^2. \qquad (18.32\text{d})$$

The last equation defines the light intensity in a transparent linear isotropic medium. Here $\mathcal{E}$ is the complex amplitude of the electric field of the optical wave, n is the index of refraction, which is real in the transparent region of the medium (see Chapter 19), and c is the velocity of light in vacuum. If $n = 1$, eqn (18.32d) transforms into eqn (3.28) for the light intensity in vacuum.

18.5.2 The frequency transmission function and the Green function for a linear isotropic medium

Earlier, we established the relationship existing between the polarization and the field when a plane monochromatic wave propagates in a linear isotropic medium. Let us now generalize the calculation to optical signals of arbitrary form.

Consider a fixed point of the medium. Assuming locality of the medium response to the optical field, we shall write the field strength and polarization in the form of the Fourier integrals:

$$E(t) = \frac{1}{2\pi} \int_{-\infty}^{\infty} E(\omega) e^{i\omega t} d\omega, \tag{18.33}$$

$$P(t) = \frac{1}{2\pi} \int_{-\infty}^{\infty} P(\omega) e^{i\omega t} d\omega. \tag{18.34}$$

Due to the linearity of the medium, the spectral amplitudes of the polarization and the field are connected by the relation

$$P(\omega) = \chi(\omega) E(\omega), \tag{18.35}$$

where $\chi(\omega)$ is the frequency transmission function of the given linear system, or, in terms of optics, the linear optical susceptibility of the medium. Substituting eqn (18.35) into eqn (18.34), and applying the Fourier inverse transformation

$$E(\omega) = \int_{-\infty}^{\infty} E(t) e^{-i\omega t} dt, \tag{18.36}$$

which follows from eqn (18.33), we obtain

$$P(t) = \int_{-\infty}^{\infty} \chi(\tau) E(t - \tau)\, d\tau, \tag{18.37}$$

where

$$\chi(\tau) = \frac{1}{2\pi} \int_{-\infty}^{\infty} \chi(\omega) e^{i\omega\tau} d\omega. \tag{18.38}$$

Thus, the problem is solved. Specifying the form of the functions $E(t)$ and $\chi(\omega)$ and making use of eqns (18.37) and (18.38), one can calculate the function $P(t)$. If, in particular, $E(t)$ has the form of a very short pulse, i.e.

$$E(t) = \text{constant}\, \delta(t), \tag{18.39}$$

where $\delta(t)$ is the delta function, we find from (18.37) that

$$P(t) = \text{constant}\, \chi(t). \tag{18.40}$$

This result shows that the function $\chi(\tau)$ in eqn (18.37) describes the response of the system to the action of an infinitely short pulse. The function $\chi(\tau)$ is called the *pulse response function* or the *Green function* of the given linear system. As can be seen from eqn (18.38), the Green function is related to the frequency transmission function by the Fourier transformation. This is a general property of all linear systems.

Applying the Fourier inverse transformation, we obtain from eqn (18.38)

$$\chi(\omega) = \int_{-\infty}^{\infty} \chi(\tau) e^{-i\omega\tau} d\tau. \tag{18.41}$$

Since the response of the system cannot precede the action (causality principle), it follows from eqns (18.39) and (18.40) that

$$\chi(t < 0) = 0, \tag{18.42}$$

i.e. the Green function is non-zero only for positive values of its argument. This allows eqns (18.37) and (18.41) to be rewritten as

$$P(t) = \int_{0}^{\infty} \chi(\tau) E(t-\tau)\, d\tau, \tag{18.43}$$

$$\chi(\omega) = \int_{0}^{\infty} \chi(\tau) e^{-i\omega\tau} d\tau. \tag{18.44}$$

According to eqn (18.43), the response of a linear system is a linear retarded functional with respect to the input signal. Consequently, eqn (18.43) takes into account the inertia of the linear system.

Thus, the material equation of a linear isotropic medium generally has the form of eqn (18.43). In this equation, $\chi(\tau)$ is the Green function, which depends only on the properties of the medium and is related to the linear optical susceptibility of the medium $\chi(\omega)$ by eqn (18.38).

18.6 The classical oscillator model of a medium

Consider an homogeneous linear isotropic medium. The optical polarization of the medium $\vec{P}$ can then be written as

$$\vec{P} = N\vec{p}, \tag{18.45}$$

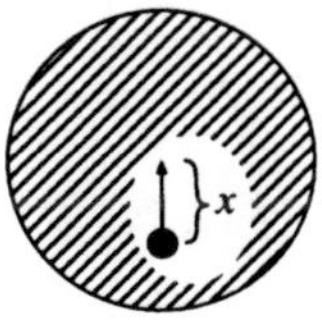

Fig. 18.1 The classical model of the atom.

where N is the number of atoms per unit volume, and $\vec{p}$ is the dipole moment of an individual atom. By definition,

$$\vec{p} = e\vec{x}, \tag{18.46}$$

where e is the electron charge, and $\vec{x}$ is the electron displacement with respect to the atomic nucleus (Fig. 18.1). The displacement of the electron relative to the nucleus is caused by the optical field and, consequently, $\vec{x} = \vec{x}(\vec{E})$. To find the specific form of this dependence, we need concretely to define the model of the atom.

Let us consider the classical oscillator model of the medium proposed by Lorentz. According to this model, the atom is an harmonic oscillator obeying the equation

$$\ddot{\vec{x}} + \Gamma\dot{\vec{x}} + \omega_0^2\vec{x} = \frac{e}{m}\vec{E}. \tag{18.47}$$

Here m is the electron mass, ω_0 is the intrinsic vibration frequency of the electron in the atom, and the parameter Γ describes the vibration decay. By substituting eqn (18.19) into eqn (18.47) we obtain the equation

$$\ddot{\vec{x}} + \Gamma\dot{\vec{x}} + \omega_0^2\vec{x} = \frac{e}{m}\tfrac{1}{2}\vec{\mathcal{E}}e^{i(\omega t - \vec{k}\vec{r})} + \text{c. c.}, \tag{18.48}$$

where $\vec{r}$ is the radius vector of the electron. Since the dimension a of the atom is much smaller than the optical wavelength λ, we can estimate $\vec{k}\vec{r} \approx ka = 2\pi a/\lambda \ll 1$. This relation allows one to consider the atom as a point object.

Let us calculate the polarization of the medium in the Lorentz model. The solution to eqn (18.48) can be written in the general form

$$\vec{x} = \tfrac{1}{2}\tilde{\vec{x}}\exp\left[i(\omega t - \vec{k}\vec{r})\right] + \text{c. c.} \tag{18.49}$$

Substituting eqn (18.49) into eqn (18.48), we obtain

$$\tilde{\vec{x}} = \frac{e}{m}\frac{1}{\omega_0^2 - \omega^2 + i\omega\Gamma}\vec{\mathcal{E}}. \tag{18.50}$$

According to eqns (18.45), (18.46), and (18.49), the polarization of the medium is

$$\vec{P} = Ne\vec{x} = Ne\tfrac{1}{2}\tilde{\vec{x}}\exp\left[i(\omega t - \vec{k}\vec{r})\right] + \text{c. c.} \tag{18.51}$$

It follows from eqns (18.51) and (18.20) that

$$\vec{\mathcal{P}} = Ne\tilde{\vec{x}} \tag{18.52}$$

or, taking into account eqn (18.50),

$$\vec{\mathcal{P}} = \frac{Ne^2}{m}\frac{1}{\omega_0^2 - \omega^2 + i\omega\Gamma}\vec{\mathcal{E}}. \tag{18.53}$$

Comparing eqns (18.28) and (18.53), we find the following expression for the optical susceptibility of the medium in the context of the Lorentz model:

$$\kappa(\omega) = \frac{Ne^2}{m}\frac{1}{\omega_0^2 - \omega^2 + i\omega\Gamma}. \tag{18.54}$$

It follows from eqns (18.12), (18.19), and (18.20) that

$$\vec{\mathcal{D}} = \vec{\mathcal{E}} + 4\pi\vec{\mathcal{P}}. \tag{18.55}$$

Taking into account eqns (18.27), (18.28), and (18.55), we can write

$$\varepsilon(\omega) = 1 + 4\pi\kappa(\omega) \tag{18.56}$$

or, in view of eqn (18.54),

$$\varepsilon(\omega) = 1 + \frac{4\pi Ne^2}{m}\frac{1}{\omega_0^2 - \omega^2 + i\omega\Gamma}. \tag{18.57}$$

Equation (18.57) defines the complex dielectric permittivity of the medium in the Lorentz model.

The expression for the linear optical susceptibility can be derived by an alternative method without recourse to the harmonic optical wave model. Indeed, multiplying eqn (18.47) by Ne, we obtain the following equation relating the polarization and the field:

$$\ddot{\vec{P}} + \Gamma\dot{\vec{P}} + \omega_0^2\vec{P} = \frac{Ne^2}{m}\vec{E}. \tag{18.58}$$

Since the medium is supposed to be isotropic, we set

$$\vec{E} = \vec{e}E, \qquad \vec{P} = \vec{e}P \tag{18.59}$$

and write the scalar equation

$$\ddot{P} + \Gamma \dot{P} + \omega_0^2 P = \frac{Ne^2}{m} E. \tag{18.60}$$

It is easy to find the solution to this equation for the arbitrary function $E(t)$. Employing the spectral approach and substituting eqns (18.33) and (18.34) into eqn (18.60), we shall obtain the following equation relating the spectral amplitudes of the polarization and the field:

$$P(\omega) = \frac{Ne^2}{m} \frac{1}{\omega_0^2 - \omega^2 + i\omega\Gamma} E(\omega). \tag{18.61}$$

Comparing eqns (18.35) and (18.61), we find that

$$\chi(\omega) = \frac{Ne^2}{m} \frac{1}{\omega_0^2 - \omega^2 + i\omega\Gamma}. \tag{18.62}$$

which coincides with eqn (18.54). Substituting eqn (18.62) into eqn (18.38) and integrating, we obtain

$$\chi(\tau) = \frac{Ne^2}{m} \begin{cases} 0, & \tau < 0, \\ \dfrac{1}{\Omega} e^{-\gamma\tau} \sin \Omega\tau, & \Omega = \sqrt{\omega_0^2 - \gamma^2}, \quad \gamma = \Gamma/2. \end{cases} \tag{18.63}$$

Equation (18.63) determines the pulse response function of the medium in the Lorentz model. Note that eqn (18.63) obeys the condition (18.42). The sought-after general solution to eqn (18.60) is given by eqns (18.43) and (18.63).

Thus, we have derived the material equation for a linear isotropic medium consisting of microscopic harmonic oscillators. It will be shown later (see Chapter 19) that this model allows one to explain a number of optical phenomena associated with light dispersion and absorption.

Bibliography

Bloembergen, N. (1965). *Nonlinear optics.* W. A. Benjamin, New York.

Born, M. and Wolf, E. (1968). *Principles of optics.* Pergamon Press, Oxford.

Zernike, F. and Midwinter, J. E. (1973). *Applied nonlinear optics*, John Wiley, New York.

19

Optical waves in linear isotropic media

The dispersion and absorption of light in a linear isotropic medium. The facts confirming Lorentz dispersion theory. Methods of investigation of optical dispersion and absorption. Optical spectroscopy. The propagation of an optical pulse in a dispersive medium.

19.1 The dispersion and absorption of light in a linear isotropic medium

The interaction of light with a linear isotropic medium brings about two main physical effects: (i) light absorption and (ii) a decrease in the velocity of light. We have considered these effects earlier, treating the medium as an ensemble of harmonic oscillators (see Chapter 7). Now we shall discuss the phenomena of dispersion and absorption of light in terms of the continuum model.

The optical properties of a linear isotropic medium can be characterized by the complex index of refraction n (see Chapter 18). Let us introduce the real and imaginary parts of the complex index of refraction, as

$$n = n' - in''. \tag{19.1}$$

Substituting eqn (19.1) into eqn (18.31), we obtain the following expression for the wave number:

$$k = k' - ik'', \tag{19.2}$$

where

$$k' = \omega n'/c, \qquad k'' = \omega n''/c \tag{19.3}$$

are the real and imaginary parts of the wave number. Let us make the physical meanings of these quantities clear. Suppose that an optical wave falls normally on the medium boundary. The optical wave in the medium can then be written in the form

$$E = \tfrac{1}{2}\mathcal{E}\exp[i(\omega t - kz)] + \text{c. c.} \tag{19.4}$$

or, in view of eqn (19.2),

$$E = \tfrac{1}{2}\mathcal{E}\exp(-k''z)\exp[i(\omega t - k'z)] + \text{c. c.} \tag{19.5}$$

It follows from eqns (19.1)–(19.5) that the imaginary part of the refractive index determines the damping and its real part the velocity of the optical wave in the medium. Indeed, writing the phase velocity of light as $v_{\text{ph}} = \omega/k'$, we obtain

$$v_{\text{ph}} = c/n'. \tag{19.6}$$

Since $n' = n'(\omega)$, the phase velocity depends on the light frequency. This phenomenon is called *dispersion*. Equation (19.5) also implies light absorption by the medium (the Bouguer law):

$$I = I_0 \exp(-\delta z), \tag{19.7}$$

where the absorption coefficient δ is

$$\delta = 2k'' = 2\omega n''/c. \tag{19.8}$$

Using eqn (18.32), we can express the index of refraction n' and the absorption coefficient δ in terms of the complex dielectric permittivity of the medium ε:

$$n'(\omega) = \text{Re}\sqrt{\varepsilon(\omega)}, \tag{19.9}$$

$$\delta(\omega) = -2(\omega/c)\text{Im}\sqrt{\varepsilon(\omega)}. \tag{19.10}$$

Here ω is the frequency of the optical wave and c is velocity of light in vacuum.

Let us calculate the refractive index and the absorption coefficient, using the expression for $\varepsilon(\omega)$ obtained in the context of the Lorentz model (see Chapter 18). We shall introduce the designation

$$\omega_p = \sqrt{4\pi N e^2/m} \tag{19.11}$$

and rewrite eqn (18.57) in the form

$$\varepsilon(\omega) = 1 + \frac{\omega_p^2}{\omega_0^2 - \omega^2 + i\omega\Gamma}. \tag{19.12}$$

Here the parameter ω_p has the dimensions of frequency and is called the *plasma frequency*. Separating the real and imaginary parts of ε, we transform eqn (19.12) into the form

$$\varepsilon(\omega) = 1 + \frac{\omega_p^2(\omega_0^2 - \omega^2 - i\omega\Gamma)}{(\omega_0^2 - \omega^2)^2 + \omega^2\Gamma^2}. \tag{19.13}$$

Assume that the medium is rarified enough, i.e. that the value of N (the number of molecules per unit volume) is sufficiently small. Then, as can be seen from eqns (19.11) and (19.12), the value of ε differs little from unity

$$\varepsilon = 1 + \mu, \qquad |\mu| \ll 1, \tag{19.14}$$

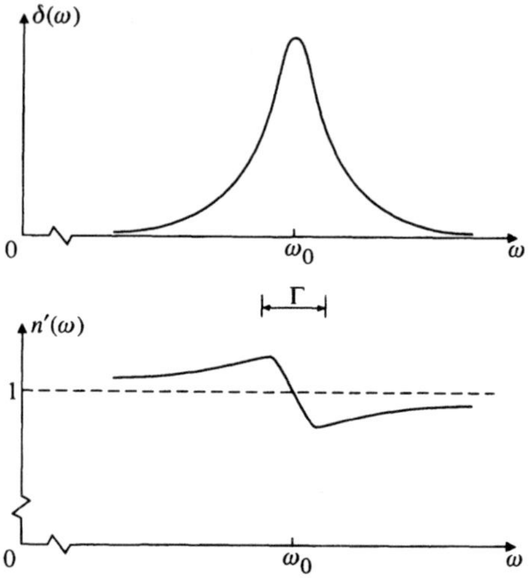

Fig. 19.1 The absorption and dispersion of light in a rarified linear isotropic medium.

and, consequently, $n = \sqrt{\varepsilon} \approx 1 + \mu/2$. In this approximation

$$n'(\omega) = 1 + \tfrac{1}{2}\frac{\omega_{\cdot}^2(\omega_0^2 - \omega^2)}{(\omega_0^2 - \omega^2)^2 + \omega^2\Gamma^2}, \tag{19.15}$$

$$\delta(\omega) = \frac{\omega_{\cdot}^2\omega^2\Gamma/c}{(\omega_0^2 - \omega^2)^2 + \omega^2\Gamma^2}. \tag{19.16}$$

Let us consider the frequency region near the resonance: $|\omega - \omega_0| \ll \omega_0$. Setting $\omega_0^2 - \omega^2 \approx 2\omega_0(\omega_0 - \omega)$, we obtain

$$n'(\omega) = 1 + \frac{\pi N e^2}{m\omega_0^2}\frac{(\omega_0 - \omega)}{(\omega_0 - \omega)^2 + \Gamma^2/4}, \tag{19.17}$$

$$\delta(\omega) = \frac{\pi N e^2}{mc}\frac{\Gamma}{(\omega_0 - \omega)^2 + \Gamma^2/4}. \tag{19.18}$$

Thus, eqns (19.17) and (19.18), obtained from the Lorentz model, describe the dispersion and absorption of light by a rarified linear isotropic medium. Note that the expression for the absorption coefficient (19.18) coincides with the expression (7.20) obtained earlier for the oscillator ensemble.

Usually the condition $\Gamma \ll \omega_0$ holds in optics. The typical dependencies $\delta(\omega)$ and $n'(\omega)$ for this case are shown in Fig. 19.1. As is evident from this figure, the spectral line of absorption has a Lorentzian form, its width is of the order of Γ, and its centre is at the frequency ω_0. The index of refraction $n'(\omega)$ differs noticeably from unity only near the absorption band.

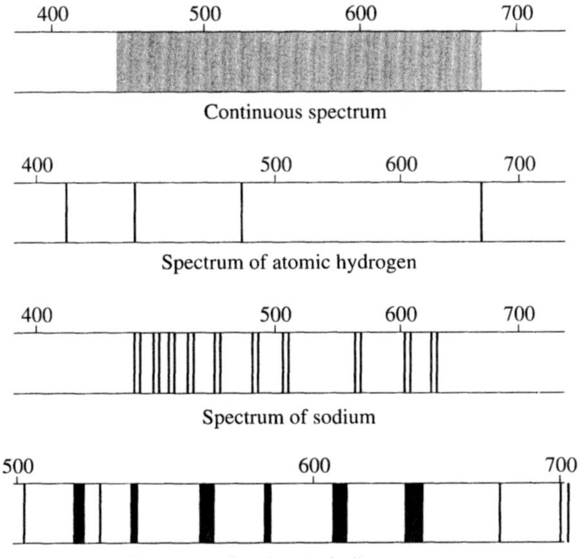

Fig. 19.2 Optical spectra.

19.1.1 Normal and anomalous dispersion

Depending on the sign of the derivative $\partial n/\partial\omega$, we can distinguish a frequency region of normal dispersion in which the index of refraction grows with frequency ($\partial n/\partial\omega > 0$), and a region of anomalous dispersion in which the index of refraction decreases with the frequency ($\partial n/\partial\omega < 0$). As can be seen from Fig. 19.1, the former region lies outside the absorption band, coinciding with the region in which the substance is transparent. By contrast, the much narrower region of anomalous dispersion is situated near the absorption line centre, i.e. in the region in which the absorption of light is a maximum.

19.1.2 Oscillator strength

As a rule, real atoms possess several spectral lines of absorption (Fig. 19.2). Denoting the frequencies and widths of these spectral lines correspondingly by ω_j and Γ_j, we can generalize eqn (18.57) for the complex dielectric permittivity of the medium:

$$\varepsilon(\omega) = 1 + \frac{4\pi N e^2}{m} \sum_j \frac{f_j}{\omega_j^2 - \omega^2 + i\omega\Gamma_j},$$

where the parameters f_j characterize the relative intensities of different spectral lines and are called the *oscillator strengths.*

Usually, the oscillator strengths are normalized so that their sum is equal to the number of electrons in the outer shell of the atom. For atoms of alkali

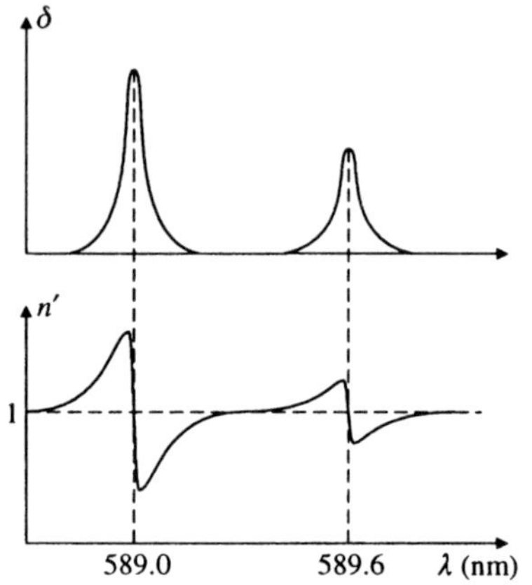

Fig. 19.3 The absorption coefficient and refractive index of sodium vapour near the D lines.

metals, having only one electron at the outer shell, the normalization condition takes the form

$$\sum_j f_j = 1.$$

For example, for two D lines of sodium, the oscillator strengths are

$$f(3S_{1/2}\text{–}3P_{1/2})=0.325,$$
$$f(3S_{1/2}\text{–}3P_{3/2})=0.65.$$

Their sum is close to unity, while the joint contribution from all other transitions is negligibly small. The dispersion and absorption of sodium vapour near the D lines are shown in Fig. 19.3. Experimental data on the characteristics of spectral lines of different atoms can be found in Corliss and Bozman (1962).

19.1.3 The refractive index of dense media

The theory of dispersion presented here assumes that the atom is subjected only to the field of the incident optical wave, while the effect of the fields generated by neighbouring atoms can be neglected. Such an approach is apparently valid only for sufficiently rarified media, in which the atoms are removed from one another. If, however, the distances between the atoms are not large, the field produced by neighbouring atoms can be comparable to the optical field. This circumstance calls for correction of the theory. We shall now generalize the theory of dispersion to media of arbitrary density.

It is known from electrodynamics that the effective field acting in a medium differs from the external field $\vec{E}$ by a value proportional to the polarization $\vec{P}$ of the medium:

$$\vec{E}_{\text{eff}} = \vec{E} + \frac{4\pi}{3}\vec{P}. \tag{19.19}$$

By replacing $\vec{E}$ with $\vec{E}_{\text{eff}}$ in eqn (18.47) describing the Lorentz model, we obtain

$$\ddot{\vec{x}} + \Gamma\dot{\vec{x}} + \omega_0^2\vec{x} = \frac{e}{m}\vec{E}_{\text{eff}}. \tag{19.20}$$

Equations (19.19) and (19.20) lead to the following equation for the polarization of the medium $\vec{P} = Ne\vec{x}$:

$$\ddot{\vec{P}} + \Gamma\dot{\vec{P}} + \omega_0^2\vec{P} = \frac{Ne^2}{m}\left(\vec{E} + \frac{4\pi}{3}\vec{P}\right). \tag{19.21}$$

Let us consider a plane monochromatic optical wave. Substituting eqns (18.19) and (18.20) into eqn (19.21), we obtain the following relation between the complex amplitudes of the polarization and the field:

$$(\omega_0^2 - \omega^2 + i\omega\Gamma)\vec{\mathcal{P}} = \frac{Ne^2}{m}\left(\vec{\mathcal{E}} + \frac{4\pi}{3}\vec{\mathcal{P}}\right). \tag{19.22}$$

From eqns (18.28) and (19.22) follows the equation for the optical susceptibility,

$$(\omega_0^2 - \omega^2 + i\omega\Gamma)\kappa(\omega) = \frac{Ne^2}{m}\left[1 + \frac{4\pi}{3}\kappa(\omega)\right], \tag{19.23}$$

while eqns (18.32), (18.56), and (19.23) yield the equation for the refractive index:

$$\frac{n^2-1}{n^2+2} = \frac{4\pi Ne^2}{3m}\frac{1}{\omega_0^2 - \omega^2 + i\omega\Gamma}. \tag{19.24}$$

Thus, eqn (19.24) takes into account that the field inside the medium differs from the field of the incident optical wave. The additional field arises in the medium due to its polarization, i.e. due to the displacement of the charged particles under the action of light. In the limiting case of an extremely thin medium, when the refractive index is close to unity, eqn (19.24) transforms into eqn (18.57). In the transmittance region defined by the condition $|\omega - \omega_0| \gg \Gamma$, eqn (19.24) takes the form

$$\frac{n^2-1}{n^2+2} = \frac{4\pi Ne^2}{3m}\frac{1}{\omega_0^2 - \omega^2} \tag{19.25}$$

and this is called the *Lorentz–Lorenz formula.*

19.2 The facts confirming Lorentz dispersion theory

The Lorentz classical theory of dispersion interprets, at least qualitatively, a number of optical effects. Let us consider the most important of them.

19.2.1 Refractivity

The Lorentz–Lorenz formula is suitable for experimental verification. Let ρ denote the density of the medium. Since $\rho \sim N$, it follows from eqn (19.25) that the quantity

Table 19.1 The molar refractivity of air at different pressures and at about 14.5°C for sodium D light.

Pressure (atm)	Refractive index	Refractivity (a.u.)
1.00	1.0002929	4.606
42.13	1.01241	4.661
96.16	1.02842	4.713
136.21	1.04027	4.743
176.27	1.05213	4.772

Table 19.2 The molar refractivity of different compounds for sodium D light.

	n (vapour)	n (liquid)	r (vapour) (a.u.)	r (liquid) (a.u.)
Oxygen	1.000271	1.221	4.05	4.00
Hydrochloric acid	1.000447	1.245	6.68	5.95
Water	1.000249	1.334	3.72	3.71
Carbon disulphide	1.00147	1.628	21.99	21.33
Acetone	1.00108	1.3589	16.16	16.14

$$r = \left(\frac{n^2 - 1}{n^2 + 2}\right)\frac{1}{\rho} \tag{19.26}$$

should not depend on density. The parameter r defined by eqn (19.26) is called the *specific refractivity of the substance*. Experiment shows that the specific refractivities of many substances (air, oxygen, water vapour, etc.) hardly depend on density. For example, the specific refractivity of air at the temperature 14.5°C varies only by 4% at the frequency of the sodium D line if the pressure grows from 1 atm to 176 atm. The specific refractivity is also found to remain practically constant when the gas (oxygen, water and acetone vapours, etc.) is condensed into liquid, i.e. when the density of the substance varies by a factor of about 1000. Thus, the dependence of refractive index on density, predicted by Lorentz theory, agrees well with experimental data (see Tables 19.1 and 19.2).

19.2.2 Normal dispersion

Lorentz theory predicts that transparent materials should have the normal dispersion

$$\frac{\partial n}{\partial \omega} > 0, \tag{19.27}$$

i.e. that the refractive index should grow with the frequency of light. In particular, at the medium boundary short waves have to be refracted more strongly than long waves. This is actually observed in all transparent media (air, water, glass, etc.). For example, in Newton's experiment (Fig. 1.6), the glass prism resolves the sunlight beam into its coloured components, with the short-wave radiation (violet colour) being refracted stronger than the long-wave radiation (red colour).

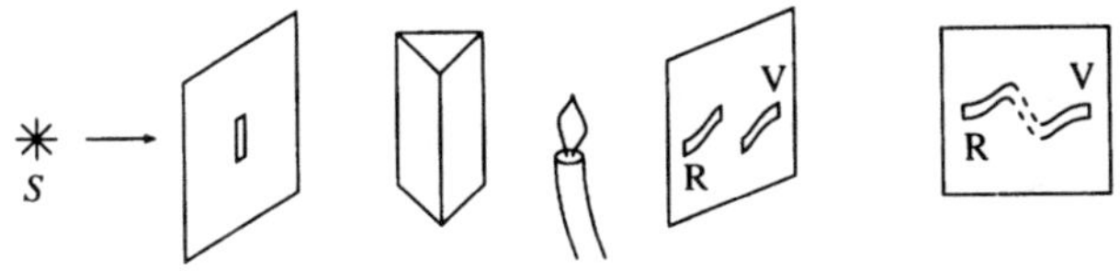

Fig. 19.4 The observation of anomalous dispersion of light in sodium vapour. The picture observed on the screen is shown on the right.

19.2.3 Anomalous dispersion

The existence of the anomalous dispersion predicted by Lorentz theory can also be confirmed by experiment. As can be seen from Fig. 19.1, anomalous dispersion takes place within the narrow frequency band near the absorption line centre. The experimental observation of anomalous dispersion is therefore a difficult problem.

In the experiment illustrated in Fig. 19.7, a white light beam from an arc lamp passes first through a glass prism with a vertical refracting edge, and then through the sodium burner flame, and falls on the screen. At the beginning of the experiment, we see a bright horizontal rainbow band on the screen, representing the spectrum of the arc lamp light. As the burner heats up and the sodium vapour density increases, a narrow dark band arises in the yellow region of the spectrum. This band corresponds to the sodium absorption line.

The spatial distribution of the sodium vapour density established in the heated sodium burner is characterized by vertical non-uniformity. The sodium vapour then starts to act like a prism with a horizontal refracting edge. As a result, the ends of the rainbow band near the absorption line bend and diverge, one upwards and the other downwards, as shown in Fig. 19.4. Such a deformation of the spectrum observed on the screen is interpreted as a manifestation of anomalous dispersion (compare Figs 19.1 and 19.4).

19.2.4 Dispersion in the low-frequency region

Let consider the dispersion of a thin gas ($n - 1 \ll 1$) in the limiting case of low frequencies ($\omega \ll \omega_0$). Using eqn (19.17), it is easy to derive the approximation

$$n - 1 = A(1 + B/\lambda^2), \tag{19.28}$$

where λ is the optical wavelength, and A and B are the constants.

Equation (19.28) is referred to as the *Cauchy dispersion formula*. Experimental measurements show that it adequately describes the dispersion of various gases in the visible and infrared regions of the spectrum. The values of the constants A and B for various gases are listed in Table 19.3. Experimentally measured values of the refractive index of air are compared in Table 19.4 with the corresponding values calculated from the Cauchy dispersion formula.

Table 19.3 The constants A and B of Cauchy's dispersion formula for various gases.

Gas	$A \times 10^5$	$B \times 10^9$ (cm^2)
Argon	27.92	5.6
Nitrogen	29.19	7.7
Helium	3.48	2.3
Hydrogen	13.6	7.7
Oxygen	26.63	5.07
Air	28.79	5.67
Ethane	73.65	9.08
Methane	42.6	14.41

Table 19.4 Observed values of the refractive index for air and the values given by Cauchy's dispersion formula.

$\lambda \times 10^5$ (cm)	$(n-1) \times 10^4$		Difference
	Observed	Calculated	
7.594	2.905	2.907	0.002
6.563	2.916	2.917	0.001
5.896	2.926	2.926	0.000
5.378	2.935	2.935	0.000
5.184	2.940	2.940	0.000
4.861	2.948	2.948	0.000
4.677	2.951	2.954	0.003
4.308	2.966	2.967	0.001
3.969	2.983	2.983	0.000
3.728	2.995	2.996	0.001
3.441	3.016	3.017	0.001
3.180	3.040	3.041	0.001
3.021	3.056	3.058	0.002
2.948	3.065	3.067	0.002

19.2.5 Dispersion in the high-frequency region

As can be seen from Fig. 19.1, the refractive index of a medium tends to unity in the limiting case of high frequencies. Assuming $n^2 + 2 = 3$ and $\omega \gg \omega_0$, we obtain, from eqn (19.25),

$$n^2 = 1 - \frac{\omega_\cdot^2}{\omega^2}, \tag{19.29}$$

where the frequency $\omega_\cdot$, called the plasma frequency, is defined by eqn (19.11): $\omega_\cdot = \sqrt{4\pi N e^2/m}$. Here e and m are the charge and mass of the electron, and N is the number of atoms per unit volume.

Let us estimate the frequency $\omega_\cdot$, assuming $N = 10^{22}$ cm^{-3}, which is typical of condensed media. Setting $e = 4.8 \times 10^{-10}$ CGSE and $m = 9.1 \times 10^{-28}$ g, we obtain the estimate $\lambda_{-} = 2\pi c/\omega_{-} = 0.3\ \mu\text{m}$. The plasma frequency for condensed medium thus lies in the ultraviolet region.

For the X-ray region, $\omega \gg \omega_{-}$ and eqn (19.29) yields the following estimate:

$$n = 1 - \delta, \qquad \delta \ll 1,$$

according to which the refractive index of the medium is close to unity. This means that in the X-ray region the medium possesses hardly any dispersion.

Experiment confirms this result. The X-rays progress through the substance and the boundaries between media almost without reflection and refraction. Physically, this is associated with the inertia of electrons, which are unable to follow the ultra-high-frequency oscillations of the electromagnetic field. The amplitude of forced vibration of electrons in such a field turns out to be very small; the radiation and the matter interact very weakly. It should be noted in this connection that this circumstance seriously complicates the development of instruments for X-ray optics to replace conventional prisms, lenses, and mirrors. In particular, it is very difficult to construct an X-ray resonator, a key element of an X-ray laser.

19.2.6 Spectral line shape

Spectral lines of real atoms and molecules are characterized by a great variety of shapes. Some experimentally observed spectral lines, however, have a simple Lorentzian shape. One such is, for example, the spectral line of Raman scattering in liquid nitrogen, shown in Fig. 19.5. This line corresponds to the vibrations of atomic nuclei in a nitrogen molecule. It has a frequency of 2326.5 cm^{-1} and a full width at half maximum of 0.058 cm^{-1}. The line is remarkable for having an almost perfect Lorentzian shape over a dynamic range of four orders of magnitude of the signal.

Thus, the primary argument in favour of the classical Lorentz model of dispersion is that its predictions agree well with experimental facts. Another argument is that this model is confirmed by rigorous quantum theory. Quantum theory argues that, for not too intense optical fields, that is, when the variation of energy level populations under the action of light is negligible, the dynamics of the atomic dipole moment are indeed described by the equation of the classical oscillator. Thus, the basic postulate of the Lorentz model proves to be justified.

19.3 Methods of investigation of optical dispersion and absorption—optical spectroscopy

The dependencies of the absorption coefficient and the refractive index on the wavelength of light convey valuable information about the structure and properties of atoms and molecules, as well as about the mechanisms of interaction of particles in the medium. The collection and analysis of this information constitutes the subject of *optical spectroscopy*.

The importance of spectral measurements in physics is well illustrated by the fact that the early spectroscopic investigations of a hydrogen atom and of

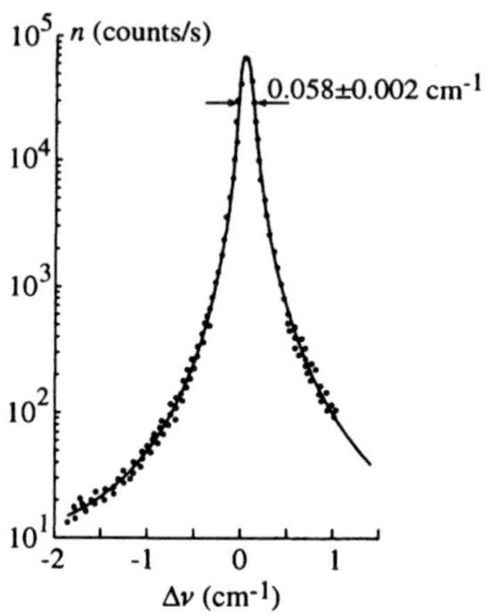

Fig. 19.5 A spectral line of Raman scattering obtained in liquid nitrogen with a high-resolution spectrometer (from Akhmanov, S. A. *et al.* (1980). *Applied optics*, **19**, 859). •, Experiment; —, the Lorentz curve.

radiation from heated bodies played a decisive part in the creation of quantum mechanics.

A simple way to measure the refractive index as a function of the wavelength, called a *crossed-prisms method*, was proposed by Newton. In this method a beam of white light passes first through a prism with a vertical refracting edge and then through a prism with a horizontal refracting edge, to form a curved coloured band on a screen (Fig. 19.6). The position and shape of this band provide information about the dependence $n(\lambda)$ for the material of both prisms.

The most sensitive methods of measurement of dispersion are based on using a spectral device (prism or diffraction grating) in combination with an interferometer. The substance under investigation is placed in the path of one of the beams of the interferometer. A schematic diagram of the measurement technique is shown in Fig. 19.7(a). A typical interference pattern observed in a similar ar-

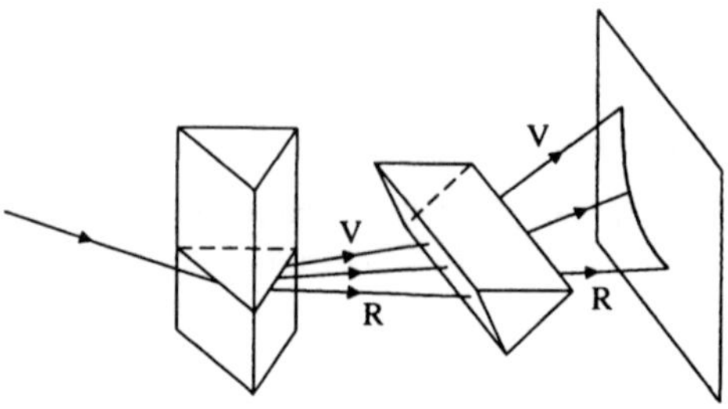

Fig. 19.6 The measurement of dispersion by the Newton method of crossed prisms.

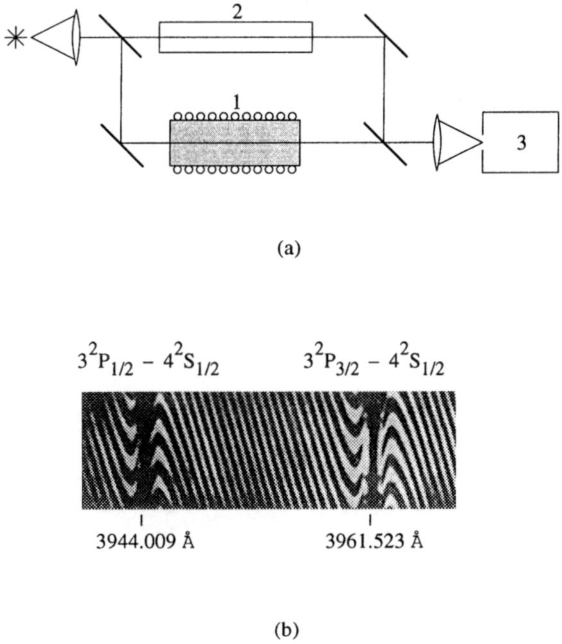

Fig. 19.7 The measurement of dispersion in atomic vapour. (a) The experimental set-up: 1, cuvette with vapour; 2, compensating cuvette; 3, spectrograph. (b) The interference pattern at different wavelengths λ near the resonant lines of aluminium.

rangement is shown in Fig. 19.7(b). In this pattern the interference fringes trace (at a certain scale) the dependence of $n - 1$ on λ, i.e. they directly produce the dispersion curve. These interference methods are especially suitable for the investigation of dispersion near the absorption bands of a substance.

An arrangement for the measurement of the absorption spectrum is schematically shown in Fig. 19.8. Its main elements are a source of light, a spectral device, and a recording system. The operation of the spectrometer can be demonstrated using the set-up shown in Fig. 19.9. A beam of light from an arc lamp passes first through a glass cuvette filled with water, then through a glass prism, and finally falls onto a screen. Initially, the observer can see on the screen a bright rainbow band in which all of the colours of the spectrum are present. If we add into the cuvette a solution of light-absorbing liquid (e.g. a solution of $KMnO_4$), part of the spectrum on the screen is replaced by dark lines corresponding to the absorption lines of the substance.

19.3.1 Laser spectroscopy

A narrow-band tunable laser is perfectly appropriate for optical spectroscopy. When such a source of light is applied, spectral devices become superfluous and measurement systems prove to be quite simple (Fig. 19.10).

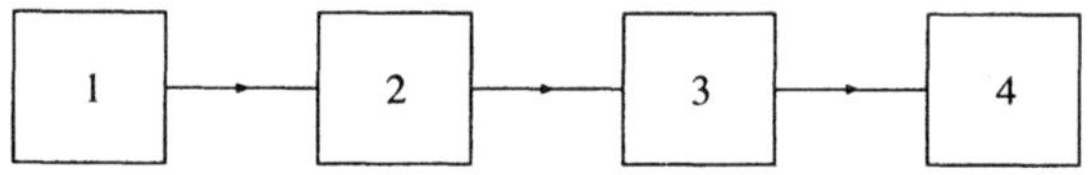

Fig. 19.8 A schematic diagram of the measurement of the absorption spectrum. 1, Source of light; 2, object under investigation; 3, spectral device; 4, recording system.

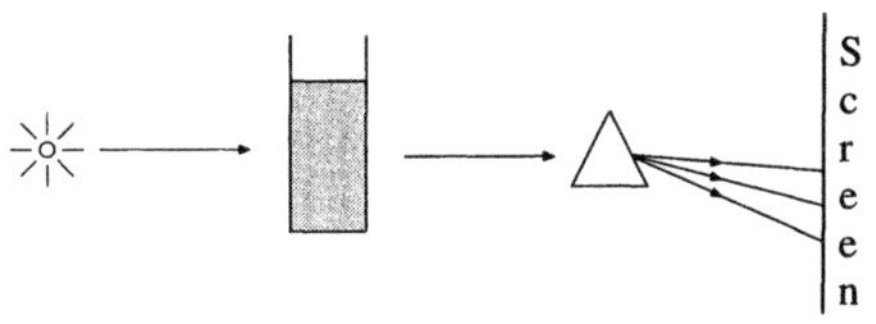

Fig. 19.9 Observation of the absorption spectrum of a $KMnO_4$–water solution.

The resolving power of a laser spectrometer is governed by the spectral bandwidth of the laser radiation. In practice, the resolving power can reach about 10^8. By way of illustration, we can refer to the spectrum of Raman scattering in liquid nitrogen shown in Fig. 19.5. This spectrum was obtained with the use of a continuous-wave tunable laser using an organic dye solution (Rhodamine 6G in ethylene glycol) as an active element. The dye laser is pumped by argon laser radiation. The radiation power is 200 mW, the spectral bandwidth of the radiation is 40 MHz, and the region of continuous tuning spreads from 570 to 620 nm. It should be noted that a laser of this type has an extremely narrow spectral bandwidth ($\sim$ 10 kHz) of radiation in the visible range and is one of the most promising instruments for high-resolution spectroscopy.

Radiation frequency control in the dye laser is attained using a dispersive element, i.e. a prism or diffraction grating, inserted in the laser resonator (Fig. 19.11).

The operation of a Rhodamine 6G dye laser can be demonstrated as follows. The second-harmonic generation of an $YAG:Nd^{3+}$ laser at a wavelength of 530 nm is used for pumping. The dye laser operates in the pulsed regime with a repetition rate of about 10 Hz. The dye laser beam is directed towards the screen. By rotating a micrometer screw, we slowly tune the laser resonator. Simultaneously, we observe the variation in the colour of the laser beam from yellow–green to orange.

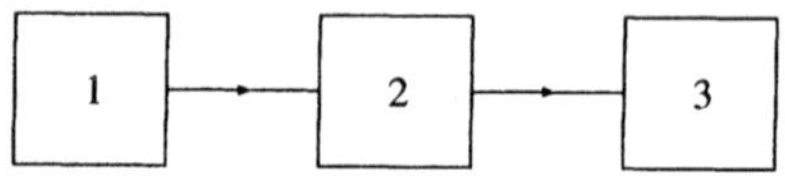

Fig. 19.10 Laser absorption spectroscopy. 1, Narrow-band tunable laser; 2, object under investigation; 3, recording system.

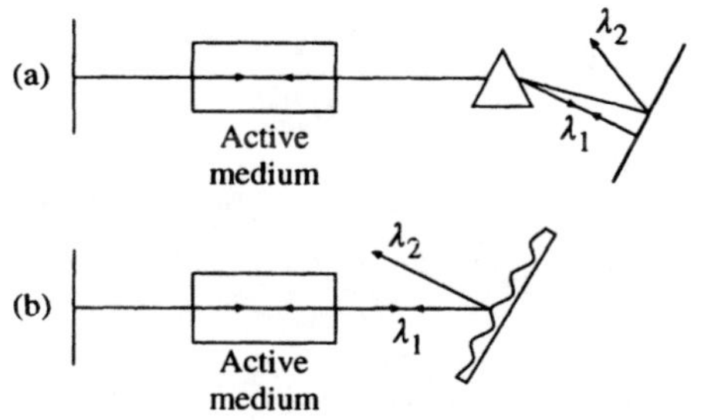

Fig. 19.11 Methods of dye laser frequency tuning: (a) with a dispersion prism; (b) with a diffraction grating.

In infrared spectroscopy tunable lasers of other types are used, as well as parametric oscillators of light; that is, special devices for light frequency control, based on non-linear optics effects (see Chapters 8, 22, and 23).

Apart from the standard arrangement shown in Fig. 19.10, there are a number of other methods based on specific properties of lasers and laser radiation. At present, one of the most sensitive spectroscopic techniques is provided by intracavity laser spectroscopy. In this method, a cuvette containing the substance under investigation is placed into the resonator of a tunable laser and the laser radiation power is measured as a function of the wavelength. Another highly sensitive method is called acousto-optical spectroscopy. In this method, the laser pulse passes through a cuvette containing a sample of the substance under investigation, in which part of the radiation is absorbed. Having absorbed the light, the atoms and molecules transfer to an excited state. Since this state is thermodynamically unstable, the particles immediately return to the ground state, contributing the energy excess to translational degrees of freedom. As a result, the laser pulse causes a rapid local heating of the absorbing medium and its expansion, accompanied by the generation of an acoustic wave. In other words, the absorption of the optical pulse is accompanied by a short click. This click is picked up by a sensitive microphone built into the cuvette. When the laser frequency is tuned, the signal from the microphone describes the absorption spectrum.

There are other methods of laser spectroscopy, such as Fourier spectroscopy (see Chapter 11), light-beat spectroscopy (see Chapter 17), time-domain (non-stationary) spectroscopy (see Chapter 7), and the methods based on non-linear optical phenomena (see, for example, Bloembergen 1977). It should be noted that laser spectroscopy is progressing rapidly and finding interesting new applications in the areas of physics, chemistry, biology, medicine, and environmental monitoring. The methods and applications of laser spectroscopy are discussed in Demtröder (1982).

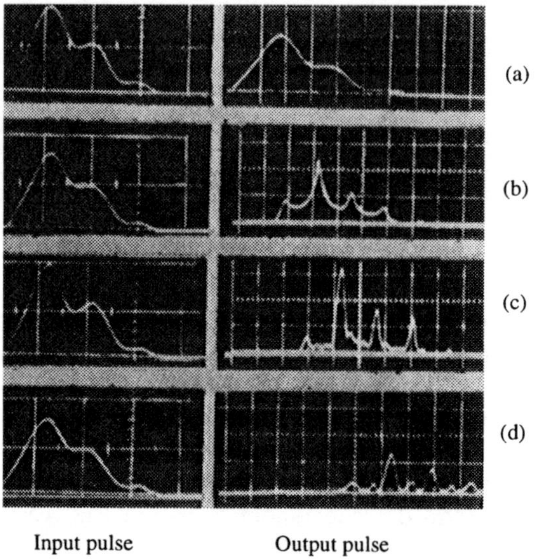

Fig. 19.12 The transformation of an envelope of short optical pulses propagating in a highly dispersive medium. The scale for input pulses is 10 ns/division and for output pulses 5 ns/division. The optical pulses generated by a tunable laser had a mean frequency close to the resonance frequency of the $^2P_{1/2}$ line in rubidium vapour. The off-resonance tuning decreases from (a) to (d) (from Grishkowsky, D. (1974). *Applied Physics Letters*, **25**, 566).

19.4 The propagation of an optical pulse in a dispersive medium

So far, we have assumed that the light propagating in the medium has the structure of a plane monochromatic wave. Now we shall generalize our considerations to an optical signal of an arbitrary form. The spatial structure of the field will be modelled, as before, by a plane wave.

From the physical point of view, the frequency dependence of the refractive index of the substance results in different spectral components of the optical field acquiring different phase shifts during the propagation of the wave through the medium. The result of summation of the optical vibrations at the exit from the medium naturally depends on the phase distribution. Consequently, the output pulse shape may differ from the input pulse shape. In other words, the optical pulse changes its shape during propagation in a dispersive medium. One may expect that the pulse deformation would be at its maximum at resonance, i.e. when the light frequency is close to the intrinsic frequency of the medium. The data presented in Fig. 19.12 confirm this conclusion.

Another example is a propagation of optical pulses in transparent optical fibres. Although the resonant effect is absent here, the pulse distortion can be significant due to the long length of the fibre. It should be noted here that the analysis of optical pulse distortion in a fibre is of practical interest in connec-

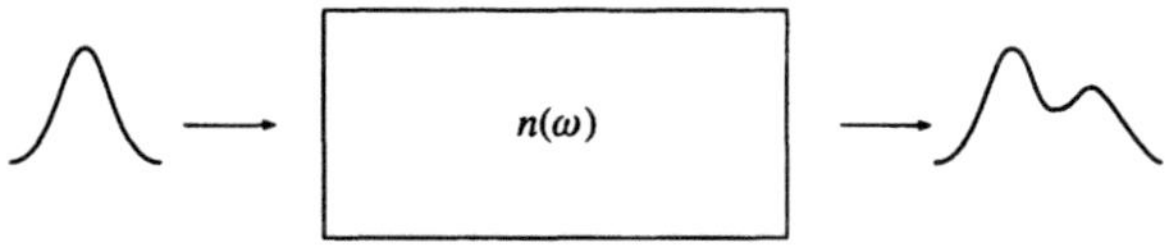

Fig. 19.13 The problem of propagation of an optical pulse in a dispersive medium.

tion with the problem of information transmission through optical fibres, the evaluation of the information capacity of communication channels, etc.

Let us calculate the shape of the optical pulse at the exit from a linear dispersive medium $E_{\text{out}}(t)$, assuming that the input pulse shape $E_{\text{in}}(t)$ and the complex refractive index of the medium $n(\omega)$ are known (Fig. 19.13).

The problem can be solved by the spectral approach. First, we shall expand the input pulse into a Fourier integral, describe the propagation through the medium of the individual monochromatic components of the field, and sum up the fields at the exit from the medium. This approach is based on the fact that the medium is linear and, consequently, each optical wave propagates through it independently of the others.

19.4.1 A monochromatic optical wave

As was shown earlier, the propagation of a plane monochromatic optical wave in a linear isotropic dispersive medium is described by the formula

$$\vec{E} = \tfrac{1}{2}\vec{\mathcal{E}}e^{i[\omega t - k(\omega)z]} + \text{c. c.}, \tag{19.30}$$

where $\vec{\mathcal{E}}$ is the complex amplitude of the field; $k(\omega)$ is the wave number depending on the light frequency ω and connected with the complex refractive index of the medium by the relation

$$k(\omega) = \frac{\omega}{c}n(\omega). \tag{19.31}$$

19.4.2 The narrow-band optical signal

Let us restrict our consideration to a narrow-band optical pulse, i.e. let us assume that the spectral bandwidth of the pulse is much less than its carrier frequency:

$$\Delta\omega \ll \omega_0, \tag{19.32}$$

as shown in Fig. 19.14. This means that the pulse is assumed to be much longer than the optical vibration period:

$$\tau_{\text{p}} \gg T_0. \tag{19.33}$$

In optics, $T_0 = 10^{-14}$–10^{-15} s and the condition (19.33) is therefore satisfied for pulses of duration $\tau_{\text{p}} \geq 10^{-13}$ s.

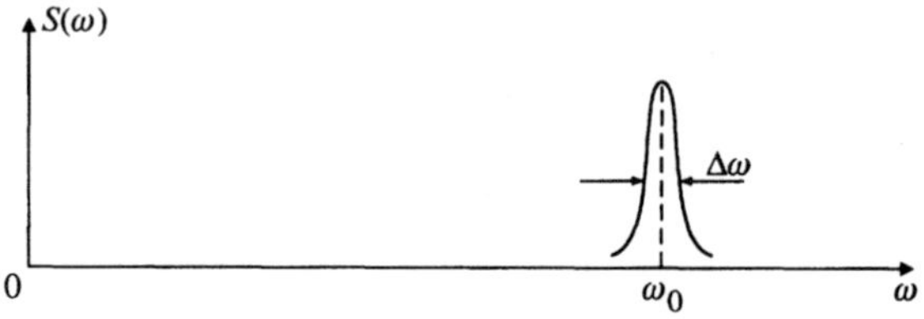

Fig. 19.14 The spectrum of a narrow-band optical signal.

Let us consider the propagation of a narrow-band optical pulse through a dispersive medium of length z. We shall write the input pulse in the form

$$\vec{E}_0(t) = \tfrac{1}{2}\vec{e}\mathcal{E}_0(t)\mathrm{e}^{i\omega_0 t} + \text{c. c.} \tag{19.34}$$

and the output signal as

$$\vec{E}(t) = \tfrac{1}{2}\vec{e}\mathcal{E}(t)\mathrm{e}^{i\omega_0 t} + \text{c. c.} \tag{19.35}$$

Expanding the amplitudes $\mathcal{E}_0(t)$ and $\mathcal{E}(t)$ into a Fourier integral, we have

$$\mathcal{E}_0(t) = \frac{1}{2\pi}\int_{-\infty}^{\infty} \mathcal{E}_0(\omega)\mathrm{e}^{i\omega t}d\omega, \tag{19.36}$$

$$\mathcal{E}(t) = \frac{1}{2\pi}\int_{-\infty}^{\infty} \mathcal{E}(\omega)\mathrm{e}^{i\omega t}d\omega. \tag{19.37}$$

The propagation process can then be described by the formula

$$\mathcal{E}(\omega) = \mathcal{E}_0(\omega)\mathrm{e}^{-ik(\omega)z}. \tag{19.38}$$

The spectral amplitude $\mathcal{E}_0(\omega)$ is expressed through the complex amplitude of the input signal $\mathcal{E}_0(t)$ by the Fourier inverse transformation

$$\mathcal{E}_0(\omega) = \int_{-\infty}^{\infty} \mathcal{E}_0(t)\mathrm{e}^{-i\omega t}dt. \tag{19.39}$$

Thus, the problem is solved. By specifying the form of the input signal amplitude $\mathcal{E}_0(t)$, one may find the amplitude of the output signal from eqns (19.37)–(19.39). Note that eqns (19.37)–(19.39) express the three successive stages of the solution: (i) the Fourier analysis of the input signal; (ii) transformation of the spectral component of the field by dispersive medium; and (iii) Fourier synthesis of the output signal.

19.4.3 The frequency transmission function and the Green function of the dispersive medium

The solution results become quite clear when expressed in terms of the general theory of linear systems.

Let us introduce the frequency transmission function of the dispersive medium:

$$\chi(\omega) = \mathrm{e}^{-ik(\omega)z}. \tag{19.40}$$

Equation (19.38) can then be rewritten in the form

$$\mathcal{E}(\omega) = \mathcal{E}_0(\omega)\chi(\omega). \tag{19.41}$$

Substituting eqns (19.41) and (19.39) into eqn (19.37), we obtain the following formula which directly relates the amplitudes of the input and output signals:

$$\mathcal{E}(t) = \int_{-\infty}^{\infty} \mathcal{E}_0(\theta)h(t-\theta)\,d\theta, \tag{19.42}$$

where

$$h(t) = \frac{1}{2\pi}\int_{-\infty}^{\infty} \chi(\omega)\mathrm{e}^{i\omega t}d\omega. \tag{19.43}$$

The function $h(t)$ in eqn (19.42) is called the *Green function* of the given linear system. Equation (19.43) shows that the frequency transmission function and the Green function are related by the Fourier transformation.

19.4.4 Models of dispersive media

Without specifying the form of the dispersion law (19.31), and assuming the condition (19.32) to be valid, we can expand the wave number in a power series of frequency in the vicinity of ω_0:

$$k(\omega) = k_0 + \omega k_0' + \tfrac{1}{2}\omega^2 k_0'' + \dots . \tag{19.44}$$

Here the frequency ω is determined with reference to the frequency ω_0, and the following designations are introduced:

$$k_0 = k(0), \quad k_0' = \left.\frac{\partial k}{\partial \omega}\right|_0, \quad k_0'' = \left.\frac{\partial^2 k}{\partial \omega^2}\right|_0. \tag{19.45}$$

Taking into account differing numbers of terms in the series (19.44), we obtain different models of the dispersive medium. The following models are of great importance.

Medium with first-order dispersion:

$$k(\omega) = k_0 + \omega k_0'. \tag{19.46}$$

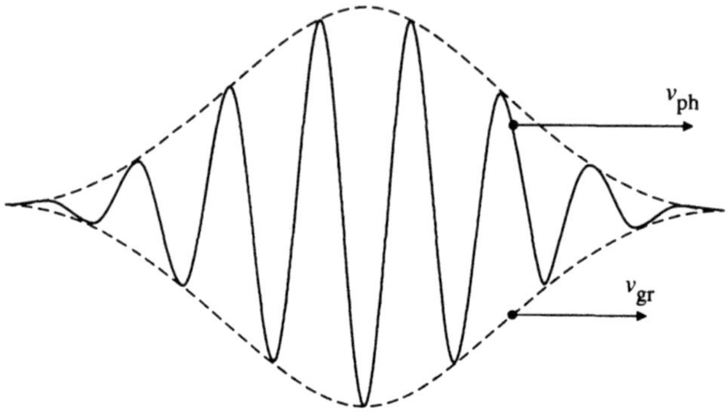

Fig. 19.15 The notion of the group velocity of light.

Medium with second-order dispersion:

$$k(\omega) = k_0 + \omega k_0' + \tfrac{1}{2}\omega^2 k_0''. \tag{19.47}$$

Let us consider each of these models separately.

19.4.5 A medium with first-order dispersion—the phase and group velocities of light

Substituting eqn (19.46) into eqns (19.40), (19.42), and (19.43), we obtain

$$h(t) = \delta(t - k_0' z)\mathrm{e}^{-ik_0 z}, \quad \mathcal{E}(t) = \mathcal{E}_0(t - k_0' z)\mathrm{e}^{-ik_0 z}$$

and

$$\vec{E}(t) = \tfrac{1}{2}\vec{e}\mathcal{E}_0(t - k_0' z)\mathrm{e}^{i(\omega_0 t - k_0 z)} + \text{c. c.} \tag{19.48}$$

Equation (19.48) shows that in a medium with first-order dispersion the optical pulse propagates without distortion, but the pulse propagation velocity

$$u = \frac{1}{k_0'} = \left.\frac{\partial \omega}{\partial k}\right|_0 \equiv v_{\text{gr}} \tag{19.49}$$

generally differs from the phase velocity of light in the given medium:

$$v = \frac{\omega_0}{k_0} = \frac{c}{n(\omega_0)} \equiv v_{\text{ph}}. \tag{19.50}$$

The carrier frequency sinusoid behaves as if being "driven" through the slower propagating envelope of the pulse (Fig. 19.15).

The velocity of motion of the pulse envelope in the dispersive medium defined by eqn (19.49) is called the *group velocity* of light. It is this quantity that determines the velocity of motion of the optical energy. The notion of the group velocity can be made clear by the simple example of a biharmonic optical wave

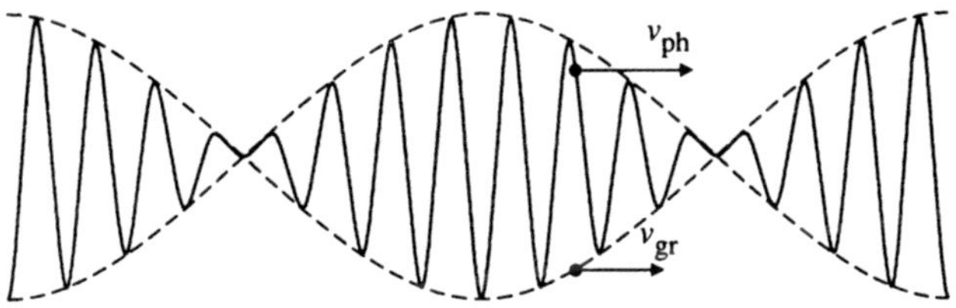

Fig. 19.16 The phase and group velocities of light in the case of propagation of a biharmonic optical wave.

$$E = A\cos(\omega_1 t - k_1 z) + A\cos(\omega_2 t - k_2 z).$$

In this case, the field can be presented in the form

$$E = F\left(t - \frac{\Delta k}{\Delta\omega} z\right)\cos(\omega t - kz), \tag{19.51}$$

where $F(t) = 2A\cos(\Delta\omega t)$, $\Delta\omega = (\omega_1 - \omega_2)/2$, $\Delta k = (k_1 - k_2)/2$, and $k = (k_1 + k_2)/2$. Equation (19.51) demonstrates that the propagation of the biharmonic wave can also be characterized by two velocities; that is, the phase velocity $v_{ph} = \omega/k$ and the group velocity $v_{gr} = \Delta\omega/\Delta k$. The former characterizes the motion of the phase front of the carrier and the latter the motion of the envelope (Fig. 19.16).

Thus, the group velocity of light is a velocity of motion of the optical pulse envelope (wave packet, or "group" of waves) in the dispersive medium. It should be emphasized that the notion of group velocity is by no means universal. It exists only for narrow-band optical signals propagating in transparent media. In resonant conditions, when the optical pulse is strongly distorted by the medium, the notion of group velocity is inapplicable.

Using eqns (19.31), (19.49), and (19.50), it is easy to obtain the following relation:

$$v_{gr} = v_{ph} \Big/ \left[1 + \frac{\omega}{n}\frac{\partial n}{\partial\omega}\right]. \tag{19.52}$$

Since the notion of the group velocity is applicable only for the regions of normal dispersion, in which the complex refractive index is real and $\partial n/\partial\omega > 0$ (see Fig. 19.1), it follows from eqn (19.52) that the group velocity of light is generally less than the phase velocity:

$$v_{gr} < v_{ph}.$$

This means that in a dispersive medium the optical wave modulation and optical energy move more slowly than the carrier. In vacuum ($n = 1$), the phase and group velocities are identical and equal to c.

19.4.6 A medium with second-order dispersion—dispersion of the group velocity

The group velocity of light in such media depends on the light frequency, or, in other words, *dispersion of the group velocity* takes place:

$$v_{\text{gr}} = v_{\text{gr}}(\omega).$$

This follows directly from eqns (19.47) and (19.49):

$$\frac{1}{v_{\text{gr}}} = \frac{\partial k}{\partial \omega} = k_0' + k_0''\omega.$$

The dispersion of the group velocity leads to specific physical effects, such as dispersive spreading of optical pulses, and the formation of "spectrons"; that is, pulses with a stable envelope shape reproducing the shape of their frequency spectra, etc. This has allowed the development of optical fibre devices for the control of optical pulse length and, specifically, of generators of ultra-short optical pulses. Inserting eqn (19.47) into eqns (19.40) and (19.43), we obtain

$$h(t) = \frac{1}{2\pi} e^{-ik_0 z} \int_{-\infty}^{\infty} e^{i\omega\theta - i\frac{1}{2}\omega^2\tau_0^2} d\omega,$$

where

$$\theta = t - k_0' z, \qquad \tau_0 = \sqrt{k_0'' z}. \tag{19.53}$$

Using the Poisson integral,

$$\int_{-\infty}^{\infty} e^{-\alpha x^2} dx = \sqrt{\pi/\alpha},$$

we find the explicit expression for $h(t)$:

$$h(t) = e^{-ik_0 z} \frac{\exp(i\theta^2/2\tau_0^2)}{\sqrt{2\pi i \tau_0^2}}. \tag{19.54}$$

Finally, substituting eqn (19.54) into eqn (19.42), we obtain

$$\mathcal{E}(t) = e^{-ik_0 z} \frac{\exp(i\theta^2/2\tau_0^2)}{\sqrt{2\pi i \tau_0^2}} \int_{-\infty}^{\infty} \mathcal{E}_0(\tau) e^{\frac{i\tau^2}{2\tau_0^2} - \frac{i\theta\tau}{\tau_0^2}} d\tau. \tag{19.55}$$

Equation (19.55) expresses the complex amplitude of the output signal through the input signal amplitude.

19.4.7 The far-field zone and the spectron

It can be seen from the structure of the integrand in eqn (19.55) that the integration domain is limited by the input pulse duration τ_{p0}; whence it follows that at sufficiently long distances z, such that

$$\frac{\tau_{p0}^2}{2\tau_0^2} = \frac{\tau_{p0}^2}{2k_0'' z} \ll \pi, \tag{19.56}$$

eqn (19.55) transforms to the simpler form

$$\mathcal{E}(t) = \mathrm{e}^{-ik_0 z} \frac{\exp(i\theta^2/2\tau_0^2)}{\sqrt{2\pi i \tau_0^2}} \int_{-\infty}^{\infty} \mathcal{E}_0(\tau) \mathrm{e}^{-i\theta\tau/\tau_0^2} \, d\tau. \tag{19.57}$$

The integral in eqn (19.57) is the Fourier transform of the complex amplitude of the input pulse. Taking into account eqn (19.39), we can rewrite eqn (19.57) as

$$\mathcal{E}(t) = \mathrm{e}^{-ik_0 z} \frac{\exp(i\theta^2/2\tau_0^2)}{\sqrt{2\pi i \tau_0^2}} \mathcal{E}_0(\omega)\big|_{\omega=\theta/\tau_0^2} \tag{19.58}$$

or, by using eqn (19.53) for τ_0, as

$$\mathcal{E}(t) = \mathrm{e}^{-ik_0 z} \frac{\exp(i\theta^2/2\tau_0^2)}{\sqrt{2\pi i \tau_0^2}} \mathcal{E}_0(\omega)\big|_{\omega=\theta/k_0'' z}, \tag{19.59}$$

where parameter θ is defined by eqn (19.53). The condition (19.56) can be rewritten as

$$z \gg z_{\mathrm{dis}}, \tag{19.60}$$

where

$$z_{\mathrm{dis}} = \frac{\tau_{p0}^2}{2\pi k_0''} \tag{19.61}$$

is the characteristic *dispersion length*. Just as in the case of optical beam diffraction, the region of space determined by the condition (19.60) is called the *far-field dispersion zone*, or simply the *far-field zone*. Using this term, we can formulate the obtained result as follows. In the far-field zone, an optical pulse acquires a stable form determined by the spectrum of the input pulse. Such a pulse is called a spectron.

19.4.8 Dispersive spreading of an optical pulse

Let $\Delta\omega$ denote the spectral bandwidth of the input pulse. This quantity is related to the input pulse duration τ_{p0} as follows:

$$\tau_{p0}\Delta\omega = 2\pi. \tag{19.62}$$

It follows from the relation $\omega = \theta/k_0'' z$ in eqn (19.59) that provided that the quantity $\Delta\omega$ characterizes the width of the distribution $\mathcal{E}_0(\omega)$, the quantity $\Delta\theta =$

$\Delta\omega k_0^{''} z$ characterizes the width of the distribution $\mathcal{E}(t)$, i.e. it denotes the pulse duration. Consequently, the optical pulse duration in the far-field zone is

$$\tau_p = \Delta\omega k_0^{''} z. \tag{19.63}$$

Equation (19.63) shows that the pulse duration grows linearly with the distance traversed by the pulse in the dispersive medium. This effect is called *dispersive spreading of the pulse.* It can be observed in media with second-order dispersion, i.e. in media possessing dispersion of the group velocity.

The effect of spectron formation in the far-field zone and its dispersive spreading can be made clear by a simple analogy. Imagine long-distance runners maintaining their differing individual constant velocities. Then, irrespective of the individual moments at which they started, there is a distance beyond which the fastest runner will finish first, the second fast will finish second, and so on, and the slowest runner will finish last. From this time on, the order of the runners will not change, although the distances (and time intervals) between them will gradually increase. An optical pulse propagating in a dispersive medium demonstrates a similar behaviour: its spectral components, propagating in the medium with differing velocities, play the part of the runners.

Another analogy of the phenomenon under consideration is the Fraunhofer diffraction of optical beams. We remind the reader that this phenomenon implies the formation of a stable angular distribution of the field, which reproduces its angular spectrum in the far-field diffraction zone, and an increase in the transverse size of the beam proportional to the distance traversed.

Taking into account eqns (19.61) and (19.62), we can rewrite eqn (19.63) as

$$\tau_p(z) = \tau_{p0} z / z_{dis}. \tag{19.64}$$

By way of example, let us consider the optical pulse propagation in an optical fibre. For wavelengths of about 1 μm, the typical optical fibre dispersion is $k_0^{''} = 3 \times 10^{-28}\ s^2\ cm^{-1}$. Thus, if the input pulse has a duration $\tau_{p0} = 10^{-12}$ s, then the dispersion length defined by eqn (19.61) is equal to $z_{dis} = 5$ m. If the fibre length equals, for example, 50 m, the output pulse has the duration $\tau_p(z) = 10^{-11}$ s, i.e. tenfold dispersive spreading takes place.

If, for example, the input pulse has a rectangular form and a duration τ_{p0}, then the intensity in the far-field zone will have the form

$$I(t) = I_{max}\, \mathrm{sinc}^2[\pi\theta/\tau_p(z)],$$

where $\theta = t - k_0' z$, $\tau_p(z)$ is the output pulse duration determined by eqn (19.64) and the normalization intensity I_{max} is independent of time (Fig. 19.17).

19.4.9 Propagation invariants

If the dispersive medium is transparent (like an optical fibre), the pulse energy

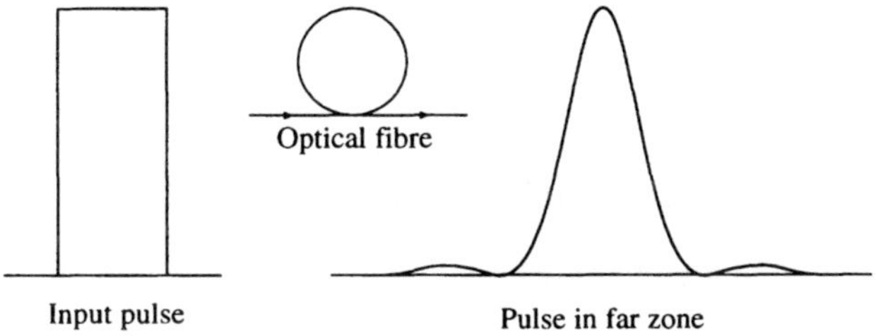

Fig. 19.17 The transformation of an optical pulse in an optical fibre.

$$W = \text{constant} \cdot \int_{-\infty}^{\infty} |\mathcal{E}(t)|^2 \, dt \tag{19.65}$$

and pulse spectrum

$$S(\omega) = |\mathcal{E}(\omega)|^2 \tag{19.66}$$

remain unchanged during in the course of propagation, i.e. the energy and spectrum of the optical pulse are propagation invariants. The spectral invariance follows from eqns (19.40) and (19.41) since, for a transparent medium, $|\chi(\omega)| = 1$. The energy invariance follows from the spectral invariance and the Parseval identity, according to which

$$\int_{-\infty}^{\infty} |\mathcal{E}(t)|^2 \, dt = \frac{1}{2\pi} \int_{-\infty}^{\infty} S(\omega) \, d\omega.$$

Of course, the spectral bandwidth $\Delta\omega$ of the pulse is also a propagation invariant.

Bibliography

Akhmanov, S. A., Vysloukh, V. A., and Chirkin, A. S. (1991). *Optics of femtosecond laser pulses.* AIP, New York.

Bloembergen, N. (1965). *Nonlinear optcis.* W. A. Benjamin, New York.

Born, M. and Wolf, E. (1968). *Principles of optics.* Pergamon Press, Oxford.

Corliss, C. H. and Bozman, W. R. (1962). *Experimental transition probabilities for spectral lines of seventy elements.* National Bureau of Standards, Washington, D.C.

Demtröder, W. (1982). *Laser spectroscopy.* Springer-Verlag, Berlin.

Moore, C. E. (1949). *Atomic energy levels.* National Bureau of Standards, Washington, D.C.

20

Optical phenomena at a boundary between two media

Reflection and refraction at a separation boundary. Boundary conditions for the electromagnetic field. The geometry of reflection and refraction—Snell's law. Total internal reflection. Energy relations for reflection and refraction—Fresnel's formulae. The Brewster effect. Variation of the optical wave phase on reflection and refraction. Reflection of light for normal incidence. The antireflective coating. The reflection of light for oblique incidence. The reflection of light by a metal surface.

20.1 Reflection and refraction at a separation boundary

It is well known from experience that a beam of light falling on a boundary between two different optical media splits into two (reflected and refracted) beams (Fig. 20.1). The characteristics of the reflected and refracted beams, such as the propagation direction, power, and polarization, depend on the composition of both media and also on the parameters of the incident optical beam; that is, its direction, frequency, polarization, and power. These properties of reflection and refraction find extensive application in light-controlling instruments (e.g. lenses, prisms, mirrors, etc.) to focus light, form optical images, perform spectral decomposition of light, obtain totally polarized light and so on. The phenomenon of light reflection is used in laser resonators and interferometers. Total internal reflection allows the transmission of light over large distances via an optical fibre. The list of similar examples could be continued.

Note another important fact. Since the optical wave possesses a momentum (see Chapter 4), the change in the beam direction caused by reflection and re-

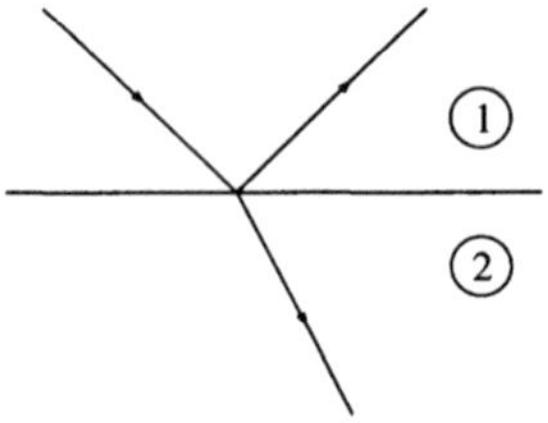

Fig. 20.1 The reflection and refraction of an optical beam at a boundary separating two media.

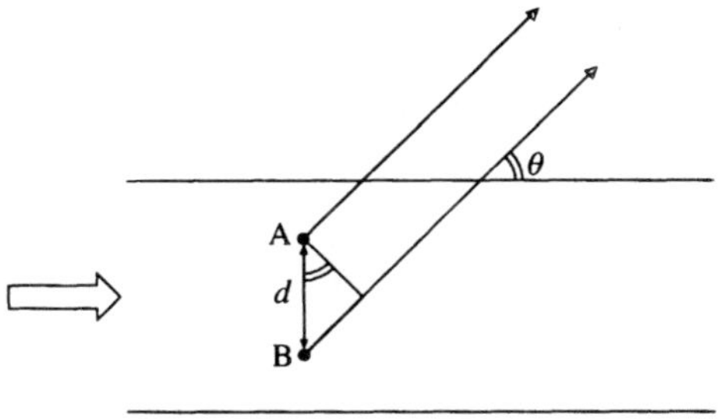

Fig. 20.2 The mechanism of rectilinear propagation of light in an homogeneous medium.

fraction triggers the forces that light exerts on material bodies. These forces are usually very small, but there are phenomena in which they play a noticeable part, e.g. optical levitation of transparent particles in the focused laser beam (see Ashkin 1972).

20.1.1 The mechanisms of reflection and refraction

The optical phenomena arising at a boundary between two media can be regarded as effects associated with the non-homogeneity of the medium. To show this, let us examine first the mechanism of rectilinear propagation of light in an homogeneous medium.

As is known from the Fresnel theory of diffraction, the rectilinear propagation of light in vacuum is associated with interference cancellation of secondary spherical optical wavelets emanating from different points in space. Such a cancellation takes place for all directions except a straight ahead one, which causes rectilinear propagation of the optical beam.

In a similar way, the spherical optical waves emitted by individual atoms in an homogeneous medium interfere, so that the optical beam also turns out to be rectilinear.

The mechanism of suppression of lateral emission of radiation and of spreading of the optical beam is illustrated in Fig. 20.2. Let A be an atom of the medium located within the optical beam. Under the action of the incident optical wave, the atomic electron makes forced oscillations, producing the secondary spherical optical wavelet. Let us consider the radiation emitted by the atom in a direction making an angle θ with the optical beam axis. It can be seen from the figure that, for any angle θ, except $\theta = 0$, there is a certain atom B emitting a wave of the opposite phase in the same direction. Indeed, if, for example, the incident radiation has the structure of a plane wave, then the atom B is located in the same cross-section of the beam at a distance d from atom A, determined by the condition $d \sin\theta = \lambda/2$, where λ is the optical wavelength. Thus, atoms A and B suppress their mutual radiations in the direction θ. Since a similar pair of atoms can be chosen for any angle $\theta \neq 0$, one can conclude that light propagates rectilinearly.

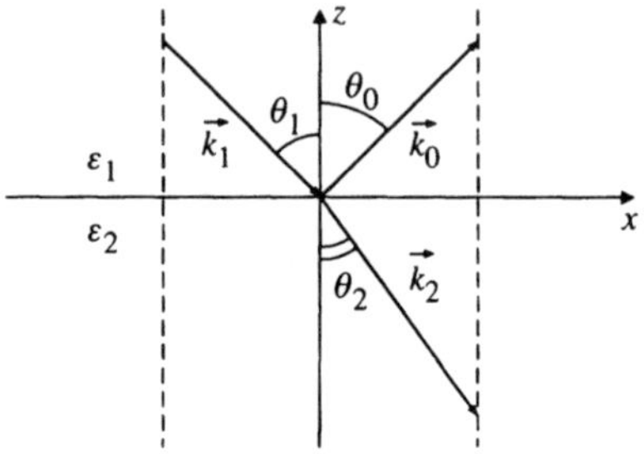

Fig. 20.3 The reflection and refraction of a plane wave at a plane boundary.

It is easy to see that the foregoing consideration implies essentially the identity of all atoms, i.e. the *homogeneity* of the medium. One may expect, however, that in a *non-homogenous* medium a beam of light will not be rectilinear. Experience shows that non-homogeneous media are characterized by strong scattering of light. For this reason, smoke or fog look turbid. From this point of view, it seems natural that the rectilinear propagation of light is violated at the boundary between two media where the reflected and refracted beams arise.

20.2 Boundary conditions for the electromagnetic field

The quantitative theory of reflection and refraction is developed from Maxwell's equations and from the boundary conditions for an electromagnetic field. As is known from electrodynamics, the tangential (i.e. parallel to the surface of discontinuity) components of electric and magnetic fields have to be continuous across the surface. Thus, the boundary conditions have the form

$$E_{1t} = E_{2t}, \qquad H_{1t} = H_{2t}. \tag{20.1}$$

Here, the subscript t denotes the tangential component and the numbers 1 and 2 the media on the opposite sides of the boundary.

20.3 The geometry of reflection and refraction—Snell's law

Let a plane monochromatic linearly polarized optical wave fall on to the plane boundary between two media:

$$\vec{E}_1 = \tfrac{1}{2}\vec{\mathcal{E}}_1 e^{i(\omega t - \vec{k}_1 \vec{r})} + \text{c. c.}, \tag{20.2}$$

where $\vec{\mathcal{E}}_1$ is the complex amplitude, ω is the frequency, $\vec{k}_1$ is the wave vector, and ε_1 and ε_2 are the complex dielectric permittivities of the media. Let θ_1 denote the angle between the normal $\vec{n}$ to the surface of discontinuity and the wave vector $\vec{k}_1$ of the incident wave. This angle is called the angle of incidence (Fig. 20.3). The plane specified by the vectors $\vec{n}$ and $\vec{k}_1$ is called the plane of incidence.

The optical fields of the reflected and refracted waves can be written in the form

$$\vec{E}_0 = \tfrac{1}{2}\vec{\mathcal{E}}_0 e^{i(\omega t - \vec{k}_0\vec{r})} + \text{c. c.},$$
$$\vec{E}_2 = \tfrac{1}{2}\vec{\mathcal{E}}_2 e^{i(\omega t - \vec{k}_2\vec{r})} + \text{c. c.}, \tag{20.3}$$

The angles θ_0 and θ_2 shown in Fig. 20.3 are called the angle of reflection and the angle of refraction, respectively.

Let us direct the z-axis normally to the separating boundary. Writing the boundary condition $E_{1t} = E_{2t}$ in the plane of the separating boundary $z = 0$, we obtain the relation

$$\mathcal{E}_{1t} \exp\left[i(\omega t - \vec{k}_1\vec{r})\right] + \mathcal{E}_{0t} \exp\left[i(\omega t - \vec{k}_0\vec{r})\right] = \mathcal{E}_{2t} \exp\left[i(\omega t - \vec{k}_2\vec{r})\right], \tag{20.4}$$

where $\mathcal{E}_{it}$ are constants. The condition (20.4) must hold for any moment in time t and for any point $\vec{r}$ lying on the boundary. Hence, we have

$$\omega t - \vec{k}_1\vec{r} = \omega t - \vec{k}_0\vec{r} = dt - \vec{k}_2\vec{r},$$

and also

$$\vec{k}_1\vec{r} = \vec{k}_0\vec{r} = \vec{k}_2\vec{r} \tag{20.5}$$

or, in Cartesian coordinates, .

$$k_{1x}x + k_{1y}y = k_{0x}x + k_{0y}y = k_{2x}x + k_{2y}y. \tag{20.6}$$

These conditions must hold for any x and y. In particular, setting $x = 0$, we obtain

$$k_{1y} = k_{0y} = k_{2y}. \tag{20.7}$$

Similarly, setting $y = 0$, we find that

$$k_{1x} = k_{0x} = k_{2x}. \tag{20.8}$$

It follows from eqns (20.7) and (20.8) that the wave vectors of the reflected and refracted waves, $\vec{k}_0$ and $\vec{k}_2$, lie in the plane of incidence, i.e. in the plane containing the wave vector $\vec{k}_1$ of the incident wave and the normal $\vec{n}$ to the separating boundary. All four vectors $\vec{n}$, $\vec{k}_1$, $\vec{k}_0$, and $\vec{k}_2$ thus lie in the same plane.

Taking the plane of incidence as the x–z plane (Fig. 20.3), we have $k_{1y} = k_{0y} = k_{2y} = 0$, $k_{1x} = k_{0x} = k_{2x}$, where

$$k_{1x} = k_1 \sin\theta_1, \quad k_{0x} = k_0 \sin\theta_0, \quad k_{2x} = k_2 \sin\theta_2, \tag{20.9}$$

and

$$k_1 = \frac{\omega}{c} n_1 = k_0, \quad k_2 = \frac{\omega}{c} n_2, \tag{20.10}$$

Here $n_1 = \sqrt{\varepsilon_1}$ and $n_2 = \sqrt{\varepsilon_2}$ are the complex refractive indices of the first and second media.

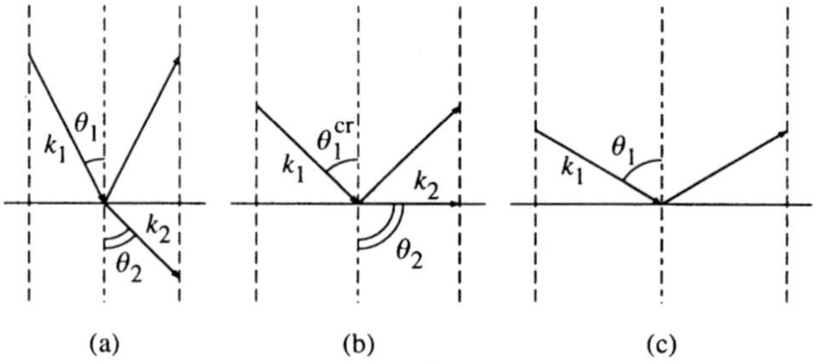

Fig. 20.4 The reflection and refraction of an optical beam passing from an optically denser medium to a less dense medium ($n_1 > n_2$). (a) The angle of incidence is less than the critical angle. (b) The angle of incidence is equal to the critical angle. (c) The angle of incidence exceeds the critical angle. In the last case the refracted ray is absent and the effect of total internal reflection takes place.

The condition $k_{1x} = k_{0x}$ with eqns (20.9) and (20.10) leads to the relation

$$\theta_1 = \theta_0, \tag{20.11}$$

according to which the angle of incidence is equal to the angle of reflection. This is the *law of light reflection.*

Now suppose that both media are transparent and that the refractive indices n_1 and n_2 are then real. Inserting eqns (20.9) and (20.10) into the relation $k_{1x} = k_{2x}$, we obtain

$$n_1 \sin\theta_1 = n_2 \sin\theta_2. \tag{20.12}$$

This equation relates the angles of incidence and refraction of light. It expresses the *law of refraction*, or *Snell's law.* Both relations (20.11) and (20.12) are well known from experiment.

20.4 Total internal reflection

Suppose that an optical beam crosses a boundary, making a transition from an optically dense medium into a less dense medium, i.e. $n_1 > n_2$. According to Snell's law, the angle of refraction will then be greater than the angle of incidence: $\theta_2 > \theta_1$. An illustration of refraction for this case is shown in Fig. 20.4(a). If the angle of incidence is

$$\theta_1^{\rm cr} = \arcsin{(n_2/n_1)}, \tag{20.13}$$

then we obtain from eqn (20.12) that $\theta_2 = 90°$ and, consequently, the refracted beam does not penetrate the second medium, but propagates along the boundary (Fig. 20.4(b)). Moreover, if the angle of incidence exceeds the critical value, the refracted beam is absent, and the light is totally reflected from the boundary (Fig. 20.4(c)). This effect is called the *total internal reflection of light.* The angle of incidence defined by eqn (20.13) is called the *limiting angle of total reflection.*

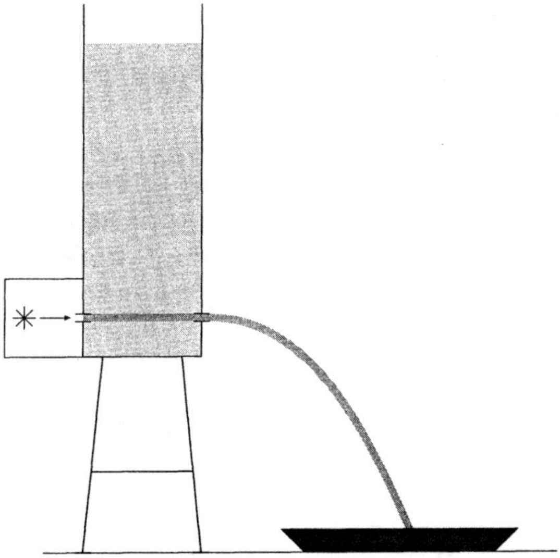

Fig. 20.5 An experimental observation of the total internal reflection of light.

For example, for the glass–air boundary ($n_1 = 1.5$, $n_2 = 1.0$) this angle is about 40°.

The effect of total internal reflection can be observed in the experiment shown schematically in Fig. 20.5. In this experiment a beam of light is introduced into a stream of water flowing out of a vessel through a hole in its wall. The stream forms a parabola, and the light beam becomes locked in the stream due to the effect of total internal reflection. The experiment is carried out in the darkroom, and the propagation of light within the stream is clearly seen due to the fluorescence of a dye added to the water.

Another experiment displays the transmission of helium–neon laser radiation through an optical fibre (Fig. 20.6). Although in this experiment the fibre is bent to form a ring, light is still not emitted from the transparent side surface of the fibre because of total internal reflection at the boundary.

Light transmission along the fibre can also be observed with a fibre braid; experiment shows that light is transmitted through the fibre braid irrespective of its shape (e.g. we can tie it in a knot).

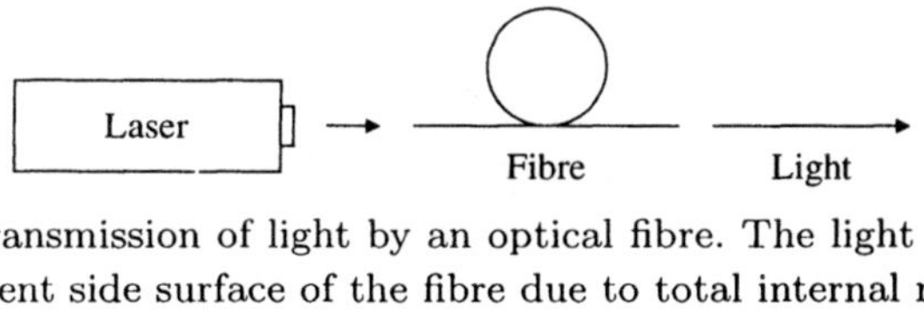

Fig. 20.6 The transmission of light by an optical fibre. The light does not come out from the transparent side surface of the fibre due to total internal reflection.

The possibility of transmitting light along optical fibres is now employed in areas such as optical communications, medical diagnostics, laser therapy, etc.

20.5 Energy relations for reflection and refraction—Fresnel's formulae

How is the energy divided between the reflected and refracted beams? How does the energy ratio depend on the angle of incidence and on the polarization of the incident beam? To answer these questions, we need to calculate the amplitudes of the reflected and refracted waves. We shall do this, using the boundary conditions (20.1) and the relation between the fields $\vec{E}$ and $\vec{H}$ in an optical wave, following from Maxwell's equations.

Assume, as before, that the incident, reflected, and refracted optical waves are plane and monochromatic. The two media forming the boundary between them are assumed to be linear and isotropic, and are characterized by the complex dielectric permittivities ε_1 and ε_2.

The propagation of light in both media is described by Maxwell's equations:

$$\operatorname{rot}\vec{E} = -\frac{1}{c}\frac{\partial \vec{H}}{\partial t}, \qquad \operatorname{rot}\vec{H} = \frac{1}{c}\frac{\partial \vec{D}}{\partial t}. \tag{20.14}$$

Let us write $\vec{E}$, $\vec{H}$, and $\vec{D}$ in the form of plane monochromatic waves:

$$\begin{aligned}\vec{E} &= \tfrac{1}{2}\vec{\mathcal{E}}\exp\left[i(\omega t - \vec{k}\vec{r})\right] + \text{c. c.},\\ \vec{H} &= \tfrac{1}{2}\vec{\mathcal{H}}\exp\left[i(\omega t - \vec{k}\vec{r})\right] + \text{c. c.},\\ \vec{D} &= \tfrac{1}{2}\vec{\mathcal{D}}\exp\left[i(\omega t - \vec{k}\vec{r})\right] + \text{c. c.}\end{aligned} \tag{20.15}$$

Substituting eqns (20.15) into eqns (20.14), we obtain the following relations between the amplitudes of electric and magnetic fields in the optical wave:

$$[\vec{k},\vec{\mathcal{E}}] = \frac{\omega}{c}\vec{\mathcal{H}}, \qquad [\vec{k},\vec{\mathcal{H}}] = -\frac{\omega}{c}\varepsilon\vec{\mathcal{E}}. \tag{20.16}$$

Here we have taken into account that in the linear isotropic medium $\vec{\mathcal{D}} = \varepsilon\vec{\mathcal{E}}$, where ε is the complex dielectric permittivity of the medium.

The value of the tangential component of the field depends on the polarization state of the incident optical wave. Let us consider separately the cases in which the vector $\vec{E}$ is perpendicular and parallel to the plane of incidence. A wave with an arbitrary polarization state can be represented as a superposition of two linearly polarized waves of the types just discussed, because in the general case

$$\vec{E} = \vec{E}_{\perp} + \vec{E}_{\parallel}. \tag{20.17}$$

Let the vector $\vec{E}$ of the incident wave be perpendicular to the plane of incidence, i.e.

$$\vec{E} = \vec{E}_{\perp}. \tag{20.18}$$

Such polarization of the wave is called "s"-polarization. Here, the reflected and refracted waves will also be polarized perpendicular to the plane of incidence. On the contrary, the vectors $\vec{H}$ of all three waves will lie in the plane of incidence.

Using the coordinate system shown in Fig. 20.3, we can write

$$E_{\mathrm{t}} = E_y = E, \qquad H_{\mathrm{t}} = H_x.$$

The boundary conditions (20.1) then take the form

$$E_1 + E_0 = E_2, \qquad H_{1x} + H_{0x} = H_{2x}$$

or, taking into account eqns (20.15), (20.2), (20.3), and (20.5),

$$\mathcal{E}_1 + \mathcal{E}_0 = \mathcal{E}_2, \tag{20.19}$$

$$\mathcal{H}_{1x} + \mathcal{H}_{0x} = \mathcal{H}_{2x}. \tag{20.20}$$

It follows from the first formula (20.16) that

$$\mathcal{H}_x = -\frac{c}{\omega} k_z \mathcal{E}. \tag{20.21}$$

Substituting eqn (20.21) into eqn (20.20), we obtain the equation

$$k_{1z}\mathcal{E}_1 + k_{0z}\mathcal{E}_0 = k_{2z}\mathcal{E}_2$$

or

$$k_{1z}(\mathcal{E}_1 - \mathcal{E}_0) = k_{2z}\mathcal{E}_2, \tag{20.22}$$

where we have taken into account that

$$k_{0z} = -k_{1z} \tag{20.23}$$

(see Fig. 20.3).

Let us introduce the complex coefficients of light reflection ($r_{\perp}$) and refraction ($t_{\perp}$):

$$r_{\perp} = \mathcal{E}_0/\mathcal{E}_1, \qquad t_{\perp} = \mathcal{E}_2/\mathcal{E}_1. \tag{20.24}$$

Using eqns (20.19) and (20.22) we then obtain

$$r_{\perp} = \frac{k_{1z} - k_{2z}}{k_{1z} + k_{2z}}, \qquad t_{\perp} = \frac{2k_{1z}}{k_{1z} + k_{2z}}. \tag{20.25}$$

Equations (20.24) and (20.25) solve the problem; they establish the relationship between the amplitudes of the incident, reflected, and refracted optical waves.

Similarly, we shall consider the case in which the vector $\vec{E}$ of the incident wave is parallel to the plane of incidence, i.e. when

$$\vec{E} = \vec{E}_{\parallel}. \tag{20.26}$$

Such polarization of the wave is called "p"-polarization. In this case,

$$H_{\rm t} = H_y = H, \qquad E_{\rm t} = E_x.$$

The boundary conditions (20.1) take the form

$$H_1 + H_0 = H_2, \qquad E_{1x} + E_{0x} = E_{2x},$$

whence follow the relations for the complex amplitudes:

$$\mathcal{H}_1 + \mathcal{H}_0 = \mathcal{H}_2, \tag{20.27}$$

$$\mathcal{E}_{1x} + \mathcal{E}_{0x} = \mathcal{E}_{2x}. \tag{20.28}$$

It follows from the second formula (20.16) that

$$\mathcal{E}_x = \frac{c}{\omega}\frac{k_z}{\varepsilon}\mathcal{H}. \tag{20.29}$$

Inserting eqn (20.29) into eqn (20.28), we obtain the equation

$$\frac{k_{1z}}{\varepsilon_1}\mathcal{H}_1 + \frac{k_{0z}}{\varepsilon_1}\mathcal{H}_0 = \frac{k_{2z}}{\varepsilon_2}\mathcal{H}_2$$

or

$$\frac{k_{1z}}{\varepsilon_1}(\mathcal{H}_1 - \mathcal{H}_0) = \frac{k_{2z}}{\varepsilon_2}\mathcal{H}_2, \tag{20.30}$$

where eqn (20.23) is taken into account.

Let us define the coefficients of reflection and refraction as

$$r_{\parallel} = \mathcal{H}_0/\mathcal{H}_1, \qquad t_{\parallel} = \mathcal{H}_2/\mathcal{H}_1. \tag{20.31}$$

Using eqns (20.27) and (20.30), we obtain

$$r_{\parallel} = \frac{(k_{1z}/\varepsilon_1) - (k_{2z}/\varepsilon_2)}{(k_{1z}/\varepsilon_1) + (k_{2z}/\varepsilon_2)}, \qquad t_{\parallel} = \frac{2k_{1z}/\varepsilon_1}{(k_{1z}/\varepsilon_1) + (k_{2z}/\varepsilon_2)}. \tag{20.32}$$

Now suppose that both media are transparent. Then the dielectric permittivities, ε_1 and ε_2, and the refractive indices of the media,

$$n_1 = \sqrt{\varepsilon_1}, \qquad n_2 = \sqrt{\varepsilon_2}, \tag{20.33}$$

are real. The normal components of the wave vectors of the incident and refracted waves are expressed by the formulae (see Fig. 20.3):

$$k_{1z} = -k_1 \cos\theta_1, \qquad k_{2z} = -k_2 \cos\theta_2, \tag{20.34}$$

where the wave numbers are

$$k_1 = (\omega/c)n_1, \qquad k_2 = (\omega/c)n_2. \tag{20.35}$$

Substituting eqns (20.33)–(20.35) into eqns (20.25) and (20.32), we obtain

$$r_\perp = \frac{n_1 \cos\theta_1 - n_2 \cos\theta_2}{n_1 \cos\theta_1 + n_2 \cos\theta_2} \tag{20.36}$$

and

$$r_\parallel = \frac{n_2 \cos\theta_1 - n_1 \cos\theta_2}{n_2 \cos\theta_1 + n_1 \cos\theta_2}. \tag{20.37}$$

Allowing for Snell's law (20.12), we can transform these expressions into the form

$$r_\perp = \frac{\sin(\theta_1 - \theta_2)}{\sin(\theta_1 + \theta_2)}, \qquad r_\parallel = -\frac{\tan(\theta_1 - \theta_2)}{\tan(\theta_1 + \theta_2)}. \tag{20.38}$$

Thus, eqns (20.38) determine the amplitude coefficients of the reflection of light at the boundary between two linear isotropic transparent media. In these formulae, θ_1 is the angle of incidence and θ_2 the angle of refraction of the optical beam at the boundary (see Fig. 20.3). These angles are interrelated by Snell's law (20.12). The expression for $r_\perp$ pertains to the case in which the incident optical wave is polarized perpendicularly to the plane of incidence, while the expression for $r_\parallel$ relates to the case of wave polarization in the plane of incidence. Equations (20.38) are called the *Fresnel formulae*. These formulae enable one to calculate the energies and phases of the reflected and refracted optical waves. In the specific case in which $n_1 = n_2$ (an homogeneous medium without a separating boundary), eqns (20.38), (20.32), and (20.25) describe the conventional rectilinear propagation of light. It is worthwhile noting that the Fresnel formulae were obtained before the discovery of the electromagnetic nature of light.

20.6 The Brewster effect

The most noticeable property of eqns (20.38) is that under certain conditions the coefficient of reflection $r_\parallel$ becomes zero:

$$r_\parallel = 0. \tag{20.39}$$

This happens when the sum of the angles of incidence and refraction is equal to $\pi/2$:

$$\theta_1 + \theta_2 = \pi/2. \tag{20.40}$$

The Fresnel formulae thus predict the total disappearance of the reflected wave provided that the incident optical wave is polarized in the plane of incidence and condition (20.40) is satisfied. Experiment confirms this result. The disappearance of the reflected wave is called the *Brewster effect*.

From eqn (20.40) and the law of refraction (20.12) it is easy to calculate the angle of incidence at which the Brewster effect takes place. This angle, called the Brewster angle, is

$$\theta_B = \arctan(n_2/n_1). \tag{20.41}$$

For example, for the air–glass boundary, when $n_1 = 1$, and $n_2 = 1.52$, we have $\theta_B = 56°40'$.

The Brewster angle is sometimes called the *total polarization angle*. When non-polarized light falls at this angle, the reflected beam is linearly polarized in the direction perpendicular to the plane of incidence. The Brewster effect can thus be used to produce linearly polarized light.

In an experimental demonstration, a beam of white non-polarized light emitted by an arc lamp falls on to a glass plate at the Brewster angle. The reflected beam of light is directed towards a screen. It has linear polarization, as can readily be checked by a polaroid. Thus, the polarizing effect really does take place.

In another experiment, the reflected beam is directed towards a second glass plate, oriented so that the beam turns out to be polarized in the plane of incidence. We observe the reflection of light from the second plate. By rotating the second plate, we change the angle of incidence and observe, at a certain angle of incidence, the total disappearance of the reflected beam; that is, the Brewster effect.

The origin of the Brewster effect can be explained as follows. The reflected beam can be treated as a non-compensated radiation, emitted by the dipoles of the second medium and propagating in the first medium. Figure 20.7 shows that when condition (20.40) is satisfied, the reflected and refracted beams turn out to be mutually perpendicular. Here, the dipoles of the second medium, oscillating parallel to the vector $\vec{E}$ of the refracted wave and, consequently, perpendicular to the refracted beam, do not emit light in the direction of the reflected beam because this direction coincides with that of the dipole oscillations. As a result, the reflected beam is absent, and all of the energy of the incident light is transmitted to the refracted beam.

It follows from the foregoing discussion that the Brewster effect is possible only when the incident beam is polarized in the plane of incidence. If, on the contrary, the incident beam is polarized perpendicularly to the plane of incidence, the reflected beam should be observed for any angle of incidence. This conclusion also agrees with eqns (20.38).

20.7 Variation of the optical wave phase on reflection and refraction

It follows from eqns (20.25) and (20.32) that the amplitude transmission coefficients $t_\perp$ and $t_\parallel$ are real positive numbers, provided that both media are

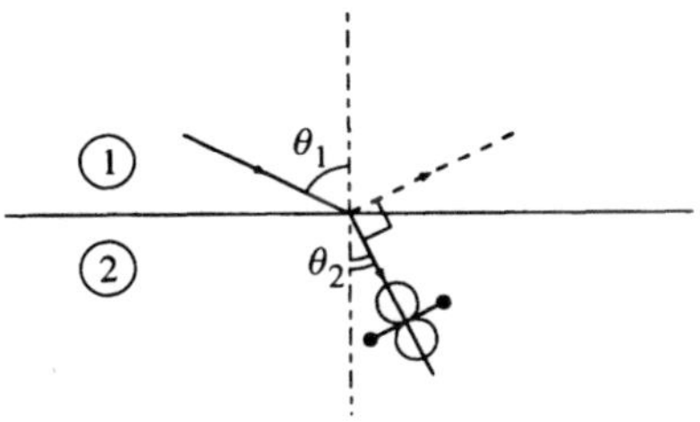

Fig. 20.7 The mechanism of the Brewster effect. The incident and refracted beams are shown by solid straight lines. Also shown is the direction of dipole vibrations in the second medium and the dipole radiation pattern (a torus).

transparent. This signifies that the phase of the refracted wave coincides with the phase of the incident wave at the boundary.

The amplitude reflection coefficients $r_\perp$ and $r_\parallel$ are also real, but they can have either positive or negative values. Consequently, depending on the conditions, the phase of the reflected wave either equals the phase of the incident wave at the boundary or differs from it by π.

20.8 Reflection of light for normal incidence

If $\theta_1 = 0$, then eqns (20.11), (20.12), and (20.38) yield $\theta_0 = 0$ and $\theta_2 = 0$:

$$r_\perp = -\frac{n_2 - n_1}{n_1 + n_2}, \qquad r_\parallel = \frac{n_2 - n_1}{n_1 + n_2}. \tag{20.42}$$

Consequently, for normal incidence,

$$r_\perp = -r_\parallel. \tag{20.43}$$

The minus sign in this formula means that reflection in the backward direction changes the direction of one of the vectors $\vec{E}$ or $\vec{H}$ of the optical wave to the opposite sense; only in this case do the vectors $\vec{E}$, $\vec{H}$, and $\vec{k}$ of both the incident and reflected waves form a right-handed triad, as required by Maxwell's equations.

Let us consider a specific example. Suppose that a plane monochromatic optical wave falls normally onto a boundary with an optically denser medium, for example, on the air–glass boundary, with $n_2 > n_1$. According to eqn (20.42), we have here $r_\perp = \mathcal{E}_0/\mathcal{E}_1 < 0$ and $r_\parallel = \mathcal{H}_0/\mathcal{H}_1 > 0$ and, consequently, the vector $\vec{H}$ retains its direction after the reflection, while the vector $\vec{E}$ changes its direction to the opposite sense (see Fig. 20.8). In other words, the phase of the wave "H" remains unchanged on reflection from the denser medium, while the phase of the wave "E" changes by π. For the reflection from the less dense medium ($n_2 < n_1$), on the contrary, the phase of the wave "E" remains unchanged, while the phase of the wave "H" changes by π.

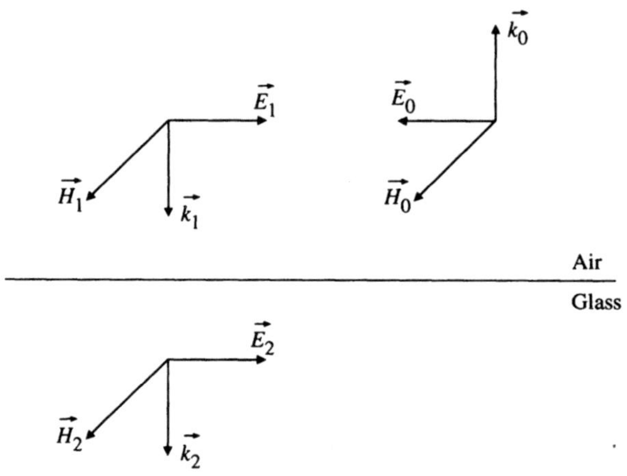

Fig. 20.8 The mutual orientation of the vectors $\vec{E}$, $\vec{H}$, and $\vec{k}$ in the incident, reflected, and transmitted waves for the normal incidence of an optical wave on an air–glass boundary ($n_2 > n_1$).

As for the optical wave transmitted across the boundary, both of its vectors $\vec{E}$ and $\vec{H}$ have the same directions at the boundary as have the vectors of the incident wave (Fig. 20.8). In other words, the phases of both waves "E" and "H" do not change when passing across the boundary.

Using eqns (20.24), (20.25), (20.29), (20.31), and (20.32), it is easy to check that the law of energy conservation actually holds in reflection and refraction processes. In particular, for normal incidence, the intensity $I_1 = (cn_1/8\pi)|\mathcal{E}_1|^2$ of the incident wave is equal to the sum of the intensities of the reflected $I_0 = (cn_1/8\pi)|\mathcal{E}_0|^2$ and transmitted $I_2 = (cn_2/8\pi)|\mathcal{E}_2|^2$ waves:

$$I_1 = I_0 + I_2. \tag{20.44}$$

The coefficients of reflection and transmission are related by the formulae

$$|r_\perp|^2 + \frac{n_2}{n_1}|t_\perp|^2 = 1, \qquad |r_\parallel|^2 + \frac{n_1}{n_2}|t_\parallel|^2 = 1.$$

Here the quantities $r_\perp$ and $r_\parallel$ are determined by eqns (20.42) and

$$t_\perp = 2n_1/(n_1 + n_2), \qquad t_\parallel = 2n_2/(n_1 + n_2),$$

according to eqns (20.25) and (20.32).

The energy reflection coefficient $R = |r|^2$ for normal incidence is

$$R = \left(\frac{n_1 - n_2}{n_1 + n_2}\right)^2. \tag{20.45}$$

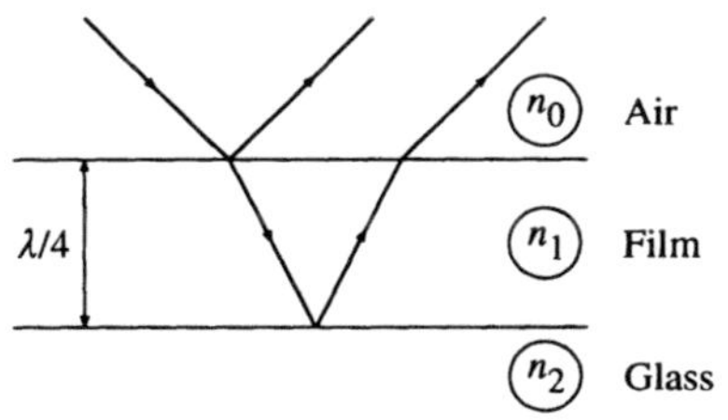

Fig. 20.9 The action of an antireflective coating.

For example, for the air–glass boundary when $n_1 = 1$ and $n_2 = 1.5$, we obtain $R = 0.04$, i.e. the reflection is 4%.

20.9 The antireflective coating

Systems forming optical images, such as camera objectives or microscope objectives, usually employ several lenses, so that the total number of air–glass boundaries is quite large. If no special care is taken, an optical beam passing through the objective will lose an appreciable fraction of its energy. Thus, the problem of reducing the Fresnel reflection of light at the boundaries becomes important.

One way of solving this problem is to use the interference suppression of the reflected wave. For this purpose, one could coat the glass surface with a transparent layer of substance (an antireflective film), so that the optical waves reflected from both boundaries of this layer will cancel each other (Fig. 20.9).

Let l be the thickness of the film, n_0 be the refractive index of air, n_1 be the refractive index of the film, and n_2 be the refractive index of the glass. It is easy to show that the Fresnel reflection will be totally cancelled if the film thickness is taken to be equal to a quarter of the optical wavelength,

$$l = \lambda/4,$$

and the refractive index of the film is chosen so that the coefficients of reflection of light at both boundaries of the film are equal. Therefore, we obtain the following equation for n_1:

$$\left(\frac{n_0 - n_1}{n_0 + n_1}\right)^2 = \left(\frac{n_1 - n_2}{n_1 + n_2}\right)^2,$$

the solution of which is

$$n_1 = \sqrt{n_0 n_2}.$$

Thus, the refractive index of the antireflective film should be equal to the geometric mean of the refractive indices of the glass and air. For example, assuming $n_0 = 1$ and $n_2 = 1.5$, we obtain $n_1 = 1.22$. For an optical wavelength in vacuum of 550 nm, the corresponding wavelength within the coating is $550/1.22 = 450$ nm. This corresponds to a coating thickness of $450/4 = 112$ nm. Such a layer can be deposited on the glass surface by evapouration of the appropriate material in a vacuum chamber.

The refractive indices for most solids lie within an interval from 1.5 to 2.2, which makes the selection of a substance with a refractive index as low as 1.22 a difficult problem. Substances such as sodium fluoride (NaF) and cryolite (Na_3AlF_6), the refractive indices of which are about 1.33, satisfy this requirement well. The reflection coefficient then becomes equal to 0.008, which is significantly less than that for conventional glass (~ 0.04). The reflection coefficient for radiation of other wavelengths is somewhat higher, but still remains sufficiently low throughout the visible range.

20.10 The reflection of light for oblique incidence

The reflection coefficient for the case of oblique incidence depends on the polarization of light. If the incident beam is polarized perpendicularly to the plane of incidence or parallel to it, the energy reflection coefficients are, respectively,

$$R_\perp = r_\perp^2, \qquad R_\parallel = r_\parallel^2, \tag{20.46}$$

where the quantities $r_\perp$ and $r_\parallel$ are defined by eqns (20.38). If the incident beam of light is linearly polarized and the vector $\vec{E}$ makes an angle ψ with the plane of incidence, then the reflection coefficient is

$$R = R_\parallel \cos^2\psi + R_\perp \sin^2\psi. \tag{20.47}$$

For non-polarized (natural) light one must average expression (20.47) over all angles ψ. This gives

$$R = \tfrac{1}{2}(R_\perp + R_\parallel). \tag{20.48}$$

The reflection coefficients for polarized and non-polarized light are shown in Fig. 20.10 as a function of the angle of incidence on the air–glass boundary ($n_1 = 1$ and $n_2 = 1.52$).

20.11 The reflection of light by a metal surface

The optical properties of metals, such as the reflection and absorption of light, are primarily determined by the presence of free electrons in the metals. The complex dielectric permittivity of a metal can be approximately calculated using the Lorentz model and equating to zero the parameters determining the elastic bond of the electron in the atom (ω_0) and electron vibration damping (Γ). Equation (19.12) then yields

$$\varepsilon(\omega) = 1 - \frac{\omega_\mathrm{p}^2}{\omega^2}, \tag{20.49}$$

where ω_p is the so-called *plasma frequency*, defined by eqn (19.11):

$$\omega_\mathrm{p} = \sqrt{\frac{4\pi N e^2}{m}}. \tag{20.50}$$

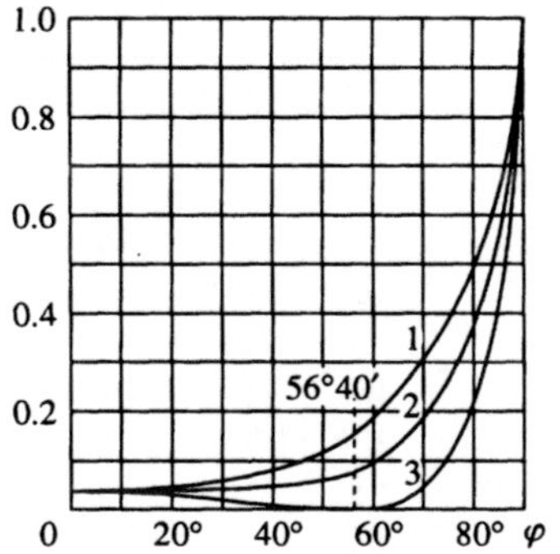

Fig. 20.10 The coefficient of light reflection from an air–glass boundary as a function of the angle of incidence. 1, Linearly polarized light with polarization perpendicular to the plane of incidence; 2, naturally polarized light; 3, linearly polarized light with polarization parallel to the plane of incidence.

As shown in Chapter 19, the plasma frequency corresponds to the wavelength $\lambda_p = 0.3\ \mu m$, i.e. it lies in the ultraviolet region, when the free electron density $N = 10^{22}\ cm^{-3}$. The inequality $\omega < \omega_p$ then holds for the visible range, and the value of ε determined by eqn (20.49) proves to be negative:

$$\varepsilon < 0. \tag{20.51}$$

The complex index of refraction $n = \sqrt{\varepsilon}$ turns out to be imaginary, because

$$\mathrm{Re}\,\sqrt{\varepsilon} = 0. \tag{20.52}$$

Consequently, the complex refractive index of a metal can be written as

$$n = in'', \tag{20.53}$$

where n'' is a real quantity.

According to eqns (20.25) and (20.32), the coefficient of light reflection by a metal surface for normal incidence can be written in the form

$$R = \left| \frac{\sqrt{\varepsilon_1} - \sqrt{\varepsilon_2}}{\sqrt{\varepsilon_1} + \sqrt{\varepsilon_2}} \right|^2, \tag{20.54}$$

where ε_1 and ε_2 are the complex dielectric permittivities of the media forming a separating boundary. For example, for an air–metal boundary, we have $\sqrt{\varepsilon_1} = 1$ and $\sqrt{\varepsilon_2} = in_2''$. Then, we find from eqn (20.54) that

$$R = 1, \tag{20.55}$$

i.e. the metal should completely reflect the light.

Table 20.1 Coefficients of light reflection from polished surfaces of metals for normal incidence (in per cent).

λ (μm)	0.188	0.2	0.251	0.305	0.357	0.5
Aluminium	—	—	80	—	84	88
Copper	—	—	26	25	27	44
Steel	22	27	38	44	50	56
Silver	22	25	34	9	75	91
Nickel	—	—	38	44	49	61
Chromium	33	36	—	37	41	55
Zinc	17	22	39	48	51	55
Silicon	64	73	75	73	60	34
λ (μm)	0.6	0.7	0.9	1.0	5.0	10.0
Aluminium	89	87	85	93	94	97
Copper	72	83	89	90	98	98
Steel	57	58	61	63	90	94
Silver	93	95	97	97	99	99
Nickel	65	69	70	72	94	—
Chromium	—	56	—	57	81	93
Zinc	58	61	62	69	97	—
Silicon	32	—	—	—	—	—

Most metals reflect light quite efficiently. It is this property that causes the characteristic lustre of pure metallic surfaces. It should be also noted that as the radiation frequency decreases, the inequality $\omega < \omega_p$ is satisfied better, and the reflection coefficient increases. Experiment confirms this conclusion. The reflection coefficients of metals increase with the radiation wavelength, reaching almost 100% at $\lambda = 10$ μm (see Table 20.1).

For some metals, the plasma frequency lies in the visible region. Since the reflection of light from a metal drops sharply at frequencies $\omega > \omega_p$, the metal surface in this case no longer looks white. For example, gold, having a plasma frequency with a wavelength of about 600 nm, weakly reflects blue–green light and thus looks yellow when illuminated by white light. The plasma frequency of aluminium lies in the ultraviolet region, and aluminium therefore efficiently reflects all visible light and looks white under usual illumination.

Bibliography

Born, M. and Wolf, E. (1968). *Principles of optics.* Pergamon Press, Oxford.

21

Optics of anisotropic media

Anisotropic materials and the basic effects of crystal optics. The structure of an optical wave in an anisotropic crystal. The material equation of an anisotropic medium. Classification of crystals. The polarization eigenstates of an optical wave in an anisotropic crystal. Ordinary and extraordinary waves. Optical birefringence at the anisotropic medium boundary. The production and analysis of polarized light. The interference of polarized rays. Induced anisotropy.

21.1 Anisotropic materials and the basic effects of crystal optics

As early as at the time of Huygens, it was known that some crystals possess unusual optical properties. For example, a crystal of Iceland spar (calcite) refracts light differently depending on the orientation of the crystal with respect to the incident beam. A light beam falling normally onto this crystal propagates straightforwardly inside the crystal only along a definite direction. For other directions, the optical beam splits into two beams, showing the effect of *optical birefringence* (Fig. 21.1). Such dependence of optical properties on the crystalline direction is called *optical anisotropy*. Let us consider some experiments illustrating anisotropic properties of crystals.

21.1.1 Optical birefringence

One of the first experiments on optical birefringence is shown schematically in Fig. 21.1. A white light beam from an arc lamp passes through an Iceland spar crystal to produce two light spots of nearly equal brightness on a screen (Fig. 21.2(a)). When we rotate the crystal about the axis of the incident beam, one light spot on the screen remains motionless, while the other moves round the first one synchronously with the rotating of the crystal (Fig. 21.2(b)). Analysis of the polarization of the beams, using a polaroid, shows that both beams emerging from the crystal are linearly polarized and that their polarization directions are mutually perpendicular (Fig. 21.2(c)). This can be seen better when we slowly rotate the crystal about the axis of the incident beam. Then, the beams of light, having emerged from the crystal and passed through the polaroid, alternately change their intensities so that when one beam fades, the other reaches its maximum intensity. A similar picture is observed when we rotate the polaroid, keeping the crystal fixed.

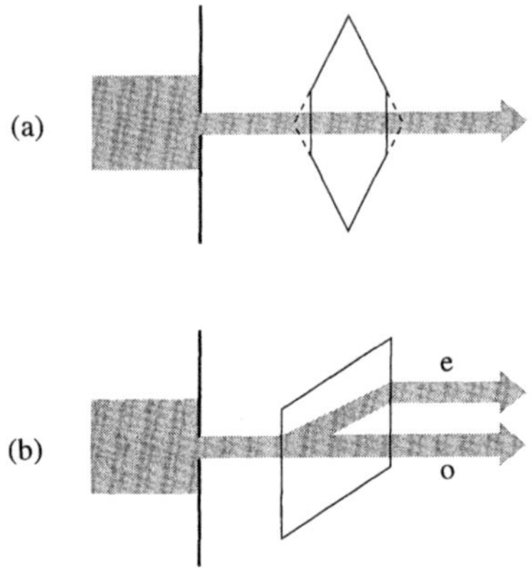

Fig. 21.1 (a) Optical birefringence does not take place when light travels along the optic axis of the Iceland spar crystal. (b) The optical birefringence of light falling normally on the natural face of an Iceland spar crystal.

21.1.2 Anisotropic absorption

Another experiment shows the anisotropic absorption of light by a tourmaline crystal. A thin plate of tourmaline is pasted into a special holder to be fixed against the objective of a projection lamp. An image of the tourmaline plate projected on a screen has the form of a darkened greenish band (Fig. 21.3(a)). Inserting a polaroid into the beam and rotating it about the beam axis, we can see that the brightness of the light band on the screen changes. We can also note that there is a certain position of the polaroid for which the image of the tourmaline plate becomes completely black, which means that the light passing through the tourmaline plate has been totally polarized and its polarization is linear.

We then remove the polaroid from the beam and fix one more holder with a tourmaline plate against the objective of the projection lamp, so that two

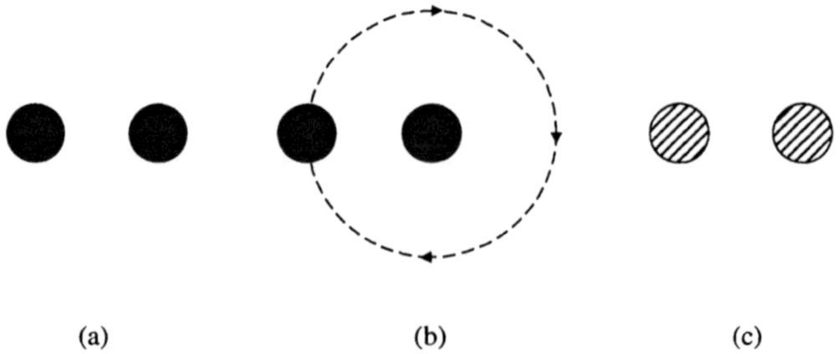

Fig. 21.2 (a,b) Pictures observed on a screen in experiments with Iceland spar crystal. (c) Polarization of the rays.

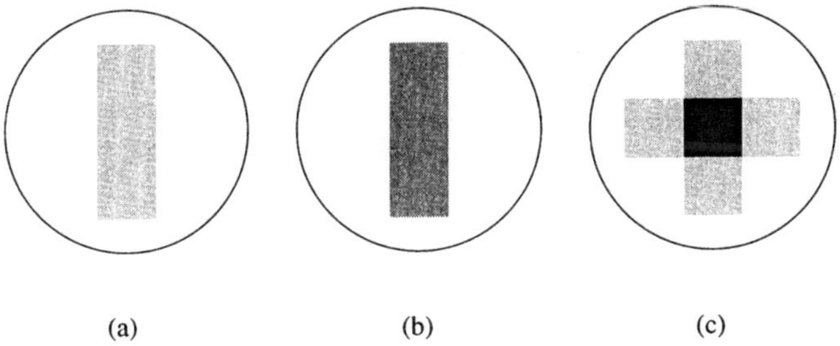

Fig. 21.3 Images of tourmaline plates on a screen: (a) single plate; (b) two plates set parallel over each other; (c) two plates set perpendicularly over each other.

tourmaline plates are placed close to each other. Now turning one plate relative to the other, we can see that for a parallel orientation of the plates the light beam passes through them, although the image on the screen becomes darker (Fig. 21.3(b)). With crossed plates, we can see a completely dark square in the region of overlap (Fig. 21.3(c)). The result of this experiment shows that the absorption of light by the tourmaline plate generally depends on its orientation. In other words, tourmaline displays the anisotropic absorption of light.

The mechanism of anisotropic absorption can be explained as follows. The anisotropy of the tourmaline structure makes the electrons move predominantly along a definite direction with respect to the crystal axis. If the polarization of the incident optical wave coincides with this direction, the optical field excites strong vibration of the electrons, lending them energy. The electrons, in turn, transfer their energy to the crystal lattice. As a result, the optical wave is absorbed by the crystal. If, conversely, the polarization of the incident optical wave is perpendicular to the allowed direction of the electron motion in the crystal, the vibration amplitude of the electrons is small and their energy is transferred only to the secondary radiation, but not to the crystal lattice. The optical wave experiences only slight absorption. This also explains why a crystal illuminated by non-polarized (natural) light produces a linearly polarized beam: tourmaline transmits only light the polarization direction of which is orthogonal to the allowed direction of the electron motion in the crystal.

In the experiment described above, the tourmaline plates were about 1 mm thick. As the experiment shows, such a plate can be used to produce linearly polarized light, i.e. as a polarizer of light. Another good polarizer employs an artificial anisotropic absorber (usually synthetic polyvinyl alcohol with iodine).

Optical anisotropy is caused by the structural anisotropy of the medium. Consequently, anisotropic properties are observed only in crystals, but not in gases, liquids (except liquid crystals), plastic materials, or glasses. Sometimes, however, optical anisotropy is exhibited by isotropic media because of mechanical stress, an external electric field, etc. For example, optical birefringence can arise in isotropic media under the action of a constant electric field (Kerr's effect). Similar phenomena are observed in an alternating electric field and in the field of a powerful laser pulse.

21.2 The structure of an optical wave in an anisotropic crystal

The structure of a plane monochromatic optical wave propagating in a linear anisotropic medium is determined by Maxwell's equations:

$$\begin{aligned} \operatorname{rot}\vec{E} &= -\frac{1}{c}\frac{\partial\vec{H}}{\partial t}, \quad \operatorname{rot}\vec{H} = \frac{1}{c}\frac{\partial\vec{D}}{\partial t}, \\ \operatorname{div}\vec{D} &= 0, \qquad\qquad \operatorname{div}\vec{H} = 0. \end{aligned} \tag{21.1}$$

Since these equations are linear, we shall assume all of the vectors $\vec{E}$, $\vec{D}$, and $\vec{H}$ to be plane monochromatic waves:

$$\begin{aligned} \vec{E} &= \tfrac{1}{2}\vec{\mathcal{E}}e^{i(\omega t-\vec{k}\vec{r})} + \text{c. c.}, \\ \vec{H} &= \tfrac{1}{2}\vec{\mathcal{H}}e^{i(\omega t-\vec{k}\vec{r})} + \text{c. c.}, \\ \vec{D} &= \tfrac{1}{2}\vec{\mathcal{D}}e^{i(\omega t-\vec{k}\vec{r})} + \text{c. c.}, \end{aligned} \tag{21.2}$$

Substituting eqns (21.2) into eqns (21.1), we obtain

$$\begin{aligned} [\vec{k},\vec{\mathcal{E}}] &= \frac{\omega}{c}\vec{\mathcal{H}}, \quad [\vec{k},\vec{\mathcal{H}}] = -\frac{\omega}{c}\vec{\mathcal{D}}, \\ (\vec{k},\vec{\mathcal{D}}) &= 0, \qquad (\vec{k},\vec{\mathcal{H}}) = 0. \end{aligned} \tag{21.3}$$

One can see from eqns (21.3) that the vectors $\vec{\mathcal{D}}$, $\vec{\mathcal{H}}$, and $\vec{k}$ form a right-handed triad of mutually perpendicular vectors (Fig. 21.4). As for the vector $\vec{\mathcal{E}}$, it is perpendicular to the vector $\vec{\mathcal{H}}$, and consequently lies in the plane of the vectors $\vec{\mathcal{D}}$ and $\vec{k}$, but generally is not parallel to the vector $\vec{\mathcal{D}}$.

The Poynting vector,

$$\vec{S} = \frac{c}{4\pi}[\vec{E},\vec{H}], \tag{21.4}$$

determining the direction of propagation of the optical energy (see Chapter 3) is perpendicular to both vectors $\vec{\mathcal{E}}$ and $\vec{\mathcal{H}}$, and, consequently, lies in the plane of the vectors $\vec{k}$, $\vec{\mathcal{E}}$, and $\vec{\mathcal{D}}$, but generally is not parallel to the vector $\vec{k}$ (Fig. 21.4). This means that the optical beam direction $\vec{S}$ in the anisotropic crystal does not generally coincide with the normal $\vec{k}$ to the wavefront of the optical wave. The angle between the vectors $\vec{k}$ and $\vec{S}$ is equal to the angle between the vectors $\vec{\mathcal{D}}$ and $\vec{\mathcal{E}}$, and is called the *angle of anisotropy*. In Fig. 21.4, this angle is denoted by β.

Later, we shall show that the specific optical properties of anisotropic crystals are caused by the fact that the vectors $\vec{\mathcal{E}}$ and $\vec{\mathcal{D}}$ of an optical wave are not collinear. First, let us consider the physical reason for this phenomenon.

The property of the anisotropic medium that we are considering can be made clear by a mechanical analogy; for example, with the help of an abacus. Take an

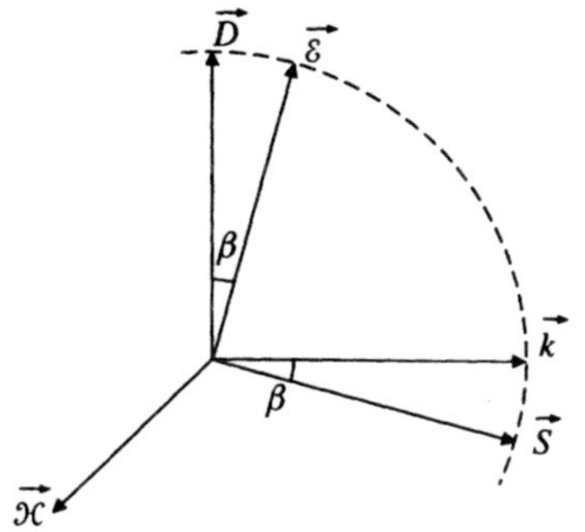

Fig. 21.4 The mutual orientation of the vectors determining the structure of an optical wave in an anisotropic crystal. The vectors $\vec{D}$, $\vec{\mathcal{E}}$, $\vec{k}$, and $\vec{S}$, the ends of which are joined by a dashed line, lie in the same plane perpendicular to the vector $\vec{\mathcal{H}}$.

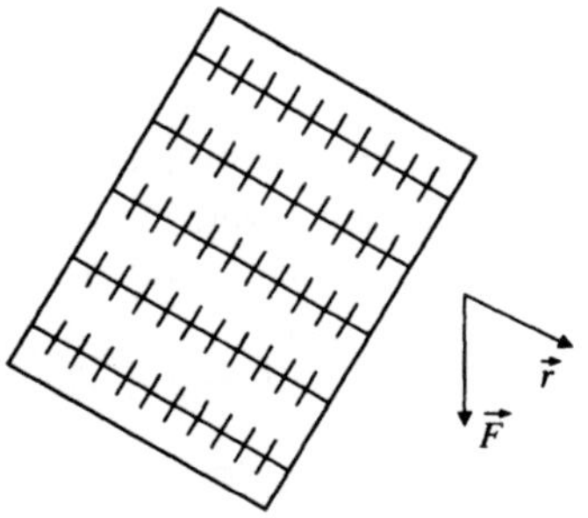

Fig. 21.5 An abacus as a simple mechanical model of an anisotropic system. The important property of such a system is that the displacement of the particles $\vec{r}$ is generally not parallel to the applied external force $\vec{F}$.

abacus and tilt it so that the stones start to move along the wires (Fig. 21.5). It can clearly be seen that the stones move not in the direction of the external force, which is here the gravitational force directed downwards, but along the wires, i.e. in the direction *prescribed by the structure of the system.*

Something similar occurs in an anisotropic crystal. When an optical wave propagates in a crystal, the electrons of the crystal experience an external force directed along the electric vector $\vec{E}$ of the wave. The electrons, however, shift not in the direction of this external force, but in the direction imposed by the *crystalline structure*. As a result, the vector of polarization $\vec{P}$ of the medium turns out to be non-parallel to the vector $\vec{E}$. The vector of the electric displacement $\vec{D} = \vec{E} + 4\pi\vec{P}$ is also non-parallel to the vector $\vec{E}$. Thus, the situation illustrated in Fig. 21.4 appears.

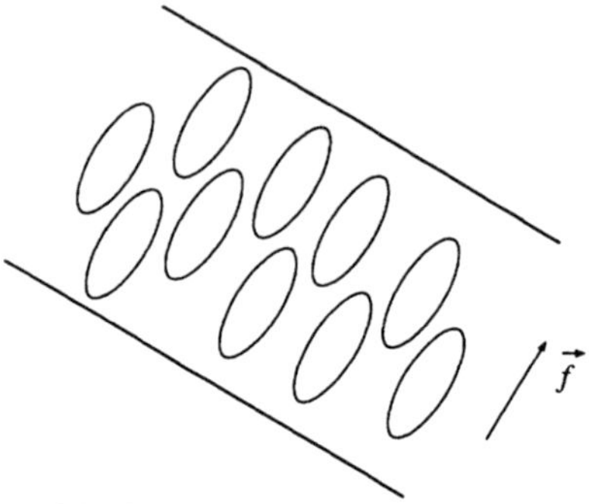

Fig. 21.6 The simplest model of an anisotropic medium, in which electrons move only in the direction determined by the vector $\vec{f}$.

21.3 The material equation of an anisotropic medium

An essential property of an anisotropic medium is that the electron mobility is different for different crystalline directions. Allowing for this, we shall consider the simplest model of an anisotropic medium, in which the electron motion is allowed only in one direction. This direction is characterized by the unit vector $\vec{f}$, called the *director* (Fig. 21.6). The given model is the simplest in terms of mathematics. It should be noted, however, that anisotropy of a similar type is actually observed in real media; specifically, in some liquid crystals.

Let us write the optical polarization of the medium in the form

$$\vec{P} = N\vec{p}, \tag{21.5}$$

where N is the number of molecules per unit volume,

$$\vec{p} = e\vec{x} \tag{21.6}$$

is the dipole moment of an individual molecule,

$$\vec{x} = x\vec{f}, \tag{21.7}$$

is the electron displacement with respect to its equilibrium position, and $\vec{f}$ is the displacement direction (the director).

Using the Lorentz model, we can write the equation of the electron motion as the oscillator equation

$$\ddot{x} + \Gamma\dot{x} + \omega_0^2 x = \frac{e}{m}(\vec{f}, \vec{E}), \tag{21.8}$$

where $(\vec{f}, \vec{E})$ is the scalar product of the vectors $\vec{f}$ and $\vec{E}$, equal to the projection of the vector $\vec{E}$ in the displacement direction.

Substituting the expression for $\vec{E}$ in the form of a plane monochromatic wave (21.2) into eqn (21.8), we shall seek a solution in the form

$$x = \tfrac{1}{2}\tilde{x}\, e^{i(\omega t - \vec{k}\vec{r})} + \text{c. c.} \tag{21.9}$$

Inserting eqn (21.9) into eqn (21.8), we can find the relation between the complex amplitudes of the electric field and the electron displacement:

$$\tilde{x} = \frac{e}{m}\frac{(\vec{f}, \vec{\mathcal{E}})}{\omega_0^2 - \omega^2 + i\omega\Gamma}. \tag{21.10}$$

According to eqns (21.5)–(21.7), and (21.10), the optical polarization of the medium can be written as

$$\vec{P} = \tfrac{1}{2}\vec{\mathcal{P}} e^{i(\omega t - \vec{k}\vec{r})} + \text{c. c.}, \tag{21.11}$$

where

$$\vec{\mathcal{P}} = \frac{Ne^2}{m}\frac{(\vec{f}, \vec{\mathcal{E}})}{\omega_0^2 - \omega^2 + i\omega\Gamma}\vec{f}. \tag{21.12}$$

The formula (21.12), relating the complex amplitudes of the polarization $\vec{\mathcal{P}}$ and the field $\vec{\mathcal{E}}$, is the material equation of the model anisotropic medium. In this formula, N is the number of molecules per unit volume, e is the electron charge, m is the electron mass, ω_0 is the natural frequency of an elementary oscillator, Γ is the damping coefficient, and $\vec{f}$ is the direction vector of the electron motion. According to the basic idea of our model, eqn (21.12) shows that the optical polarization vector $\vec{P}$ of the medium is generally not parallel to the electric field vector $\vec{E}$ of the optical wave:

$$\vec{P} \nparallel \vec{E}.$$

As we mentioned earlier, this is caused by the fact that the displacement of the electrons in an anisotropic crystal occurs not in the direction of the electric field of an optical wave, but in a direction prescribed by the structure of the crystal.

21.3.1 The optical susceptibility tensor

Let us find the relationship between the Cartesian components of the polarization and the field. It is convenient here to introduce tensor notation, according to which the Cartesian axes are denoted by x_1, x_2, and x_3, and the Cartesian components of the vectors are numbered by the subscripts 1, 2, and 3. For example, the components of the director are f_1, f_2, and f_3.

Writing the scalar product in Cartesian components as

$$(\vec{f}, \vec{\mathcal{E}}) = f_1\mathcal{E}_1 + f_2\mathcal{E}_2 + f_3\mathcal{E}_3 \tag{21.13}$$

and substituting eqn (21.13) into eqn (21.12), we obtain

$$\mathcal{P}_1 = \frac{Ne^2}{m}\frac{f_1f_1\mathcal{E}_1 + f_1f_2\mathcal{E}_2 + f_1f_3\mathcal{E}_3}{\omega_0^2 - \omega^2 + i\omega\Gamma},$$
$$\mathcal{P}_2 = \frac{Ne^2}{m}\frac{f_2f_1\mathcal{E}_1 + f_2f_2\mathcal{E}_2 + f_2f_3\mathcal{E}_3}{\omega_0^2 - \omega^2 + i\omega\Gamma}, \tag{21.14}$$
$$\mathcal{P}_3 = \frac{Ne^2}{m}\frac{f_3f_1\mathcal{E}_1 + f_3f_2\mathcal{E}_2 + f_3f_3\mathcal{E}_3}{\omega_0^2 - \omega^2 + i\omega\Gamma}.$$

Here $\mathcal{P}_1$, $\mathcal{P}_2$, and $\mathcal{P}_3$ are the Cartesian components of the vector $\vec{\mathcal{P}}$, and $\mathcal{E}_1$, $\mathcal{E}_2$, and $\mathcal{E}_3$ are the Cartesian components of the vector $\vec{\mathcal{E}}$. We then see from eqn (21.14) that each Cartesian component of the vector $\vec{\mathcal{P}}$ is generally expressed through all three Cartesian components of the vector $\vec{\mathcal{E}}$. Such a relation between two vectors is called a tensor relation. The tensor expresses the operation of transformation of one vector into the other, when these vectors are not parallel.

The matrix relating the Cartesian components of the complex amplitudes of the polarization and the field is called a linear optical susceptibility tensor of the anisotropic medium. Let us denote this matrix by $\kappa_{\alpha\beta}$. Then, by definition,

$$\mathcal{P}_\alpha = \sum_{\beta=1}^{3} \kappa_{\alpha\beta}(\omega)\mathcal{E}_\beta. \tag{21.15}$$

According to eqns (21.14), in our model this matrix has the form

$$\kappa_{\alpha\beta}(\omega) = \frac{Ne^2}{m}\frac{f_\alpha f_\beta}{\omega_0^2 - \omega^2 + i\omega\Gamma}. \tag{21.16}$$

The summation sign in eqn (21.15) can be omitted if summation is taken over the repeating subscript, giving

$$\mathcal{P}_\alpha = \kappa_{\alpha\beta}(\omega)\mathcal{E}_\beta. \tag{21.17}$$

A more compact notation can be used when all the subscripts are omitted and the matrix (operator) is denoted by the hat "^" sign:

$$\vec{\mathcal{P}} = \hat{\kappa}(\omega)\vec{\mathcal{E}}. \tag{21.18}$$

Thus, eqns (21.15) and (21.17) express the tensor character of the relation between the polarization and the field in an anisotropic crystal.

21.3.2 The general form of the material equation for a linear anisotropic medium

Generalizing eqns (18.43) and (18.38), we can write the material equation for a linear anisotropic medium as

$$P_\alpha(t) = \int_0^\infty \kappa_{\alpha\beta}(\tau) E_\beta(t-\tau)\, d\tau. \tag{21.19}$$

Here $P_\alpha(t)$ is the Cartesian component of the polarization of the medium, $E_\beta(t)$ is the Cartesian component of the electric field of an optical wave, and $\kappa_{\alpha\beta}(\tau)$ is a tensor Green function depending only on the properties of the medium and related to the linear optical susceptibility tensor $\kappa_{\alpha\beta}(\omega)$ by Fourier transformation:

$$\kappa_{\alpha\beta}(\tau) = \frac{1}{2\pi}\int_{-\infty}^{\infty} \kappa_{\alpha\beta}(\omega) \mathrm{e}^{i\omega\tau}\, d\omega. \tag{21.20}$$

In particular, the tensor $\kappa_{\alpha\beta}(\omega)$ for the Lorentz model is described by eqn (21.16).

By making the Fourier inverse transformation, we obtain from eqn (21.20)

$$\kappa_{\alpha\beta}(\omega) = \int_{-\infty}^{\infty} \kappa_{\alpha\beta}(\tau) \mathrm{e}^{-i\omega\tau}\, d\tau \tag{21.20a}$$

or, considering that $\kappa_{\alpha\beta}(\tau < 0) = 0$ (the causality principle),

$$\kappa_{\alpha\beta}(\omega) = \int_0^{\infty} \kappa_{\alpha\beta}(\tau) \mathrm{e}^{-i\omega\tau}\, d\tau. \tag{21.20b}$$

The material equation of the form (21.19) thus considers not only the anisotropic, but also the dispersive properties of a linear anisotropic medium.

21.3.3 The dielectric permittivity tensor

By the definition of the electric displacement vector, $\vec{D} = \vec{E} + 4\pi\vec{P}$. Substituting eqns (21.2) and (21.11) in this formula, we find the following relationship between the complex amplitude vectors:

$$\vec{\mathcal{D}} = \vec{\mathcal{E}} + 4\pi\vec{\mathcal{P}}.$$

In Cartesian components,

$$\mathcal{D}_\alpha = \mathcal{E}_\alpha + 4\pi\mathcal{P}_\alpha. \tag{21.21}$$

Substituting eqn (21.17) into eqn (21.21), we obtain

$$\mathcal{D}_\alpha = \mathcal{E}_\alpha + 4\pi\kappa_{\alpha\beta}(\omega)\mathcal{E}_\beta \tag{21.22}$$

or

$$\mathcal{D}_\alpha = \varepsilon_{\alpha\beta}(\omega)\mathcal{E}_\beta, \tag{21.23}$$

where

$$\varepsilon_{\alpha\beta}(\omega) = \delta_{\alpha\beta} + 4\pi\kappa_{\alpha\beta}(\omega) \tag{21.24}$$

is the *dielectric permittivity tensor* of a linear anisotropic medium and $\delta_{\alpha\beta}$ is the Kronecker symbol. Equation (21.23) implies summation over the repeating subscript. As seen from eqns (21.16) and (21.24), the linear optical susceptibility tensor and the dielectric permittivity tensor are symmetric:

$$\kappa_{\alpha\beta} = \kappa_{\beta\alpha}, \quad \varepsilon_{\alpha\beta} = \varepsilon_{\beta\alpha}; \quad \alpha, \beta = 1, 2, 3. \tag{21.25}$$

21.3.4 The principal coordinate axes of an anisotropic crystal

So far, we have assumed the coordinate system x_1, x_2, x_3 to be arbitrary. It can be shown, however, that for any anisotropic crystal there is a coordinate system x, y, z, in which the dielectric permittivity tensor is diagonal:

$$\varepsilon_{\alpha\beta} = \begin{pmatrix} \varepsilon_{xx} & 0 & 0 \\ 0 & \varepsilon_{yy} & 0 \\ 0 & 0 & \varepsilon_{zz} \end{pmatrix}. \tag{21.26}$$

Such axes x, y, z are called the *principal coordinate axes*. For example, in the system shown in Fig. 21.6, one principal axis is directed along the director $\vec{f}$; the directions of the other two axes are arbitrary. Hereinafter, we shall use only the principal coordinate axes. The relations between Cartesian components of the electric displacement and the field (21.23) take the following form:

$$\mathcal{D}_x = \varepsilon_{xx}\mathcal{E}_x, \quad \mathcal{D}_y = \varepsilon_{yy}\mathcal{E}_y, \quad \mathcal{D}_z = \varepsilon_{zz}\mathcal{E}_z. \tag{21.27}$$

Here the parameters ε_{xx}, ε_{yy}, and ε_{zz} are called *principal dielectric permittivities* of the anisotropic crystal.

21.4 Classification of crystals

Depending on the relation between the principal dielectric permittivities ε_{xx}, ε_{yy}, and ε_{zz}, all crystals can be divided into three groups: isotropic, uniaxial, and biaxial. Crystals the three principal dielectric permittivities of which are equal, i.e.

$$\varepsilon_{xx} = \varepsilon_{yy} = \varepsilon_{zz}. \tag{21.28}$$

are called *isotropic*. If only two of the three principal permittivities are equal, i.e.

$$\varepsilon_{xx} = \varepsilon_{yy} \neq \varepsilon_{zz}, \tag{21.29}$$

the crystal is called *uniaxial*. For a *biaxial* crystal, all three principal dielectric permittivities are different:

$$\varepsilon_{xx} \neq \varepsilon_{yy} \neq \varepsilon_{zz}. \tag{21.30}$$

In the region of transparency, the crystals are also characterized by *refractive indices*. An isotropic crystal is characterized by one refractive index only:

$$n = \sqrt{\varepsilon}. \tag{21.31}$$

Table 21.1 Refractive indices of isotropic crystals.

Crystal	n
CdTe	2.69
NaCl	1.544
Diamond	2.417
GaAs	3.40

An uniaxial crystal has two principal refractive indices:

$$n_o = \sqrt{\varepsilon_{xx}} = \sqrt{\varepsilon_{yy}}, \qquad n_e = \sqrt{\varepsilon_{zz}}. \tag{21.32}$$

Finally, a biaxial crystal has three principal refractive indices:

$$n_x = \sqrt{\varepsilon_{xx}}, \qquad n_y = \sqrt{\varepsilon_{yy}}, \qquad n_z = \sqrt{\varepsilon_{zz}}. \tag{21.33}$$

Depending on the relation between the principal refractive indices, uniaxial crystals are subdivided into positive and negative ones. The crystals for which

$$n_o < n_e \tag{21.34}$$

are called *positive*, while the crystals for which

$$n_o > n_e \tag{21.35}$$

are called *negative*. Details of refractive indices for several crystals are given in Tables 21.1–21.3.

21.5 The polarization eigenstates of an optical wave in an anisotropic crystal

The main feature of light propagation in anisotropic crystals is that an optical wave with an arbitrary polarization state splits into two linearly polarized waves with orthogonal polarization directions, each travelling at a different velocity. To make this clear, let us first consider a simple case.

Suppose that a plane monochromatic optical wave propagates along one of the principal axes of a crystal, e.g. along the z-axis (Fig. 21.7). Denoting the unit vectors directed along the principal axes of the crystal by $\vec{x}_0$, $\vec{y}_0$, and $\vec{z}_0$, and the wave vector of the optical wave by $\vec{k}$, we can write

$$\vec{k} \parallel \vec{z}_0. \tag{21.36}$$

The optical wave is transverse, i.e. the relation $(\vec{k}, \vec{\mathcal{D}}) = 0$ holds. According to this relation and the condition (21.36), the vector $\vec{\mathcal{D}}$ lies in the x–y plane. Let us consider the following alternatives.

Table 21.2 Refractive indices of uniaxial crystals.

Crystal	n_o	n_e
Positive		
Ice	1.309	1.310
Quartz	1.544	1.553
BeO	1.717	1.732
Zircon	1.923	1.968
Rutile	2.616	2.903
ZnS	2.354	2.358
Negative		
$(NH_4)H_2PO_4$ (ADP)	1.552	1.478
Beryl	1.598	1.590
KH_2PO_4 (KDP)	1.507	1.467
$NaNO_3$	1.587	1.336
Calcite	1.658	1.486
Tourmaline	1.638	1.618
$LiNbO_3$	2.300	2.208
$BaTiO_3$	2.416	2.364
Proustite	3.019	2.739

Table 21.3 Refractive indices of biaxial crystals.

Crystal	n_x	n_y	n_z
Gypsum	1.520	1.523	1.530
Feldspar	1.522	1.526	1.530
Mica	1.522	1.582	1.588
Topaz	1.619	1.620	1.627
$NaNO_2$	1.344	1.411	1.651
SbSi	2.7	3.2	3.8
$YAlO_3$	1.923	1.938	1.947

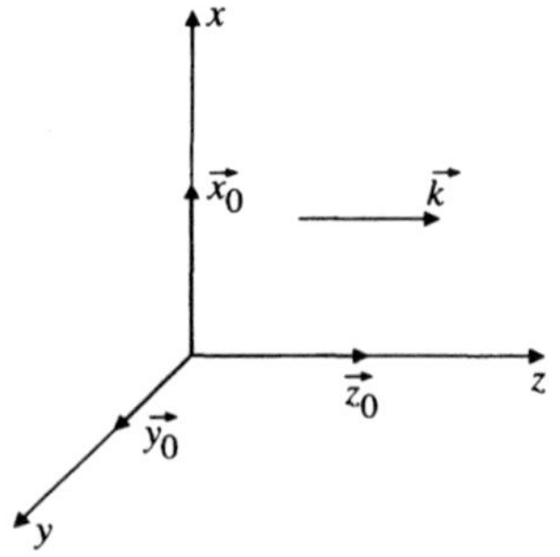

Fig. 21.7 The polarization eigenstates of an optical wave in an anisotropic crystal. The principal axes x, y, and z of the crystal and the optical wave vector $\vec{k}$ are shown.

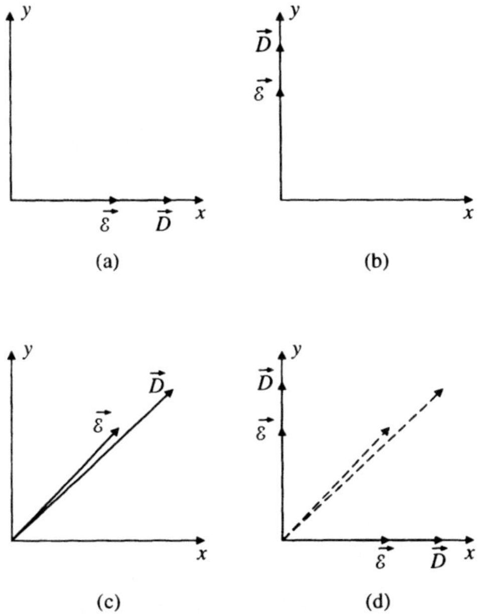

Fig. 21.8 The emergence of polarization eigenstates of an optical wave in an anisotropic crystal.

(a) Suppose that

$$\vec{\mathcal{D}} \parallel \vec{x}_0, \tag{21.37}$$

i.e. the wave is polarized along the principal x-axis of the crystal. Then $\mathcal{D}_y = \mathcal{D}_z = 0$, and eqns (21.27) yield $\mathcal{E}_y = \mathcal{E}_z = 0$, $\mathcal{D}_x = \varepsilon_{xx}\mathcal{E}_x$, whence it follows that the vector $\vec{\mathcal{D}}$ is parallel to the vector $\vec{\mathcal{E}}$:

$$\vec{\mathcal{D}} \parallel \vec{\mathcal{E}}. \tag{21.38}$$

This means that the medium behaves as though it is isotropic with respect to the given wave. The wave propagates with a velocity

$$v_1 = c/\sqrt{\varepsilon_{xx}}. \tag{21.39}$$

The wave polarization is stable (Fig. 21.8(a)).

(b) As before, let condition (21.36) hold, but this time let

$$\vec{\mathcal{D}} \parallel \vec{y}_0, \tag{21.40}$$

i.e. the optical wave is linearly polarized along the principal y-axis of the crystal. Using eqns (21.27), it is easy to show that the vectors $\vec{\mathcal{D}}$ and $\vec{\mathcal{E}}$ are parallel to each other again,

$$\vec{\mathcal{D}} \parallel \vec{\mathcal{E}} \parallel \vec{y}_0, \tag{21.41}$$

but that the wave velocity,

$$v_2 = c/\sqrt{\mathcal{E}_{yy}}, \tag{21.42}$$

differs from that determined by eqn (21.39). The wave polarization is stable (Fig. 21.8(b)).

(c) As before, let $\vec{k} \parallel \vec{z}_0$, but let the vector $\vec{\mathcal{D}}$ no longer be parallel to any of the principal axes of the crystal, i.e.

$$\vec{\mathcal{D}} \nparallel \vec{x}_0, \qquad \vec{\mathcal{D}} \nparallel \vec{y}_0. \tag{21.43}$$

Using eqns (21.27) and allowing for the anisotropy of the crystal, i.e. that

$$\varepsilon_{xx} \neq \varepsilon_{yy}, \tag{21.44}$$

we conclude that both of the vectors $\vec{\mathcal{D}}$ and $\vec{\mathcal{E}}$ lie in the x–y plane, but that they are no longer parallel (Fig. 21.8(c)). Here, however, the vectors $\vec{\mathcal{D}}$, $\vec{\mathcal{E}}$, and $\vec{k}$ are not coplanar and, consequently, a wave with such a configuration cannot propagate in the crystal because of Maxwell's equations (see Fig. 21.4).

What happens when a wave with polarization (21.43) enters the crystal? The wave will apparently break down into two waves with stable polarizations, travelling with different velocities:

$$\vec{\mathcal{D}} = \vec{x}_0\mathcal{D}_x + \vec{y}_0\mathcal{D}_y \tag{21.45}$$

(Fig. 21.8(d)). As the waves propagate in the crystal, the phase difference between the orthogonal components of the field will change. The wave in the crystal will therefore have an elliptical polarization, the parameters of which will vary with the distance covered by the wave in the crystal.

Thus, via this specific example, we have shown that an optical wave in an anisotropic crystal is generally unstable; it splits into two linearly polarized waves with orthogonal polarization directions. Both waves travel in the same direction but with different phase velocities.

In the example just considered, the optical wave propagating along the principal z-axis of the crystal has only two stable polarization states; that is, linear polarization along the x-axis and linear polarization along the y-axis. These polarization states can be called the *polarization eigenstates* of the optical wave in the anisotropic crystal.

It can be shown that in the general case, when the optical wave travels in an arbitrary direction with respect to the crystal, a similar effect takes place, i.e. the optical wave decomposes into waves with polarization eigenstates (Fig. 21.9).

For any direction $\vec{k}$ in the crystal there are only two allowed (inherent) polarization directions of the optical wave, which are orthogonal to each other. An

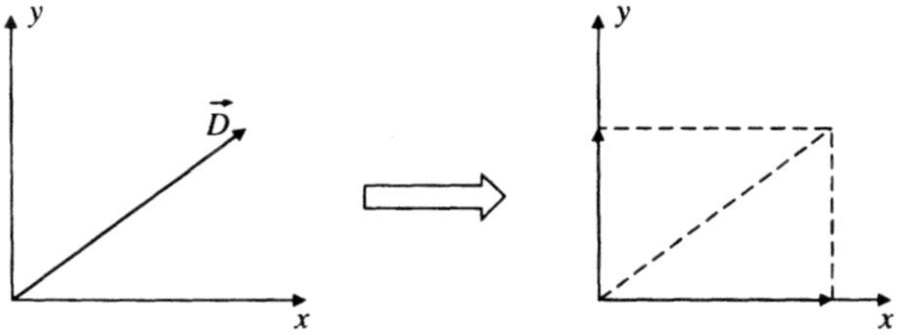

Fig. 21.9 The breakdown of a linearly polarized optical wave into waves with polarization eigenstates in an anisotropic crystal.

arbitrarily polarized wave breaks down into two linearly polarized intrinsic waves travelling with different velocities.

How can one find the polarization eigenstates, when the optical wave propagates in an arbitrary direction with respect to the crystal's principal axes? The simplest answer to this question can be provided, if the anisotropic crystal is uniaxial.

21.5.1 The uniaxial crystal

In view of condition (21.29), the principal z-axis is physically distinguished in a uniaxial crystal. This axis is called the optic axis of the crystal. Using the reasoning of the preceding section, it is easy to show that any optical wave travelling along the optic axis of an uniaxial crystal keeps its polarization invariable. The propagation velocity is independent of the wave polarization. There is only one such direction in a uniaxial crystal. This can be used to find the direction of the optic axis in the crystal.

Let a plane monochromatic optical wave propagate in a uniaxial anisotropic crystal in the direction characterized by the wave vector $\vec{k}$. The allowed polarization states of the wave are governed by two conditions: (i) the vectors $\vec{\mathcal{D}}$ and $\vec{\mathcal{E}}$ of the optical wave are generally not collinear due to anisotropy of the crystal, and (ii) the vectors $\vec{\mathcal{D}}$, $\vec{\mathcal{E}}$, and $\vec{k}$ must lie in the same plane, according to Maxwell's equations.

The first of these conditions (about the vectors $\vec{\mathcal{D}}$ and $\vec{\mathcal{E}}$ being not collinear) is expressed mathematically by eqns (21.27), which can be rewritten for the uniaxial crystal as

$$\mathcal{D}_x = n_0^2 \mathcal{E}_x, \qquad \mathcal{D}_y = n_0^2 \mathcal{E}_y, \qquad \mathcal{D}_z = n_e^2 \mathcal{E}_z \tag{21.46}$$

(see eqn (21.32)). The second condition (that the vectors $\vec{\mathcal{D}}$, $\vec{\mathcal{E}}$, and $\vec{k}$ are coplanar) can be written as

$$\left(\vec{k}, [\vec{\mathcal{E}}, \vec{\mathcal{D}}]\right) = 0. \tag{21.47}$$

where $(\vec{k}, [\vec{\mathcal{E}}, \vec{\mathcal{D}}])$ is the triple scalar product.

It should be stressed that eqns (21.46) and (21.47) are independent: eqns (21.46) are the material equations of the medium, while eqn (21.47) follows from Maxwell's equations. They are jointly valid only for some specific polarizations of the wave, which are called the polarization eigenstates of the optical wave in an anisotropic crystal.

Let us construct a vectorial product of the vectors $\vec{\mathcal{E}}$ and $\vec{\mathcal{D}}$. Denoting by $\vec{x}_0$, $\vec{y}_0$, and $\vec{z}_0$ the unit vectors directed along the principal axes of the uniaxial anisotropic crystal (with the optic z-axis) and allowing for the material equations (21.46), we obtain

$$[\vec{\mathcal{E}}, \vec{\mathcal{D}}] = \begin{vmatrix} \vec{x}_0 & \vec{y}_0 & \vec{z}_0 \\ \mathcal{E}_x & \mathcal{E}_y & \mathcal{E}_z \\ n_o^2 \mathcal{E}_x & n_o^2 \mathcal{E}_y & n_e^2 \mathcal{E}_z \end{vmatrix}. \tag{21.48}$$

One can see from eqn (21.48) that the vector $[\vec{\mathcal{E}}, \vec{\mathcal{D}}]$ is perpendicular to the z-axis. This can be written as the identity

$$\left(\vec{z}_0, [\vec{\mathcal{E}}, \vec{\mathcal{D}}]\right) = 0. \tag{21.49}$$

It follows from eqn (21.49) that the vectors $\vec{z}_0, \vec{\mathcal{E}}$, and $\vec{\mathcal{D}}$ should be coplanar. The conditions (21.47) and (21.49) specify the possible polarization states of the optical wave in the uniaxial crystal.

Both of the indicated conditions are satisfied in the following two cases. First, the vector $\vec{\mathcal{D}}$ is parallel to the vector $\vec{\mathcal{E}}$ and

$$[\vec{\mathcal{E}}, \vec{\mathcal{D}}] = 0. \tag{21.50}$$

Second, the vector $\vec{\mathcal{D}}$ lies in the plane of the vectors $\vec{z}_0$ and $\vec{k}$, which can be written mathematically as the identity

$$\left(\vec{z}_0, [\vec{k}, \vec{\mathcal{D}}]\right) = 0. \tag{21.51}$$

21.6 Ordinary and extraordinary waves

Both conditions can be expressed in the simplest way by introducing the notion of a *principal plane*, defined as a plane containing the vectors $\vec{k}$ and $\vec{z}_0$ (Fig. 21.10).

It is easy to see that the condition (21.50) is satisfied for a wave *polarized perpendicular to the principal plane*. Such a wave is called an *ordinary wave* and is denoted by the subscript "o". The polarization vector of this wave is perpendicular to the optic axis of the crystal, the vectors $\vec{\mathcal{D}}_o$ and $\vec{\mathcal{E}}_o = \vec{\mathcal{D}}_o/n_o^2$ are parallel, and the velocity of the ordinary wave, $v_o = c/n_o$, is independent of the propagation direction in the crystal.

The condition (21.51) holds for a wave *polarized in the principal plane*. Such a wave is called an *extraordinary wave* and is denoted by the subscript "e". The polarization vector $\vec{\mathcal{D}}_e$ of this wave is not perpendicular to the optic axis of the crystal, the vectors $\vec{\mathcal{D}}_e$ and $\vec{\mathcal{E}}_e$ are not parallel to each other (Fig. 21.10), and

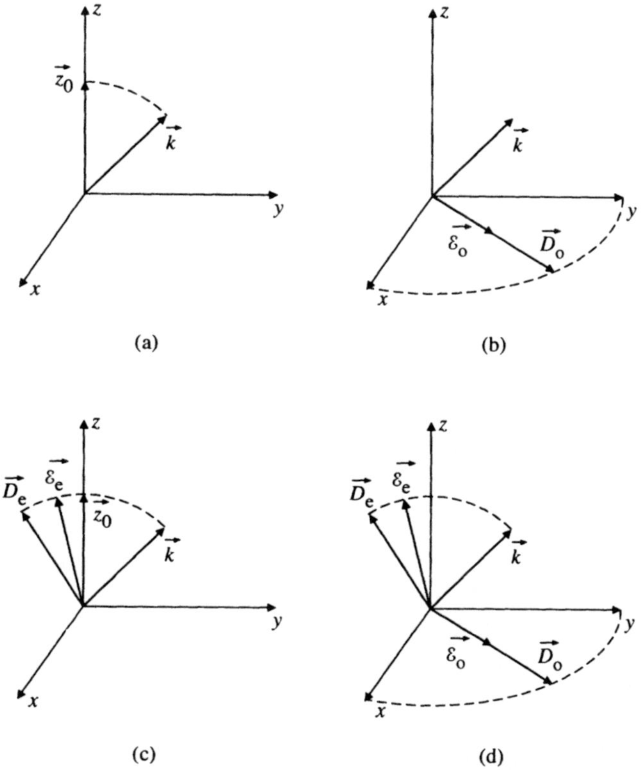

Fig. 21.10 Polarizations of ordinary and extraordinary optical waves in an uniaxial anisotropic crystal. (a) The principal plane. (b) Polarization of the ordinary wave. (c) Polarization of the extraordinary wave. (d) The mutual orientation of polarizations of the ordinary and extraordinary waves. A dashed line joins the ends of the vectors lying in the same plane.

the velocity of the extraordinary wave, $v_e = c/n$, depends on the propagation direction in the crystal.

The polarization vectors of ordinary and extraordinary waves, $\vec{\mathcal{D}}_o$ and $\vec{\mathcal{D}}_e$, are mutually perpendicular:

$$\vec{\mathcal{D}}_o \perp \vec{\mathcal{D}}_e. \tag{21.52}$$

We have thus shown that there are only two allowed (inherent) polarization directions of an optical wave for any direction $\vec{k}$ in uniaxial anisotropic crystals. One of these is perpendicular and the other parallel to the principal plane. Having entered the crystal, a wave with an arbitrary polarization state breaks down into two linearly polarized waves with mutually orthogonal (inherent) polarization directions. The propagation velocities of these waves differ. The velocity of the ordinary wave, $v_o = c/n_o$, is independent of the propagation direction. The

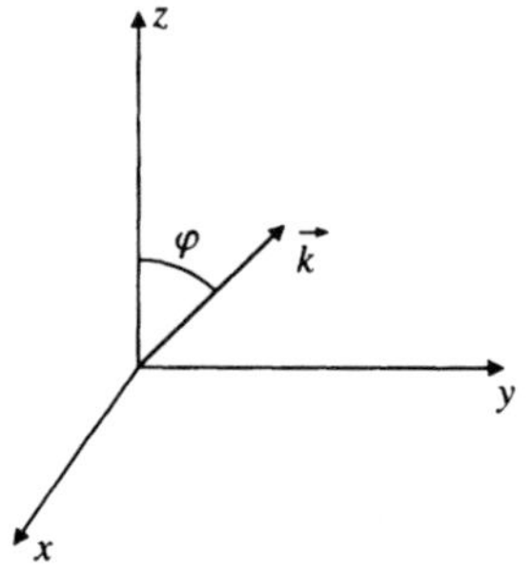

Fig. 21.11 The calculation of the velocity of the extraordinary wave in an anisotropic crystal. $\vec{k}$ is the wave vector of the optical wave and z is the optic axis of the crystal.

velocity of the extraordinary wave depends on the propagation direction in the crystal, falling within the range from c/n_o to c/n_e.

Let us now calculate the velocity of the extraordinary wave for a given direction of propagation in the crystal.

21.6.1 The propagation velocity of an extraordinary wave and the refractive index ellipsoid

Suppose that the extraordinary wave propagates in a direction $\vec{k}$, which does not coincide with any principal axis of the uniaxial anisotropic crystal (Fig. 21.11). Let us calculate the wave velocity as a function of the propagation direction $v = v(\vec{k})$.

We shall use the dispersion equation that follows from eqn (21.3):

$$\left[\vec{k}, [\vec{k}, \vec{\mathcal{E}}]\right] + \frac{\omega^2}{c^2}\vec{\mathcal{D}} = 0$$

or

$$(\vec{k}, \vec{\mathcal{E}})\vec{k} - k^2\vec{\mathcal{E}} + \frac{\omega^2}{c^2}\vec{\mathcal{D}} = 0. \tag{21.53}$$

For the ordinary wave, we have $\vec{\mathcal{D}} = n_0^2\mathcal{E}$ and $(\vec{k}, \vec{\mathcal{E}}) = 0$. According to eqn (21.53) $k = (\omega/c)n_0$ and $v_0 = c/n_0$ irrespective of the direction $\vec{k}$. For the extraordinary wave, we have $(\vec{k}, \vec{\mathcal{E}}) \neq 0$ and, consequently, we need to solve the general equation (21.53).

Let us introduce the unit vector $\vec{m}$ directed along the vector $\vec{k}$ of the optical wave,

$$\vec{m} = \vec{k}/k, \tag{21.54}$$

and the refractive index for the extraordinary wave n related to the wave number k by the formula

$$k = \frac{\omega}{c}n. \tag{21.55}$$

Substituting eqns (21.54) and (21.55) into eqn (21.53), we obtain the equation

$$\vec{\mathcal{E}} - \frac{1}{n^2}\vec{\mathcal{D}} = (\vec{m}, \vec{\mathcal{E}})\vec{m}. \tag{21.56}$$

For the uniaxial anisotropic crystal, the components of the electric displacement vector are related to the components of the electric field vector of the optical wave by the formulae (21.46). Writing eqn (21.56) in Cartesian components and allowing for eqn (21.46), we obtain three scalar equations:

$$\mathcal{D}_x = \frac{(\vec{m}, \vec{\mathcal{E}})m_x}{\dfrac{1}{n_o^2} - \dfrac{1}{n^2}}, \quad \mathcal{D}_y = \frac{(\vec{m}, \vec{\mathcal{E}})m_y}{\dfrac{1}{n_o^2} - \dfrac{1}{n^2}}, \quad \mathcal{D}_z = \frac{(\vec{m}, \vec{\mathcal{E}})m_z}{\dfrac{1}{n_e^2} - \dfrac{1}{n^2}}, \tag{21.57}$$

where n_o and n_e are the principal refractive indices of the uniaxial crystal. Now we shall use the equation

$$(\vec{m}, \vec{\mathcal{D}}) = 0, \tag{21.58}$$

following from the equation $(\vec{k}, \vec{\mathcal{D}}) = 0$ (see eqns (21.3)) and eqn (21.54). We obtain, from eqn (21.58),

$$m_x\mathcal{D}_x + m_y\mathcal{D}_y + m_z\mathcal{D}_z = 0. \tag{21.59}$$

Substituting eqn (21.57) into eqn (21.59) and considering that in the given case $(\vec{m}, \vec{\mathcal{E}}) \neq 0$), we obtain the equation

$$\frac{m_x^2 + m_y^2}{\dfrac{1}{n_o^2} - \dfrac{1}{n^2}} + \frac{m_z^2}{\dfrac{1}{n_e^2} - \dfrac{1}{n^2}} = 0. \tag{21.60}$$

In principle, this equation solves the problem: it allows one to calculate the refractive index for an extraordinary wave travelling in an arbitrary direction $\vec{m}$ within the crystal. This equation is called *Fresnel's equation of the wave normals.*

The Fresnel equation can be transformed into a more compact form if we use the angle φ between the wave vector and the optic axis of the crystal (see Fig. 21.11), instead of the vector $\vec{m}$ characterizing the propagation direction of the optical wave in the crystal. Considering that $|\vec{m}| = 1$ and, consequently,

$$m_x^2 + m_y^2 + m_z^2 = 1,$$

we obtain $m_z = \cos\varphi$, $m_x^2 + m_y^2 = \sin^2\varphi$. Equation (21.60) then takes the form

$$\frac{\sin^2\varphi}{\dfrac{1}{n_o^2} - \dfrac{1}{n^2}} + \frac{\cos^2\varphi}{\dfrac{1}{n_e^2} - \dfrac{1}{n^2}} = 0$$

or

$$\frac{\sin^2\varphi}{n_e^2} + \frac{\cos^2\varphi}{n_o^2} = \frac{1}{n^2}. \tag{21.61}$$

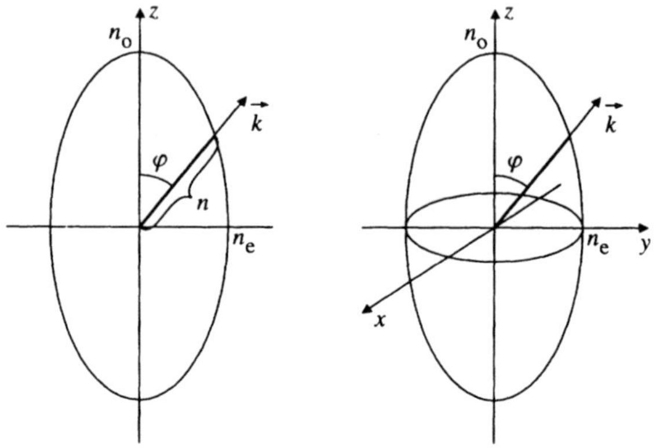

Fig. 21.12 The ellipse and ellipsoid of the refractive index.

The solution to eqn (21.61) is

$$n(\varphi) = \frac{n_o n_e}{\sqrt{n_e^2 \cos^2 \varphi + n_o^2 \sin^2 \varphi}}. \tag{21.62}$$

Thus, using eqn (21.62) and knowing the principal refractive indices n_o and n_e of the uniaxial anisotropic crystal, one may calculate the refractive index for the extraordinary wave n and its phase velocity

$$v = c/n \tag{21.63}$$

for any direction φ in the crystal. The quantities n_o and n_e for several crystals are given in Table 21.2.

The dependence $n(\varphi)$ is illustrated graphically in Fig. 21.12, which shows an ellipse with semi-axes equal to the principal refractive indices of the crystal n_o and n_e. The value of the refractive index for the extraordinary wave n propagating in direction $\vec{k}$ is equal to the length of the bold line section.

Figure 21.12 shows a drawing of the principal plane section. Due to the axial symmetry of the uniaxial crystal, the three-dimensional surface of the refractive index has the form of an ellipsoid of revolution with semi-axes n_o and n_e (see Fig. 21.12(b)). This ellipsoid is called an ellipsoid of the wave normals or a refractive index ellipsoid.

The refractive index surface for the ordinary wave is a sphere of radius n_o. The sphere and ellipsoid of refractive indices make contact in the direction of the optic axis of the crystal. For a positive uniaxial crystal ($n_o < n_e$), the sphere is inscribed within the ellipsoid, and, conversely, for a negative crystal ($n_o > n_e$), the ellipsoid is inscribed within the sphere (Fig. 21.13).

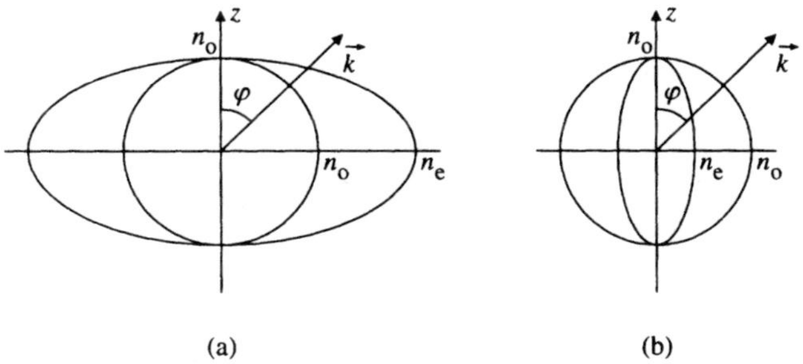

Fig. 21.13 The sphere and ellipsoid of the refractive index for (a) positive and (b) negative uniaxial anisotropic crystals.

21.7 Optical birefringence at the anisotropic medium boundary

Experiment shows that light refraction at the boundary of an anisotropic crystal has several characteristic features. The most important are listed below:

1. Having entered the crystal, an optical ray splits into two rays, demonstrating the effect of *optical birefringence*. Irrespective of the polarization state of the incident light, both refracted beams turn out to be linearly polarized, the polarization directions being mutually perpendicular (see Figs 21.1 and 21.2).
2. One of the refracted rays may not lie in the plane of incidence.
3. The refraction may take place even for normal incidence of the ray on the anisotropic crystal boundary.

To explain all of these facts, we shall consider the refraction of light at the boundary of a uniaxial anisotropic crystal (Fig. 21.14). The boundary conditions for the electromagnetic field require that the tangential, i.e. parallel to the boundary, components of the electric and magnetic fields be continuous across the boundary:

$$E_{t1} = E_{t2}, \qquad H_{t1} = H_{t2}.$$

As with isotropic media, it follows that the tangential components of the wave vectors of the incident, reflected, and refracted waves should be equal to each other:

$$k_{1x} = k_{0x} = k_{2x}. \tag{21.64}$$

These equations, in turn, lead to the law of reflection (which says that the angle of incidence is equal to the angle of reflection),

$$\theta_1 = \theta_0, \tag{21.65}$$

and to the law of refraction (Snell's law),

$$n_1 \sin\theta_1 = n_2 \sin\theta_2. \tag{21.66}$$

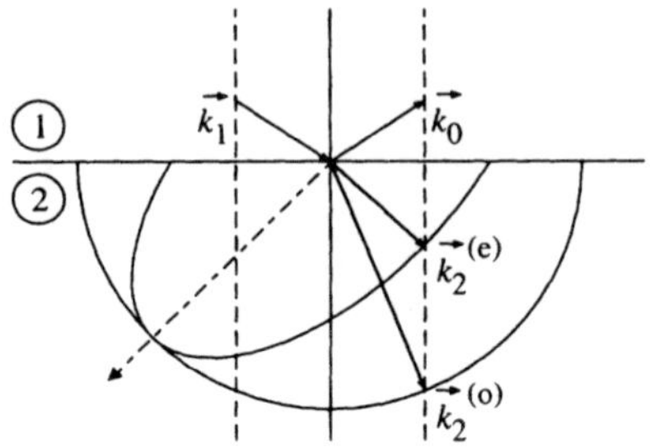

Fig. 21.14 The refraction of light at the boundary of an uniaxial anisotropic crystal. The optic axis of the crystal (shown dashed) is assumed to lie in the plane of incidence.

However, since the refractive indices for the ordinary and extraordinary waves in the anisotropic crystal are generally different, the law of refraction (21.66) should be written separately for the ordinary and extraordinary waves:

$$n_1 \sin\theta_1 = n_2^{(o)} \sin\theta_2^{(o)}, \tag{21.67}$$

$$n_1 \sin\theta_1 = n_2^{(e)} \sin\theta_2^{(e)}. \tag{21.68}$$

Construction of the wave vectors for refracted waves is shown in Fig. 21.14. One can see that the angles of refraction for the ordinary and extraordinary waves are different.

Note also that the ray direction (the direction of the energy flow),

$$\vec{S} = \frac{c}{4\pi}[\vec{E}, \vec{H}],$$

for the extraordinary wave does not generally coincide with the direction of the wave vector (see Figs 21.4 and 21.10(c)). It is this circumstance that clarifies the light refraction observed in an anisotropic crystal for normal incidence.

A picture of double refraction for normal incidence is shown in Fig. 21.15. A plane monochromatic optical wave with wave vector $\vec{k}_1$ falls normally on to the boundary of a uniaxial anisotropic crystal so that the crystal optic axis, shown dashed, lies in the plane of incidence. According to eqns (21.64) we have for normal incidence $k_{1x} = k_{0x} = k_{2x} = 0$ and, consequently, the wave vectors of the optical waves in the crystal are directed normally to the boundary. The ordinary ray $\vec{S}_o$ has the same direction. Simultaneously, the extraordinary ray deflects from the normal, producing the optical birefringence effect, which we have observed experimentally (Fig. 21.1).

How can one find the direction of the extraordinary ray? Let us show that in the uniaxial anisotropic crystal the extraordinary ray is directed normally to the surface of the refractive index ellipsoid.

Since the ray $\vec{S}$ is orthogonal to the electric field vector $\vec{\mathcal{E}}$ of the optical wave, it will suffice to show that the vector $\vec{\mathcal{E}}$ of the extraordinary wave is parallel to the tangent to the refractive index ellipsoid.

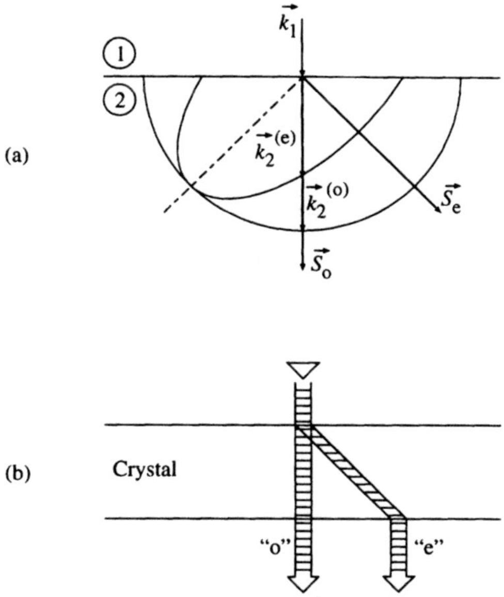

Fig. 21.15 Optical birefringence for the normal incidence of light. (a) The wave vectors. (b) The optical beams.

Let $\vec{n}$ denote the vector drawn from the centre of the refractive index ellipsoid to the point of intersection of the ellipsoid with the wave vector $\vec{k}$ of the extraordinary wave (Fig. 21.16). The modulus of the vector $\vec{n}$ is equal to the refractive index n for the extraordinary wave propagating in the given direction $\vec{k}$. The direction of the tangent to the refractive index ellipsoid is determined by the vector $d\vec{n}$ that denotes the increment of the vector $\vec{n}$ due to the increase in the angle φ by the value $d\varphi$.

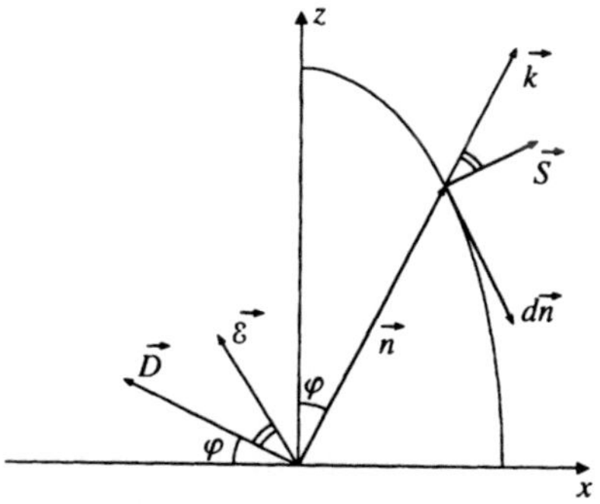

Fig. 21.16 The derivation of the direction of the extraordinary ray.

Thus, we need to prove that the vector $\vec{\mathcal{E}}$ is parallel to the vector $d\vec{n}$. To do this, it will suffice to show that their vector product is equal to zero, i.e.

$$[\vec{\mathcal{E}}, d\vec{n}] = 0. \tag{21.69}$$

Suppose, for simplicity, that the principal plane of the crystal shown in Fig. 21.16 is perpendicular to one of the principal axes, e.g. to the y-axis. The vector product (21.69) can then be rewritten in Cartesian coordinates as

$$[\vec{\mathcal{E}}, d\vec{n}] = \begin{vmatrix} \vec{x}_0 & \vec{y}_0 & \vec{z}_0 \\ \mathcal{E}_x & 0 & \mathcal{E}_z \\ dn_x & 0 & dn_z \end{vmatrix} = \vec{y}_0(\mathcal{E}_z dn_x - \mathcal{E}_x dn_z).$$

Let us prove that

$$\mathcal{E}_z dn_x - \mathcal{E}_x dn_z = 0. \tag{21.70}$$

Using the material equation (21.46) and the expressions

$$\mathcal{D}_x = -\mathcal{D}\cos\varphi, \qquad \mathcal{D}_z = \mathcal{D}\sin\varphi,$$

that follow from Fig. 21.16, we rewrite eqn (21.70) in the form

$$n_{\text{o}}^2 \sin\varphi\, dn_x + n_{\text{e}}^2 \cos\varphi\, dn_z = 0. \tag{21.71}$$

It can be seen from Fig. 21.16 that

$$n_x = n\sin\varphi, \qquad n_z = n\cos\varphi, \tag{21.72}$$

where $n = n(\varphi)$ is the function determined by the Fresnel equation (21.61). It follows from eqns (21.72) and (21.61) that

$$dn_x = n\cos\varphi\, d\varphi + \sin\varphi\, dn, \qquad dn_z = -n\sin\varphi\, d\varphi + \cos\varphi\, dn, \tag{21.73}$$

where

$$dn = \left(\frac{1}{n_{\text{o}}^2} - \frac{1}{n_{\text{e}}^2}\right) n^3 \sin\varphi\, \cos\varphi\, d\varphi. \tag{21.74}$$

Substituting eqns (21.73) and (21.74) into eqn (21.71) and considering eqn (21.61), we obtain an identity. This means that the relation (21.69) holds and, consequently, the optical ray is indeed directed normally to the refractive index ellipsoid.

In the examples just considered, the optic axis of the uniaxial anisotropic crystal was supposed to lie in the plane of incidence of the optical wave on the crystal boundary (see Figs 21.14 and 21.15). Here, the refracted ray also lies in the plane of incidence. If, however, the plane of incidence does not coincide with the principal plane, the refracted extraordinary ray no longer lies in the plane of

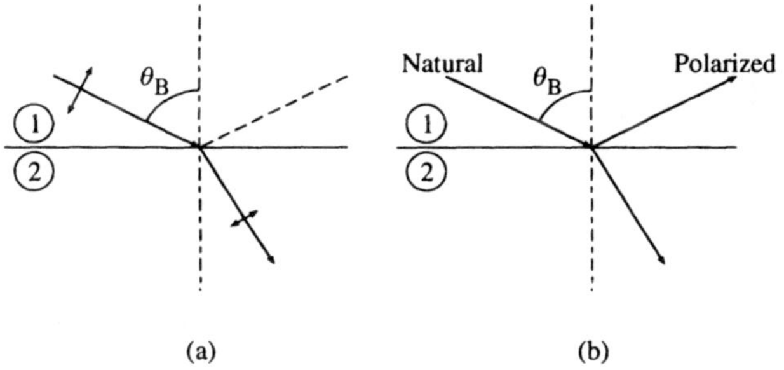

Fig. 21.17 The anisotropic reflection of light. Reflection of the light incident at the Brewster angle allows one (a) to analyse its polarization and (b) to produce polarized light.

incidence because the deflection of this ray from the wave vector occurs in the principal plane, i.e. in the direction of the optic axis of the crystal. Thus, the refraction of light in the anisotropic crystal does not necessarily take place in the plane of incidence.

21.8 The production and analysis of polarized light

Any physical phenomenon sensitive to the polarization of light can be used for the production and analysis of polarized light. Anisotropic reflection, absorption, and refraction all belong to the range of such phenomena.

21.8.1 Anisotropic reflection

It is known that the reflection of light at the boundary between two isotropic media depends essentially on the polarization of the incident optical wave (see Chapter 20). If the incident light is linearly polarized, then under certain conditions reflection can be totally suppressed. This happens when the optical wave polarized in the plane of incidence falls onto the boundary at the angle equal to the Brewster angle:

$$\theta_B = \arctan(n_2/n_1).$$

Using this phenomenon, one can confirm that the polarization is linear and determine the polarization direction (Fig. 21.17). If the non-polarized (natural) light is directed onto the glass plate at the Brewster angle, then the light reflected from the plate is linearly polarized and its polarization direction is perpendicular to the plane of incidence (Fig. 21.17(b)).

21.8.2 Anisotropic absorption

There are anisotropic crystals that are characterized by different absorption coefficients with respect to the ordinary and extraordinary waves. For example, a tourmaline crystal strongly absorbs an ordinary wave. Directing a ray of non-polarized light onto a crystal plate of sufficient thickness, we can obtain linearly

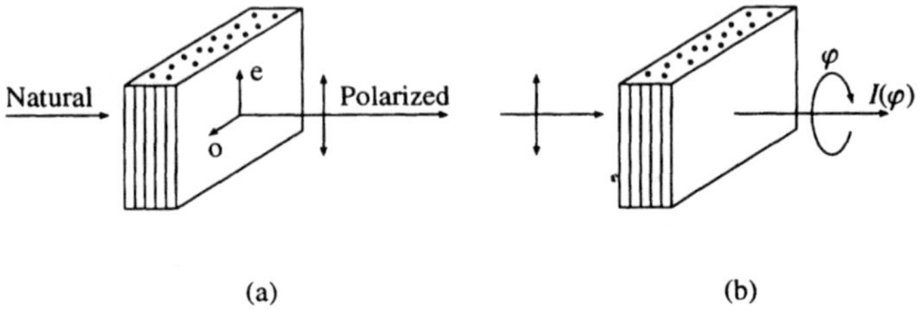

Fig. 21.18 The anisotropic absorption of light. The production (a) and analysis (b) of polarized light using a tourmaline crystal.

polarized light at the crystal exit, with the extraordinary wave passing through the crystal and the ordinary wave being absorbed. Here, the crystal acts as a polarizer of light (Fig. 21.18).

If we direct a ray of linearly polarized light at the crystal, then the light transmission will depend on the mutual orientation of the crystal and the light polarization direction. By rotating the crystal around the axis of the optical beam and observing the variation in the intensity of the transmitted light, we can confirm that the polarization is linear and find the polarization direction (Fig. 21.18(b)). Here, the crystal acts as an analyser. Film polaroids using other anisotropic materials operate in a similar way.

21.8.3 Anisotropic refraction

The phenomenon of optical birefringence can also be applied to produce polarized light (Fig. 21.19). Both optical beams (ordinary and extraordinary) are known to be linearly polarized. As can be seen from Fig. 21.15, the angle of optical birefringence is a maximum when the crystalline plate is cut at an angle of 45° to the optic axis.

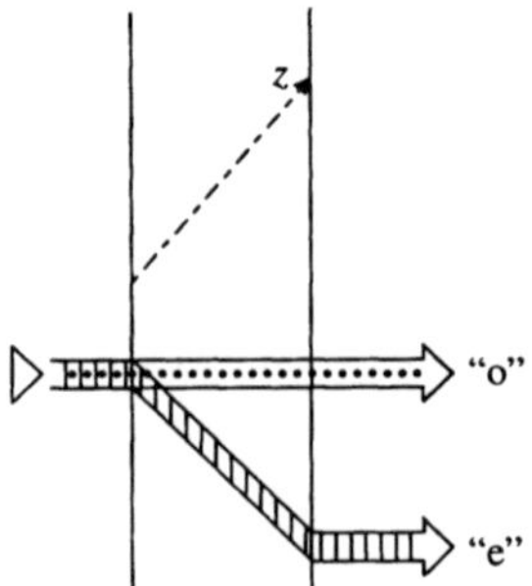

Fig. 21.19 The anisotropic refraction of light. The production of polarized light using the phenomenon of optical birefringence.

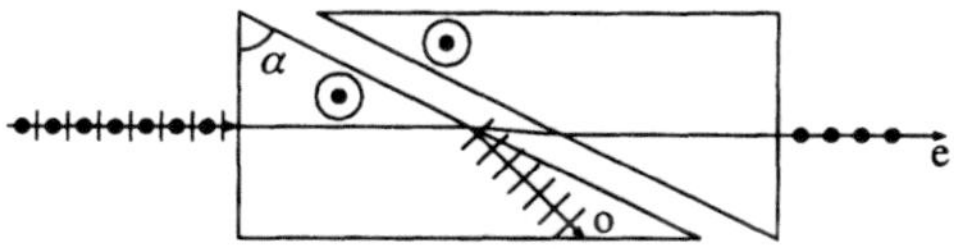

Fig. 21.20 The Glan prism.

21.8.4 The Glan prism

Polarization prisms such as the Glan prism and the Nicol prism are most frequently used for the production and analysis of polarized light. A schematic diagram of the Glan prism is shown in Fig. 21.20. The prism consists of two calcite crystals separated by an air gap. The principal refractive indices of calcite are $n_o = 1.658$ and $n_e = 1.486$. The optic axis of the crystal is perpendicular to the plane of the figure. The original optical beam is incident normally on the face of the prism. Since the beam is perpendicular to the optic axis of the crystal, optical birefringence does not take place (see Fig. 21.15). The optical wave propagating in the crystal maintains the direction of the incident wave, but breaks down into ordinary and extraordinary waves, with the difference between the refractive indices for these waves reaching its maximum value:

$$\Delta n = n_o - n = n_o - n_e = \Delta n_{\max}$$

(see Fig. 21.13). The angle α of the prism is chosen so that the angle of incidence of the light beam at the crystal–air boundary is close to the limiting angle of total internal reflection. Since the jump in the refractive index across this boundary is greater for the ordinary wave, this wave undergoes total internal reflection while the extraordinary wave crosses the boundary. The extraordinary wave then passes through the air gap and through the second crystal, leaving the prism as a linearly polarized beam of light. In this arrangement, the prism operates as a polarizer of light. The Glan prism can apparently be used as an analyser as well. The transmission coefficient of the prism with respect to linearly polarized light depends on the mutual orientation of the wave polarization vector and the optic axis of the crystals.

21.8.5 The compensator

This is an optical device intended gradually to vary the phase difference between the orthogonal components of the optical field. Using the compensator, one can transform elliptically polarized light into linearly polarized light and measure the parameters of a polarization ellipse.

Figure 21.21 shows a schematic diagram of the *Babinet compensator*. The compensator consists of two quartz wedges, together forming a plane-parallel plate. The principal refractive indices of quartz are $n_o = 1.544$ and $n_e = 1.553$. The optic axes of the wedges are directed perpendicular to each other and to the incident optical beam. In this arrangement, optical birefringence does not

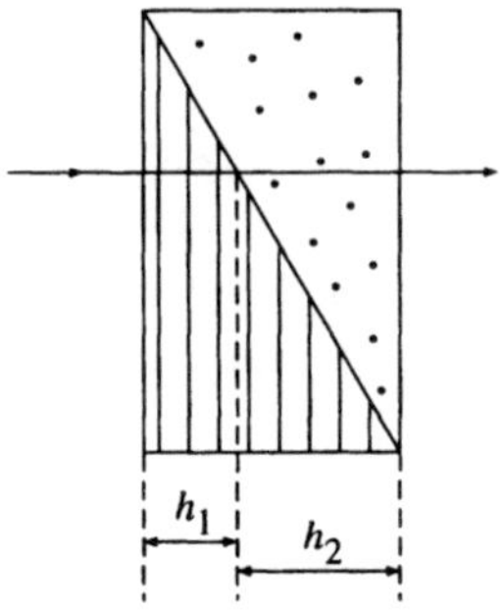

Fig. 21.21 The Babinet compensator.

occur, but the incident optical wave splits into ordinary and extraordinary waves travelling with different velocities. The phase difference between the orthogonal components of the field, introduced by the compensator, is determined by the expression

$$\Delta\varphi = \frac{2\pi}{\lambda}(n_e - n_o)(h_1 - h_2),$$

where λ is the optical wavelength, n_o and n_e are the principal refractive indices of the crystal, and h_1 and h_2 are the distances traversed by the optical beam in the first and in the second crystals, respectively. The phase difference $\Delta\varphi$ can be varied gradually by displacing the compensator across the optical beam.

Figure 21.22 shows an experimental arrangement for measuring the degree of light polarization. A broad beam of light passes successively through the compensator and the analyser to produce a system of alternating dark and light bands on the screen. The visibility (degree of contrast) of this pattern,

$$\gamma = \frac{I_{\text{max}} - I_{\text{min}}}{I_{\text{max}} + I_{\text{min}}},$$

characterizes the degree of light polarization. The limiting cases $\gamma = 1$ and $\gamma = 0$ correspond to totally polarized and totally non-polarized light, respectively.

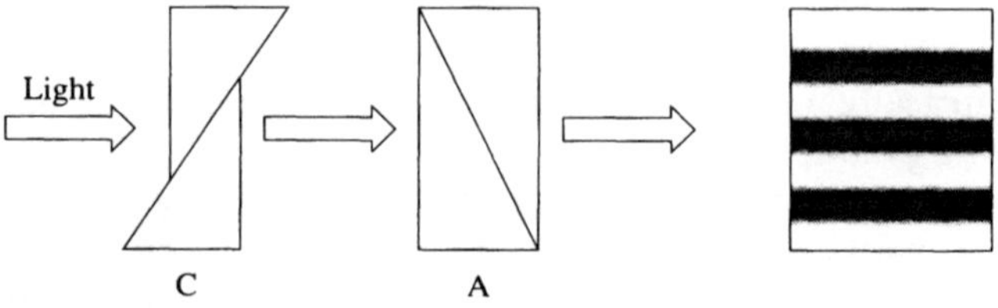

Fig. 21.22 Measuring the degree of light polarization. C, Compensator; A, analyser. The pattern observed on the screen is shown on the right.

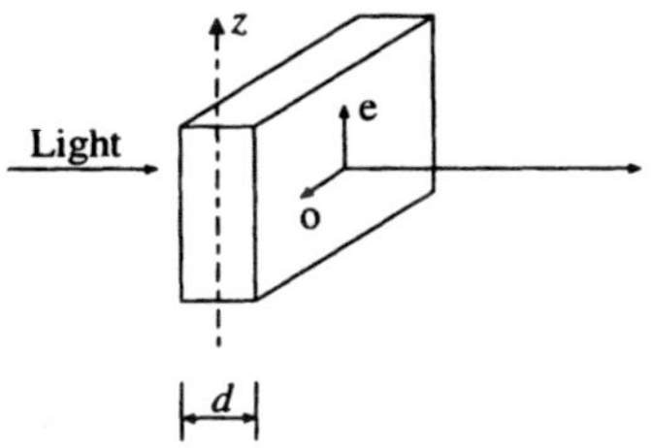

Fig. 21.23 Light polarization control with a crystalline plate. z is the optic axis of the crystal.

21.8.6 Light polarization control

The physical phenomena accompanying light propagation in an anisotropic crystal can be employed for light polarization control. For example, a compensator allows one to control the polarization ellipse parameters of a polarized optical beam. In practice, one often needs to convert linear polarization into circular polarization or vice versa, and also to vary the direction of the linear polarization or the direction of rotation in a circularly polarized optical wave. Special plates of anisotropic crystals, so-called quarter-wave and half-wave plates, are used for these purposes.

Let a plane monochromatic optical wave fall normally onto a crystalline plate cut of uniaxial anisotropic crystal parallel to its optic axis, i.e. so that the optic axis of the crystal is parallel to the face illuminated by the light (Fig. 21.23). Having entered the crystal, the optical wave breaks down into two linearly polarized waves with orthogonal directions of polarization; that is, into ordinary and extraordinary waves. The refractive index of the crystal for the ordinary wave is equal to n_o and for the extraordinary wave to n_e. This causes a phase shift between the ordinary and extraordinary waves:

$$\Delta\varphi = \frac{2\pi}{\lambda}(n_o - n_e)d, \tag{21.75}$$

where λ is the wavelength and d is the thickness of the plate. Variation of the phase difference between the orthogonal components of the optical field changes its polarization state. It is this effect that underlies the operation of the above-mentioned plates.

21.8.7 The quarter-wave plate

Such a plate transforms the linearly polarized light into circularly polarized light and vice versa. The plate thickness d is chosen so that $\Delta\varphi = \pi/2$, or

$$(n_o - n_e)d = \lambda/4.$$

Thus, the path difference between the ordinary and extraordinary waves in the plate is equal to a quarter of a wavelength.

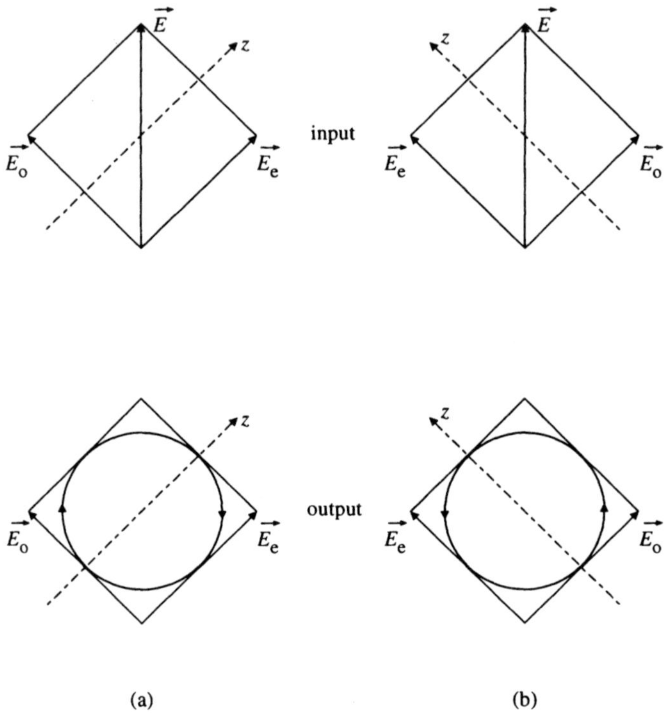

Fig. 21.24 The conversion of linearly polarized light into circularly polarized light by a quarter-wave plate. The incident optical wave is linearly polarized in the vertical direction. The optic axis of the crystal is oriented at an angle of 45° to the vertical. The cases of right-polarized and left-polarized light are shown.

To convert linearly polarized light into circularly polarized light, we orient the plate so that the optic axis of the crystal makes an angle of 45° with the polarization direction of the incident wave (Fig. 21.24). In this arrangement, the orthogonal vibrations of the optical field in the ordinary and extraordinary waves have the same amplitudes and phases at the front face of the plate. At the back face of the plate, these vibrations turn out to have phases shifted by $\pi/2$ and, consequently, the output beam has circular polarization. By turning the plate through 90° about the axis of the optical beam, we can reverse the direction of the circular polarization.

The inverse operation, i.e. conversion of circular polarization to linear polarization, can be done in a like manner.

21.8.8 The half-wave plate

The half-wave plate rotates the polarization plane through 90°. Then $\Delta\varphi = \pi$ and

$$(n_o - n_e)d = \lambda/2.$$

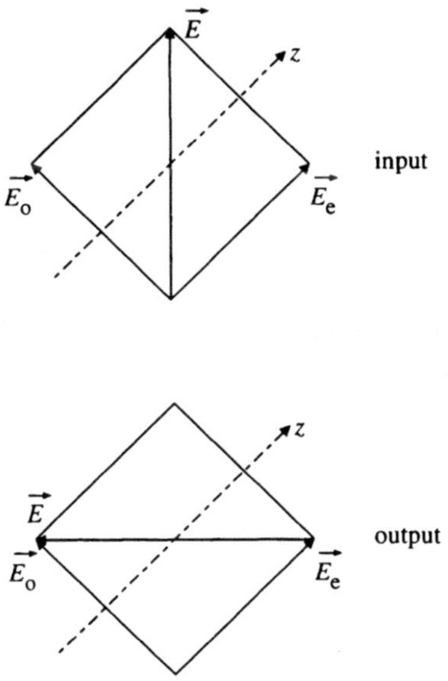

Fig. 21.25 The conversion of a vertically polarized optical wave into an horizontally polarized wave by a half-wave plate.

The operation of the half-wave plate is illustrated in Fig. 21.25.

The performance of quarter-wave and half-wave plates can be demonstrated with the help of a helium–neon laser. A block diagram of the experiment is shown in Fig. 21.26. First, we project the laser beam directly onto the screen, on which a bright red spot is observed. Then we insert a polaroid into the laser beam to observe the variation in the brightness of the spot when the polaroid is rotated about the beam axis. At a certain position of the polaroid, the spot on the screen vanishes completely. This observation proves that the laser radiation has linear polarization.

Second, we insert a quarter-wave plate into the laser beam and orient it so that the optic axis of the crystal makes an angle of 45° with the polarization plane of the laser beam. We can note that the plate does not change the brightness of the spot on the screen. The polarization of the light leaving the plate is

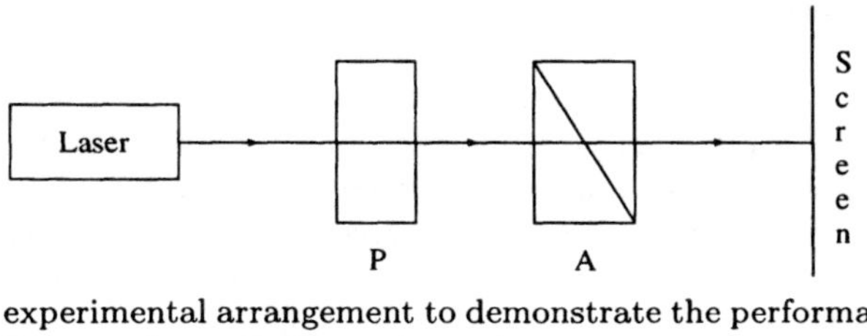

Fig. 21.26 An experimental arrangement to demonstrate the performance of the quarter-wave and half-wave plates. P, Plate; A, analyser.

then studied using the polaroid. This experiment shows that the rotation of the polaroid does not vary the brightness of the spot, which indicates that the light emerging from the plate has circular polarization.

If the angle between the optic axis of the crystal and the polarization direction of the laser beam differs from 45°, the rotation of the polaroid causes periodic variations of light intensity on the screen. However, the spot on the screen does not vanish altogether, which means that the light emerging from the plate has elliptical polarization.

In a similar fashion, we explain the performance of the half-wave plate. The experiment shows that the half-wave plate rotates the polarization plane of the laser beam through 90°. First, we place an analyser into the laser beam and orient it so that the screen becomes totally dark. Then we insert the half-wave plate into the beam in front of the analyser and see a bright spot of light appearing on the screen. Turning the analyser through 90°, we again observe total darkness on the screen.

21.9 The interference of polarized rays

Destructive interference of two light waves is obviously possible only if the waves have the same polarization direction. If the waves are polarized in mutually perpendicular planes, they cannot cancel each other for any phase difference and, consequently, cannot produce an interference pattern. Moreover, it is easy to show that the light intensity is independent of the phase difference between the orthogonal components of the field. Indeed, setting

$$\vec{E} = \vec{x}_0 A_1 \cos(\omega t - kz) + \vec{y}_0 A_2 \cos(\omega t - kz + \varphi),$$

we obtain

$$I = I_1 + I_2,$$

where

$$I = \frac{c}{4\pi}\langle E^2 \rangle, \qquad I_1 = \frac{c}{8\pi} A_1^2, \qquad I_2 = \frac{c}{8\pi} A_2^2,$$

and the angular brackets denote averaging over the optical vibration period.

Therefore, it follows that the ordinary and extraordinary waves propagating in a crystal cannot interfere with each other. Experiment confirms this conclusion. The optical beam retains its original intensity after crossing a plate of transparent anisotropic crystal. The intensity remains invariable irrespective of the phase difference arising between the ordinary and extraordinary waves in the crystal.

However, the phase shift arising between the ordinary and extraordinary waves in an anisotropic crystal affects the polarization of the light. The change in the polarization can be transformed into a change in intensity by passing the optical beam through a polaroid. Thus, using a polarizing device, one can notice that the light intensity varies with the phase difference between the orthogonal components of the optical field. Such phenomena are caused by *interference of polarized rays*.

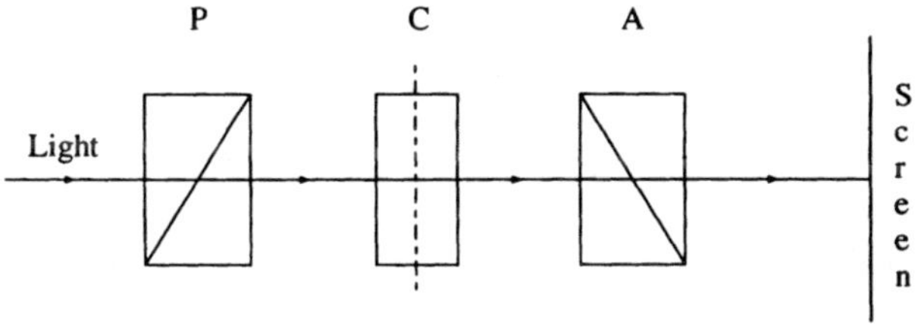

Fig. 21.27 An experimental set-up for the observation of interference of polarized rays. P, Polarizer; C, crystal; A, analyser.

A block diagram of the experimental arrangement intended for observation of the interference of polarized rays is shown in Fig. 21.27. A source of light, a polarizer, and an analyser are the basic elements of the arrangement. First, we set the polarizer and analyser in the crossed position, i.e. so that the system as a whole does not transmit any light. Then, we insert a plate of anisotropic crystal into the optical beam between the polarizer and the analyser. A light spot appears on the screen, proving that the system transmits the light.

If the positions of the polarizer and analyser are fixed, then the light intensity on the screen depends on the mutual orientation of the polarizer and crystal, and on the phase shift $\Delta\varphi$, arising between the ordinary and extraordinary waves in the crystal and determined by eqn (21.75). In particular, if the optic axis of the crystal makes an angle of 45° with the polarization plane of the incident light and $\Delta\varphi = \pi$, i.e. the crystalline plate operates as a half-wave plate (see Fig. 21.25), then the polarization plane of the light is rotated through 90° within the plate and the light leaving the plate is totally transmitted by the analyser. Here, the light on the screen reaches its maximum intensity corresponding to the intensity of the original optical beam. For the same position of the crystal, but for $\Delta\varphi = 2\pi$, the rotation of the polarization plane does not take place and the light spot does not appear on the screen.

The phase shift arising between the ordinary and extraordinary waves in the crystalline plate depends on the material and thickness of the plate, and also on the wavelength of the incident light (see eqn (21.75)). If we use white light in the arrangement shown in Fig. 21.27, the waves of different frequencies, while propagating through the crystal, accumulate different phase shifts $\Delta\varphi$. These waves experience different changes in their polarization states and, consequently, are transmitted differently by the analyser. Thus, the transmission function of the arrangement depends on the optical wavelength. As a result, the light passing through the arrangement acquires colour.

21.9.1 The colours of crystalline plates

The colours of crystalline plates can be demonstrated using the experimental set-up shown in Fig. 21.27. We use white light from an arc lamp and crystalline plates composed of small sheets of gypsum or mica of various thicknesses. The following phenomena are demonstrated.

By setting the polarizer and analyser in the crossed position, we make the screen totally dark. Then we introduce a crystalline plate between the polarizer and the analyser. A bright coloured light appears on the screen.

Having turned the analyser through 90° about the beam axis, we observe the colours on the screen changing to the complementary ones. For example, red changes to light blue, green to violet, and blue to yellow. If we turn the analyser through 45° about the optical beam axis, the colours vanish.

With the polarizer and analyser set in the crossed position, we turn the plate to observe darkening of the screen followed by maximum brightness, this effect occurring four times during one complete revolution.

Then we carry out the experiment using the plates of different thicknesses. We observe different colours on the screen. In particular, a gypsum plate detached along the cleavage plane produces an attractive mottled pattern with different colours corresponding to sites with different thicknesses.

Note that the coloured pattern on the screen can only be observed when both polarization instruments, that is, a polarizer and an analyser, are applied (see Fig. 21.27). If we remove any of these instruments, the light on the screen turns white.

21.10 Induced anisotropy

Isotropic materials can acquire anisotropic properties when subjected to an "anisotropic" external action. For example, liquids, plastic materials, and glasses show optical anisotropy under the action of mechanical stresses, external electric fields, and intense polarized laser pulses. These phenomena caused by induced anisotropy find extensive practical use.

Isotropic Plexiglass becomes anisotropic under mechanical stress. For example, if we replace an anisotropic crystal in the arrangement for observation of the interference of polarized rays (Fig. 21.27) with a transparent cube of Plexiglass squeezed in a vice, we can observe on the screen an interference pattern associated with the internal anisotropic stresses induced in the Plexiglass cube. Increasing the squeezing force, we may observe an increase in the pattern contrast, and also a shift and deformation of the interference bands. Similar observations can be used for the modelling and investigation of internal stresses and deformations in various materials and constructions, e.g. in hooks, shafts, and rollers.

21.10.1 The Kerr effect

Some fluids and glasses acquire anisotropy under the action of an external electric field. This effect, called the *Kerr effect*, is caused by the external electric field orienting the elementary electric dipoles (molecules) and thus simulating an optic axis of the medium (see Fig. 21.6).

Figure 21.28 shows a *Kerr cell* employed as an electro-optical modulator of an optical beam. The cell consists of a transparent cuvette filled with a liquid (usually with nitrobenzene), and a capacitor providing the electric field. The Kerr cell is placed between a polarizer and an analyser of light, e.g. between polaroids selecting optical waves the polarization directions $\vec{E}'$ and $\vec{E}''$ of which

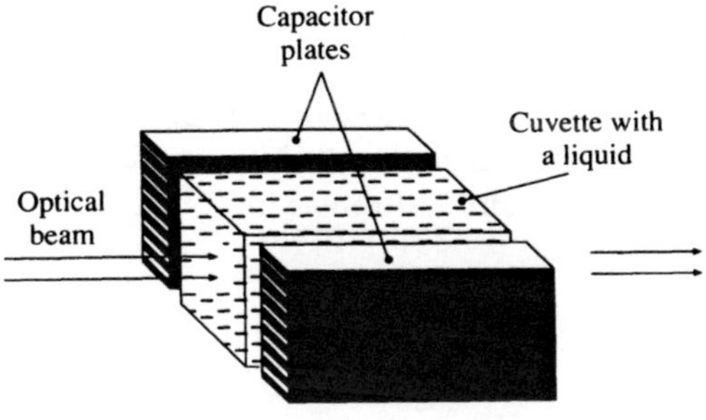

Fig. 21.28 The Kerr cell.

are mutually perpendicular and make an angle of 45° with the direction of the electric field created by the capacitor. Without the electric field, the system does not transmit the light. When applied, the field induces anisotropy in the liquid. The Kerr cell changes the polarization state of the optical wave so that a fraction of the incident light passes through the device (Fig. 21.29). If an alternating electric field of frequency ω is applied, the output optical beam intensity turns out to be modulated with the same frequency. The Kerr cell can thus be used as an electro-optical modulator in optical communications systems. The cell is also used as a shutter in lasers to produce powerful ("giant") optical pulses.

High-frequency Kerr effect

The Kerr effect caused by the electric field of an intense polarized optical wave is called a high-frequency Kerr effect. Because of this effect, the refractive index of a liquid depends on the light intensity, i.e. the medium becomes *non-linear* (see Chapter 22).

The Kerr effect induced by a short laser pulse can be used to construct an *ultra-fast optical shutter* (Duguay 1971).

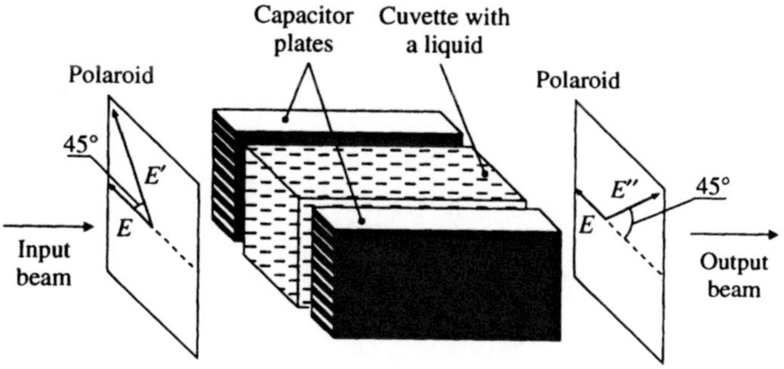

Fig. 21.29 A modulator of light.

Bibliography

Azzam, R. M. A. and Bashara, N. M. (1977). *Ellipsometry and polarized light.* North-Holland, Amsterdam.

Born, M. and Wolf, E. (1968). *Principles of optics.* Pergamon Press, Oxford.

Duguay, M. A. (1971). *American Scientist,* **59**, 551–556.

Yariv, A. and Yeh, P. (1984). *Optical waves in crystals.* John Wiley, New York.

22

Non-linear optics

Fundamental effects of non-linear optics and the mechanisms of optical non-linearity. Violation of the superposition principle for intense optical waves in a medium. The material equation of a non-linear medium. Non-linear polarization. Non-linear susceptibility. An ensemble of non-linear oscillators as a classical model of a non-linear medium. Optical rectification.

This chapter is devoted to the physics of the interaction of intense laser radiation with matter. Fundamental effects of non-linear optics are considered, theoretical foundations are presented, and applications of non-linear optical phenomena are discussed.

22.1 Fundamental effects of non-linear optics and the mechanisms of optical non-linearity

Non-linear optics deals with the processes of the interaction of light and matter that depend on the intensity of light. Let us consider some of these processes.

22.1.1 Optical second-harmonic generation

This phenomenon consists of the doubling of the light frequency in a crystal. The mechanism of the process is linked to the non-linearity of an elementary atomic oscillator.

Historically, light frequency doubling in a crystal was the first non-linear optical effect to be discovered soon after the invention of the laser. A drawing of the Franken experiment in which this effect was observed is shown in Fig. 22.1.

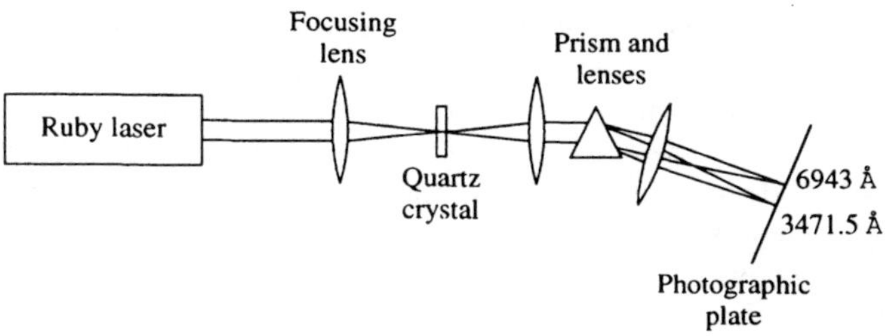

Fig. 22.1 Franken's experiment on the generation of the optical second harmonic.

The initial ruby laser radiation of wavelength $\lambda_1 = 694.3$ nm was focused into a quartz crystal. The radiation emerging from the crystal was decomposed into a spectrum with the help of a dispersion prism and then focused onto a photographic plate. The experiment showed that the light emerging from the crystal contained not only radiation at the laser frequency, but also radiation at the doubled frequency (second harmonic), having a wavelength of $\lambda_2 = \lambda_1/2 = 347.2$ nm. Although the optical second-harmonic radiation in the Franken experiment was extremely weak, this experiment has played an important role by initiating the development of non-linear optics.

Subsequent experiments showed that the efficiency of the second-harmonic generation can be sharply increased by employing other crystals. To date, methods have been developed that ensure the conversion of a significant fraction of the initial laser radiation into radiation at the optical second harmonic; sometimes, the conversion efficiency is close to 100%.

Optical second-harmonic generation can be demonstrated by using a neodymium laser and a crystal of lithium niobium ($LiNbO_3$) or KDP. The experimental set-up is similar to that shown in Fig. 22.1. The laser radiation has a wavelength of 1.06 μm, while that for the second-harmonic is 0.53 μm. We observe bright flashes of green light (second harmonic) on a white screen in the darkroom. The efficiency of the second-harmonic generation in this experiment exceeds 10%. The experiment shows that the power of the second-harmonic radiation is at a maximum at some definite orientation of the crystal. If we turn the crystal from this position, the efficiency of the harmonic generation decreases sharply. It should be also noted that the second-harmonic arises only at sufficiently high field intensities attained by focusing the laser radiation into the crystal.

Optical second-harmonic generation was first demonstrated by Franken in 1961. At present, this effect is widely used for frequency conversion of laser radiation.

22.1.2 Stimulated Raman scattering (SRS) of light

This effect consists of the following: a medium illuminated by high-powered laser light produces intense radiation, the frequency of which is shifted with respect to the laser generation frequency by a value equal to the frequency of molecular vibration of the medium. The mechanism of the process is the same as in spontaneous Raman scattering, i.e. the modulation of light by molecular vibrations. However, unlike the spontaneous scattering, which is very weak and has an isotropic directivity pattern, the stimulated scattering looks like a laser emission. The power and directivity of the stimulated scattering are comparable to the analogous parameters of the laser beam. This is caused by the fact that the stimulated scattering involves not the thermal molecular vibrations, as in the spontaneous scattering, but also the vibrations excited by light. The transition from spontaneous to stimulated scattering occurs when the intensity of the exciting light exceeds a certain value, called the *threshold* of SRS.

SRS was first observed by Woodbury and Ng (1962) during experiments with a Q-switched ruby laser. They used a nitrobenzene cell for Q-switching

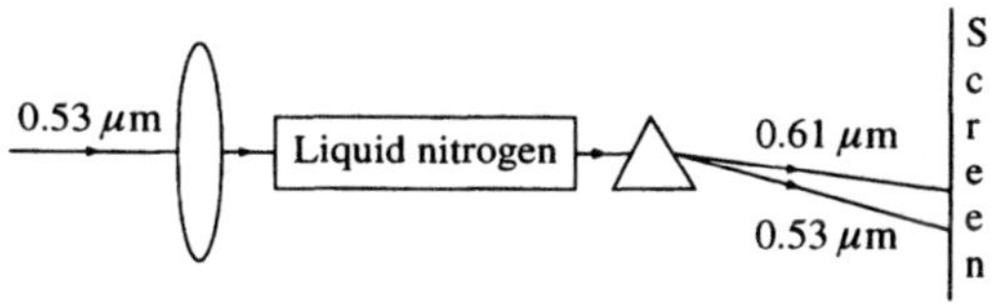

Fig. 22.2 Stimulated Raman scattering of light in liquid nitrogen.

and noticed the appearance of an infrared component in the laser emission, the frequency of which was less than the frequency of the basic laser radiation by 1345 cm^{-1}. Since the frequency shift coincided with the frequency of one of the normal vibrations of the nitrobenzene molecule, a suggestion was made that the infrared component was due to the Raman scattering of light in nitrobenzene. The high intensity of the infrared radiation was attributed to the stimulated character of the process, in which the molecular vibrations were strongly excited by the laser light. This suggestion was then confirmed by subsequent experiments with various liquids, gases, and solids.

SRS in liquid nitrogen can be demonstrated using the experimental set-up shown in Fig. 22.2. The second-harmonic radiation of a neodymium laser at a wavelength of 530 nm is focused into a glass cuvette containing liquid nitrogen. The radiation emerging from the cuvette passes through a prism and falls on to the screen. During the laser pulse, we can see spots of green and orange on the screen. The green spot corresponds to the second-harmonic radiation and the orange spot to SRS at a wavelength of 610 nm. The frequency shift between these waves equals 2326 cm^{-1}, which is the frequency of molecular vibrations in liquid nitrogen. The experiment shows that SRS can be observed only at sufficiently high light intensities achieved by focusing the laser beam into the vessel. Unlike the spontaneous Raman scattering process characterized by very low intensities, the SRS intensity is comparable to the laser beam intensity. This allows the development of effective converters of laser radiation, based on the SRS process. Such converters are now used for radiation frequency conversion, for improving the spatial coherence of light, and for compressing laser pulses (by reducing their duration and increasing their power). In addition, the SRS process can be used for coherent summation of the radiation of several laser modules.

The techniques of driving coherent molecular vibrations with the help of a pair of optical waves ("biharmonic pumping") is applied in the spectroscopy of coherent anti-Stokes Raman scattering (CARS).

22.1.3 Self-focusing of light

Through this effect, a medium illuminated by intense laser light acquires focusing (lens-like) properties. As a result, the laser beam in the medium can sharply collapse to transform into a thin shining filament, or break down into several filaments.

Self-focusing is linked to a variation of the refractive index of a medium under the action of an intense optical wave. The reasons for this variation can be diverse. For example, electrostriction in the optical field leads to a pressure growth that changes the density and, consequently, the refractive index of the medium in the region occupied by the optical beam. In liquids, the intense polarized optical field orients anisotropically polarized molecules by interacting with their induced dipole moments. The medium becomes anisotropic, and its mean refractive index increases. This effect is usually referred to as the high-frequency Kerr effect. Variation of the refractive index here is caused by alignment of the molecules along the field, just as in the well-known static Kerr effect. The refractive index increment depending on the intensity of the optical wave can also be attributed to the non-linearity of electronic polarization. Finally, the density variation and, consequently, the variation of the refractive index, can be associated with heating of the medium by the intense optical wave.

Self-focusing of light was predicted theoretically by Askaryan in 1962 and was first observed by Pilipetskii and Rustamov in 1965. They photographed narrow shining filaments in organic liquids illuminated by a focused beam from a ruby laser.

Figure 22.3 shows a single-mode laser beam at the exit window of a cuvette filled with toluene. Figure 22.3(a) corresponds to a short cuvette, in which the laser beam did not have enough time to become focused (the beam diameter is 700 μm). In Fig. 22.3(b), the cuvette was close to the self-focusing length and the output beam diameter was about 50 μm. In Fig. 22.3(c), the cuvette was long enough for the beam self-focusing to develop; the output beam has the form of a filament with a limiting diameter of 10 μm. Figure 22.4 shows the self-focusing effects for a multiple-mode laser beam in CS_2.

The thermal self-action of an argon laser beam in a weakly absorbing liquid can be demonstrated using the experimantal set-up shown in Fig. 22.5. A laser beam passes through a cuvette filled with alcohol and falls on the screen. The self-action effect can be emphasized by adding a light-absorbing dye, e.g. fuchsin, to the alcohol.

With the laser operating in a low-power mode, we can see a light spot the dimensions of which are governed by the usual diffraction divergence of the laser beam. Switching the laser into a full-power mode, we can see that the beam divergence and the spot size on the screen increase sharply. This is caused by the action of a thermal (here, defocusing) lens induced in the liquid by the laser beam. Self-defocusing develops as the light-absorbing liquid heats up more strongly near the axis of the laser beam, where the light intensity is at a maximum. The heating leads to thermal expansion of the liquid, decreasing its density and refractive index, and eventually forms a thermal defocusing lens that increases the angular divergence of the laser beam. The mechanism of the effect is clear from Fig. 22.6.

Similar mechanisms cause another non-linear self-action effect; that is, self-modulation of an optical pulse. Due to the self-modulation of an optical pulse that takes place, for example, in an optical fibre, the frequency spectrum of the pulse expands sharply, which makes it possible to use the pulse compression

(a)

(b)

(c)

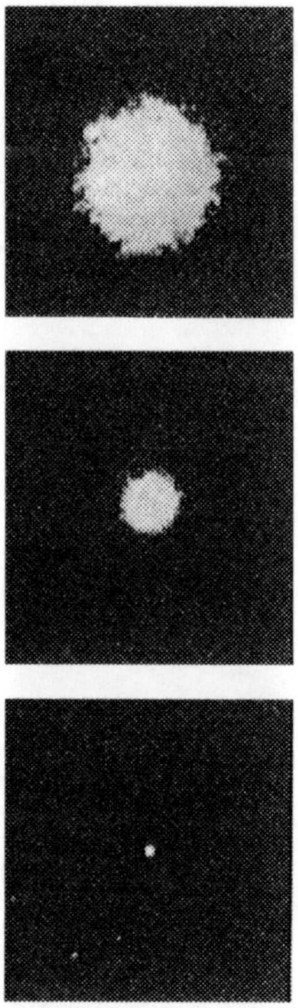

Fig. 22.3 The self-focusing of light in toluene (from Shen, Y. R. (1975). *Progress in Quantum Electronics*, **4**, 3).

technique to form extremely short optical pulses. This effect is applied to produce femtosecond laser pulses (see Akhmanov *et al.* 1991).

A non-linear material placed into an optical resonator exhibits the properties of a bistable element, and can thus be used in an all-light computer (Gibbs 1985).

A non-linear element in an adaptive optical system with two-dimensional feedback allows the generation of such structures of the optical field as helical and rotating waves, and also the observation of "optical turbulence". Adaptive non-linear optical systems are now used to model the dynamics of neuronal networks (Figs 22.7 and 22.8) (Vorontsov 1991).

The basic effects of non-linear optics and their applications are briefly listed in Table 22.1. Note that non-linear optics today, after 30 years of development, continues to make progress, and the number of applications is still growing.

22.2 Violation of the superposition principle for intense optical waves in a medium

Despite the great variety of non-linear optical phenomena, we can distinguish some typical features that are inherent to all of them.

Table 22.1 Fundamental effects of non-linear optics and their applications.

Effect	Material	Mechanism	Application
Generation of optical harmonics. Parametric generation of light	Crystals, semiconductors, noble gases, metal vapours	Non-linearity of atomic oscillator	Frequency conversion of laser radiation. Continuous tuning of laser frequency. Generation of coherent radiation in the visible, IR, UV, and X-ray ranges. Visualization of IR image. Diagnostics of semiconductor materials
Stimulated light scattering	Gases, liquids, crystals, metal vapours, optical fibre	Non-linearity of atomic or molecular oscillator. Light scattering by electronic, molecular, acoustic vibrations, etc. Coherent excitation of vibrations in matter by optical waves	Frequency conversion of laser radiation. Compression of laser pulses. Wave front conjugation and wavefront correction. Improvement of laser spatial coherence. Coherent summation of radiation of several laser modules. Non-linear spectroscopy; in particular, coherent anti-Stokes Raman scattering (CARS)
Self-action of light	Liquids, gases, crystals, liquid crystals, optical fibres	Non-linearity of atomic oscillator. Orientation of anisotropic molecules by optical wave. Electrostriction. Heating of the medium	Self-modulation of optical pulses. Generation of femtosecond laser pulses. Bistable optical elements. Elements of optical computer. Modelling of neural networks

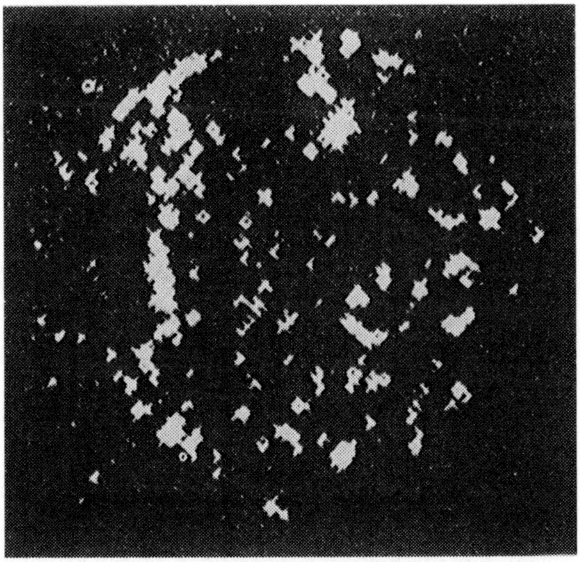

Fig. 22.4 The self-focusing of a multimode laser beam in CS_2 (from Shen 1984).

The first is a strong dependence on the intensity of light. As a rule, a non-linear optical effect becomes perceptible only at sufficiently high intensities of light. It is not accidental, therefore, that non-linear optics appeared only after the advent of the laser. Non-linear optics is the optics of intense optical fields; specifically, the optics of high-powered laser beams.

Second, violation of the superposition principle is typical of non-linear effects. According to the superposition principle, optical waves of different frequencies, propagation directions, and polarizations interact with a medium independently of one another. In non-linear optics, however, this is not true. As we have mentioned earlier, non-linear optical processes are accompanied by the appearance of new spectral components of the optical field, and a strong interaction of different optical waves, including energy interchange and complete conversion of one wave into another. A typical example is optical second-harmonic generation.

Thus, in non-linear optical processes, we come up against a violation of the superposition principle. The inverse statement can also be made: violation of

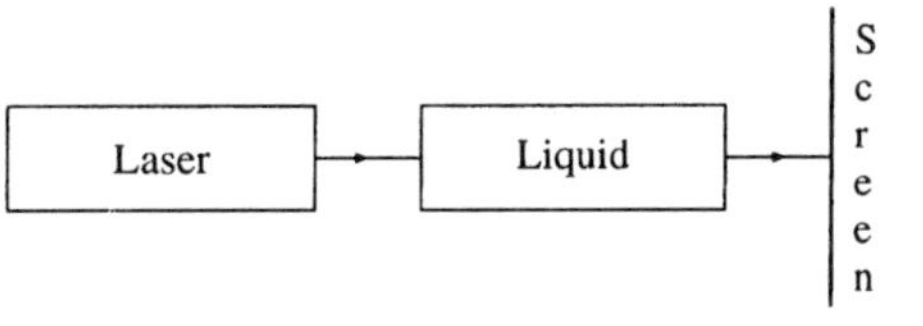

Fig. 22.5 A demonstration of thermal self-defocusing of light in a liquid.

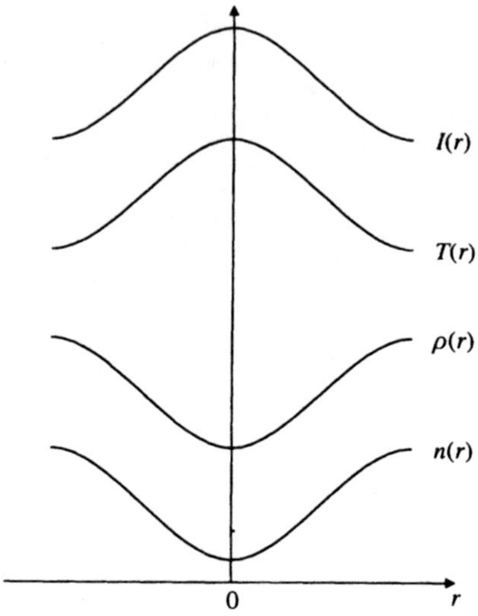

Fig. 22.6 A mechanism of thermal self-defocusing of light in a weakly absorbing liquid. r is the radial coordinate in the beam cross-section, I is the intensity of light, T is the temperature of the medium, ρ is the density of the medium, and n is the refractive index of the medium.

the superposition principle is directly or indirectly associated with a non-linear effect.

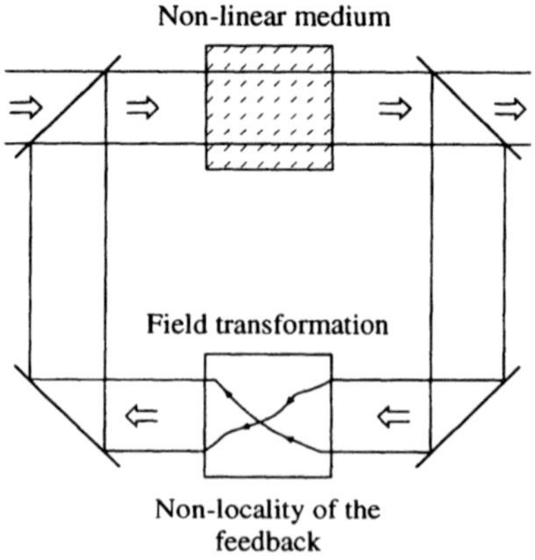

Fig. 22.7 A non-linear optical resonator with two-dimensional feedback.

Fig. 22.8 The structures of the optical field observed in a non-linear resonator with two-dimensional feedback: (a,b) rotating waves (optical reverberators); (c,d) spiral waves; (e) the coexistence of structures; (f) optical turbulence.

22.3 The material equation of a non-linear medium

The theory of non-linear optical phenomena is based on the material equations and the Maxwell equations.

The Maxwell equations for a dielectric non-magnetic medium have the form (see Chapter 18)

$$\operatorname{rot} \vec{E} = -\frac{1}{c}\frac{\partial \vec{H}}{\partial t}, \quad \operatorname{div} \vec{D} = 0,$$

$$\operatorname{rot} \vec{H} = \frac{1}{c}\frac{\partial \vec{D}}{\partial t}, \quad \operatorname{div} \vec{H} = 0,$$

where $\vec{D} = \vec{E} + 4\pi\vec{P}$. The Maxwell equations lead to the wave equation,

$$\mathrm{rot\,rot}\,\vec{E} + \frac{1}{c^2}\frac{\partial^2 \vec{E}}{\partial t^2} = -\frac{4\pi}{c^2}\frac{\partial^2 \vec{P}}{\partial t^2}, \tag{22.1}$$

which in an isotropic medium takes the form

$$\Delta\vec{E} - \frac{1}{c^2}\frac{\partial^2 \vec{E}}{\partial t^2} = \frac{4\pi}{c^2}\frac{\partial^2 \vec{P}}{\partial t^2}. \tag{22.2}$$

Here $\vec{E}$ is the electric field strength of an optical wave, and $\vec{P}$ is the polarization of the medium.

Equations (22.1) and (22.2) are valid for both linear and non-linear media. According to these equations, the polarization of the medium is a source of the optical field. The polarization, in turn, arises under the action of the incident optical wave. The polarization induced by an optical field is described by the material equation

$$\vec{P} = \vec{P}(\vec{E}),$$

which allows for the structure and properties of the medium.

The simplest material equation of a non-linear medium is

$$P = \kappa E + \chi^{(2)}E^2 + \chi^{(3)}E^3 + \cdots . \tag{22.3}$$

According to this equation, the polarization of the medium is a *non-linear function* of the optical field strength. From a mathematical point of view it is this that causes the violation of the superposition principle for optical waves in a non-linear medium. The possibility of optical harmonics generation and other non-linear optical effects follows directly from eqns (22.1)–(22.3). Note that the relative value of the non-linear terms in eqn (22.3) increases with the optical field strength, i.e. with the light intensity. This explains why non-linear effects can be observed mostly in intense laser fields.

The coefficients κ, $\chi^{(2)}$, $\chi^{(3)}$, ... in eqn (22.3) are called *optical susceptibilities* and depend on the properties of the medium. In particular, κ is the linear optical susceptibility, $\chi^{(2)}$ is the non-linear susceptibility of the second order, $\chi^{(3)}$ is the non-linear susceptibility of the third order, etc.

22.4 Non-linear polarization

A fraction of the polarization of the medium, which depends non-linearly on the optical field strength, is called the *non-linear polarization*. Separating the linear and non-linear components of the polarization, we can write

$$\vec{P} = \vec{P}_{\mathrm{L}} + \vec{P}_{\mathrm{NL}}. \tag{22.4}$$

Substituting eqn (22.4) into eqns (22.1) and (22.2), we obtain the wave equation,

$$\mathrm{rot\,rot}\,\vec{E} + \frac{1}{c^2}\frac{\partial^2 \vec{E}}{\partial t^2} + \frac{4\pi}{c^2}\frac{\partial^2 \vec{P}_{\mathrm{L}}}{\partial t^2} = -\frac{4\pi}{c^2}\frac{\partial^2 \vec{P}_{\mathrm{NL}}}{\partial t^2}, \tag{22.5}$$

for a non-linear anisotropic medium and

$$\Delta\vec{E} - \frac{1}{c^2}\frac{\partial^2\vec{E}}{\partial t^2} - \frac{4\pi}{c^2}\frac{\partial^2\vec{P}_{\rm L}}{\partial t^2} = \frac{4\pi}{c^2}\frac{\partial^2\vec{P}_{\rm NL}}{\partial t^2} \tag{22.6}$$

for a non-linear isotropic medium. It can be seen from eqns (22.3)–(22.6) that the non-linear polarization of the medium is a source of new spectral components of the field (e.g. optical harmonics, combination frequencies, etc.).

The material equation of the form (22.3) describes an isotropic non-linear medium with an instantaneous localized response to the optical field. The analogous equation for an anisotropic dispersive medium has the form

$$\begin{aligned} P_\alpha(t) &= \int_0^\infty \kappa_{\alpha\beta}(\tau)E_\beta(t-\tau)\,d\tau \\ &+ \int_0^\infty\int_0^\infty \chi^{(2)}_{\alpha\beta\gamma}(\tau_1,\tau_2)E_\beta(t-\tau_1)E_\gamma(t-\tau_2)\,d\tau_1\,d\tau_2 \\ &+ \int_0^\infty\int_0^\infty\int_0^\infty \chi^{(3)}_{\alpha\beta\gamma\delta}(\tau_1,\tau_2,\tau_3)E_\beta(t-\tau_1)E_\gamma(t-\tau_2)E_\delta(t-\tau_3)\,d\tau_1\,d\tau_2\,d\tau_3 + \cdots . \end{aligned} \tag{22.7}$$

Here, the indices α, β, $\gamma, \ldots$ adopt values that label the Cartesian coordinate axes. Equation (22.7) implies summation over the repeating indices. The functions

$$\kappa_{\alpha\beta}(\tau), \quad \chi^{(2)}_{\alpha\beta\gamma}(\tau_1,\tau_2), \quad \chi^{(3)}_{\alpha\beta\gamma\delta}(\tau_1,\tau_2,\tau_3), \ldots$$

are in fact tensor Green functions characterizing the linear and non-linear responses of the medium to the impact excitation.

The Green functions for media with a non-localized response depend not only on time, but also on the coordinates:

$$\kappa_{\alpha\beta} = \kappa_{\alpha\beta}(\tau;\vec{r}), \qquad \chi^{(2)}_{\alpha\beta\gamma} = \chi^{(2)}_{\alpha\beta\gamma}(\tau_1,\tau_2;\vec{r}_1,\vec{r}_2)$$

and so on. Integration over the spatial variables should be then added to eqn (22.7). The non-locality of the medium response should be allowed for when the elementary oscillators of the medium, located at different points in space, are bonded and interact with one another. Media that possess this property—including some types of crystal— are regarded as having what is termed *spatial dispersion*.

22.5 Non-linear susceptibility

The material equation (22.7) is a generalization of eqn (21.19) to the case of a non-linear medium. Generalizing eqn (21.20b),

$$\kappa_{\alpha\beta}(\omega) = \int_0^\infty \kappa_{\alpha\beta}(\tau)\mathrm{e}^{-i\omega\tau}\,d\tau, \tag{22.8}$$

defining the linear optical susceptibility $\kappa_{\alpha\beta}(\omega)$, we shall introduce the *non-linear susceptibilities of the medium*:

$$\chi^{(2)}_{\alpha\beta\gamma}(\omega_1,\omega_2) = \int_0^\infty\int_0^\infty \chi^{(2)}_{\alpha\beta\gamma}(\tau_1,\tau_2)e^{-i(\omega_1\tau_1+\omega_2\tau_2)}\,d\tau_1\,d\tau_2,$$
$$\chi^{(3)}_{\alpha\beta\gamma\delta}(\omega_1,\omega_2,\omega_3) = \int_0^\infty\int_0^\infty\int_0^\infty \chi^{(3)}_{\alpha\beta\gamma\delta}(\tau_1,\tau_2,\tau_3)e^{-i(\omega_1\tau_1+\omega_2\tau_2+\omega_3\tau_3)}\,d\tau_1\,d\tau_2\,d\tau_3, \tag{22.9}$$

and so on. The analogous quantities for media with spatial dispersion depend not only on frequencies, but also on the wave vectors $\vec{k}$, $\vec{k}_1$, $\vec{k}_2$, etc. Note that the susceptibilities defined by eqns (22.8) and (22.9) have the same dimensions as the susceptibilities appearing in eqn (22.3).

22.5.1 Intensity-dependent refractive index

Usually, the cubic non-linearity, i.e. the non-linearity of the third order, is the lowest non-zero non-linearity in isotropic non-linear media. In the approximation of the instantaneous localized response, such a medium can be described by a material equation of the form

$$P = \kappa E + \chi^{(3)}E^3. \tag{22.10}$$

In the same approximation, the refractive index n of the medium is determined by the formulae $D = E + 4\pi P = \varepsilon E = n^2 E$, from which

$$n = \sqrt{1 + 4\pi P/E}. \tag{22.11}$$

Substituting eqn (22.10) into eqn (22.11) and taking into account that the non-linear term in eqn (22.10) is relatively small, we obtain

$$n = n_0 + \frac{2\pi}{n_0}\chi^{(3)}E^2, \tag{22.12}$$

where

$$n_0 = \sqrt{1 + 4\pi\kappa}. \tag{22.13}$$

The quantity E^2 in eqn (22.12) is related to the light intensity I as $I = cE^2/8\pi$. This yields

$$n = n_0 + n_2 I, \tag{22.14}$$

where

$$n_2 = \frac{16\pi^2}{n_0 c}\chi^{(3)}. \tag{22.15}$$

n_0 is the linear refractive index of the medium, and c is the velocity of light.

The formula (22.14) shows that media with a cubic non-linearity can be characterized by a refractive index that depends on the light intensity. This effect leads to the self-action of optical waves; and, in particular, to such effects as self-focusing of an optical beam, self-phase modulation of an optical pulse, bistability of a resonator filled with a non-linear medium, etc. The quantity n_2, that has the dimensions of the inverse of light intensity, is a convenient characteristic of the cubic non-linearity of the medium. For example, for the quartz crystal $n_2 = 3 \times 10^{-16}\ \mathrm{cm^2\,W^{-1}}$.

A specific mechanism leading to non-linearity of the type (22.14) can be associated, for example, with the alignment of anisotropic molecules in liquid in the field of an intense polarized optical wave. Since the same mechanism causes optical birefringence, induced by the constant electric field (Kerr's effect), the variation of the refractive index due to the light intensity is called a *high-frequency Kerr effect.* The non-linearity described by eqn (22.14) is called a *Kerr-type non-linearity.*

22.5.2 Estimates

Let us estimate the linear and non-linear optical susceptibilities of a medium, e.g. a crystal. The linear optical susceptibility κ is related to the refractive index n of the medium by $n^2 = 1 + 4\pi\kappa$. Typically, the refractive index in a crystal is $n = 1.5$. Using this value, we obtain the estimate

$$\kappa = 0.1. \tag{22.16}$$

To evaluate the quadratic susceptibility $\chi^{(2)}$, we can use the material equation (22.3). According to this equation, the dimensionality of $\chi^{(2)}$ is determined by

$$\chi^{(2)} = \kappa/E. \tag{22.17}$$

Let us use eqn (22.17) to make a quantitative estimate. Since the susceptibility is a material property, the electric field strength E in eqn (22.17) should be replaced with a certain quantity characterizing the substance. The natural scale of the electric field strength in the medium is the intra-atomic electric field strength, which can be estimated from the formula

$$E_{\mathrm{at}} = e/a^2, \tag{22.18}$$

where e is the electron charge and a is the radius of the atom. Setting $e = 4.8 \times 10^{-10}$ CGSE and $a = 0.5 \times 10^{-8}$ cm (the Bohr radius), we obtain

$$E_{\mathrm{at}} = 2 \times 10^7\,\mathrm{CGSE}. \tag{22.19}$$

From eqns (22.16), (22.17), and (22.19), we obtain

$$\chi^{(2)} = 5 \times 10^{-9}\,\mathrm{CGSE}. \tag{22.20}$$

The obtained estimate correctly predicts the order of magnitude of the quadratic non-linear susceptibility of a crystal. For example, the KDP crystal used extensively in non-linear optics has $\chi^{(2)} = 3 \times 10^{-9}$ CGSE. As seen from eqn (22.17), $\chi^{(2)}$ has the dimensions of the inverse of the electric field strength:

$$\left[\chi^{(2)}\right] = \mathrm{g}^{-1/2}\,\mathrm{cm}^{1/2}\,\mathrm{s} = \mathrm{erg}^{\ 1/2}\,\mathrm{cm}^{3/2}. \tag{22.21}$$

Table 22.2 Non-linear optical susceptibilities of the third order ($n = n_0 + n_2 I$).

Material	n_2 ($cm^2\,W^{-1}$)	$\chi_3^{eff}(-\omega;\omega,\omega,-\omega)$ ($cm^3\,erg^{-1}$)	T (K)	Notes
CS_2	3×10^{-14}	10^{-12}	300	Non-resonant
Quartz	3×10^{-16}	10^{-14}	300	Non-resonant
Si	10^{-10}	10^{-8}	300	Non-resonant
Liquid crystal	10^{-4}	10^{-3}	300	Oriented NLQ oriented mesophase, very slow (seconds)
InSb	3×10^{-3}	1	77	Resonant amplification
AlGaAs	2×10^{-4}	6×10^{-2}	300	Resonant amplification, quantum-well structures

Similarly, we can estimate the cubic susceptibility $\chi^{(3)}$. Using the material equation (22.3) and putting $E = E_{at}$, we obtain

$$\chi^{(3)} = \kappa/E_{at}^2. \tag{22.22}$$

The numerical estimation by eqns (22.16), (22.19), and (22.22) yields

$$\chi^{(3)} = 2.5\times10^{-16}\,\text{CGSE} = 2.5\times10^{-16}\,\text{cm}^3\,\text{erg}^{-1}. \tag{22.23}$$

The real value of $\chi^{(3)}$, for example, for the quartz crystal, is 10^{-14} $cm^3\,erg^{-1}$. There are, however, materials possessing a cubic susceptibility of a much greater magnitude. Several examples are listed in Table 22.2.

It is also interesting to compare the values of the linear and non-linear optical polarizations of a medium. For example, restricting ourselves to second-order non-linearity, we can write

$$P = P_L + P_{NL}^{(2)} = \kappa E + \chi^{(2)} E^2,$$

$$\frac{P_{NL}^{(2)}}{P_L} = \frac{\chi^{(2)} E^2}{\kappa E} = \frac{\chi^{(2)} E}{\kappa}.$$

Using the estimate

$$\chi^{(2)} = \kappa/E_{at}, \tag{22.24}$$

we obtain

$$P_{NL}^{(2)}/P_L = E/E_{at}. \tag{22.25}$$

The ratio between the quadratic and linear polarizations of a medium is thus equal to the ratio of the optical wave electric field strength to the intra-atomic field strength. In a similar way, we obtain

$$P_{\rm NL}^{(3)}/P_{\rm L} = (E/E_{\rm at})^2 \tag{22.26}$$

and so on.

Let us estimate the value of $E/E_{\rm at}$. The light intensity I is related to the electric field strength E of the optical wave by the formula $I = cE^2/8\pi$. Hence, $E = \sqrt{8\pi I/c}$. Putting, for example, $I = 10^9\ \mathrm{W\,cm^{-2}} = 10^{16}$ CGSE and $c = 3 \times 10^{10}\ \mathrm{cm\,s^{-1}}$, we obtain $E = 3 \times 10^3$ CGSE. Using eqn (22.19), we obtain

$$E/E_{\rm at} = 10^{-4}. \tag{22.27}$$

Thus, for the given light intensity ($10^9\ \mathrm{W\,cm^{-2}}$), the relative value of the non-linear polarization turns out to be very small. Nevertheless, the non-linear effect can be appreciable due to its ability to *accumulate* in the course of propagation of the optical wave through the non-linear medium.

22.6 An ensemble of non-linear oscillators as a classical model of a non-linear medium

The mechanisms of optical non-linearity are multifarious (see Table 22.1). It is believed, however, that one of the most universal is the mechanism associated with the non-linearity of an elementary oscillator of the medium; that is, an atom or a molecule. Let us calculate the non-linear polarization and non-linear susceptibility, treating the medium as an ensemble of non-linear oscillators.

22.6.1 The non-linear oscillator

Using Newton's second law, we can write the equation of motion for an oscillator in the form

$$m\ddot{x} = F_{\rm res} + eE. \tag{22.28}$$

Here m and e are the mass and charge of the electron, x is the displacement of the electron with respect to the atomic nucleus (Fig. 22.9), E is the electric field strength of the optical wave, and $F_{\rm res}$ is the restoring force due to attraction of the electron to the nucleus. This force is related to the potential energy $U(x)$ of the electron in the field of the nucleus by

$$F_{\rm res} = -\frac{\partial U}{\partial x}. \tag{22.29}$$

The potential energy $U(x)$ can be expanded in a power series of x near the point of equilibrium of the electron ($x = 0$):

$$U(x) = \tfrac{1}{2}\alpha x^2 + \tfrac{1}{3}\beta x^3 + \dots . \tag{22.30}$$

The potential energy as a function of the electron coordinate is shown schematically in Fig. 22.10. The first term in eqn (22.30) corresponds to a parabolic approximation, shown in Fig. 22.10 as a dashed line. The other terms describe

Fig. 22.9 The classical model of the atom.

the difference between the real potential well and the parabolic one. These terms have to be allowed for if the electron vibration amplitude is sufficiently large, as can happen in an intense optical field.

Substituting eqn (22.30) into eqn (22.29), we obtain

$$F_{\text{res}} = -\alpha x - \beta x^2 + \dots . \tag{22.31}$$

Thus, the restoring force turns out to be a non-linear function of the electron displacement. Substituting eqn (22.25) into eqn (22.22) and allowing only for the first non-linear correction, we obtain the equation

$$\ddot{x} + \omega_0^2 x + \gamma x^2 = \frac{e}{m} E, \tag{22.32}$$

where $\omega_0^2 = \alpha/m$ is the natural frequency of the oscillator and $\gamma = \beta/m$ is the non-linearity parameter. To allow for the damping of electron vibrations, let us introduce the term $\Gamma\dot{x}$ into the left-hand side of eqn (22.32). We then obtain the equation

$$\ddot{x} + \Gamma\dot{x} + \omega_0^2 x + \gamma x^2 = \frac{e}{m} E. \tag{22.33}$$

This equation describes the vibration of an atomic oscillator under the action of an optical field. It allows for the oscillator non-linearity, which becomes appreciable for sufficiently large vibration amplitudes. Anharmonicity of the elementary oscillator gives rise to the non-linear polarization of the medium.

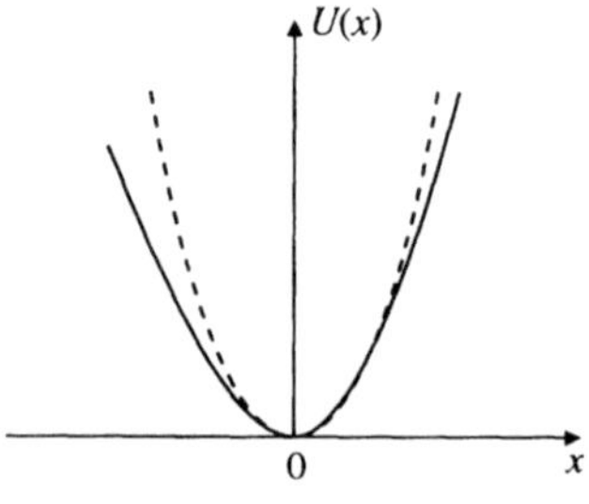

Fig. 22.10 The potential energy of a non-linear oscillator (solid curve) and its parabolic approximation (dashed curve).

22.6.2 Polarization of the medium

The polarization P of the medium is defined as the electric dipole moment of the unit volume of the medium. Assuming the medium to be homogeneous, we can write

$$P = Np, \tag{22.34}$$

where N is the number of atoms per unit volume,

$$p = ex \tag{22.35}$$

is the dipole moment of the elementary oscillator (atom), e is the charge of the electron, and x is the electron displacement governed by eqn (22.33). To calculate the polarization, we thus need to solve eqn (22.33).

An exact solution of this equation is unknown. Here we come across a typical situation in non-linear optics, in which the equation describing a non-linear effect either has no exact solution at all or the solution is too complicated for easy application. Under these conditions, one needs to use various approximate methods to analyse the non-linear system, and the choice of a specific method requires an individual approach for every problem.

22.6.3 The perturbation method

One of the most universal methods for the analysis of non-linear systems is the perturbation method. The main idea of the method is to describe first the motion of the system in the *linear approximation*, and then consider a non-linear effect as a *small correction.*

Suppose that the oscillator vibration amplitude is so small that the non-linear term in eqn (22.33) at any moment in time remains much less than the linear terms. Then

$$|\gamma x^2| \ll |\omega_0^2 x| \tag{22.36}$$

or

$$|\gamma x| \ll \omega_0^2. \tag{22.37}$$

The approximate solution of eqn (22.33) can be then written as

$$x = x_{\mathrm{L}} + x_{\mathrm{NL}}, \tag{22.38}$$

where x_{L} is the solution to the linear equation

$$\ddot{x}_{\mathrm{L}} + \Gamma \dot{x}_{\mathrm{L}} + \omega_0^2 x_{\mathrm{L}} = \frac{e}{m} E, \tag{22.39}$$

and x_{NL} is the non-linear correction, which is small compared to x_{L}:

$$|x_{\mathrm{NL}}| \ll |x_{\mathrm{L}}|. \tag{22.40}$$

22.6.4 The linear approximation

Let us write the optical field in the form of a plane monochromatic wave

$$E = \tfrac{1}{2}\mathcal{E}e^{i(\omega t - \vec{k}\vec{r})} + \text{c. c.} \tag{22.41}$$

and substitute eqn (22.41) into eqn (22.39). The solution to eqn (22.39) will be sought in the form

$$x_\text{L} = \tfrac{1}{2}\tilde{x}_\text{L}e^{i(\omega t - \vec{k}\vec{r})} + \text{c. c.} \tag{22.42}$$

Substituting eqn (22.42) into eqn (22.39), we find that

$$\tilde{x}_\text{L} = \frac{e}{m}\frac{1}{\omega_0^2 - \omega^2 + i\omega\Gamma}\mathcal{E} \tag{22.43}$$

or

$$\tilde{x}_\text{L} = \alpha(\omega)\mathcal{E}, \tag{22.44}$$

where the quantity

$$\alpha(\omega) = \frac{e}{m}\frac{1}{\omega_0^2 - \omega^2 + i\omega\Gamma} \tag{22.45}$$

is called the *linear atomic polarizability*. The problem is thus solved in the linear approximation.

22.6.5 Calculation of the non-linear correction

Substituting eqn (22.38) into eqn (22.33), we obtain the equation

$$\ddot{x}_\text{L} + \ddot{x}_\text{NL} + \Gamma\dot{x}_\text{L} + \Gamma\dot{x}_\text{NL} + \omega_0^2 x_\text{L} + \omega_0^2 x_\text{NL} + \gamma(x_\text{L}^2 + x_\text{NL}^2 + 2x_\text{L}x_\text{NL}) = \frac{e}{m}E. \tag{22.46}$$

By virtue of eqns (22.39), some terms in this equation cancel, to yield

$$\ddot{x}_\text{NL} + \Gamma\dot{x}_\text{NL} + \omega_0^2 x_\text{NL} + \gamma(x_\text{L}^2 + x_\text{NL}^2 + 2x_\text{L}x_\text{NL}) = 0. \tag{22.47}$$

Let us retain in this equation only the terms of the lowest order of magnitude with respect to x_NL. The inequalities (22.37) and (22.40) allow us to write

$$x_\text{L}^2 + x_\text{NL}^2 \approx x_\text{L}^2 \tag{22.48}$$

and

$$\omega_0^2 x_\text{NL} + 2\gamma x_\text{L}x_\text{NL} = (\omega_0^2 + 2\gamma x_\text{L})x_\text{NL} \approx \omega_0^2 x_\text{NL}. \tag{22.49}$$

Finally, we obtain the equation

$$\ddot{x}_\text{NL} + \Gamma\dot{x}_\text{NL} + \omega_0^2 x_\text{NL} = -\gamma x_\text{L}^2. \tag{22.50}$$

We have thus linearized the original non-linear eqn (22.33) with respect to the small parameter x_NL.

Equation (22.50) is a linear equation of forced vibrations, with the term $\gamma x_{\rm L}^2$ representing a driving force. The quantity $x_{\rm L}$ is determined by eqns (22.42), (22.44), and (22.45). Let us now find the solution to eqn (22.50). According to eqn (22.42),

$$\tilde{x}_{\rm L}^2 = \tfrac{1}{4}x_{\rm L}^2 e^{i(2\omega t - 2\vec{k}\vec{r})} + \tfrac{1}{4}|x_{\rm L}|^2 + \text{c. c.} \tag{22.51}$$

Thus, the driving force in eqn (22.50) contains a constant component and an alternating component oscillating at the second harmonic frequency. Due to the linearity of eqn (22.50), the quantity $x_{\rm NL}$ would have a similar structure. Therefore, we shall seek a solution to eqn (22.50) in the form

$$x_{\rm NL}(t) = \tfrac{1}{2}x_2 e^{i(2\omega t - 2\vec{k}\vec{r})} + \tfrac{1}{2}x_0 + \text{c. c.} \tag{22.52}$$

Substituting eqns (22.51) and (22.52) into eqn (22.50), we obtain the equation

$$\tfrac{1}{2}(\omega_0^2 - 4\omega^2 + 2i\omega\Gamma)x_2 e^{i(2\omega t - 2\vec{k}\vec{r})} + \tfrac{1}{2}\omega_0^2 x_0 + \text{c. c.} = -\tfrac{1}{4}\gamma\tilde{x}_{\rm L}^2 e^{i(2\omega t - 2\vec{k}\vec{r})} - \tfrac{1}{4}\gamma|\tilde{x}_{\rm L}|^2 + \text{c. c.} \tag{22.53}$$

Equating the pre-exponential factors and also the constant terms, we find

$$x_0 = -\frac{\gamma}{2\omega_0^2}|\tilde{x}_{\rm L}|^2, \quad x_2 = -\frac{\gamma}{2(\omega_0^2 - 4\omega^2 + 2i\omega\Gamma)}\tilde{x}_{\rm L}^2. \tag{22.54}$$

The non-linear correction is thus calculated.

We have obtained an approximate solution to eqn (22.33). Now it is easy to calculate the polarization of the medium.

Substituting eqns (22.38), (22.42), and (22.52) into eqns (22.34) and (22.35), we obtain

$$P = P_{\rm L} + P_{\rm NL}, \tag{22.55}$$

where

$$P_{\rm L} = \tfrac{1}{2}\mathcal{P}_{\rm L} e^{i(\omega t - \vec{k}\vec{r})} + \text{c. c.} \tag{22.56}$$

is the linear polarization and

$$P_{\rm NL} = \tfrac{1}{2}\mathcal{P}_2 e^{i(2\omega t - 2\vec{k}\vec{r})} + \tfrac{1}{2}\mathcal{P}_0 + \text{c. c.} \tag{22.57}$$

is the non-linear polarization. The quantities $\mathcal{P}_{\rm L}$, $\mathcal{P}_2$, and $\mathcal{P}_0$ are determined by the formulae

$$\mathcal{P}_{\rm L} = Ne\tilde{x}_{\rm L}, \qquad \mathcal{P}_2 = Nex_2, \qquad \mathcal{P}_0 = Nex_0 \tag{22.58}$$

or, in view of eqns (22.44) and (22.54),

$$\mathcal{P}_{\rm L} = Ne\alpha(\omega)\mathcal{E},$$
$$\mathcal{P}_2 = -\frac{1}{2}\gamma Ne\frac{\alpha^2(\omega)}{\omega_0^2 - 4\omega^2 + 2i\omega\Gamma}\mathcal{E}^2, \quad \mathcal{P}_0 = -\frac{1}{2}\gamma Ne\frac{|\alpha(\omega)|^2}{\omega_0^2}|\mathcal{E}|^2, \tag{22.59}$$

where $\alpha(\omega)$ is the linear atomic polarizability defined by eqn (22.45).

(a)

(b)

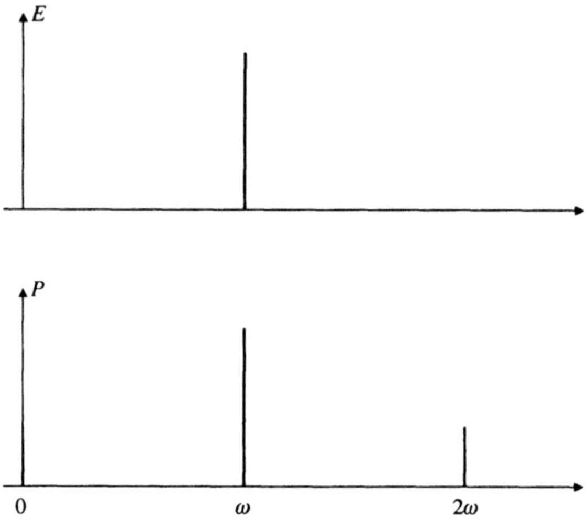

Fig. 22.11 Spectra of the incident optical wave (a) and polarization (b) in a medium with quadratic non-linearity.

Equations (22.55)–(22.59) show that the polarization of the non-linear medium considered here contains three spectral components: (i) a component with the frequency of the incident optical wave (the linear polarization); (ii) a component with frequency 2ω; and (iii) a constant component (Fig. 22.11). The polarization component with the frequency 2ω is responsible for the optical second-harmonic generation.

22.6.6 Optical susceptibilities

The linear optical susceptibility $\kappa(\omega)$ of the medium is defined as a proportionality coefficient between the complex amplitudes of the field and the polarization at the frequency ω:

$$\mathcal{P}_{\mathrm{L}} = \kappa(\omega)\mathcal{E}. \tag{22.60}$$

Let us define the quadratic non-linear susceptibility $\chi^{(2)}(2\omega)$ of the medium for the second-harmonic frequency as a proportionality coefficient between the complex amplitude $\mathcal{P}_2$ of the polarization at this frequency and the square of the field complex amplitude:

$$\mathcal{P}_2 = \chi^{(2)}(2\omega)\mathcal{E}^2. \tag{22.61}$$

Finally, the quadratic non-linear susceptibility $\chi^{(2)}(0)$ of the medium at zero frequency is defined as

$$\mathcal{P}_0 = \chi^{(2)}(0)|\mathcal{E}|^2. \tag{22.62}$$

It follows from eqns (22.59)–(22.62) that

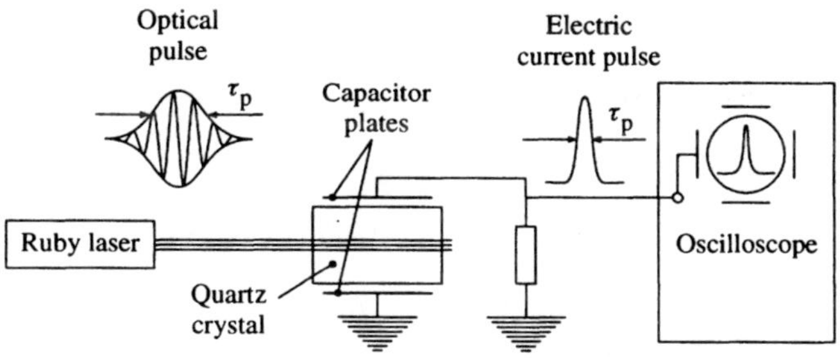

Fig. 22.12 The experimental observation of optical rectification. A ruby laser beam passes through a quartz crystal placed into a capacitor. Due to optical rectification, the laser pulse generates an electric current pulse in the capacitor circuit.

$$\begin{aligned}\kappa(\omega) &= Ne\alpha(\omega),\\ \chi^{(2)}(2\omega) &= -\tfrac{1}{2}\gamma Ne\frac{\alpha^2(\omega)}{\omega_0^2 - 4\omega^2 + 2i\omega\Gamma},\\ \chi^{(2)}(0) &= -\tfrac{1}{2}\gamma Ne\frac{|\alpha(\omega)|^2}{\omega_0^2},\end{aligned} \tag{22.63}$$

where

$$\alpha(\omega) = \frac{e}{m}\frac{1}{\omega_0^2 - \omega^2 + i\omega\Gamma} \tag{22.64}$$

is the linear atomic polarizability.

Hence, we have calculated the linear and non-linear optical susceptibilities for the ensemble of classical non-linear oscillators. The derived eqns (22.63) show that the non-linear optical susceptibility of the medium depends, first, on the anharmonicity of the elementary oscillator, which is characterized by the parameter γ, and, second, on the linear atomic polarizability $\alpha(\omega)$. Consequently, non-linear media can be identified primarily by a large refractive index (liquids, crystals). Additionally, it follows from eqns (22.63) and (22.64) that the non-linear susceptibility of the medium increases under *resonant conditions*, when the frequency ω of the incident light wave, or the second-harmonic frequency 2ω, is close to the natural frequency of the elementary oscillator.

22.7 Optical rectification

Our calculation shows that the intense light wave propagating in a quadratically non-linear medium, i.e. in a medium possessing a non-linear susceptibility of the second order, produces a constant polarization, the magnitude of which is proportional to the light intensity. This effect is called *optical rectification*. The constant polarization, in turn, gives rise to a constant electric field, which can be registered and measured.

Figure 22.12 shows a schematic diagram of the observation of optical rectification in a quartz crystal. A ruby laser beam passes through a quartz crystal placed into a capacitor. Due to the optical rectification effect, the laser pulse generates a pulse of electric current in the capacitor circuit. It should be noted that similar arrangements can be used for the measurement of the quadratic optical non-linearity of various media.

Bibliography

Akhmanov, S. A. and Khokhlov, R. V. (1972). *Problems of nonlinear optics.* Gordon and Breach, New York.

Akhmanov, S. A., Vysloukh, V. A., and Chirkin, A. S. (1991). *Optics of femtosecond laser pulses.* AIP, New York.

Allen, L. and Eberly, J. H. (1975). *Optical resonance and two-level atoms.* John Wiley, New York.

Bloembergen, N. (1965). *Nonlinear Optics.* W. A. Benjamin, New York.

Bloembergen, N. (ed.) (1975). *Spectroskopia nonlineare.* North-Holland, Amsterdam.

Fisher, R. and Müller, R. (1989). *Qvantovaya Elektronika,* **16**, 1723–1730. (in Russian).

Franken, P. A., Hill, A. E., Peters, C. W., and Weinreich, G. (1961). *Physical Review Letters,* **7**, 118–119.

Gibbs, H. M. (1985). *Optical bistability: controling light with light.* Academic Press, New York.

Ready, J. F. (1971). *Effects of high-power laser radiation.* Academic Press, New York.

Reintjes, G. R. (1987). *Nonlinear optical parametric processes in liquids and gases.* Academic Press, New York.

Shapiro, S. L. (ed). (1977). *Ultrashort light pulses.* Springer-Verlag, Berlin.

Shen, Y. R. (1984). *The principles of nonlinear optics.* John Wiley, New York.

Vorontsov, M. A. (1991). *Proceedings, SPIE,* **1402**, 116–144.

Vorontsov, M. A. and Miller, W. B. (eds.) (1994). *Self-organization in nonlinear optics and some applications to optical processing.* Springer-Verlag, Berlin.

Woodbury, E. J. and Ng, W. K. (1962). *Proceedings of the Institute of Radio Engineers,* **50**, 2347.

Yariv, A. and Yeh, P. (1984). *Optical waves in crystals.* John Wiley, New York.

Yariv, A. (1982). *An introduction to theory and applications of quantum mechanics.* John Wiley, New York.

Zernike, F. and Midwinter, J. E. (1973). *Applied nonlinear optics.* John Wiley, New York.

23

Theoretical non-linear optics

Optical second-harmonic generation. Stimulated Raman scattering. Self-focusing of light.

The theory of non-linear optical effects is based on Maxwell's equations and the material equations. From Maxwell's equations one can obtain the wave equation (see Chapter 22)

$$\operatorname{rot}\operatorname{rot}\vec{E}+\frac{1}{c^2}\frac{\partial^2\vec{D}_{\rm L}}{\partial t^2}=-\frac{4\pi}{c^2}\frac{\partial^2\vec{P}_{\rm NL}}{\partial t^2}, \tag{23.1}$$

which, for an isotropic medium, takes the form

$$\Delta\vec{E}-\frac{1}{c^2}\frac{\partial^2\vec{D}_{\rm L}}{\partial t^2}=\frac{4\pi}{c^2}\frac{\partial^2\vec{P}_{\rm NL}}{\partial t^2}. \tag{23.2}$$

Here $\vec{D}_{\rm L}=\vec{E}+4\pi\vec{P}_{\rm L}$ is the linear component of the electric displacement, $\vec{E}$ is the electric field vector of the optical wave, $\vec{P}_{\rm L}$ is the linear polarization of the medium, $\vec{P}_{\rm NL}$ is the non-linear polarization, and Δ is the Laplace operator, which in Cartesian coordinates has the form

$$\Delta=\frac{\partial^2}{\partial x^2}+\frac{\partial^2}{\partial y^2}+\frac{\partial^2}{\partial z^2}.$$

According to eqns (23.1) and (23.2), the polarization of the medium is a source of the optical field. The polarization, in turn, is induced by the incident optical wave and described by equation

$$\vec{P}=\vec{P}(\vec{E})=\vec{P}_{\rm L}+\vec{P}_{\rm NL}. \tag{23.3}$$

Specific examples of the material equation are given in Chapter 22. We shall now apply this theory to the solution of specific non-linear optical problems.

23.1 Optical second-harmonic generation

The estimates made in Chapter 22 show that the local non-linear effect is usually small. For example, for optical second-harmonic generation we have

$$P_{\rm NL}/P_{\rm L}\simeq E/E_{\rm at}\approx 10^{-4}. \tag{23.4}$$

Here E is the electric field strength of the optical wave corresponding to a light intensity of 1 GW cm^{-2}, and E_{at} is the intra-atomic electric field strength.

23.1.1 Wave synchronism

The estimate (23.4) shows that the radiation power re-emitted by the atom at the second-harmonic frequency is much less than that re-emitted at the incident wave frequency. Nevertheless, the harmonic generation efficiency can be quite high. This is explained by the fact that the non-linear effect can *accumulate* while the optical wave propagates through the non-linear medium. Such an accumulation takes place provided that the *wave synchronism* condition is satisfied.

For second-harmonic generation, the wave synchronism condition requires that the velocity of the second-harmonic wave should be equal to the incident wave velocity:

$$v(\omega) = v(2\omega). \tag{23.5}$$

With this condition satisfied, an optimal phase relation is attained between the elementary second-harmonic waves emitted by individual atoms and the wave arriving from the previous region of the crystal. As a result, constructive interference of these waves takes place and, provided that the crystal size is sufficiently large, the incident optical wave can be totally converted into the second harmonic wave. The power of the second-harmonic wave grows monotonously along the crystal length. Conversely, if the wave synchronism condition is not met, the elementary second-harmonic waves emitted by individual atoms interfere destructively with incoming waves. The power of the second-harmonic wave then varies periodically along the crystal length and the resultant generation efficiency does not significantly exceed the value of the local effect.

The normal dispersion of the crystal, implying that the refractive index grows with the frequency of light (see Chapter 19), prevents the synchronism condition from being satisfied. In 1962, Giordmaine and Terhune suggested using the phenomenon of optical birefringence in anisotropic crystals to solve this problem. In a negative uniaxial anisotropic crystal, the velocities of the wave with the fundamental (pump) frequency and of the wave of the second harmonic can be the same if these waves have different polarizations. The pump wave should be ordinary and the harmonic wave extraordinary. Refractive index surfaces for this case are shown in Fig. 23.1. Figure 23.1(a) shows the refractive index surfaces for a KDP crystal and ruby laser pumping. One can see that there is a direction in the crystal for which the refractive index ellipsoid of the second harmonic (extraordinary wave) intersects the refractive index sphere of the pump (ordinary wave). This direction is called the direction of synchronism. The intersection of the refractive index surfaces is shown separately in Fig. 23.1(b). Figure 23.2 shows a similar picture for quartz crystal. No direction of synchronism apparently exists in quartz. This explains why the second-harmonic radiation was very weak in quartz (Franken's experiment, 1961).

The curve in Fig. 23.3 shows that the angular width of the synchronism is very small: in the KDP crystal it is about 0.1°.

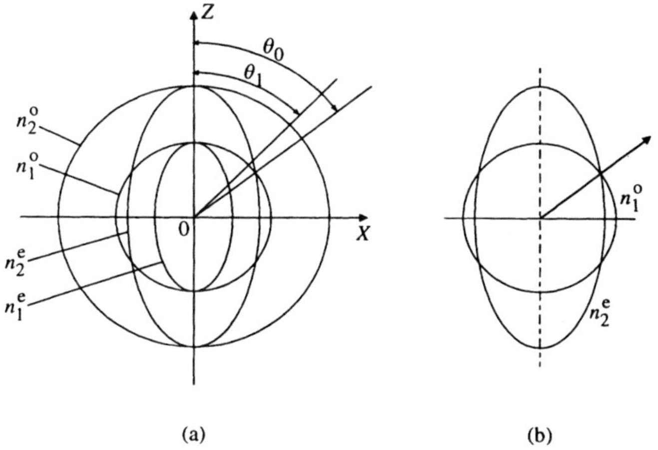

Fig. 23.1 Wave synchronism in an anisotropic crystal. (a) Sections of refractive index surfaces of a KDP crystal (a sphere for an ordinary wave n^{o}, and an ellipsoid for an extraordinary wave n^{e}) for the ruby laser frequency (subscript 1) and its second harmonic (subscript 2). (b) The direction of synchronism.

23.1.2 The theory of the generation of the optical second-harmonic in a KDP crystal

Let us consider in detail the generation of the optical second-harmonic in a non-linear anisotropic crystal. For definiteness, we shall consider the KDP crystal pumped by neodymium laser radiation ($\lambda_1 = 1.06\ \mu$m).

23.1.3 Non-linear polarization

Considering only the linear and quadratic susceptibilities of the crystal and allowing for its dispersion and anisotropy, we can write the material equation in

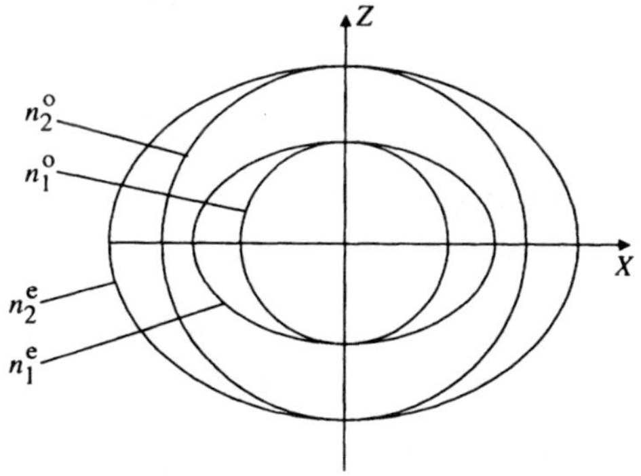

Fig. 23.2 Sections of the refractive index surfaces of a quartz crystal. No direction of synchronism is available in this crystal.

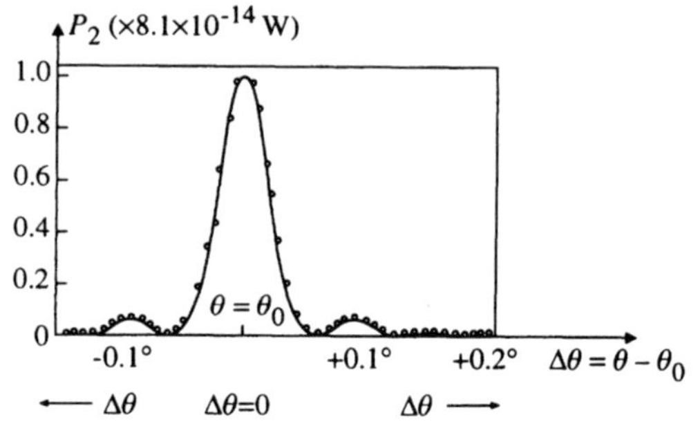

Fig. 23.3 The generation of the second harmonic of helium–neon laser radiation ($\lambda = 1.15$ microns) in a KDP crystal. The harmonic power P_2 is plotted as a function of the angle $\Delta\theta$ between the laser beam and direction of synchronism. The length of the crystal $l = 1.23$ cm.

the form

$$P_i(t) = \int_0^\infty \kappa_{ij}(\tau) E_j(t-\tau)\, d\tau + \iint_0^\infty \chi_{ijk}(\tau_1,\tau_2) E_j(t-\tau_1) E_k(t-\tau_2)\, d\tau_1\, d\tau_2. \quad (23.6)$$

Here, the indices i, j, and k adopt the values x, y, and z, designating the principal axes of the crystal. As usual, our notation implies summation over the repeating indices.

23.1.4 The KDP crystal

Potassium dihydrophosphate, KH_2PO_4, abbreviated to KDP, is a negative uniaxial anisotropic crystal with an optic z-axis. It has six non-zero components of the quadratic non-linear susceptibility tensor:

$$\chi_{xyz} = \chi_{xzy} = \chi_{yzx} = \chi_{yxz} \neq 0, \qquad \chi_{zxy} = \chi_{zyx} \neq 0. \quad (23.7)$$

23.1.5 Representation of the field: notation

Suppose that the optical field in the crystal is a superposition of two plane monochromatic waves: (i) the wave of the fundamental (pump) frequency and (ii) the second-harmonic wave. Denoting the quantities pertaining to these waves by subscripts 1 and 2, respectively, we write

$$\vec{E} = \vec{E}_1 + \vec{E}_2 = \tfrac{1}{2}\vec{e}_1 A_1(\vec{r}) e^{i(\omega_1 t - \vec{k}_1\vec{r})} + \tfrac{1}{2}\vec{e}_2 A_2(\vec{r}) e^{i(\omega_2 t - \vec{k}_2\vec{r})} + \text{c. c.} \quad (23.8)$$

Here, the symbols $\vec{e}$, A, ω, and $\vec{k}$ denote the unit polarization vector, the complex amplitude, the frequency and the wave vector, respectively. The frequencies ω_1 and ω_2 are related by

$$2\omega_1 = \omega_2. \tag{23.9}$$

23.1.6 The condition of synchronism

As we mentioned earlier, the wave synchronism condition in the KDP crystal holds provided that the fundamental harmonic wave is ordinary and the second-harmonic wave is extraordinary. We remind the reader that the ordinary wave is polarized perpendicular to the optic z-axis of the crystal (see Chapter 21), and, consequently,

$$E_{1z} = 0. \tag{23.10}$$

According to eqns (23.6) and (23.7), there is only one non-zero Cartesian component of the non-linear polarization at the second harmonic frequency; that is, the component parallel to the optic z-axis of the crystal:

$$P_{\mathrm{NL},z} = 2 \iint_0^\infty \chi_{zxy}(\tau_1, \tau_2) E_{1x}(t - \tau_1) E_{1y}(t - \tau_2)\, d\tau_1\, d\tau_2. \tag{23.11}$$

23.1.7 Non-linear susceptibility

Substituting eqn (23.8) into eqn (23.11), we obtain the following expression for the non-linear polarization at the second-harmonic frequency:

$$P_{\mathrm{NL},z} = \tfrac{1}{2}\mathcal{P}_{\mathrm{NL},z} e^{i(2\omega_1 t - 2\vec{k}_1\vec{r})} + \mathrm{c.\,c.}, \tag{23.12}$$

where

$$\mathcal{P}_{\mathrm{NL},z} = e_{1x} e_{1y} \chi_{zxy}(\omega_1, \omega_1) A_1^2 \tag{23.13}$$

is the non-linear polarization amplitude, and

$$\chi_{zxy}(\omega_1, \omega_2) = \iint_0^\infty \chi_{zxy}(\tau_1, \tau_2) e^{-i\omega_1\tau_1 - i\omega_2\tau_2} d\tau_1\, \tau_2 \tag{23.14}$$

is the non-linear susceptibility of the crystal.

Since the pump wave is ordinary, its polarization vector $\vec{e}_1$ lies in the x–y plane. Let φ denote the angle between the x-axis and the vector $\vec{e}_1$. We then have

$$e_{1x} = \cos\varphi, \qquad e_{1y} = \sin\varphi \tag{23.15}$$

and

$$\mathcal{P}_{\mathrm{NL},z} = \tfrac{1}{2}\sin 2\varphi \chi^{(2)} A_1^2, \tag{23.16}$$

where the designation

$$\chi^{(2)} = \chi_{zxy}(\omega_1, \omega_1) \tag{23.17}$$

is used for brevity. The diagram of second-harmonic generation in a KDP crystal is shown in Fig. 23.4.

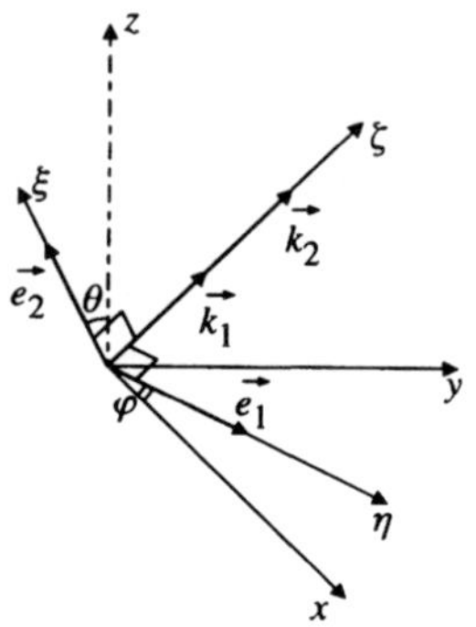

Fig. 23.4 A diagram of optical second-harmonic generation in a KDP crystal. x, y, and z are the principal axes of the crystal, z is the optic axis of the crystal, $\vec{e}_1$ and $\vec{e}_2$ are the polarization vectors of the pump wave and of the second harmonic, $\vec{k}_1$ and $\vec{k}_2$ are the wave vectors of the pump wave and of the second harmonic, and ξ, η, and ζ is the auxiliary coordinate system. The ζ-axis is directed along the optical beam, the ξ-axis lies in the z–ζ plane (the principal plane), and the η-axis lies in the x–y plane.

23.1.8 The wave equation

Let us now derive the equation for the second-harmonic amplitude $A_2(\vec{r})$. We shall use the wave equation

$$\operatorname{rot}\operatorname{rot}\vec{E} + \frac{1}{c^2}\frac{\partial^2\vec{E}}{\partial t^2} + \frac{4\pi}{c^2}\frac{\partial^2\vec{P}_\mathrm{L}}{\partial t^2} = -\frac{4\pi}{c^2}\frac{\partial^2\vec{P}_\mathrm{NL}}{\partial t^2} \tag{23.18}$$

or

$$\operatorname{rot}\operatorname{rot}\vec{E} + \frac{1}{c^2}\frac{\partial^2\vec{D}_\mathrm{L}}{\partial t^2} = -\frac{4\pi}{c^2}\frac{\partial^2\vec{P}_\mathrm{NL}}{\partial t^2}, \tag{23.19}$$

where

$$\vec{D}_\mathrm{L} = \vec{E} + 4\pi\vec{P}_\mathrm{L} \tag{23.20}$$

is the linear component of the electric displacement, and $\vec{P}_\mathrm{L}$ is the linear component of the polarization of the medium.

23.1.9 The equation for the second-harmonic amplitude

Substituting eqn (23.8) into eqn (23.19) and making use of the known formulae of vector analysis,

$$\begin{gathered}\operatorname{rot}\vec{\Phi} = [\vec{\nabla}, \vec{\Phi}],\\ [\vec{\nabla}, \Phi_1\vec{\Phi}_2] = \Phi_1[\vec{\nabla}, \vec{\Phi}_2] + [\vec{\nabla}\Phi_1, \vec{\Phi}_2],\\ [\vec{\nabla}, \vec{e}\Phi_1\Phi_2] = \Phi_1[\vec{\nabla}, \vec{e}\Phi_2] + \Phi_2[\vec{\nabla}, \vec{e}\Phi_1]\end{gathered} \tag{23.21}$$

(one can prove these formulae using the definition of the operator $\vec{\nabla}$: $\vec{\nabla} = \vec{x}_0\partial/\partial x + \vec{y}_0\partial/\partial y + \vec{z}_0\partial/\partial z$), we obtain the following equation for the amplitude A_2 of the second-harmonic wave:

$$\tfrac{1}{2}e^{i(\omega_2 t-\vec{k}_2\vec{r})}\left\{\left[-i\vec{k}_2,[-i\vec{k}_2,\vec{e}_2A_2]\right]-\frac{\omega_2^2}{c^2}\vec{\mathcal{D}}_{L2}+\left[\vec{\nabla},[-i\vec{k}_2,\vec{e}_2]A_2\right]\right.$$
$$\left.+\left[-i\vec{k}_2,[\vec{\nabla},\vec{e}_2A_2]\right]+\left[\vec{\nabla},[\vec{\nabla},\vec{e}_2A_2]\right]\right\}+\text{c. c.}=-\frac{4\pi}{c^2}\frac{\partial\vec{P}_{NL}}{\partial t^2}. \quad (23.22)$$

Here, $\vec{\mathcal{D}}_{L2}$ is the amplitude of the linear component of the electric displacement at the frequency ω_2, defined by the formula

$$\vec{D}_{L2}=\tfrac{1}{2}\vec{\mathcal{D}}_{L2}e^{i(\omega_2 t-\vec{k}_2\vec{r})}+\text{c. c.} \quad (23.23)$$

Let us set

$$\left[-i\vec{k}_2,[-i\vec{k}_2,\vec{e}_2A_2]\right]-\frac{\omega_2^2}{c^2}\vec{\mathcal{D}}_{L2}=0 \quad (23.24)$$

and consider eqn (23.24) as an equation defining $\vec{k}_2$. If we denote $\vec{e}_2A_2=\vec{\mathcal{E}}_2$, the given equation coincides with the equation relating the vectors $\vec{\mathcal{E}}$ and $\vec{\mathcal{D}}$ of the optical wave in a linear anisotropic medium (see eqn (21.53)).

Equation (23.24) reduces eqn (23.22) to

$$\tfrac{1}{2}e^{i(\omega_2 t-\vec{k}_2\vec{r})}\left\{\left[\vec{\nabla},[-i\vec{k}_2,\vec{e}_2]A_2\right]+\left[-i\vec{k}_2,[\vec{\nabla},\vec{e}_2A_2]\right]\right.$$
$$\left.+\left[\vec{\nabla},[\vec{\nabla},\vec{e}_2A_2]\right]\right\}+\text{c. c.}=-\frac{4\pi}{c^2}\frac{\partial\vec{P}_{NL}}{\partial t^2}. \quad (23.25)$$

Let us now derive the scalar equation for A_2. To do this, we shall project the vector equation (23.25) on the ξ-axis shown in Fig. 23.4. Taking into account that the vector $\vec{P}_{NL}$ is directed along the z-axis, that is, along the optic axis of the crystal, and its magnitude is determined by eqns (23.12) and (23.16), we obtain

$$\left\{\left[\vec{\nabla},[-i\vec{k}_2,\vec{e}_2]A_2\right]+\left[-i\vec{k}_2,[\vec{\nabla},\vec{e}_2A_2]\right]\right.$$
$$\left.+\left[\vec{\nabla},[\vec{\nabla},\vec{e}_2A_2]\right]\right\}_\xi=\frac{4\pi\omega_2^2}{c^2}\mathcal{P}_{NL,\xi}e^{i\Delta\vec{k}\vec{r}}, \quad (23.26)$$

where

$$\mathcal{P}_{NL,\xi}=\mathcal{P}_{NL,z}\sin\theta, \quad (23.27)$$

θ being the angle between the propagation direction of the optical wave and the optic z-axis of the crystal (see Fig. 23.4), and

$$\Delta\vec{k}=\vec{k}_2-2\vec{k}_1 \quad (23.28)$$

is the difference between the wave vectors of the non-linear polarization wave at the second-harmonic frequency and the optical wave at the same frequency ("wave mismatch").

Suppose that the pump wave and the second-harmonic wave propagate in crystal in the same direction ζ shown in Fig. 23.4. Then, we can write

$$\vec{E} = \vec{E}_1 + \vec{E}_2 = \tfrac{1}{2}\vec{e}_1 A_1(\zeta)e^{i(\omega_1 t - k_1\zeta)} + \tfrac{1}{2}\vec{e}_2 A_2(\zeta)e^{i(\omega_2 t - k_2\zeta)} + \text{c.c.}, \qquad (23.29)$$

$$\vec{k}_1 = \vec{\zeta}_0 k_1, \qquad \vec{k}_2 = \vec{\zeta}_0 k_2, \qquad k_1 = \frac{\omega_1}{c} n_1, \qquad k_2 = \frac{\omega_2}{c} n_2,$$

$$\Delta\vec{k}\vec{r} = \Delta k\zeta, \qquad \Delta k = k_2 - 2k_1 = \frac{2\omega_1}{c}(n_2 - n_1), \qquad \vec{\nabla} = \vec{\zeta}_0 \frac{\partial}{\partial\zeta},$$

where $\vec{\zeta}_0$ is the unit vector directed along the ζ-axis, and n_1 and n_2 are the refractive indices of the crystal for the pump wave and the second-harmonic wave, respectively.

Allowing for eqn (23.29), we rewrite eqn (23.26) as

$$\left[\vec{\zeta}_0, [\vec{\zeta}_0, \vec{e}_2]\right]_\xi \cdot \left(-2ik_2 \frac{\partial A_2}{\partial\zeta} + \frac{\partial^2 A_2}{\partial\zeta^2}\right) = \frac{4\pi\omega_2^2}{c^2} \mathcal{P}_{\mathrm{NL},\xi} e^{i\Delta k\zeta}. \qquad (23.30)$$

In experiments on optical second-harmonic generation, the following condition is usually well satisfied:

$$\left|-2ik_2 \frac{\partial A_2}{\partial\zeta}\right| \gg \left|\frac{\partial^2 A_2}{\partial\zeta^2}\right|, \qquad (23.31)$$

signifying that the gain of the second-harmonic is relatively small within the crystal region the length of which is of the order of an optical wavelength. Neglecting the term $\partial^2 A_2/\partial\zeta^2$, we then rewrite eqn (23.3) as

$$\left[\vec{\zeta}_0, [\vec{\zeta}_0, \vec{e}_2]\right]_\xi \frac{\partial A_2}{\partial\zeta} = i\frac{2\pi\omega_2}{cn_2} \mathcal{P}_{\mathrm{NL},\xi} e^{i\Delta k\zeta}. \qquad (23.32)$$

Figure 23.5 illustrates the principal plane of the crystal. We see from this figure that

$$\left[\vec{\zeta}_0, [\vec{\zeta}_0, \vec{e}_2]\right]_\xi = -\cos\beta, \qquad (23.33)$$

where β is the angle between the vectors $\vec{\mathcal{E}}$ and $\vec{\mathcal{D}}$ of the second-harmonic wave, the so-called angle of anisotropy. Since here $\beta \ll 1$, we shall set

$$\cos\beta = 1 \qquad (23.34)$$

and rewrite eqn (23.32) in the form

$$\frac{\partial A_2}{\partial\zeta} = -i\frac{2\pi\omega_2}{cn_2} \mathcal{P}_{\mathrm{NL},\xi} e^{i\Delta k\zeta}. \qquad (23.35)$$

This is the desired equation for the second-harmonic amplitude. Using eqns (23.16) and (23.27), we can present it in the form

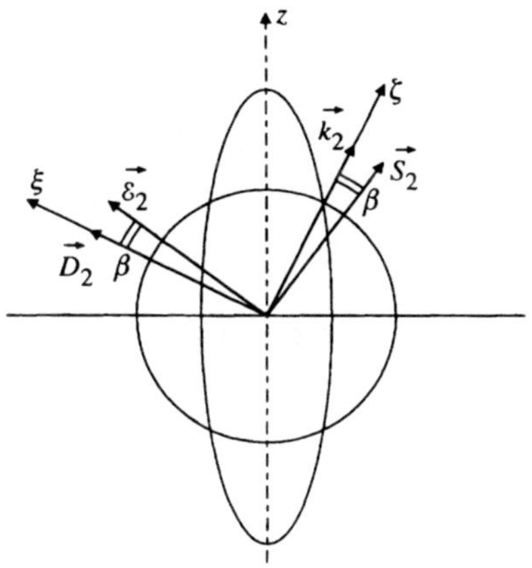

Fig. 23.5 Optical second-harmonic generation in a KDP crystal. The principal plane of the crystal is shown. β is the angle of anisotropy (here $\beta \ll 1$).

$$\frac{\partial A_2}{\partial \zeta} = -i\frac{2\pi\omega_2}{cn_2}\chi^{(2)}_{\text{eff}} A_1^2 e^{i\Delta k\zeta}, \tag{23.36}$$

where

$$\chi^{(2)}_{\text{eff}} = (\tfrac{1}{2}\sin 2\varphi \sin\theta)\chi^{(2)}, \tag{23.37}$$

the angles θ and φ are shown in Fig. 23.4, the quantity $\chi^{(2)}$ is defined by eqn (23.17), and the quantity Δk is determined from eqn (23.29).

23.1.10 The efficiency of second-harmonic generation

Integrating eqn (23.36) and allowing for the boundary condition

$$A_2(\zeta = 0) = 0, \tag{23.38}$$

we obtain

$$A_2(\zeta) = -iKA_1^2\zeta \operatorname{sinc}\left(\frac{\Delta k\zeta}{2}\right) e^{i\Delta k\zeta/2}, \tag{23.39}$$

where

$$K = \frac{2\pi\omega_2}{cn_2}\chi^{(2)}_{\text{eff}}, \qquad \operatorname{sinc} x \equiv \frac{\sin x}{x}. \tag{23.40}$$

It is now easy to find the second-harmonic intensity:

$$I_2 = \frac{cn_2}{8\pi}|A_2|^2 = \frac{8\pi n_2}{cn_1^2}|K|^2 I_1^2 \zeta^2 \operatorname{sinc}^2\left(\frac{\Delta k\zeta}{2}\right), \tag{23.41}$$

where

$$I_1 = \frac{cn_1}{8\pi}|A_1|^2 \tag{23.42}$$

is the pump wave intensity.

Let us define the efficiency of second-harmonic generation as the ratio of the harmonic intensity to the pump wave intensity:

$$\eta = I_2/I_1. \tag{23.43}$$

We find from eqns (23.41) and (23.43) that

$$\eta = \frac{8\pi n_2}{cn_1^2}|K|^2 I_1 \zeta^2 \operatorname{sinc}^2\left(\frac{\Delta k\zeta}{2}\right) \tag{23.44}$$

or, in view of eqns (23.40),

$$\eta = \kappa I_1 \zeta^2 \operatorname{sinc}^2\left(\frac{\Delta k\zeta}{2}\right), \tag{23.45}$$

where

$$\kappa = \frac{32\pi^3\omega_2^2}{c^3 n_1^2 n_2}\left|\chi_{\text{eff}}^{(2)}\right|^2. \tag{23.46}$$

Since we have neglected the pump depletion, eqn (23.45) is valid for

$$\eta \ll 1. \tag{23.47}$$

To estimate $\eta \approx 1$ for the regime of high conversion efficiency, one needs to derive the equation for the pump amplitude and to solve the system of coupled non-linear differential equations for the amplitudes of the pump and second-harmonic waves.

23.1.11 Numerical estimates

According to the experimental conditions for optical second-harmonic generation in the KDP crystal (see, e.g., Shen 1984), we put

$$\lambda_1 = 1.06\,\mu\text{m} \qquad \theta = 40.5° \text{ (angle of synchronism)}. \tag{23.48}$$

Then $\Delta k = 0$,

$$\begin{gathered} \omega_2 = 3.56 \times 10^{15}\,\text{s}^{-1}, \\ \left|\chi^{(2)} \sin\theta\right| = 1,5 \times 10^{-9}\,\text{erg}^{-1/2}\,\text{cm}^{3/2}, \\ \eta = \kappa I_1 \zeta^2. \end{gathered} \tag{23.49}$$

Furthermore, let

$$\varphi = 45°, \qquad n_1 = n_2 = 1.5. \tag{23.50}$$

In view of eqns (23.37), (23.46), and (23.49), we obtain

$$\begin{gathered} \left|\chi_{\text{eff}}^{(2)}\right| = 0.75 \times 10^{-9}\,\text{erg}^{-1/2}\,\text{cm}^{3/2}, \\ \kappa = 0.77 \times 10^{-16}\,\text{s}\,\text{erg}^{-1} = 0.77 \times 10^{-9}\,\text{W}^{-1}. \end{gathered} \tag{23.51}$$

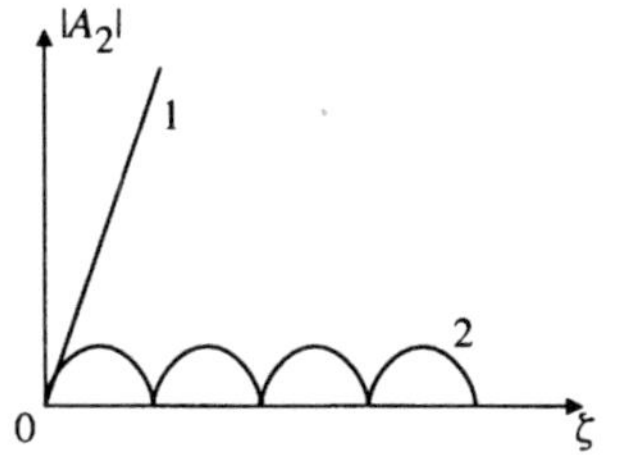

Fig. 23.6 Variation of the second-harmonic amplitude along the crystal. 1, The condition of wave synchronism is satisfied; 2, the condition of wave synchronism is not satisfied.

Finally, let

$$\zeta = 1\,\text{cm}, \qquad I_1 = 10^8\,\text{W}\,\text{cm}^{-2}. \tag{23.52}$$

Then

$$\eta = 0.077. \tag{23.53}$$

Note that the condition (23.47) is satisfied. Thus, our estimate shows that a pump intensity of the order of 100 MW cm^{-2} is required to obtain an efficiency of second-harmonic generation in the KDP crystal of about 10%. Such an intensity can be obtained by focusing a laser pulse into the crystal.

The given estimate is obtained on the assumption that the wave synchronism condition is met, i.e. the wave mismatch $\Delta k = 0$. One can see from eqn (23.39) that here the harmonic amplitude grows linearly with the crystal length ζ (see Fig. 23.6). However, if $\Delta k \neq 0$, then the harmonic amplitude varies periodically along the ζ coordinate and the conversion efficiency decreases sharply (see eqn (23.39) and curve 2 in Fig. 23.6).

23.2 Stimulated Raman scattering

The stimulated Raman scattering (SRS) of light is usually observed in isotropic media, a gas or a liquid. The process is described by the wave equation

$$\Delta E - \frac{1}{c^2}\frac{\partial^2 D_{\text{L}}}{\partial t^2} = \frac{4\pi}{c^2}\frac{\partial^2 P_{\text{NL}}}{\partial t^2}, \tag{23.54}$$

where $D_{\text{L}} = E + 4\pi P_{\text{L}}$ is the linear component of the electric displacement and P_{L} is the linear component of polarization of the medium. Let us calculate the non-linear polarization P_{NL} for the Raman-active medium.

23.2.1 The classical model

In the SRS process the frequency ω_{p} of the exciting (pump) radiation and the frequency ω_{s} of the scattered radiation ("Stokes component") are related by the relation

$$\omega_{\text{s}} = \omega_{\text{p}} - \omega_0, \tag{23.55}$$

where ω_0 is the frequency of molecular vibrations of the medium. Consequently, we can treat this process as a *modulation* of the incident light by the molecular vibration. The process can be described mathematically, having introduced the electron polarizability α of the molecule, depending on the nuclear coordinate q:

$$\alpha = \alpha(q) = \alpha_0 + \alpha' q. \tag{23.56}$$

Here α_0 is the electron polarizability of the molecule the nuclei of which are located at the equilibrium positions, i.e. when $q = 0$, and $\alpha' = (\partial\alpha/\partial q)_0$ is its derivative with respect to the nuclear coordinate, also taken for $q = 0$.

Writing the polarization of the medium in the form

$$P = Np, \qquad p = \alpha E, \tag{23.57}$$

where N is the number of molecules per unit volume, p is the electric dipole moment of the molecule, and E is electric field strength of the optical wave, we obtain

$$P = P_{\rm L} + P_{\rm NL}, \tag{23.58}$$

where

$$P_{\rm L} = N\alpha_0 E, \qquad P_{\rm NL} = N\alpha' q E. \tag{23.59}$$

The polarization of the medium is thus calculated.

23.2.2 Light frequency shift in SRS

The formulae (23.54) and (23.59) provide an explanation of why the light frequency shifts during the SRS process. Indeed, if the incident optical field $E = E_{\rm p}$ oscillates at the frequency $\omega_{\rm p}$, i.e.

$$E = A_{\rm p} \cos \omega_{\rm p} t,$$

and the molecular vibrational coordinate q oscillates at the frequency ω_0, i.e.

$$q = Q \cos \omega_0 t,$$

then the product

$$qE = \tfrac{1}{2} A_{\rm p} Q \left[\cos(\omega_{\rm p} - \omega_0)t + \cos(\omega_{\rm p} + \omega_0)t\right]$$

will oscillate at the frequencies $\omega_{\rm p} - \omega_0$ and $\omega_{\rm p} + \omega_0$. In view of eqns (23.54) and (23.59), the non-linear polarization $P_{\rm NL}$ and the generated optical field will oscillate at the same frequencies.

23.2.3 The driving and phasing of molecular vibration by the optical wave

The high intensity and directivity of SRS are explained by the fact that the light is scattered not by thermal molecular vibrations, as happens in spontaneous scattering, but by molecular vibrations driven and phase-matched by the optical waves in a large volume of the medium. This inverse effect of light on the molecular vibrations is quite essential for the SRS process and is also described by the model given by eqns (23.56) and (23.57).

The molecular vibrations are driven by the force exerted by the optical waves:

$$F = -\partial W/\partial q, \tag{23.60}$$

where W is the energy of the molecule (dipole) in the optical field and q is the molecular vibrational coordinate, e.g. the internuclear distance in a hydrogen or a nitrogen molecule. The energy W is linked to the dipole moment p of the molecule by the relationship

$$p = -\partial W/\partial E. \tag{23.61}$$

Inserting $p = \alpha E$ in this formula and integrating, we obtain

$$W = -\tfrac{1}{2}\alpha E^2. \tag{23.62}$$

Substituting eqn (23.62) into eqn (23.60) and allowing for eqn (23.56), we obtain the formula

$$F = \tfrac{1}{2}\alpha' E^2. \tag{23.63}$$

This formula shows that a pair of optical waves with frequencies $\omega_{\rm p}$ and $\omega_{\rm s} = \omega_{\rm p} - \omega_0$, i.e. an optical field of the form

$$E = E_{\rm p} + E_{\rm s} = \tfrac{1}{2}A_{\rm p}\exp\left[i(\omega_{\rm p}t - k_{\rm p}z)\right] + \tfrac{1}{2}A_{\rm s}\exp\left[i(\omega_{\rm s}t - k_{\rm s}z)\right] + \text{c. c.} \tag{23.64}$$

can cause the resonant excitation of molecular vibrations because the quantity E^2 contains a component oscillating at the frequency ω_0:

$$E^2 = \tfrac{1}{2}A_{\rm p}A_{\rm s}^{*}{\rm e}^{i(\omega_0 t - k_0 z)} + \text{c. c.} + \ldots, \tag{23.65}$$

where $\omega_0 = \omega_{\rm p} - \omega_{\rm s}$, $k_0 = k_{\rm p} - k_{\rm s}$.

23.2.4 The equation for the molecular vibrations

The equation for forced molecular vibrations in an optical field has the form

$$\ddot{q} + \Gamma\dot{q} + \omega_0^2 q = F/M, \tag{23.66}$$

where q is the coordinate, F is the force, M is the reduced mass of the oscillator, ω_0^2 is the natural frequency of molecular vibrations, and Γ is the damping coefficient; a dot above a letter denotes the differentiation with respect to time.

Substituting eqn (23.63) into eqn (23.66), we obtain

$$\ddot{q} + \Gamma\dot{q} + \omega_0^2 q = \frac{\alpha'}{2M}E^2. \tag{23.67}$$

Thus, in the classical model, the SRS is described by the equations

$$\Delta E - \frac{1}{c^2}\frac{\partial^2 D_{\rm L}}{\partial t^2} = \frac{4\pi}{c^2}\frac{\partial^2 P_{\rm NL}}{\partial t^2}, \qquad D_{\rm L} = E + 4\pi P_{\rm L}, \tag{23.68}$$

$$P_{\rm NL} = N\alpha' q E, \tag{23.69}$$

$$\ddot{q} + \Gamma\dot{q} + \omega_0^2 q = \frac{\alpha'}{2M}E^2. \tag{23.70}$$

Here E is the electric field strength of the optical wave, $P_{\rm L}$ is the linear polarization of the medium, $P_{\rm NL}$ is the non-linear polarization of the medium, $D_{\rm L}$ is the linear component of the electric displacement, q is the molecular vibrational coordinate, ω_0 is the natural frequency of molecular vibration, Γ is the damping coefficient, M is the reduced mass of the molecular oscillator, α' is the derivative of the electron polarizability of the molecule with respect to the coordinate q taken for the equilibrium position ($q = 0$), N is the number of molecules per unit volume of the medium, t is the time, Δ is the Laplace operator, and c is the velocity of light.

23.2.5 The quantum-mechanical model

Using the general equations of a quantum-mechanical system in the external field (see, e.g., Yariv 1967), the SRS equations can be presented in the form

$$\Delta E - \frac{1}{c^2}\frac{\partial^2 D_{\rm L}}{\partial t^2} = \frac{4\pi}{c^2}\frac{\partial^2 P_{\rm NL}}{\partial t^2}, \tag{23.71}$$

$$P_{\rm NL} = N\alpha'\langle q\rangle E, \tag{23.72}$$

$$\langle q\rangle = {\rm Tr}(\rho q) = \sum_m \sum_n \rho_{mn} q_{nm}, \tag{23.73}$$

$$\frac{d\rho}{dt} \equiv \dot{\rho} = \frac{i}{\hbar}[\rho, H] + R. \tag{23.74}$$

Here ρ is the density matrix, H is the Hamiltonian, R is the relaxation matrix, and $\hbar$ is Planck's constant. The other designations are the same as in eqns (23.68)–(23.70).

23.2.6 A two-level approximation

Treating the molecule as a two-level quantum-mechanical system and neglecting the relaxation processes, we obtain from eqn (23.74) the following equations for the density matrix elements:

$$
\begin{aligned}
\dot{\rho}_{11} &= \frac{i}{\hbar}(\rho_{12}H_{21} - H_{12}\rho_{21}), \\
\dot{\rho}_{22} &= \frac{i}{\hbar}(\rho_{21}H_{12} - H_{21}\rho_{12}), \\
\dot{\rho}_{12} &= \frac{i}{\hbar}(H_{22} - H_{11})\rho_{12} + \frac{i}{\hbar}H_{12}n, \\
\dot{\rho}_{21} &= -\frac{i}{\hbar}(H_{22} - H_{11})\rho_{21} - \frac{i}{\hbar}H_{21}n,
\end{aligned}
\tag{23.75}
$$

where

$$n = \rho_{11} - \rho_{22}. \tag{23.76}$$

The Hamiltonian of the molecule in the optical field can be represented in the form

$$H = H^{(0)} + V, \tag{23.77}$$

where $H^{(0)}$ is Hamiltonian of the free molecule, and

$$V = -\tfrac{1}{2}(\alpha_0 + \alpha' q)E^2 \tag{23.78}$$

is the energy of interaction between the molecule and the optical field. Equation (23.78) follows from eqns (23.56) and (23.62).

The matrix elements of the Hamiltonian

$$H_{mn} = H_{mn}^{(0)} + V_{mn} \tag{23.79}$$

are calculated in the system of eigenfunctions of the operator $H^{(0)}$. Here

$$H_{11}^{(0)} = E_1, \quad H_{22}^{(0)} = E_2, \quad H_{12}^{(0)} = H_{21}^{(0)} = 0, \tag{23.80}$$

where E_1 and E_2 are the energies of a steady-state molecule, related to the frequency ω_0 of the vibrational transition by

$$\hbar\omega_0 = E_2 - E_1. \tag{23.81}$$

The matrix elements of the perturbation operator (23.78) are

$$V_{mn} = -\tfrac{1}{2}\alpha_0 E^2 \delta_{mn} - \tfrac{1}{2}\alpha' E^2 q_{mn}, \tag{23.82}$$

where δ_{mn} is the Kronecker symbol, and

$$q_{mn} = \int \varphi_m^* q \varphi_n \, dq \tag{23.83}$$

are the matrix elements of the vibrational coordinate.

Suppose that the wave functions of the steady states φ_n of the molecule satisfy the conditions

$$\varphi_n = \varphi_n^*, \tag{23.84}$$

$$\varphi_n(q) = \pm\varphi_n(-q). \tag{23.85}$$

(These conditions are satisfied, in particular, for the wave functions of an harmonic oscillator.) Then

$$q_{11} = q_{22} = 0, \qquad q_{12} = q_{21}. \tag{23.86}$$

We thus have

$$V_{11} = V_{22} = -\tfrac{1}{2}\alpha_0 E^2, \qquad V_{12} = V_{21} = -\tfrac{1}{2}\alpha' E^2 q_{12}. \tag{23.87}$$

Neglecting the energy $-\frac{1}{2}\alpha_0 E^2$ in comparison with E_1 and E_2, we obtain

$$H_{11} = E_1, \quad H_{22} = E_2, \quad H_{12} = H_{21} = -\tfrac{1}{2}\alpha' E^2 q_{12}. \tag{23.88}$$

It follows from eqns (23.73) and (23.86) that

$$\langle q \rangle = (\rho_{12} + \rho_{21}) q_{12}. \tag{23.89}$$

Substituting eqn (23.88) into eqn (23.75) and allowing for eqns (23.76) and (23.81), we obtain

$$\dot{n} = \frac{2i}{\hbar}(\rho_{12} - \rho_{21}) H_{12}, \tag{23.90}$$

$$\dot{\rho}_{12} = i\omega_0 \rho_{12} + \frac{i}{\hbar} H_{12} n, \tag{23.91}$$

$$\dot{\rho}_{21} = -i\omega_0 \rho_{21} - \frac{i}{\hbar} H_{12} n. \tag{23.92}$$

Summing up eqns (23.91) and (23.92) and then subtracting eqn (23.92) from eqn (23.91), we obtain

$$(\dot{\rho}_{12} + \dot{\rho}_{21}) = i\omega_0(\rho_{12} - \rho_{21}), \tag{23.93}$$

$$(\dot{\rho}_{12} - \dot{\rho}_{21}) = i\omega_0(\rho_{12} + \rho_{21}) + \frac{2i}{\hbar} H_{12} n. \tag{23.94}$$

It follows from eqns (23.93) and (23.94) that

$$(\ddot{\rho}_{12} + \ddot{\rho}_{21}) + \omega_0^2(\rho_{12} + \rho_{21}) = -\frac{2\omega_0}{\hbar} H_{12} n. \tag{23.95}$$

Equations (23.90) and (23.93) yield

$$\dot{n} = \frac{2}{\hbar\omega_0}(\dot{\rho}_{12} + \dot{\rho}_{21}) H_{12}. \tag{23.96}$$

Considering eqn (23.89), we can rewrite eqns (23.95) and (23.96) as

$$\langle\ddot{q}\rangle + \omega_0^2\langle q\rangle = -\frac{2\omega_0}{\hbar}H_{12}q_{12}n, \tag{23.97}$$

$$\dot{n} = \frac{2}{\hbar\omega_0}\frac{H_{12}}{q_{12}}\langle\dot{q}\rangle. \tag{23.98}$$

Substituting eqn (23.88) into eqns (23.97) and (23.98), we obtain

$$\langle\ddot{q}\rangle + \omega_0^2\langle q\rangle = \frac{\omega_0\alpha'}{\hbar}q_{12}^2E^2n, \tag{23.99}$$

$$\dot{n} = -\frac{\alpha'}{\hbar\omega_0}E^2\langle\dot{q}\rangle. \tag{23.100}$$

Let us consider the molecule as an harmonic oscillator, with states 1 and 2 being its ground and first excited states, respectively. The matrix element q_{12} of the vibrational coordinate is then given by the formula (see, e.g., Yariv 1982)

$$q_{12}^2 = \frac{\hbar}{2m\omega_0}, \tag{23.101}$$

where m is the mass of the oscillator, ω_0 is the natural vibration frequency, and $\hbar$ is Planck's constant. Substituting eqn (23.101) into eqn (23.99), we obtain

$$\langle\ddot{q}\rangle + \omega_0^2\langle q\rangle = \frac{\alpha'}{2m}E^2n. \tag{23.102}$$

Finally, adding the relaxation terms to eqns (23.100) and (23.102), we obtain

$$\langle\ddot{q}\rangle + \frac{2}{T_2}\langle\dot{q}\rangle + \omega_0^2\langle q\rangle = \frac{\alpha'}{2m}E^2n, \tag{23.103}$$

$$\dot{n} + \frac{n-n_0}{T_1} = -\frac{\alpha'}{\hbar\omega_0}E^2\langle\dot{q}\rangle. \tag{23.104}$$

Here n_0 is the equilibrium population difference in the absence of the field, T_1 is the population relaxation time, and T_2 is the polarization relaxation time (or the dephasing time of molecular vibrations).

The complete set of equations describing the interaction of light with a Raman-active medium has the form

$$\begin{aligned}
\Delta E - \frac{1}{c^2}\frac{\partial^2 D_{\mathrm{L}}}{\partial t^2} &= \frac{4\pi}{c^2}\frac{\partial^2 P_{\mathrm{NL}}}{\partial t^2},\\
P_{\mathrm{NL}} &= N\alpha'\langle q\rangle E,\\
\langle\ddot{q}\rangle + \frac{2}{T_2}\langle\dot{q}\rangle + \omega_0^2\langle q\rangle &= \frac{\alpha'}{2m}E^2n,\\
\dot{n} + \frac{n-n_0}{T_1} &= -\frac{\alpha'}{\hbar\omega_0}E^2\langle\dot{q}\rangle.
\end{aligned} \tag{23.105}$$

Here n is the population difference of the Raman transition, n_0 is the equilibrium value of n, T_1 is the population relaxation time, T_2 is the dephasing time of molecular vibrations, and $\hbar$ is Planck's constant. The other designations are identical to those used in eqns (23.68)–(23.70). Note that if

$$n = n_0 = \text{constant} \tag{23.106}$$

eqns (23.105) transform into eqns (23.68)–(23.70) describing the classical model. The condition (23.106) means that the populations of the quantum energy levels do not vary essentially under the action of the optical field. This is true unless the optical field is too intense.

23.2.7 Reduced equations of SRS

Two waves play the most important roles in SRS: the pumping wave and the Stokes component. In the plane wave approximation the optical field can be represented as

$$E = \tfrac{1}{2}A_{\mathrm{p}}\mathrm{e}^{i(\omega_{\mathrm{p}}t - k_{\mathrm{p}}z)} + \tfrac{1}{2}A_{\mathrm{s}}\mathrm{e}^{i(\omega_{\mathrm{s}}t - k_{\mathrm{s}}z)} + \text{c. c.}, \tag{23.107}$$

where

$$\omega_{\mathrm{s}} = \omega_{\mathrm{p}} - \omega_0. \tag{23.108}$$

The optical field drives coherent molecular vibrations

$$\langle q \rangle = \tfrac{1}{2}Q\mathrm{e}^{i(\omega_0 t - k_0 z)} + \text{c. c.}, \tag{23.109}$$

where

$$k_0 = k_{\mathrm{p}} - k_{\mathrm{s}}. \tag{23.110}$$

Inserting eqns (23.107)–(23.110) into eqn (23.105) and using the approximation of the slowly varying amplitudes A_{p}, A_{s}, and Q, we obtain

$$\begin{aligned}
\frac{\partial A_{\mathrm{p}}}{\partial z} + \frac{1}{v_{\mathrm{p}}}\frac{\partial A_{\mathrm{p}}}{\partial t} &= \gamma_{\mathrm{p}} A_{\mathrm{s}} Q, \\
\frac{\partial A_{\mathrm{s}}}{\partial z} + \frac{1}{v_{\mathrm{s}}}\frac{\partial A_{\mathrm{s}}}{\partial t} &= \gamma_{\mathrm{s}} A_{\mathrm{p}} Q^*, \\
\frac{\partial Q}{\partial t} + \frac{Q}{T_2} &= \gamma_q A n, \quad A = A_{\mathrm{p}} A_{\mathrm{s}}^*, \\
\frac{\partial n}{\partial t} + \frac{n - n_0}{T_1} &= \gamma_n (AQ^* - A^*Q),
\end{aligned} \tag{23.111}$$

where v_{p} and v_{s} are the velocities of the pump wave and of the Stokes component. The constant coefficients in eqns (23.111) are determined by the formulae

$$\gamma_{\rm p} = \frac{\pi N\alpha'\omega_{\rm p}}{in_{\rm p}c}, \qquad \gamma_{\rm s} = \frac{\pi N\alpha'\omega_{\rm s}}{in_{\rm s}c},$$
$$\gamma_q = \frac{\alpha'}{4iM\omega_0}, \qquad \gamma_n = \frac{i\alpha'}{4\hbar}, \tag{23.112}$$

where $n_{\rm p}$ and $n_{\rm s}$ are the refractive indices at the frequencies $\omega_{\rm p}$ and $\omega_{\rm s}$, respectively.

23.2.8 Specific Raman gain

A general solution to eqns (23.111) is impossible to otain. We can, however, simplify the problem by making some additional assumptions, which are often justified in experiment. In the steady-state approximation, eqns (23.111) transform, for a given pump field and without the population motion, to a single equation

$$\frac{\partial A_{\rm s}}{\partial z} = \tfrac{1}{2} g I_{\rm p} A_{\rm s}, \tag{23.113}$$

where $I_{\rm p}$ is the pump intensity and

$$g = \frac{4\pi^2 N(\alpha')^2 T_2\omega_{\rm s}}{M\omega_0 c^2 n_{\rm p} n_{\rm s}} \tag{23.114}$$

is the specific Raman gain, having the dimensions of cm W^{-1}. It follows from eqn (23.113) that the intensity of the Stokes wave grows exponentially along the Raman medium:

$$I_{\rm s} = I_{\rm s}\exp(gI_{\rm p}z). \tag{23.115}$$

The formula (23.115) can be used for the experimental measurement of the parameter g. For many Raman media (compressed hydrogen, liquid nitrogen, etc.) the value of g lies in the range 10^{-9}–10^{-8} cm W^{-1}.

23.3 Self-focusing of light

As was mentioned above (see Chapter 22), the cubic non-linearity of the medium provides an addition to the refractive index, proportional to the light intensity. In the theory of self-focusing, the refractive index of a non-linear medium is usually written as

$$n = n_0 + n_2|A|^2, \tag{23.116}$$

where A is the optical wave amplitude.

Since the refractive index affects the velocity of light in the medium, eqn (23.116) can be interpreted as a dependence of the velocity of light on its intensity. Depending on the sign of n_2, an increase or decrease in the phase velocity of an intense optical wave can occur in the medium and thus influence the efficiency of wave interactions. The effects associated with the non-linear addition to the refractive index are of particular interest for confined optical beams. Here, a new physical effect arises, which is called the non-linear refraction of optical rays.

Indeed, according to eqn (23.116), an originally homogeneous medium becomes optically inhomogeneous in the field of the confined optical beam; the refractive index of the medium is affected by the intensity distribution of the optical wave. The non-linear character of the refraction is governed by the sign of n_2. A positive sign is usually observed for the high-frequency Kerr effect, for electrostriction in the field of an intense optical wave, and sometimes because of heating of the medium by the optical wave. In media with a positive $n_2 > 0$, the regions of maximum light intensity are optically denser. The non-linear refraction should then apparently lead to a concentration of optical energy, because the peripheral rays will be deflected to the region in which the field is a maximum. This effect is called self-focusing of an optical beam. An important circumstance that distinguishes self-focusing from other non-linear optical effects is its avalanche-type character. Indeed, even a weak increase in the intensity in some region of the optical beam propagating in a medium with $n_2 > 0$ leads to a concentration of the rays in this region and, consequently, to an additional intensity increase; this, in turn, enhances the effect of non-linear refraction, and so on.

Let us consider the self-focusing of a Gaussian optical beam. First, neglecting diffraction, we can write the wave field in the form

$$E = \tfrac{1}{2}A\mathrm{e}^{i(\omega t - kz)} + \mathrm{c.\,c.}, \tag{23.117}$$

where

$$A = A_0 \exp(-r^2/2\rho_0^2), \tag{23.118}$$

r is the distance of a given point in the cross-section of the beam from its axis, and ρ_0 is the beam radius. The wave number can be represented as

$$k = \omega n/c, \tag{23.119}$$

where the refractive index n is defined by eqn (23.116).

Using eqns (23.116)–(23.119) and assuming $r \ll \rho_0$ and $n_2|A_0|^2 \ll n_0$, it is easy to obtain the formula

$$kz = k_0 z - k_0 r^2/2R_{\mathrm{NL}}, \tag{23.120}$$

where $k_0 = \omega n_0/c$, $k_2 = \omega n_2/c$, and

$$R_{\mathrm{NL}} = \frac{k_0 \rho_0^2}{2k_2|A_0|^2 z} \tag{23.121}$$

is the curvature radius of the wavefront, caused by non-linear refraction. The minus sign in eqn (23.120) shows that for $n_2 > 0$ the non-linearity of the medium leads to focusing (convergence) of the beam (Fig. 23.7).

We shall consider now the diffraction divergence of the beam. In the near-field diffraction zone, $z \ll z_{\mathrm{d}}$, the wavefront curvature of a Gaussian beam, caused by diffraction, is characterized by the radius

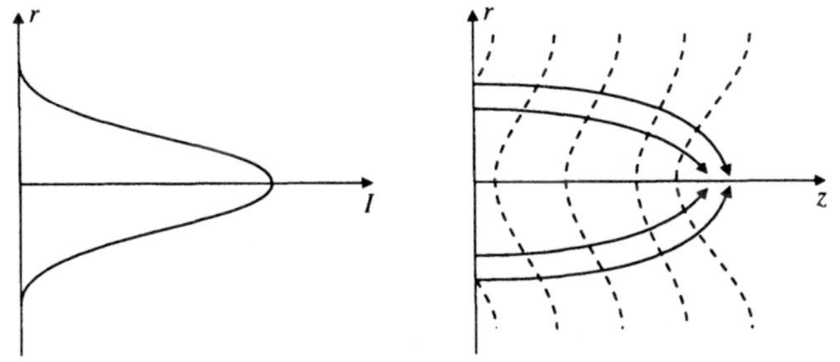

(a) (b)

Fig. 23.7 The self-focusing of light. (a) A Gaussian optical beam intensity profile. (b) Transformation of the wavefront of a Gaussian beam (dashed curves) due to non-linearity of the medium for $n_2 > 0$.

$$R_{\mathrm{d}} = z_{\mathrm{d}}^2/z, \tag{23.122}$$

where

$$z_{\mathrm{d}} = k_0\rho_0^2 \tag{23.123}$$

is the diffraction length of the beam.

If

$$R_{\mathrm{d}} = R_{\mathrm{NL}}, \tag{23.124}$$

the non-linear refraction offsets the diffraction divergence. In this regime, called the *self-channelling regime*, the optical beam propagates in the medium, keeping its wavefront plane and its radius constant.

The power of a Gaussian optical beam is given by the formulae

$$P = \pi\rho_0^2 I_0, \quad I_0 = \frac{cn_0}{8\pi}|A_0|^2. \tag{23.125}$$

It follows from eqns (23.121)–(23.125) that a self-channelling regime is expected for a beam power $P = P_{\mathrm{cr}}$, where

$$P_{\mathrm{cr}} = \frac{c\lambda^2}{64\pi^2 n_2} \tag{23.126}$$

is the critical power of self-focusing and λ is the optical wavelength.

If $P < P_{\mathrm{cr}}$, the optical beam diverges, with the divergence rate being less than in the linear medium. If $P > P_{\mathrm{cr}}$, the rays deflect to the beam axis and self-focusing sets in.

Let us estimate the critical power of self-focusing for liquid CS_2, for which $n_2 = 1.45 \times 10^{-11}\ \mathrm{cm^3\,erg^{-1}}$. Taking $\lambda = 532$ nm, we obtain $P_{\mathrm{cr}} = 0.9$ kW.

A comprehensive discussion of self-focusing can be found, for example, in Shen (1984).

Bibliography

Akhmanov, S. A., Vysloukh, V. A., and Chirkin A. S. (1991). *Optics of femtosecond laser pulses.* AIP, New York.

Bloembergen, N. (ed.) (1975). *Spectroscopia nonlineare.* North-Holland, Amsterdam.

Giordmaine, J. A. and Kaiser, W. (1966). *Physical Review*, **144**, 676–688.

Laubereau, A. and Kaiser, W. (1978). *Review of Modern Physics*, **50**, 607–665.

Shapiro, S. L. (ed.) (1977). *Ultrashort light pulses.* Springer-Verlag, Berlin.

Shen, Y. R. (1984). *The principles of nonlinear optics.* John Wiley, New York.

Yariv, A. (1967). *Quantum electronics.* John Wiley, New York.

Yariv, A. (1982). *An introduction to theory and applications of quantum mechanics.* John Wiley, New York.

Yariv, A. and Yeh, P. (1984). *Optical waves in crystals.* John Wiley, New York.

Index

Lightning Source UK Ltd.
Milton Keynes UK
10 March 2010

151146UK00001B/53/A

9 780198 517955